Adapted Wavelet Analysis from Theory to Software

Adapted Wavelet Analysis from Theory to Software

Mladen Victor Wickerhauser

Washington University
St. Louis, Missouri

A K Peters
Wellesley, Massachusetts

Editorial, Sales, and Customer Service Office

A K Peters, Ltd.
289 Linden Street
Wellesley, MA 02181

Library of Congress Cataloging-in-Publication Data

Wickerhauser, Mladen Victor
Adapted wavelet analysis from theory to software / Mladen Victor Wickerhauser.
p, cm.
Includes bibliographical references and index.
ISBN 1-56881-041-5
1. Wavelets. I. Title.
QA403.3.W53 1993
621.382 ' 2 ' 015152433--dc20 94-14995
CIP

Printed in the United States of America
98 97 96 95 94 10 9 8 7 6 5 4 3 2 1

Preface

In the past decade, wavelet analysis has grown from a mathematical curiosity into a major source of new signal processing algorithms. The subject has branched out to include wavelet packets, and has merged with other methods of transient signal analysis such as Gabor expansions, Wilson bases, and adapted lapped orthogonal transforms. Dozens of conferences and workshops on wavelet applications have been sponsored on five continents. The 1993 *Wavelet Literature Survey* [89] contains 976 articles and books, a number which has doubled annually since 1991. Both the March 1992 *IEEE Transactions on Information Theory* and the December 1993 *IEEE Transactions on Signal Processing* were special issues dedicated to applications of wavelet transforms. Seminal articles on the subject have appeared in a remarkable variety of publications, ranging from *Dr. Dobb's Journal* to the *Journal of Chemical Information and Computer Science* to *Revista Matemática Iberoamericana.* The *Wavelet Digest* electronic mailing list currently boasts 3255 subscribers, including engineers, scientists, and mathematicians.

In spite of all this activity, there have been few works devoted to implementation. Apart from a short book chapter ([90], §13.10) and three popular articles [16, 17, 18], there is no wavelet programming guide. This may be due to the commercial value of wavelet transform software, which has kept it hidden beneath a proprietary cloak. Sources for discrete and lapped trigonometric transforms are easier to find ([92] appendices; [75]), and there are innumerable implementations of the discrete Fourier transform. But if we collect all these deeply interrelated transforms into a single toolbox, then we can profit by choosing the right tool for each job. In adapted wavelet analysis, choice makes the method more powerful than the sum of its parts.

This text goes beyond the existing literature to aid the engineer and applied mathematician in writing computer programs to analyze real data. It addresses the properties of wavelet and related transforms, to establish criteria by which the proper analysis tool may be chosen, and then details software implementations to perform the needed computation. It will also be useful to the pure mathematician who is familiar with some parts of wavelet theory but has questions about the applications. The worked exercises make this a useful textbook for self-study, or for a course in the theory and practice of wavelet analysis.

Beginning with an overview of the mathematical prerequisites, successive chapters rigorously examine the properties of the waveforms used in adapted wavelet analysis: discrete "fast" Fourier transforms, orthogonal and biorthogonal wavelets, wavelet packets, and localized trigonometric or lapped orthogonal functions. Other chapters discuss the "best basis" method, time-frequency analysis, and combinations of these algorithms useful for signal analysis, de-noising, and compression.

Each chapter discusses the technicalities of implementation, giving examples in pseudocode backed up with machine-readable Standard C source code available on the optional diskette. Each chapter finishes with a list of worked exercises in both the mathematics and the programming of adapted wavelet algorithms. Especially emphasized are the pitfalls and limitations of the algorithms, with examples and suggestions given to show how to avoid them.

Most of the adapted wavelet algorithms described here are the product of research conducted at Yale University from 1989–1991, and at Washington University in St. Louis from 1991–1994. Some of the algorithms have been reduced to practice and patented by Aware, Inc., FMA&H Corporation, Positive Technologies, Inc., and Yale University.

The author gratefully acknowledges the support provided by the National Science Foundation (NSF), the Air Force Office of Scientific Research (AFOSR), the Office of Naval Research (ONR), and the Defense Advanced Research Projects Agency (DARPA) during the past several years. The author also wishes to thank Professor Ronald R. Coifman of Yale University, Professor Yves Meyer of the University of Paris–Dauphine, and Professor Alexander Grossmann of the Centre de Physique Théorique in Luminy for many fruitful visits and conversations.

This book was written at a time when brutal war raged against the author's ancestral Croatian homeland. It is dedicated to fellow Croatians throughout the world, in appreciation for many contributions to science and culture, and in honor of the spirit which prevails against savagery.

University City, Missouri
9 May 1994

Contents

1 Mathematical Preliminaries 1

1.1 Basic analysis . . . 1

1.1.1 Convergence of series and products . . . 2

1.1.2 Measurability . . . 3

1.1.3 Integrability . . . 5

1.1.4 Metrics, norms, and inner products . . . 6

1.2 Function spaces . . . 8

1.2.1 Continuous functions on the circle . . . 8

1.2.2 Lebesgue spaces . . . 9

1.2.3 Spaces of test functions . . . 10

1.2.4 Dual spaces . . . 11

1.2.5 Frames, bases, and orthonormality . . . 13

1.3 Fourier analysis . . . 18

1.3.1 Fourier integrals . . . 19

1.3.2 Fourier series . . . 20

1.3.3 General orthogonal transformations . . . 22

1.3.4 Discrete Fourier transforms . . . 23

1.3.5 Heisenberg's inequality . . . 24

1.3.6 Convolution . . . 25

1.3.7 Dilation, decimation, and translation . . . 31

1.4 Approximation . . . 32

1.4.1 Averaging and sampling . . . 32

1.4.2 Band-limited functions . . . 35

1.4.3 Approximation by polynomials . . . 36

1.4.4 Smooth functions and vanishing moments . . . 37

1.5 Exercises . . . 38

2 Programming Techniques **41**
2.1 Computation in the real world 41
2.1.1 Finiteness 41
2.1.2 Validity 43
2.1.3 Pseudocode 44
2.1.4 Common utilities 46
2.2 Structures 47
2.2.1 Complex arithmetic 47
2.2.2 Intervals 48
2.2.3 Binary trees 51
2.2.4 Hedges 55
2.2.5 Atoms 57
2.3 Manipulation 58
2.3.1 Hedges and trees 58
2.3.2 Atoms and trees 61
2.3.3 Hedges to atoms 63
2.4 Exercises 66

3 The Discrete Fourier Transform **67**
3.1 The Fourier transform on $\mathbf{C}^N$ 68
3.1.1 The "fast" Fourier transform 69
3.1.2 Implementation of DFT 73
3.2 The discrete Hartley transform 77
3.2.1 The "fast" discrete Hartley transform 78
3.2.2 Implementation of DHT 81
3.3 Discrete sine and cosine transforms 83
3.3.1 DCT-I and DST-I 85
3.3.2 DCT-II, DCT-III, DST-II, and DST-III 90
3.3.3 DCT-IV and DST-IV 94
3.3.4 Implementations of DCT and DST 96
3.4 Exercises 101

4 Local Trigonometric Transforms **103**
4.1 Ingredients and examples 104
4.1.1 Unitary folding and unfolding 104
4.1.2 Smooth orthogonal projections 112
4.1.3 Periodization 118
4.1.4 Some analytic properties 122
4.2 Orthogonal bases 126

4.2.1 Compatible partitions . . . 127
4.2.2 Orthonormal bases on the line . . . 130
4.2.3 Discrete orthonormal bases . . . 134
4.3 Basic implementation . . . 137
4.3.1 Rising cutoff functions . . . 137
4.3.2 Midpoint folding and unfolding functions . . . 140
4.3.3 Midpoint local trigonometric transforms . . . 143
4.3.4 Midpoint local periodization . . . 145
4.4 Implementation of adapted transforms . . . 146
4.4.1 Adapted local cosine analysis . . . 146
4.4.2 Extraction of coefficients . . . 149
4.4.3 Adapted local cosine synthesis . . . 151
4.5 Exercises . . . 151

5 Quadrature Filters 153
5.1 Definitions and basic properties . . . 154
5.1.1 Action on sequences . . . 154
5.1.2 Biorthogonal QFs . . . 156
5.1.3 Orthogonal QFs . . . 158
5.1.4 Action on functions . . . 160
5.2 Phase response . . . 163
5.2.1 Shifts for sequences . . . 164
5.2.2 Shifts in the periodic case . . . 171
5.3 Frequency response . . . 176
5.3.1 Effect of a single filter application . . . 177
5.3.2 Effect of iterated filter applications . . . 182
5.4 Implementing convolution-decimation . . . 191
5.4.1 General assumptions . . . 191
5.4.2 Aperiodic convolution-decimation . . . 194
5.4.3 Adjoint aperiodic convolution-decimation . . . 197
5.4.4 Periodic convolution-decimation . . . 198
5.4.5 Adjoint periodic convolution-decimation . . . 207
5.4.6 Tricks . . . 208
5.5 Exercises . . . 210

6 The Discrete Wavelet Transform 213
6.1 Some wavelet basics . . . 214
6.1.1 Origins . . . 214
6.1.2 The DWT family . . . 215

6.1.3 Multiresolution analysis . . . 216
6.1.4 Sequences from functions . . . 217
6.2 Implementations . . . 218
6.2.1 Periodic DWT and iDWT . . . 220
6.2.2 Aperiodic DWT and iDWT . . . 226
6.2.3 Remarks . . . 234
6.3 Exercises . . . 234

7 Wavelet Packets **237**
7.1 Definitions and general properties . . . 238
7.1.1 Fixed-scale wavelet packets on **R** . . . 238
7.1.2 Multiscale wavelet packets on **R** . . . 242
7.1.3 Numerical calculation of wavelet packet coefficients . . . 245
7.1.4 The discrete wavelet packet analysis family . . . 251
7.1.5 Orthonormal bases of wavelet packets . . . 254
7.2 Implementations . . . 257
7.2.1 Generic algorithms . . . 258
7.2.2 Periodic DWPA and DWPS . . . 260
7.2.3 Aperiodic DWPA and DWPS . . . 265
7.2.4 Biorthogonal DWPA and DWPS . . . 271
7.3 Exercises . . . 272

8 The Best Basis Algorithm **273**
8.1 Definitions . . . 274
8.1.1 Information cost and the best basis . . . 274
8.1.2 Entropy, information, and theoretical dimension . . . 276
8.2 Searching for the best basis . . . 282
8.2.1 Library trees . . . 283
8.2.2 Fast searches for minimum information cost . . . 283
8.2.3 Adapted waveform analysis meta-algorithm . . . 286
8.3 Implementation . . . 287
8.3.1 Information cost functionals . . . 287
8.3.2 Extracting a basis subset . . . 289
8.3.3 Extracting a branch . . . 294
8.4 Exercises . . . 297

9 Multidimensional Library Trees **299**
9.1 Multivariable splitting operators . . . 300
9.1.1 Tensor products of CQFs . . . 301
9.1.2 Tensor products of DTTs and LTTs . . . 304
9.1.3 Complexity of the d-dimensional best basis algorithm . . . 307
9.1.4 Anisotropic dilations in multidimensions . . . 308
9.2 Practical considerations . . . 310
9.2.1 Labeling the bases . . . 310
9.2.2 Saving memory . . . 313
9.3 Implementations . . . 313
9.3.1 Transposition . . . 314
9.3.2 Separable convolution-decimation . . . 317
9.3.3 Separable adjoint convolution-decimation . . . 319
9.3.4 Separable wavelet packet bases . . . 322
9.3.5 Separable folding and unfolding . . . 326
9.4 Exercises . . . 327

10 Time-Frequency Analysis **329**
10.1 The time-frequency plane . . . 329
10.1.1 Waveforms and time-frequency atoms . . . 329
10.1.2 The idealized time-frequency plane . . . 333
10.1.3 Bases and tilings . . . 336
10.1.4 Analysis and compression . . . 339
10.1.5 Time-frequency analysis with library trees . . . 340
10.2 Time-frequency analysis of basic signals . . . 345
10.2.1 Benefits of adaption . . . 345
10.2.2 Wavelets versus wavelet packets . . . 348
10.2.3 Chirps . . . 349
10.2.4 Speech signals . . . 351
10.3 Implementation . . . 353
10.3.1 Drawing primitives . . . 354
10.3.2 Plotting the signal samples . . . 355
10.3.3 Plotting the time-frequency plane . . . 356
10.3.4 Computing the atoms . . . 358
10.4 Exercises . . . 359

11 Some Applications **361**
11.1 Picture compression 361
11.1.1 Digital pictures 362
11.1.2 Transform coding image compression 366
11.2 Fast approximate factor analysis 377
11.2.1 The approximate KL transform 382
11.2.2 Classification in large data sets 385
11.2.3 Jacobians of complicated maps 389
11.3 Nonstandard matrix multiplication 398
11.3.1 Two-dimensional best basis sparsification 398
11.3.2 Applying operators to vectors 399
11.3.3 Composing operators 408
11.4 Speech signal segmentation 409
11.4.1 Adapted local spectral analysis 410
11.4.2 Voiced-unvoiced segmentation 411
11.4.3 Experimental results 413
11.5 Speech scrambling 415
11.5.1 Objects to be scrambled 416
11.5.2 Feature-preserving permutations 416
11.6 Adapted waveform de-noising 417
11.6.1 Coherency and noise 417
11.6.2 Experimental results 419

A Solutions to Some of the Exercises **425**

B List of Symbols **441**

C Quadrature Filter Coefficients **443**
C.1 Orthogonal quadrature filters 443
C.1.1 Beylkin filters 444
C.1.2 Coifman or “Coiflet” filters 444
C.1.3 Standard Daubechies filters 449
C.1.4 Vaidyanathan filters 454
C.2 Biorthogonal quadrature filters 455
C.2.1 Symmetric/antisymmetric, one moment 456
C.2.2 Symmetric/symmetric, two moments 457
C.2.3 Symmetric/antisymmetric, three moments 459

Bibliography *463*
Index *475*

Chapter 1

Mathematical Preliminaries

The behavior of the algorithms discussed in this volume cannot be understood without some knowledge of the mathematical properties of the underlying functions. Thus, we will go over a few facts from real and harmonic analysis. This chapter can be skipped by those who are skimming the text, but the reader is advised to try the exercises as preparation for the mathematical development later.

1.1 Basic analysis

Analysis is the mathematical theory of infinite algorithms: evaluation of infinite sums, limits of arithmetic operations iterated infinitely many times, and so on. While no such algorithm can ever be implemented, some of its properties can be determined a priori without ever completing the calculation. What is more, a large finite number of arithmetic operations from a convergent infinite algorithm will produce a result close to the limit, whose properties will be similar to those of the limit. We will have a *truncated infinite algorithm* if we simply stop after a sufficiently large number of steps.

If a finite algorithm is a truncated infinite algorithm which has no limit, then the finite algorithm will be *unstable*, *i.e.*, the result will vary greatly with differing truncations. On the other hand, truncating a convergent infinite process at any sufficiently large number of steps will produce essentially the same result. This is the idea behind *Cauchy's criterion* for a sequence $\{a(n)\}$: For each $\epsilon > 0$ we can find a sufficiently large N such that if both n and m are greater than N, then $|a(n) - a(m)| < \epsilon$. If $a(n)$ is the output of an infinite algorithm after n operations, then after sufficiently many (N) operations all the outputs will be within ϵ of each other. If $|\epsilon/a(N)|$ is smaller than the machine precision, then all outputs after $a(N)$ are indistinguishable and we might as well truncate the algorithm.

Algorithms whose outputs are themselves infinite sets, like solution functions for differential equations, must be truncated still further. The output will always be a finite set, such as a finite set of samples for a function or a finite list of coefficients in an expansion for the function, and we may imagine that the rank of this set is fixed in advance. For each such rank M there will be an error between the finite-rank approximation and the actual infinite-rank solution, and we may ask how this error behaves as $M \to \infty$. At the very least the error must go to zero, but as a practical matter we should also know how fast it goes to zero in comparison with various functions of M, like M^{-2} or $1/\log M$ or 2^{-M}.

It has become abundantly clear that the successful interpretation of the numbers produced by numerical and signal processing transformations requires some basic results from measure theory and functional analysis.

1.1.1 Convergence of series and products

A sequence $\{a(n) : n = 0, 1, \ldots\}$ is *absolutely summable* if the following limit exists and is finite:

$$\lim_{N\to\infty} \sum_{n=0}^{N} |a(n)| < \infty.$$

An absolutely summable sequence has the advantage that its elements can be added up in any order and will yield the same result, up to roundoff error on a finite precision machine. The sequence is *square-summable* if we have the following:

$$\lim_{N\to\infty} \sum_{n=0}^{N} |a(n)|^2 < \infty.$$

Then $\{|a(n)|^2\}$ is absolutely summable. This is a weaker condition than absolute summability for $\{a(n)\}$.

Doubly infinite sequences $\{a(n) : n = 0, \pm1, \pm2, \ldots\}$ are absolutely summable if both the positive-indexed and the negative-indexed sides are absolutely summable. They are square-summable if both sides are square-summable.

A sequence of functions $\{u_n = u_n(t)\}$ is said to *converge at a point* t_0 if the sequence of numbers $\{u_n(t_0) : n = 1, 2, \ldots\}$ converges as $n \to \infty$. If $\{u_n(t)\}$ converges at each t, then the limits define a function $u(t) = \lim_{n\to\infty} u_n(t)$. The convergence is said to be *uniform* if $u_n(t) \to u(t)$ at comparable rates for each t, *i.e.*, if for every $\epsilon > 0$ there is an $N > 0$ such that we have $n > N \Rightarrow |u_n(t) - u(t)| < \epsilon$ for every t.

Proposition 1.1 *If u_n is a continuous function for each n and $u_n(t) \to u(t)$ converges uniformly in t, then $u = u(t)$ is a continuous function.* □

This result is proved as Theorem 9.2 of [3]. It implies that if $\{v_n = v_n(t)\}$ is a sequence of bounded continuous functions and $\{c_n\}$ is an absolutely summable sequence, then $\sum c_n v_n$ is a continuous function.

The *infinite product* of a sequence $\{b(n) : n = 0, 1, \dots\}$ is defined by

$$\prod_{n=0}^{\infty} b(n) \stackrel{\text{def}}{=} \lim_{N\to\infty} \prod_{n=0}^{N} b(n), \tag{1.1}$$

whenever this limit exists and has a finite, nonzero value.

Lemma 1.2 (Weierstrass Product Test) *If the sequence $a(n) = b(n) - 1$ is absolutely summable, then $\prod_{n=0}^{\infty} b(n)$ exists and is no larger than $\exp(\sum_n |a(n)|)$.*

Proof: First note that $0 \le \log(1 + |x|) \le |x|$ for all real numbers x. Then, writing $|b(n)| = |1 + a(n)| \le 1 + |a(n)|$, observe that

$$\log\left(\prod_{n=0}^{N} |b(n)|\right) = \sum_{n=0}^{N} \log|b(n)| \le \sum_{n=0}^{N} \log(1 + |a(n)|) \le \sum_{n=0}^{\infty} |a(n)| < \infty.$$

This also gives the upper bound on the size of the product.

Since $\{a(n)\}$ is absolutely summable its elements must satisfy $|a(n)| < 1/2$ for all sufficiently large n, so we may assume (extracting finitely many factors if necessary) that $a(n) > -1/2$ for all n. Notice that $\log|1 + x| \ge -(2\log 2)\,|x|$ for all $x > -1/2$. We may thus write $\log|b(n)| = \log|1 + a(n)| \ge -(2\log 2)|a(n)|$, and observe that

$$\log\left(\prod_{n=0}^{N} |b(n)|\right) \ge -(2\log 2) \sum_{n=0}^{\infty} |a(n)| > -\infty.$$

This implies that $\prod_{n=0}^{\infty} |b(n)| > 0$. □

1.1.2 Measurability

Define *Lebesgue measure* on $\mathbf{R}$ to be the map $E \mapsto |E|$ defined on all subsets $E \subset \mathbf{R}$ by

$$|E| = \inf\left\{ \sum_{k=0}^{\infty} |b_k - a_k| : E \subset \bigcup_{k=0}^{\infty} (a_k, b_k) \right\}. \tag{1.2}$$

Here we are taking the infimum over all countable covers of E by open intervals of the form (a_k, b_k), where possibly some of the intervals are empty. This is often called *Lebesgue outer measure*, but following Evans and Gariepy [44] we will not treat the restriction of outer measure to well-behaved ("measurable") subsets as a different object. We will also treat the qualifier "Lebesgue" as optional. It is elementary to prove that $|\emptyset| = 0$ and $0 \le |E| \le \infty$ for each $E \subset \mathbf{R}$, and that the following properties hold:

- *Countable subadditivity:* If E_n for $n = 0, 1, 2, \ldots$ is a countable sequence of subsets of $\mathbf{R}$, then $|\bigcup_{n=0}^{\infty} E_n| \le \sum_{n=0}^{\infty} |E_n|$;

- *Monotonicity:* If E and F are any two subsets of $\mathbf{R}$, then $|F| \le |F \cap E| + |F \cap E^c|$, where E^c is the complement of E in $\mathbf{R}$;

- *Measure of an interval:* $|[a,b]| = |]a,b[| = |[a,b[| = |]a,b]| = b - a$.

More generally, a *positive measure* on $\mathbf{R}$ is a map from subsets of $\mathbf{R}$ to $\mathbf{R}^+$ which is countably subadditive and monotonic, but which measures intervals $[a,b]$ using some other nondecreasing function than $b - a$. A *measure* is the difference between two positive measures.

We can construct Lebesgue measure on $\mathbf{R}^n$, $n > 1$, simply by using products of intervals. We will also denote the Lebesgue measure of a set $E \subset \mathbf{R}^n$ by $|E|$. The properties described below continue to hold if we replace $\mathbf{R}$ with $\mathbf{R}^n$. The construction is completely abstract and applies to other measures as well.

A subset $E \subset \mathbf{R}$ has *zero (Lebesgue) measure* if for every $\epsilon > 0$ there is a countable collection of open intervals I_k, $k \in \mathbf{Z}$, such that $\sum_{k \in \mathbf{Z}} |I_k| < \epsilon$ and $E \subset \bigcup_{k \in \mathbf{Z}} I_k$. If a property holds for all points in $\mathbf{R}$ except possibly for a set of measure 0, then we will say that it holds *almost everywhere*, or *for almost every* $x \in \mathbf{R}$, or abbreviate *a.e.x.*

A subset $E \subset \mathbf{R}$ is *(Lebesgue) measurable* if for every $F \subset \mathbf{R}$,

$$|F| = |E \cap F| + |E^c \cap F|,$$

i.e., if equality holds in the monotonicity property. Lebesgue measure is *countably additive* on measurable sets: If $\{E_n\}_{n \in \mathbf{Z}}$ is a sequence of disjoint measurable sets, then $|\bigcup_{n \in \mathbf{Z}} E_n| = \sum_{n \in \mathbf{Z}} |E_n|$. Intervals are measurable, and measurability is preserved by countable unions, intersections and complements. But not all subsets of $\mathbf{R}$ are Lebesgue measurable; see [3], p.304 for a counterexample.

A real-valued function $u = u(x)$ of one real variable is said to be *(Lebesgue) measurable* if for every $a \in \mathbf{R}$, the set $E_a = \{x : u(x) > a\}$ is measurable. A complex-valued function is said to be measurable if both its real and imaginary parts are measurable.

1.1.3 Integrability

The *Riemann integral* of a function $u = u(x)$ on an interval $[a, b]$ is the limit of the *Riemann sums*:

$$\int_a^b u(x)\,dx \stackrel{\text{def}}{=} \lim_{\Delta \to 0} \sum_{k=0}^{N-1} u(x_k)(a_{k+1} - a_k). \tag{1.3}$$

Here $a = a_0 < a_1 < \cdots < a_{N-1} < a_N = b$ is an arbitrary partition, $x_k \in [a_k, a_{k+1}]$ is an arbitrary point in the kth subinterval of the partition, and $\Delta = \max\{|a_{k+1} - a_k| : 0 \leq k < N\}$ is called the *mesh* of the partition. As $\Delta \to 0$, it forces $N \to \infty$. A function u is *Riemann integrable* if this limit exists regardless of the exact manner in which we choose the partition or the points of evaluation x_k in the subintervals. Riemann sums provide an infinite algorithm for computing the integral of a function; Riemann integrability is just the guarantee that this algorithm can be truncated.

The following basic fact, proved for example as Theorem 7.48 in [3], shows that most reasonable functions are Riemann integrable:

Proposition 1.3 *If u is a bounded function defined on $[a, b]$ whose discontinuities form a set of zero measure, then u is Riemann integrable on $[a, b]$.* □

We also define the *improper Riemann integral* to be $\lim_{b\to\infty} \int_a^b u(x)\,dx$ in the case $b = \infty$, and to be $\lim_{c\to a+} \int_c^b u(x)\,dx$ in the case that $|u(x)| \to \infty$ as $x \to a+$. However, Riemann sums are unwieldy in calculations, so we turn to a more general notion of integral.

If $u = u(x)$ is a measurable function on $\mathbf{R}$ taking only nonnegative real values, then the function

$$\alpha_u(r) \stackrel{\text{def}}{=} |\{x \in \mathbf{R} : u(x) > r\}| \geq 0 \tag{1.4}$$

defined for $r \geq 0$ will be called the *distribution function* of u. Then $r \leq s \Rightarrow \alpha_u(r) \geq \alpha_u(s)$, so α_u is monotonic and decreasing and can therefore have at most countably many discontinuities. Hence α_u is Riemann integrable on $]0, \infty[$, though the improper Riemann integral may be $+\infty$. We now define

$$u_+(x) = \begin{cases} u(x), & \text{if } u(x) > 0; \\ 0, & \text{if } u(x) \leq 0; \end{cases} \tag{1.5}$$

$$u_-(x) = \begin{cases} 0, & \text{if } u(x) \geq 0; \\ -u(x), & \text{if } u(x) < 0. \end{cases} \tag{1.6}$$

If u is measurable then so are both u_+ and u_-, hence so is $|u| = u_+ + u_-$. Also, $u = u_+ - u_-$.

The *Lebesgue integral* of a measurable function u is the difference of the two Riemann integrals below, in the case that both are finite:

$$\int u(x)\,dx = \int_0^\infty \alpha_{u_+}(r)\,dr - \int_0^\infty \alpha_{u_-}(r)\,dr. \tag{1.7}$$

Measurable functions u for which both α_{u_+} and α_{u_-} have finite Riemann integral will thus be called *Lebesgue integrable.*

These notions of integrability can be compared using a smooth positive function u. The Riemann integral divides the domain of u into small simple pieces $I_k = [a_k, a_{k+1}]$, each of known size $|I_k| < \Delta$, then picks a random representative $u(x_k)$ for the value of u in I_k. For smooth u these representatives are close to all values of u in I_k, so that $\sum_k u(x_k)|I_k|$ converges to $\int_a^b u(x)\,dx$ as $\Delta \to 0$. The Lebesgue integral, on the other hand, fixes a small range $[u_k, u_{k+1}[$ of values of u, and looks at the approximate measure $\alpha(r_k)$, $u_k \leq r_k < u_{k+1}$ of the set of points x where u takes on such values. We can also use the analogy of counting coins. Lebesgue integration first arranges the coins into stacks by value and then measures the heights of the stacks. Riemann integration, on the other hand, scatters the coins equally among many little boxes and then picks one "average" coin from each box.

A measurable function f is Lebesgue integrable if and only if it is *absolutely integrable*, meaning that $|f|$ is Lebesgue integrable. Thus oscillation or cancellation of negative and positive parts must not be essential to the existence of the integral. Also, f is *locally integrable* if $\int_a^b |f(x)|\,dx$ exists and is finite for all finite numbers a, b, or equivalently if $\mathbf{1}_{[a,b]}(x)f(x)$ is Lebesgue integrable for every bounded interval $[a, b]$. Localizing distinguishes the two ways that the integral of a function might diverge: there may be large values on small intervals, or there may be insufficient decrease to zero at infinity.

A basic result is that we can exchange integration and differentiation under certain conditions:

Proposition 1.4 *If both $u = u(x,t)$ and $du(x,t)/dx$ are Lebesgue integrable functions of t, then*

$$\frac{d}{dx}\int u(x,t)\,dt = \int \frac{d}{dx}u(x,t)\,dt.$$

□

1.1.4 Metrics, norms, and inner products

A *linear space* or *vector space* in our context will be a set of vectors with scalars taken from either **R** or **C**. We will consider both finite-dimensional and infinite dimensional vector spaces.

A *metric* on a set S is a function $dist : S \times S \to \mathbf{R}^+$ satisfying the following conditions, for all $x, y, z \in S$:

- *Symmetry:* $dist(x, y) = dist(y, x)$;
- *Nondegeneracy:* $dist(x, y) = 0 \iff x = y$;
- *Triangle inequality:* $dist(x, z) \leq dist(x, y) + dist(y, z)$.

The pair $(S, dist)$ is called a *metric space*. If S is a linear space and the metric $dist$ satisfies $dist(x + z, y + z) = dist(x, y)$, or more generally if S is a group and $dist$ satisfies $dist(xz, yz) = dist(x, y)$, then we will say that $dist$ is *invariant under translation.*

A *Cauchy sequence* in a metric space S is an ordered subset $\{u_n : n \in \mathbf{N}\}$ such that for every $\epsilon > 0$ there is a sufficiently large number M which will guarantee that $dist(u_i, u_j) < \epsilon$ whenever both $i > M$ and $j > M$. The *limit* of a Cauchy sequence is an element u_∞ satisfying $dist(u_n, u_\infty) \to 0$ as $n \to \infty$. The metric space S is *complete* if every Cauchy sequence has a limit in S.

A *norm* on a vector space S is a map $\|\cdot\| : S \to \mathbf{R}^+$ satisfying the following properties for all $x, y, z \in S$ and $a \in \mathbf{C}$:

- *Scaling:* $\|ax\| = |a|\|x\|$;
- *Nondegeneracy:* $\|x\| = 0 \iff x = 0$;
- *Triangle inequality:* $\|x - z\| \leq \|x - y\| + \|y - z\|$.

The function $dist(x, y) = \|x - y\|$ is thus an invariant metric.

A *(Hermitean) inner product* on a linear space S is a function $\langle \cdot, \cdot \rangle : S \times S \to \mathbf{C}$ satisfying the following properties for all $f, g, h \in S$ and $a, b \in \mathbf{C}$:

- *Linearity:* $\langle f, ag + bh \rangle = a\langle f, g \rangle + b\langle f, h \rangle$;
- *Hermitean symmetry:* $\langle f, g \rangle = \overline{\langle g, f \rangle}$;
- *Positivity:* $\langle f, f \rangle = 0 \iff f = 0$.

The first and second properties imply that $\langle af + bg, h \rangle = \bar{a}\langle f, g \rangle + \bar{b}\langle g, h \rangle$, which is sometimes called *sesquilinearity.* Also, the third property implies that if $\langle f, g \rangle = 0$ for all $g \in S$, then $f = 0$, which is sometimes called *nondegeneracy.*

We can always define a norm from an inner product by the formula

$$\|x\| \stackrel{\text{def}}{=} \langle x, x \rangle^{1/2}. \tag{1.8}$$

Thus an inner product space is a normed linear space is a metric space with an invariant metric. Some simple examples of inner product spaces are $\mathbf{C}^N$ for each $N > 0$. A complete inner product space is also called a *Hilbert space*. Some familiar examples of Hilbert spaces are the N-dimensional real (or complex) vector spaces $\mathbf{R}^N$ (or $\mathbf{C}^N$), and the infinite-dimensional function space $L^2(\mathbf{R})$.

We can estimate the inner product from the norm using the *Cauchy–Schwarz inequality*:

$$|\langle f, g\rangle| \leq \|f\| \, \|g\|. \tag{1.9}$$

If the norm is derived from an inner product, then we can recover the inner product from the norm using *polarization*:

$$4\Re\{\langle f, g\rangle\} = \|f+g\|^2 - \|f-g\|^2; \qquad 4i\Im\{\langle f, g\rangle\} = \|f+ig\|^2 - \|f-ig\|^2. \tag{1.10}$$

1.2 Function spaces

1.2.1 Continuous functions on the circle

We will denote the collection of all continuous, one-periodic real (or complex-valued) functions by $C(\mathbf{T})$. If we wish to indicate that we have relaxed the periodicity condition so as to allow continuous functions u for which $u(0) \neq u(1)$, then we will use the notation $C([0,1])$. Either of these collections forms an *inner product space* with the Hermitean inner product defined by the integral:

$$\langle f, g\rangle \stackrel{\text{def}}{=} \int_0^1 \bar{f}(t) g(t)\, dt. \tag{1.11}$$

This is also a normed linear space with the derived norm $\|f\| = \langle f, f\rangle^{1/2}$, and a metric space with the derived invariant metric $dist(f,g) = \|f-g\|$. Neither $(C(\mathbf{T}), dist)$ nor $(C([0,1]), dist)$ is a complete metric space, however, because the metric permits Cauchy sequences of continuous functions $\{f_n : n = 1, 2, \ldots\}$ whose limits are not continuous functions. We will remedy this defect in two ways: by changing the metric to impose better convergence on Cauchy sequences, or by enlarging the space of functions to include all the limits.

We can alternatively define the *uniform norm* $\|\cdot\|_\infty$, which gives a different invariant metric on the continuous functions:

$$\|u\|_\infty \stackrel{\text{def}}{=} \sup\{|u(t)| : 0 \leq t \leq 1\}. \tag{1.12}$$

The normed spaces $(C(\mathbf{T}), \|\cdot\|_\infty)$ and $(C([0,1]), \|\cdot\|_\infty)$ are complete metric spaces.

1.2.2 Lebesgue spaces

The *(Lebesgue) L^p-norms* $\|\cdot\|_p$, for $1 \leq p \leq \infty$, are defined for measurable functions u by

$$\|u\|_p = \begin{cases} \left(\int |u(t)|^p \, dt\right)^{1/p} = \int_0^\infty pr^{p-1}\alpha_{|u|}(r)\, dr, & \text{if } 1 \leq p < \infty; \\ \text{ess sup } |u| \stackrel{\text{def}}{=} \inf\{r : \alpha_u(r) = 0\}, & \text{if } p = \infty. \end{cases} \tag{1.13}$$

It can be shown (see [3]) that if $\|f - g\|_p = 0$ in L^p (for any $p \in [1, \infty]$) and f and g are continuous functions, then $f(x) = g(x)$ for every x. But measurable functions need not be continuous, so to preserve the essential nondegeneracy property of the L^p norm we must change the notion of equality. All that we can conclude from $\|f\| = 0$ is that the points x where $f(x) \neq 0$ form a set of measure zero. Thus the L^p "norms" fail to satisfy the nondegeneracy condition, but this can be remedied by identifying as equal any two functions u and v for which $\|u\|_p$ and $\|v\|_p$ are both finite and $\|u - v\|_p = 0$. Any pair u, v which are thus identified can be shown to satisfy $u(t) = v(t)$ at almost every $t \in \mathbf{R}$. The collection of equivalence classes of measurable functions u satisfying $\|u\|_p < \infty$ is called the *Lebesgue space* $L^p(\mathbf{R})$. By substituting $\mathbf{T}$ for $\mathbf{R}$ we similarly define $L^p(\mathbf{T})$.

Each L^p-norm provides a derived invariant metric $dist_p(x, y) = \|x - y\|_p$. For each p, the metric space $(L^p, dist_p)$ is complete. L^2 in particular is a Hilbert space, where the inner product can be defined by polarization or directly in terms of the Lebesgue integral:

$$\|f\|_{L^2(\mathbf{R})} \stackrel{\text{def}}{=} \left(\int |f(x)|^2 \, dx\right)^{1/2} < \infty; \tag{1.14}$$

$$\langle f, g\rangle_{L^2(\mathbf{R})} \stackrel{\text{def}}{=} \int \bar{f}(x) g(x)\, dx. \tag{1.15}$$

We may also define $L^2(\mathbf{R}^n)$ for any $n > 1$ just by integrating over $\mathbf{R}^n$ rather than $\mathbf{R}$ in the definition of $\|f\|_{L^2(\mathbf{R}^n)}$ and $\langle f, g\rangle_{L^2(\mathbf{R}^n)}$. Likewise, we can define $L^2([0,1]^n)$ to be the Hilbert space of functions of n real variables which are one-periodic in each variable.

Square-summable sequences form the Hilbert space $\ell^2 = \ell^2(\mathbf{Z})$, with inner product and norm defined as below:

$$\|a\|_{\ell^2} \stackrel{\text{def}}{=} \left(\sum_{n=-\infty}^{\infty} |a(n)|^2\right)^{1/2} < \infty; \tag{1.16}$$

$$\langle a, b\rangle_{\ell^2} \stackrel{\text{def}}{=} \sum_{n=-\infty}^{\infty} \bar{a}(n) b(n). \tag{1.17}$$

Likewise, we can define $\ell^2(\mathbf{Z}^n)$ to be the Hilbert space of square-summable sequences indexed by n integers.

Henceforth, we will use $\|\cdot\|$ and $\langle\cdot,\cdot\rangle$ without subscripts to denote the norm and inner product in L^2 or ℓ^2, whenever the space is clear from the context.

1.2.3 Spaces of test functions

The *support* of a function u, denoted $\operatorname{supp} u$, is the complement of the largest open set E with the property that $t \in E \Rightarrow u(t) = 0$. Thus the support of a function is always a closed subset of the domain of the function. We say that the function u is *compactly supported* if $\operatorname{supp} u$ is compact. If the domain is $\mathbf{N}$, $\mathbf{Z}$, or $\mathbf{Z}^n$, then the sequence u is compactly supported if and only if all but finitely many of its elements are 0; we will call such u *finitely supported.* If the domain is $\mathbf{R}^n$, then u is compactly supported if and only if the support set is bounded: there exists some $R < \infty$ such that $s, t \in \operatorname{supp} u \Rightarrow |t - s| < R$. The smallest R for which the preceding implication holds is called the *width* or *diameter* of the support of u and will be denoted $\operatorname{diam} \operatorname{supp} u$.

The support of a measurable function is a measurable set. It should be noted that if $E \subset \mathbf{R}^n$, then $|E|^{1/n} \leq \operatorname{diam} E$. However, $\operatorname{diam} E$ and $|E|$ are not comparable quantities: for every $\epsilon > 0$ there is a set $E \subset \mathbf{R}^n$ for which $|E|^{1/n} \leq \epsilon \operatorname{diam} E$. In $\mathbf{R} = \mathbf{R}^1$, an easy example is the set $[0, \epsilon/2] \cup [1 - \epsilon/2, 1]$.

We will say that a function u is *smooth* if $d^n u/dt^n$ is a continuous function for all $n \in \mathbf{N}$. We will use the weaker notion of *smooth to degree d* if $d^n u/dt^n$ is a continuous function for all $0 \leq n \leq d$.

We will say that a function u has *rapid decrease* at infinity if for every $n \in N$ we have a finite constant $K_n > 0$ such that for all $t \in \mathbf{R}$, $|t^n u(t)| < K_n$. This is also called *superalgebraic decay* at infinity. It is not quite as strong a notion as *exponential decay* at infinity, for which we need to have some constants $K, \sigma > 0$ for which the estimate $|u(t)| < Ke^{-\sigma|t|}$ holds. To prove this, consider the example $f(t) = e^{-[\log|t|]^2}$, which is the subject of one of the exercises. We will say that u has *degree d decay* at infinity if $K_n < \infty$ at most for $0 \leq n \leq d$. It is evident that compactly supported functions have all possible kinds of decay at infinity.

The *Schwartz class* S is the set of measurable smooth functions of superalgebraic decay at infinity, *i.e.*, the functions $u = u(t)$ satisfying an inequality like the following for each $n, m \in \mathbf{N}$:

$$\sup \left\{ \left| t^n \frac{d^m}{dt^m} u(t) \right| : t \in \mathbf{R} \right\} < K_{m,n} < \infty. \tag{1.18}$$

This condition is equivalent to satisfying the inequality

$$\sup\left\{\left|\frac{d^m}{dt^m}\left(t^n u(t)\right)\right| : t \in \mathbf{R}\right\} < K'_{m,n} < \infty \tag{1.19}$$

for all $n, m \in \mathbf{N}$. Those Schwartz functions which are also compactly supported are called *test functions*; the class of such functions is denoted by $\mathcal{D}$. Schwartz functions (and test functions) are dense in all the L^p spaces with $p < \infty$, and in larger classes of spaces as well. For calculations, it is often useful to first define formulas on test functions or Schwartz functions, and then extend them to a larger class of functions by continuity.

1.2.4 Dual spaces

If X is a function space we can define its *dual space* X' to be the collection of continuous linear functions T from X to $\mathbf{R}$ or $\mathbf{C}$. Such maps themselves form a normed linear space using the *operator norm*:

$$\|T\|_{X'} = \|T\|_{op} \stackrel{\text{def}}{=} \sup\left\{\frac{|T(u)|}{\|u\|_X} : u \in X, u \neq 0\right\}. \tag{1.20}$$

If $X \subset Y$, then $Y' \subset X'$, since there are fewer continuous functions on a larger space. Thus a highly restricted space like $\mathcal{S}$ has a very large dual space which contains all sorts of interesting things. For example, it contains arbitrary derivatives of certain discontinuous functions.

Duals of L^p spaces

If $X = L^p$ for some $1 \leq p < \infty$, then the following theorem characterizes X':

Theorem 1.5 (Riesz Representation) *Let T be a continuous linear function on L^p, for $1 \leq p < \infty$. Then there is a function $\theta = \theta(x)$ in $L^{p'}$, such that*

$$T(u) = \int \theta(x) u(x)\, dx.$$

Here $p' = p/(1-p)$ if $p > 1$ and $p' = \infty$ if $p = 1$. □

A proof of this may be found in [96], p.132. Note that if $T(u) = \int \theta(x)u(x)\,dx = \int \mu(x)u(x)\,dx$, then $\int (\mu(x) - \theta(x))\,u(x) = 0$ for all $u \in L^p$. This implies that $\theta(x) = \mu(x)$ for almost every x, so they are equal in the sense of $L^{p'}$. Thus $T \leftrightarrow \theta$ is a one-to-one correspondence, and we can identify $(L^p)' = L^{p'}$.

The *dual index* p' satisfies the equation $\frac{1}{p} + \frac{1}{p'} = 1$, showing how we are victims of a bad choice of notation (p rather than $1/p$) made sometime in the dim past.

Measures and the dual of C

If $X = C([0,1])$ is the space of continuous functions on $[0,1]$ with the uniform norm, then X' can be identified with the space of bounded measures μ on the interval $[0,1]$, namely the measures which satisfy $\mu([0,1]) < \infty$. The identification is through the distribution function: for every bounded measure μ, we get a linear function $T = T_\mu$ defined on positive functions u by the formula:

$$T(u) = \int u(x)\, d\mu(x) \stackrel{\text{def}}{=} \int_{r=0}^{\infty} \mu(\{x : u(x) > r\})\, dr.$$

For real-valued continuous functions u we use $T(u) \stackrel{\text{def}}{=} T(u_+) - T(u_-)$, and for complex-valued u we separately compute and then add up the real and imaginary parts.

Every locally integrable function $m = m(x)$ generates a measure μ by the formula $\mu([a,b]) = \int_a^b m(x)\, dx$, but there are also some bounded measures which cannot be realized in this way. Thus the dual space $C([0,1])'$ contains more objects than just functions. In particular, it contains the *Dirac mass at* 0, δ, which is defined by the following formula:

$$\delta u = u(0). \tag{1.21}$$

The Dirac mass is a bounded measure for which there is no bounded function m. It is very common to abuse notation and write $\delta = \delta(x)$ as the *Dirac delta function* and to "evaluate the integral" $\delta u = \int u(x)\delta(x)\, dx = u(0)$. It is also common to write the Dirac mass at c as a translate of the Dirac mass at zero, namely $\delta(x-c)$.

Distributions

The dual space $\mathcal{D}'$ of the test functions contains what are called *distributions*, and the dual space $\mathcal{S}'$ of the Schwartz functions contains *tempered distributions*. Since $\mathcal{D} \subset \mathcal{S}$, we have $\mathcal{S}' \subset \mathcal{D}'$; both are proper inclusions.

Every measurable and locally integrable function θ defines a distribution by the formula $T(\phi) = \int \theta(x)\phi(x)\, dx$, where $\phi \in \mathcal{D}$. This gives a natural inclusion of the Lebesgue spaces into $\mathcal{S}'$; we have $\mathcal{S} \subset L^p \subset \mathcal{S}'$ for every $1 \le p \le \infty$.

Every distribution has a derivative which makes sense as a distribution, and every tempered distribution has a derivative among the tempered distributions. This derivative is evaluated by integration by parts: to compute the nth derivative of the distribution T, we use the formula

$$T^{(n)}(\phi) \stackrel{\text{def}}{=} (-1)^n T(\phi^{(n)}). \tag{1.22}$$

For example, the derivative of $\mathbf{1}_{\mathbf{R}^+}(x)$, also known as the *Heaviside function* or the unit step at zero, is the Dirac mass δ. Also, every measurable and locally integrable function θ has as many derivatives as we like among the distributions, given by the formula

$$T^{(n)}(\phi) \stackrel{\text{def}}{=} (-1)^n \int \theta(x)\phi^{(n)}(x)\,dx.$$

We can define the *Fourier transform of a tempered distribution*, which is itself a tempered distribution, by "passing the hat": for all $\phi \in \mathcal{S}$,

$$\hat{T}(\phi) = T(\hat{\phi}).$$

This works because $\mathcal{S}$ is preserved by the Fourier transform; it does not hold for $\mathcal{D}'$ since the compact support of $\phi \in \mathcal{D}$ is destroyed by the Fourier transform.

A distribution T has *compact support* if there is a bounded interval I such that $T(\phi) = 0$ whenever $\operatorname{supp}\phi$ is in the complement of I. The Dirac mass at a point has compact support. A measurable, locally integrable function of compact support corresponds to a compactly supported distribution. A compactly supported distribution is tempered, and its Fourier transform is given by integration against a smooth function.

Convolution is defined for two tempered distributions by the formula

$$(T_1 * T_2)(\phi) \stackrel{\text{def}}{=} T_1(\psi); \qquad \text{where } \begin{cases} \psi = \psi(y) = T_2(\phi_y), & y \in \mathbf{R}, \\ \phi_y(x) = \phi(x+y), & x \in \mathbf{R}. \end{cases} \tag{1.23}$$

It is left as an exercise to show that $\phi \in \mathcal{S} \Rightarrow \psi \in \mathcal{S}$.

If T_1 and T_2 are defined by integration against functions θ_1 and θ_2, then $T_1 * T_2$ will be defined by integration against the function $\theta_1 * \theta_2(x) \stackrel{\text{def}}{=} \int \theta_1(y)\theta_2(x-y)\,dy$. We note that the Dirac mass at a convolved with the mass at b is the Dirac mass supported at $a+b$.

1.2.5 Frames, bases, and orthonormality

If H is a complete inner product space, then we say that a collection $\{\phi_n : n \in \mathbf{Z}\}$ is an *orthonormal basis* (or *Hilbert basis*) for H if the following conditions are satisfied:

- *Orthogonality:* If $n, m \in \mathbf{Z}$ and $n \neq m$, then $\langle \phi_n, \phi_m \rangle = 0$;
- *Normalization:* For each $n \in \mathbf{Z}$, $\|\phi_n\| = 1$;
- *Completeness:* If $f \in H$ and $\langle f, \phi_n \rangle = 0$ for all $n \in \mathbf{Z}$, then $f = 0$.

Any collection satisfying the first two conditions, but not necessarily satisfying the third, will be called simply *orthonormal.* If it satisfies only the first condition, it will be called simply *orthogonal.* Note that our Hilbert basis has (at most) a countable infinity of elements; any Hilbert space with a countable basis is called *separable,* and all of the examples we have mentioned are separable.

An alternative notion to completeness in separable Hilbert spaces is *density.* The collection $\{\phi_n : n \in \mathbf{Z}\}$ is dense in H if for every $f \in H$ and $\epsilon > 0$, we can find a sufficiently large N and constants $a_{-N}, a_{-N+1}, \ldots, a_{N-1}, a_N$ such that $\|f - \sum_{k=-N}^{N} a_k \phi_k\| < \epsilon$. Another way of saying this is that finite linear combinations from the system $\{\phi_n : n \in \mathbf{Z}\}$ can be used to approximate functions in H arbitrarily well, as measured by the norm. An orthonormal system $\{\phi_n\}$ is dense in H if and only if it is complete.

If $\{\phi_n\}$ is a complete orthonormal system, then we can compute the norm of a function from its inner products with the functions ϕ_n using *Parseval's formula:*

$$\|f\|^2 = \sum_{n=-\infty}^{\infty} |\langle f, \phi_n \rangle|^2. \tag{1.24}$$

Parseval's formula implies that the inner products $\langle f, \phi_n \rangle$ form a square-summable sequence.

Note that orthogonality implies linear independence. We may wish to construct a collection $\{\phi_n : n \in \mathbf{Z}\}$ which is neither orthogonal, nor even linearly independent, but which can still be used to approximate functions. One important property to preserve is comparability between $\|f\|$ and the sum of the squares of its inner products with the collection. We say that an arbitrary collection of functions $\{\phi_n \in H : n \in \mathbf{Z}\}$ is a *frame* if there are two constants A and B, with $0 < A \leq B < \infty$, such that for every $f \in H$ we have:

$$A\|f\|^2 \leq \sum_{n=-\infty}^{\infty} |\langle f, \phi_n \rangle|^2 \leq B\|f\|^2. \tag{1.25}$$

A and B are called *frame bounds.* If $A = B$ then we have a *tight frame.* An orthonormal basis is a tight frame for which $A = B = 1$, but not all such tight frames are orthonormal bases. If the collection $\{\phi_n : n \in \mathbf{Z}\}$ is a frame and is also linearly independent, then it is called a *Riesz basis.* Frames and Riesz bases are complete by virtue of the left inequality in Equation 1.25: if $\langle f, \phi_n \rangle = 0$ for all n, then $0 \leq \|f\| \leq \sum 0 = 0 \Rightarrow f = 0$. Any Riesz basis can be made into a Hilbert basis by *Gram orthogonalization,* which is described in [37].

Standard orthonormal bases

If the Hilbert space H is $\mathbf{C}^n$, then the Hermitean inner product is

$$\langle u, v\rangle \stackrel{\text{def}}{=} \sum_{k=1}^{n} \bar{u}(k)v(k), \tag{1.26}$$

where $u(k)$ is the element in the kth row of the column vector u. The *standard basis* consists of the vectors

$$e_1 = \begin{pmatrix} 1 \\ 0 \\ \vdots \\ 0 \end{pmatrix}, \quad \ldots, \quad e_n = \begin{pmatrix} 0 \\ \vdots \\ 0 \\ 1 \end{pmatrix}. \tag{1.27}$$

Here e_k is the column vector with a single "1" in the kth row and zeroes elsewhere. These are evidently orthonormal with respect to the inner product.

If H is an infinite-dimensional but separable Hilbert space, then H is isomorphic to the space $\ell^2(\mathbf{Z})$ of infinite, square-summable sequences. The Hermitean inner product is

$$\langle u, v\rangle \stackrel{\text{def}}{=} \sum_{k=-\infty}^{\infty} \bar{u}(k)v(k). \tag{1.28}$$

There is a standard basis of *elementary sequences* $e_k = \{e_k(n)\}$ consisting of a single "1" at index k and zeroes elsewhere. How these look in the original space H depends on the isomorphism, which is not always easy to compute.

Bases of eigenvectors

A straightforward calculation shows that for any $n \times n$ matrix $A = (a_{ij})$, and any pair of vectors $u, v \in \mathbf{C}^n$, we have the identity

$$\langle u, Av\rangle = \langle A^* u, v\rangle, \tag{1.29}$$

where $A^* = a^*_{jk}$ is the *adjoint* or conjugated transpose of A, defined by $a^*_{jk} \stackrel{\text{def}}{=} \bar{a}_{kj}$. We say that the $n \times n$ matrix M is *Hermitean* or *selfadjoint* if for any pair of vectors $u, v \in \mathbf{C}^n$ we have

$$\langle u, Mv\rangle = \langle Mu, v\rangle, \tag{1.30}$$

or equivalently that $M^* = M$. A real-valued symmetric matrix is Hermitean.

An *eigenvalue* for a matrix A is any number λ such that there is a nonzero solution vector y to the equation $Ay = \lambda y$. Such a y is then called the *eigenvector* associated to λ. One way to find eigenvalues is to solve for the roots of the polynomial in λ that we obtain by expanding the equivalent condition

$$\det(A - \lambda I) = 0. \tag{1.31}$$

This method is not the best one if n is large because it is both computationally complex and ill-conditioned.

Eigenvalues of a Hermitean matrix $A = A^*$ must be real numbers, since

$$\bar{\lambda}\|y\|^2 = \langle Ay, y\rangle = \langle y, A^*y\rangle = \langle y, Ay\rangle = \lambda\|y\|^2, \tag{1.32}$$

and $\|y\| \neq 0$. Furthermore, if there are two distinct eigenvalues $\lambda_1 \neq \lambda_2$, then their associated eigenvectors y_1 and y_2 must be orthogonal:

$$0 = \langle Ay_1, y_2\rangle - \langle y_1, Ay_2\rangle = (\lambda_1 - \lambda_2)\langle y_1, y_2\rangle, \quad \Rightarrow \langle y_1, y_2\rangle = 0. \tag{1.33}$$

An important consequence of these elementary results is that we can make an orthonormal basis out of the eigenvectors of a selfadjoint matrix:

Theorem 1.6 *An $n \times n$ Hermitean matrix has n linearly independent eigenvectors $y_1, \ldots, y_n$ which, if suitably chosen and normalized so that $\|y_k\| = 1$ for all $k = 1, \ldots, n$, form an orthonormal basis for $\mathbf{C}^n$.* □

A proof of this theorem may be found, for example, on page 120 of [2] .

A result from Sturm–Liouville theory

The *Sturm–Liouville differential operator* acting on a twice-differentiable function $y = y(x)$, $0 < x < 1$, is

$$Dy \stackrel{\text{def}}{=} (py')' + qy, \tag{1.34}$$

where $q = q(x)$ is a continuous real-valued function, and $p = p(x)$ is a differentiable real-valued function. If p is never zero in the interval $[0, 1]$, then D is called a *regular* Sturm–Liouville operator. We can of course replace $[0, 1]$ with any bounded interval $[a, b]$.

Notice that Dy makes sense as a distribution even for $y \in L^2([0, 1])$. If $y \in L^2$ is regular enough so that $Dy \in L^2$, then $qy \in L^2$ so $(py')' \in L^2([0, 1])$ and is thus integrable. Integrating twice shows that y must be continuous on $[0, 1]$. We can therefore impose additional conditions on such functions by specifying their values at the boundary points 0 and 1.

The *Sturm–Liouville boundary conditions* for y on the interval $[0,1]$ are

$$ay(0) + by'(0) = 0; \qquad cy(1) + dy'(1) = 0. \tag{1.35}$$

Here a, b, c, d are four real numbers with not both $a = 0$ and $b = 0$, and not both $c = 0$ and $d = 0$. Notice that if the differentiable functions $u = u(x)$ and $v = v(x)$ both satisfy these conditions, then

$$u(0)v'(0) - u'(0)v(0) = 0 \qquad \text{and} \qquad u(1)v'(1) - u'(1)v(1) = 0. \tag{1.36}$$

This is because the following two matrices must be singular, since they have the nonzero vectors $(a,b)^T$ and $(c,d)^T$ in their null spaces:

$$\begin{pmatrix} u(0) & u'(0) \\ v(0) & v'(0) \end{pmatrix}; \qquad \begin{pmatrix} u(1) & u'(1) \\ v(1) & v'(1) \end{pmatrix}.$$

Let H be the set of functions $y \in L^2([0,1])$ satisfying $Dy \in L^2([0,1])$ and also satisfying the boundary conditions in Equation 1.35. All the conditions are linear, so the result is a subspace of $L^2([0,1])$ with the same Hermitean inner product. It is not hard to show that H is a dense subspace, so any Hilbert basis for H will automatically be a Hilbert basis for $L^2([0,1])$.

Now D is a selfadjoint operator in the domain H. We compute the adjoint through integration by parts:

$$\langle u, Dv \rangle = \langle Du, v \rangle + p(x)\left[u(x)v'(x) - u'(x)v(x)\right]\big|_{x=0}^{1} = \langle Du, v \rangle. \tag{1.37}$$

The boundary term vanishes because of Equation 1.36. In fact, D is the most general form of a selfadjoint second-order linear differential operator.

The *Sturm–Liouville eigenvalue problem* for the operator D and the domain H is to find eigenvalues $\lambda \in \mathbf{R}$ and *eigenfunctions* $y \in H$ such that

$$Dy(x) = \lambda y(x), \qquad \text{for all } 0 < x < 1. \tag{1.38}$$

The eigenvalues must be real numbers because of Equation 1.32 and the eigenfunctions associated to distinct eigenvalues must be orthogonal because of Equation 1.33, since both of those equations are valid for every Hermitean inner product.

Not only are there solutions to this problem, but there will be enough of them to construct a Hilbert basis. We have the following theorem about regular Sturm–Liouville boundary value problems:

Theorem 1.7 *Each regular Sturm–Liouville eigenvalue problem on $[0,1]$ has an infinite sequence of eigenvalues $\{\lambda_k : k = 0,1,2,\ldots\} \subset \mathbf{R}$ with $|\lambda_k| \to \infty$ as $k \to \infty$. The associated eigenfunctions $\{y_k : k = 0,1,2,\ldots\}$, when normalized so that $\|y_k\| = 1$ for all k, form an orthonormal basis for $L^2([0,1])$.* □

A proof of this theorem may be found, for example, in [93]. Notice how it generalizes Theorem 1.6.

1.3 Fourier analysis

By *Fourier transform* we shall mean any one of several mathematical transformations, with the specific transformation being unambiguously chosen from the context. The function (or sequence, or vector) produced by the Fourier transform contains *spectral information* about the function, in the sense that each Fourier coefficient is an inner product with an oscillating basis function of well-defined frequency and phase.

Fourier's representation of functions as superpositions of sines and cosines has become ubiquitous for both the analytic and numerical solution of differential equations, and for the analysis and treatment of communication signals. Fourier's original idea was to write an "arbitrary" one-periodic function $f = f(x)$ as a sum:

$$f(x) \approx a(0) + \sqrt{2}\sum_{n=1}^{\infty} a(n)\cos 2\pi nx + \sqrt{2}\sum_{n=1}^{\infty} b(n)\sin 2\pi nx, \tag{1.39}$$

where the constants $a(0), a(1), \dots$ and $b(1), b(2), \dots$, called the *Fourier coefficients* of f, are computed from f via integrals:

$$a(0) = a_f(0) \stackrel{\text{def}}{=} \int_0^1 f(x)\,dx; \tag{1.40}$$

$$a(n) = a_f(n) \stackrel{\text{def}}{=} \sqrt{2}\int_0^1 f(x)\cos 2\pi nx\,dx, \qquad \text{for } n \geq 1; \tag{1.41}$$

$$b(n) = b_f(n) \stackrel{\text{def}}{=} \sqrt{2}\int_0^1 f(x)\sin 2\pi nx\,dx, \qquad \text{for } n \geq 1. \tag{1.42}$$

There has been a tremendous amount of literature written on the question of what "arbitrary" means, and on the interpretation of the '$\approx$' symbol in this context. Some of the deepest results in mathematics concern the convergence of such "Fourier series" for nonsmooth or possibly discontinuous functions. However, if f is a continuous one-periodic function with a continuous derivative, then the two infinite series in Equation 1.39 will converge at every point $x \in \mathbf{R}$; in fact, the series of functions will converge uniformly on $\mathbf{R}$, and we can legitimately replace the $\approx$ with $=$. This elementary fact is proved, for example, in [3], which also gives several refinements. It was known even to Riemann [94] that there were continuous but not differentiable functions whose Fourier series diverged at some point x, but then it was shown by

Carleson [13], using much deeper analysis, that even without the differentiability hypothesis, we still get pointwise convergence in Equation 1.39, only now it holds just for almost every point $x \in \mathbf{R}$.

1.3.1 Fourier integrals

Following Stein and Weiss [103], we define the *Fourier integral* of a function on the line by

$$\hat{f}(\xi) = \int_{-\infty}^{\infty} f(x) e^{-2\pi i x \xi} \, dx. \tag{1.43}$$

Since the exponential $e^{-2\pi i x \xi}$ is a bounded function of x, this integral converges absolutely for all absolutely integrable functions f and for every $\xi \in \mathbf{R}$. In addition we have a few results from functional analysis.

Lemma 1.8 (Riemann–Lebesgue) *If f is absolutely integrable, then $\hat{f}$ is continuous, and $\hat{f}(\xi) \to 0$ as $|\xi| \to \infty$.* □

Absolutely integrable functions are dense in $L^2(\mathbf{R})$, and we can show that the Fourier integral preserves the L^2 norm for this dense subset so that by the Hahn-Banach theorem there is a unique continuous extension of $f \mapsto \hat{f}$ to all of $L^2(\mathbf{R})$. The Plancherel theorem shows that the Fourier integral behaves like a unitary matrix.

Theorem 1.9 (Plancherel) *If $f \in L^2(\mathbf{R})$, then $\hat{f} \in L^2(\mathbf{R})$ and $\|f\| = \|\hat{f}\|$. Also, $\langle f, g \rangle = \langle \hat{f}, \hat{g} \rangle$ for any two functions $f, g \in L^2$.* □

Since the Fourier transform is a unitary operator on L^2, its inverse is its adjoint. The adjoint is defined by the integral

$$\check{g}(x) = \int_{-\infty}^{\infty} g(\xi) e^{2\pi i x \xi} \, d\xi. \tag{1.44}$$

It is evident that $\hat{f}(y) = \check{f}(-y)$, and thus like the Fourier integral this transform also has a unique continuous extension to $L^2(\mathbf{R})$. We also have the identity $\hat{\bar{f}}(y) = \overline{\check{f}}(y)$.

Theorem 1.10 (Fourier Inversion) *If $f \in L^2(\mathbf{R})$ and $g = \hat{f}$, then $g \in L^2(\mathbf{R})$ and $\check{g} = f$. Likewise, if $h = \check{f}$, then $h \in L^2(\mathbf{R})$ and $\hat{h} = f$.* □

The Fourier integral, Plancherel's theorem, the Riemann-Lebesgue lemma, and the Fourier inversion theorem can all be extended to functions of n variables just by taking integrals over $\mathbf{R}^n$ and replacing $x\xi$ with $x \cdot \xi$ in the exponents.

The Fourier integral transforms functions of rapid decay into smooth functions, and transforms smooth functions into functions of rapid decay. In short, it maps the Schwartz class into itself. We state this result for $\mathbf{R}$; it holds for higher dimensions, too.

Theorem 1.11 *If u belongs to the Schwartz class on $\mathbf{R}$, then so do $\hat{u}$ and $\check{u}$.*

Proof: We know that $\frac{d^n}{dx^n}u(x)$ is Lebesgue integrable because it is continuous and decreases rapidly as $x \to \pm\infty$. But then $\int [\frac{d^n}{dx^n}u(x)]e^{-2\pi i x\xi}\,dx$ is bounded, and we can integrate by parts to conclude that $(2\pi i\xi)^n \int u(x)e^{-2\pi i x\xi}\,dx = (2\pi i\xi)^n \hat{u}(\xi)$ is bounded as $\xi \to \pm\infty$.

We can interchange integral and derivative: $\frac{d^n}{d\xi^n}\hat{u}(\xi) = \frac{d^n}{d\xi^n}\int u(x)e^{-2\pi i x\xi}\,dx = \int (-2\pi i x)^n u(x)e^{-2\pi i x\xi}\,dx$, because the integrand remains absolutely integrable due to the rapid decrease of u. This is true for all $n \in \mathbf{N}$, so we conclude that $\hat{u}$ is a smooth function of ξ.

We can combine the previous two results by Leibniz' rule to conclude that $\hat{u}$ belongs to the Schwartz class. The same conclusion evidently holds for $\check{u}(\xi) = \hat{u}(-\xi)$ as well. □

1.3.2 Fourier series

If $f = f(t)$ is a one-periodic function, then its *Fourier series* (in the modern sense) is the infinite sum

$$f(x) \approx \sum_{k=-\infty}^{\infty} c(k)e^{2\pi i k x}. \tag{1.45}$$

The numbers in the sequence $\{c(k) : k \in \mathbf{Z}\}$ are called the *Fourier coefficients* of the function f; they are computed via the formula

$$c(k) = \hat{f}(k) \stackrel{\text{def}}{=} \int_0^1 f(x)e^{-2\pi i k x}\,dx. \tag{1.46}$$

We have conserved notation by reusing the symbol for the Fourier transform of a function on the line. In a similar way, we can refer to the *inverse Fourier transform* of the sequence $\{c\}$:

$$\check{c}(\xi) \stackrel{\text{def}}{=} \sum_{k=-\infty}^{\infty} c(k)e^{2\pi i k\xi}. \tag{1.47}$$

This is a one-periodic function of the real variable ξ. If $\{c\}$ is absolutely summable, then $\check{c}$ will be a continuous function because of Proposition 1.1. We may also define the *Fourier transform of a sequence* or a *trigonometric series* by

$$\hat{c}(\xi) \stackrel{\text{def}}{=} \sum_{k=-\infty}^{\infty} c(k) e^{-2\pi i k \xi}. \tag{1.48}$$

This is to be considered a periodic function defined on the whole real line. Evidently $\hat{c}(\xi) = \check{c}(-\xi)$. Again we conserve notation by making its meaning dependent on the context.

The coefficients c are related to the coefficients a, b of Equation 1.39:

$$a(0) = c(0); \quad a(k) = \frac{1}{\sqrt{2}}\left[c(-k) + c(k)\right]; \quad b(k) = \frac{1}{\sqrt{2}}\left[c(-k) - c(k)\right]. \tag{1.49}$$

In return, the coefficients a, b are related to the coefficients c by these formulas:

$$c(0) = a(0); \quad c(-k) = \frac{1}{\sqrt{2}}\left[a(k) + ib(k)\right]; \quad c(k) = \frac{1}{\sqrt{2}}\left[a(k) - ib(k)\right]. \tag{1.50}$$

In both cases $k = 1, 2, 3, \ldots$.

It is important to note that f is assumed to be one-periodic, which implies among other things that $\hat{f}(k) = \int_z^{z+1} f(x) e^{-2\pi i k x}\, dx$ for any real z. The properties of the Fourier coefficients depend on the behavior of f as a periodic function on the whole line, and not just on its behavior in the interior of one period interval such as $[0, 1]$. In particular, one should beware of discontinuities hiding at the endpoints of a period interval.

There are analogs to the Riemann-Lebesgue lemma, the Plancherel theorem, and the Fourier Inversion theorem for Fourier series as well:

Lemma 1.12 (Riemann–Lebesgue) *Suppose f is one-periodic. If f is absolutely integrable on the interval $[0, 1]$, then $\hat{f}(k) \to 0$ as $|k| \to \infty$.*

Proof: Let $f = f(t)$ be a periodic function which is absolutely integrable on $[0, 1]$. Then

$$\begin{aligned}
\hat{f}(k) &= \int_0^1 f(t) e^{-2\pi i k t}\, dt = \int_0^1 f\left(t + \frac{1}{2k}\right) \exp\left(-2\pi i k t - i\pi\right) dt \\
&= -\int_0^1 f\left(t + \frac{1}{2k}\right) e^{-2\pi i k t}\, dt \\
\Rightarrow \quad 2\hat{f}(k) &= \int_0^1 \left[f(t) - f\left(t + \frac{1}{2k}\right)\right] e^{-2\pi i k t}\, dt \\
\Rightarrow \quad 2|\hat{f}(k)| &\leq \int_0^1 \left| f(t) - f\left(t + \frac{1}{2k}\right)\right| dt \to 0 \qquad \text{as } |k| \to \infty,
\end{aligned}$$

by the continuity of the L^1 norm with respect to translation. □

Corollary 1.13 *If f is one-periodic and absolutely integrable on the interval $[0,1]$, and $f'(x_0)$ exists, then the Fourier series for f converges at the point x_0.*

Proof: We may assume without loss that $x_0 = 0$ and that $f(0) = 0$, since if not we can just translate f and subtract a constant. We now write $f(t) = \left(e^{-2\pi it} - 1\right) g(t)$ and note that the one-periodic function g is absolutely integrable, since the quotient $f(t)/\left(e^{-2\pi it} - 1\right)$ has a finite limit as $t \to 0$. But then $\hat{f}(k) = \hat{g}(k+1) - \hat{g}(k)$, so that we obtain a telescoping Fourier series for $f(0)$:

$$\sum_{k=-n}^{m-1} \hat{f}(k) = \hat{g}(m) - \hat{g}(-n).$$

By the Riemann-Lebesgue lemma, this series converges to zero as $n, m \to \infty$ in any manner. □

Remark. These two elegant proofs were brought to my attention by Guido Weiss, who believes that they are both many years old.

Theorem 1.14 (Plancherel) *If $f \in L^2(\mathbf{T})$, then $\hat{f} \in \ell^2(\mathbf{Z})$ and*

$$\|f\| = \left(\int_0^1 |f(x)|^2\, dx\right)^{\frac{1}{2}} = \left(\sum_{k\in\mathbf{Z}} |\hat{f}(k)|^2\right)^{\frac{1}{2}} = \|\hat{f}\|.$$

Also, $\langle g, f\rangle = \int_0^1 \bar{g}(x) f(x)\, dx = \sum_{k\in\mathbf{Z}} \overline{\hat{g}(k)}\hat{f}(k) = \langle \hat{g}, \hat{f}\rangle$ for any $f, g \in L^2(\mathbf{T})$. □

Theorem 1.15 (Fourier Inversion) *If $f \in L^2([0,1])$ and we put*

$$f_N(x) = \sum_{k=-N}^{N-1} \hat{f}(k) e^{2\pi ikx}, \qquad N = 1, 2, 3, \ldots,$$

then $f_N \to f$ in the sense of $L^2(\mathbf{T})$ as $N \to \infty$. □

Proofs of these fundamental results may be found in [3], for example.

1.3.3 General orthogonal transformations

The representation of f by Equation 1.39 may be put into a more general context as an expansion in an orthonormal basis: $\{1, \sqrt{2}\cos 2\pi nx, \sqrt{2}\sin 2\pi nx : n = 1, 2, \ldots\}$ is orthonormal with respect to the (Hermitean) inner product of Equation 1.11. If

$\{\phi_k : k \in \mathbf{Z}\}$ is any orthonormal basis for $L^2([0,1])$, we can get an expansion of $f \in L^2([0,1])$ in these functions as follows:

$$f_N = f_N(x) = \sum_{k=-N}^{N-1} \langle f, \phi_k \rangle \phi_k(x). \tag{1.51}$$

The completeness of the basis implies that $\|f - f_N\| \to 0$ as $N \to \infty$.

The functions $\{\phi_j : j = 1, 2, \ldots\}$ defined by $\phi_j(x) = \sqrt{2} \sin \pi j x$ form an orthonormal basis of $L^2([0,1])$ and can be used to compute the *sine transform* in a manner identical to that of the Fourier transform, as in Equation 1.51. Likewise, we can use the functions $\phi_j(x) = \sqrt{2} \cos \pi j x$, $j = 1, 2, \ldots$ to get the *cosine transform.*

Another orthogonal transformation is obtained using $\phi_j(x) = \sqrt{2} \sin \pi (j + \frac{1}{2}) x$, $j = 0, 1, 2, \ldots$, or alternatively $\phi_j(x) = \sqrt{2} \cos \pi (j + \frac{1}{2}) x$, $j = 0, 1, 2, \ldots$. Still more variations of sine and cosine are possible, such as the *Hartley transform* given by the functions $\phi_j(x) = \sin \pi j x + \cos \pi j x$, $j = 0, 1, 2, \ldots$. The orthonormality of these sets can be shown directly, or else we can use the Sturm–Liouville theorem with a suitable operator and boundary condition. Likewise, the Sturm–Liouville theorem gives their completeness, which can also be derived directly from the completeness of the set $\{e^{2\pi i j x}\}$ in $L^2([0,1])$ by taking appropriate linear combinations.

1.3.4 Discrete Fourier transforms

We now turn to the discrete and finite-rank case. For a vector $v \in \mathbf{C}^N$, $v = \{v(k) : k = 0, 1, \ldots, N-1\}$, we define the *discrete Fourier transform* or *DFT* of v to be the vector $\hat{v} \in \mathbf{C}^N$ given by the following formula:

$$\hat{v}(k) \stackrel{\text{def}}{=} \frac{1}{\sqrt{N}} \sum_{j=0}^{N-1} v(j) e^{-2\pi i j k/N}, \qquad k = 0, 1, \ldots, N-1. \tag{1.52}$$

The identity $e^{-2\pi i j(N-k)/N} = e^{-2\pi i j(-k)/N}$ implies that large positive frequencies $N-k$ are indistinguishable from small negative frequencies $-k$, a phenomenon sometimes called *aliasing.* If N is an even number, we can identify half the coefficients $\hat{v}(k) : k = \frac{N}{2}, \frac{N}{2}+1, \ldots, N-1$ with the negative frequencies $-\frac{N}{2}, -\frac{N}{2}+1, \ldots, -1$. The other half of the coefficients gives the positive frequencies $0, 1, 2, \ldots, \frac{N}{2}-1$. This is what we will mean when we write the coefficients of $\hat{v}$ as

$$\left(\hat{v}(-\frac{N}{2}), \hat{v}(-\frac{N}{2}+1), \ldots, \hat{v}(-1), \hat{v}(0), \hat{v}(1), \ldots, \hat{v}(\frac{N}{2}-1)\right). \tag{1.53}$$

The vector in Equation 1.53 is simply $\hat{v}$ with its coefficients rearranged to better suit certain purposes; we will use it when convenient.

The vectors $\omega_j \in \mathbf{C}^N$, $j = 0, 1, \ldots, N-1$ defined by $\omega_j(k) = \frac{1}{\sqrt{N}} e^{2\pi ijk/N}$ form an orthonormal basis with respect to the Hermitean inner product on $\mathbf{C}^N$. This means that the map $v \mapsto \hat{v}$ is a unitary transformation, so we have an easy proof of the Plancherel theorem in this context:

Theorem 1.16 (Plancherel) *$\langle v, w \rangle = \langle \hat{v}, \hat{w} \rangle$ for any two vectors $v, w \in \mathbf{C}^N$. In particular, $\|v\| = \|\hat{v}\|$.* □

1.3.5 Heisenberg's inequality

A square-integrable function u defines two probability density functions: $x \mapsto |u(x)|^2/\|u\|^2$ and $\xi \mapsto |\hat{u}(\xi)|^2/\|\hat{u}\|^2$. It is not possible for both of these densities to be arbitrarily concentrated, as we shall see from the inequalities below.

Suppose that $u = u(x)$ belongs to the Schwartz class $\mathcal{S}$. Then $x\frac{d}{dx}|u(x)|^2 = x\left[u(x)\bar{u}'(x) + \bar{u}(x)u'(x)\right]$ is integrable and tends to 0 as $|x| \to \infty$. We can therefore integrate by parts to get the following formula:

$$\int_{\mathbf{R}} -x\frac{d}{dx}|u(x)|^2\,dx = \int_{\mathbf{R}} |u(x)|^2\,dx = \|u\|^2. \tag{1.54}$$

But also, the Cauchy–Schwarz inequality and the triangle inequality imply that

$$\begin{aligned} \left|\int_{\mathbf{R}} -x\frac{d}{dx}|u(x)|^2\,dx\right| &\leq 2\int_{\mathbf{R}} |xu(x)u'(x)|\,dx \qquad (1.55)\\ &\leq 2\left(\int_{\mathbf{R}} |xu(x)|^2\,dx\right)^{1/2}\left(\int_{\mathbf{R}} |u'(x)|^2\,dx\right)^{1/2}. \end{aligned}$$

Combining the last two inequalities gives $\|xu(x)\|\cdot\|u'(x)\| \geq \frac{1}{2}\|u(x)\|^2$. Now $\widehat{u'}(\xi) = 2\pi i\xi\hat{u}(\xi)$, and $\|\hat{v}\| = \|v\|$ by Plancherel's theorem, so we can rewrite the inequality as follows:

$$\frac{\|xu(x)\|}{\|u(x)\|}\cdot\frac{\|\xi\hat{u}(\xi)\|}{\|\hat{u}(\xi)\|} \geq \frac{1}{4\pi}.$$

Since the right-hand side is not changed by translation $u(x) \mapsto u(x - x_0)$ or modulation $\hat{u}(\xi) \mapsto \hat{u}(\xi - \xi_0)$, we have proved

$$\inf_{x_0}\left(\frac{\|(x - x_0)u(x)\|}{\|u(x)\|}\right)\cdot\inf_{\xi_0}\left(\frac{\|(\xi - \xi_0)\hat{u}(\xi)\|}{\|\hat{u}(\xi)\|}\right) \geq \frac{1}{4\pi}. \tag{1.56}$$

Equation 1.56 is called *Heisenberg's inequality*. We mention the usual names

$$\triangle x = \triangle x(u) \stackrel{\text{def}}{=} \inf_{x_0}\left(\frac{\|(x - x_0)u(x)\|}{\|u(x)\|}\right); \tag{1.57}$$

$$\triangle \xi = \triangle \xi(u) \stackrel{\text{def}}{=} \inf_{\xi_0}\left(\frac{\|(\xi - \xi_0)\hat{u}(\xi)\|}{\|\hat{u}(\xi)\|}\right). \tag{1.58}$$

The quantities $\triangle x$ and $\triangle\xi$ are called the *uncertainties* in position and momentum respectively, and they provide an inverse measure of how well u and $\hat{u}$ are localized. Then Heisenberg's inequality assumes the guise of the *uncertainty principle*:

$$\triangle\, x \cdot \triangle\xi \geq \frac{1}{4\pi}. \tag{1.59}$$

It is not hard to show that the infima 1.57 and 1.58 are attained at the points x_0 and ξ_0 defined by the following expressions:

$$x_0 = x_0(u) \quad = \quad \frac{1}{\|u\|^2}\int_{\mathbf{R}} x|u(x)|^2\, dx; \tag{1.60}$$

$$\xi_0 = \xi_0(u) \quad = \quad \frac{1}{\|\hat{u}\|^2}\int_{\mathbf{R}} \xi|\hat{u}(\xi)|^2\, d\xi. \tag{1.61}$$

The Dirac mass $\delta(x - x_0)$ is perfectly localized at position x_0, with zero position uncertainty, but both its frequency and frequency uncertainty are undefined. Likewise, the exponential $e^{2\pi i\xi_0 x}$ is perfectly localized in momentum (since its Fourier transform is $\delta(\xi - \xi_0)$), but both its position and position uncertainty are undefined. Equality is obtained in Equations 1.56 and 1.59 if we use the *Gaussian function* $u(x) = e^{-\pi x^2}$. It is possible to show, using the uniqueness theorem for solutions to linear ordinary differential equations, that the only functions which minimize Heisenberg's inequality are scaled, translated, and modulated versions of the Gaussian function.

If $\triangle x$ and $\triangle\xi$ are both finite, then the quantities x_0 and ξ_0 can be used to assign a nominal position and momentum to an imperfectly localized function.

1.3.6 Convolution

Convolution is a multiplication-like operation between two functions. When the functions depend upon one real variable, then it is implemented as an integral. When the functions take values only at the integers, *i.e.*, if they are sampled signals, then convolution is implemented as a (possibly) infinite sum. If however the sampled functions are periodic, then convolution reduces to a finite sum and we can also show that the convolution of two periodic functions is itself periodic.

The abstract definition of convolution requires two measurable functions u, v taking complex values on a locally compact group G.

$$u : G \to \mathbf{C}; \qquad v : G \to \mathbf{C}.$$

We also need a left-invariant or Haar measure μ on the group; see [21] for a definition of this measure, and for some of its properties. The convolution of u and v, denoted

$u * v$, is another complex-valued function on the group defined by the following abstract integral:

$$u * v(x) = \int_G u(y)v(y^{-1}x)\, d\mu(y). \tag{1.62}$$

In this integral, the property of "left-invariance" means that $d\mu(xy) = d\mu(y)$ for all $x \in G$. Thus by replacing y with xz^{-1}, $y^{-1}x$ with z, and $d\mu(y)$ with $d\mu(xz^{-1}) = d\mu(z^{-1})$ we get

$$v * u(x) = \int_G v(y)u(x^{-1}y)\, d\mu(y) = \int_G u(z)v(x^{-1}z)\, d\mu(z^{-1}).$$

This is a different convolution $u \check{*} v$, taken with respect to the right-invariant measure $\check{\mu}(z) \stackrel{\text{def}}{=} \mu(z^{-1})$. If $\check{\mu} = \mu$, then convolution is commutative: $\check{*} = *$, and $u*v = v*u$. All Abelian groups and all compact groups have this property.

We will be interested in four particular cases of G, all of which are Abelian. We shall first establish some basic mathematical properties of convolution for functions on $\mathbf{R}$ and $\mathbf{T}$, then we will transfer these to the discrete case of $\mathbf{Z}$ and $\mathbf{Z}/q\mathbf{Z}$.

Convolution on the line

Here $G = \mathbf{R}$, the real number line with addition; u and v are real-valued (or complex-valued) functions; the convolution of u and v is defined by the integral

$$u * v(x) = \int_{-\infty}^{\infty} u(y)v(x-y)\, dy.$$

Functions of a real variable represent a useful limiting case whose properties can be conveniently described within the theory of classical harmonic analysis. We begin by stating a few facts about convolution.

Theorem 1.17 (Fourier Convolution) *If $u = u(x)$ and $v = v(x)$ are Schwartz functions, then $\widehat{u * v}(\xi) = \hat{u}(\xi)\,\hat{v}(\xi)$.*

Proof:

$$\begin{aligned}
\widehat{u * v}(\xi) &= \iint u(y)v(x-y)e^{-2\pi i x\xi}\, dydx \\
&= \iint u(y)e^{-2\pi i y\xi} v(x-y)e^{-2\pi i(x-y)\xi}\, dyd(x-y) \\
&= \left(\int u(y)e^{-2\pi i y\xi}\, dy\right)\left(\int v(z)e^{-2\pi i z\xi}\, dz\right) = \hat{u}(\xi)\,\hat{v}(\xi).
\end{aligned}$$

The integrals converge since u and v are continuous and rapidly decreasing. □

In other words, the Fourier transform converts convolution into pointwise multiplication. We can use this result together with Plancherel's theorem to prove that convolution with integrable functions preserves square-integrability.

Corollary 1.18 *Suppose that $u = u(x)$ is integrable. If $v = v(x)$ belongs to L^p, $1 \le p \le \infty$, then so does the function $u * v$.*

Proof: We will prove the easy case $p = 2$; the other cases are treated in Theorem 2 of [103]. Suppose that u and v are Schwartz functions. Then by Plancherel's theorem and the convolution theorem we have $\|u * v\| = \|\widehat{u * v}\| = \|\hat{u}\,\hat{v}\|$. This gives the estimate

$$\|u * v\| \le \|\hat{u}\|_\infty \|\hat{v}\| = \|\hat{u}\|_\infty \|v\| \le \|u\|_{L^1} \|v\|. \tag{1.63}$$

The last inequality follows since $\|\hat{u}\|_\infty$ is bounded by $\|u\|_{L^1}$. The result for integrable u and square-integrable v follows since Schwartz functions are dense in L^1 and L^2. □

Convolution with integrable u is a bounded linear operator on L^2, and we will have occasion to estimate this bound with the following proposition:

Proposition 1.19 *If $u = u(x)$ is absolutely integrable on $\mathbf{R}$, then the convolution operator $v \mapsto u * v$ as a map from L^2 to L^2 has operator norm* $\sup\{|\hat{u}(\xi)| : \xi \in \mathbf{R}\}$.

Proof: By Equation 1.63, $\|u * v\| \le \sup\{|\hat{u}(\xi)|\,\|v\| : \xi \in \mathbf{R}\}$. By the Riemann–Lebesgue lemma, $\hat{u}$ is bounded and continuous and $|\hat{u}(\xi)| \to 0$ as $|\xi| \to \infty$, so $\hat{u}$ achieves its maximum amplitude $\sup\{|\hat{u}(\xi)| : \xi \in \mathbf{R}\} < \infty$ at some point $\xi_* \in \mathbf{R}$. We may assume without loss that $\xi_* = 0$. To show that the operator norm inequality is sharp, let $\epsilon > 0$ be given and find $\delta > 0$ such that $|\xi - \xi_*| < \delta \Rightarrow |\hat{u}(\xi) - \hat{u}(\xi_*)| < \epsilon$. If we take $v(x) = \frac{\sin 2\pi\delta x}{\pi x}$, then $\hat{v}(\xi) = \mathbf{1}_{[-\delta,\delta]}(\xi)$, and $\|u * v\| = \|\hat{u}\hat{v}\| > (1 - \epsilon)\,|\hat{u}(\xi_*)|\,\|\hat{v}\| = (1 - \epsilon)\,|\hat{u}(\xi_*)|\,\|v\|$. □

Lemma 1.20 (Smoothing) *If u is integrable on $\mathbf{R}$ and v is bounded and has a bounded continuous derivative on $\mathbf{R}$, then $(u * v)' = u * v'$.*

Proof: $(u * v)' = \frac{d}{dx} \int u(y) v(x - y)\, dy = \int u(y) v'(x - y)\, dy = u * v'$. □

Thus a convolution has at least as many derivatives as the smoother of the two functions.

Convolution widens the support of a function. Suppose that u and v are compactly supported functions of $\mathbf{R}$. Then they both have finite support width and satisfy the following relation:

Lemma 1.21 (Support) $\operatorname{diam}\operatorname{supp} u * v \leq \operatorname{diam}\operatorname{supp} u + \operatorname{diam}\operatorname{supp} v$. □

This inequality is sharp, in the sense that we can find two compactly supported functions u and v for which equality holds: for example, $u = \mathbf{1}_{[a,b]}$ and $v = \mathbf{1}_{[c,d]}$. It is also sharp in the sense that support diameter is the right notion rather than measure of the support set, since for every $\epsilon > 0$ we can find two functions u and v with $|\operatorname{supp} u|, |\operatorname{supp} v| < \epsilon \ll 1 = \operatorname{diam}\operatorname{supp} u = \operatorname{diam}\operatorname{supp} v$, but $\operatorname{diam}\operatorname{supp} u * v = |\operatorname{supp} u * v| = 2$. One such example is constructed in the exercises.

Convolution on the circle

Here $G = \mathbf{T}$, the circle of unit circumference with angle addition; $u = u(e^{2\pi i x})$ and $v = v(e^{2\pi i x})$ are real-valued (or complex-valued) periodic functions of $x \in \mathbf{T}$; the convolution of u and v is given by the integral

$$u * v(e^{2\pi i x}) \stackrel{\text{def}}{=} \int_0^1 u(e^{2\pi i y}) v(e^{2\pi i(x-y)})\, dy, \tag{1.64}$$

and as before we have

Proposition 1.22 *If u and v are functions on the circle $\mathbf{T}$ such that $\hat{u}$ and $\hat{v}$ exist, then $\widehat{u * v}(n) = \hat{u}(n)\hat{v}(n)$ for $n \in \mathbf{Z}$.* □

Proposition 1.23 *If $u \in L^1(\mathbf{T})$ and $v \in L^p(\mathbf{T})$, $1 \leq p \leq \infty$, then $u * v \in L^p(\mathbf{T})$.* □

Proposition 1.24 *If $u \in L^1(\mathbf{T})$, then the operator norm of the map $v \mapsto u * v$ from $L^2(\mathbf{T})$ to $L^2(\mathbf{T})$ is $\sup\{|\hat{u}(n)| : n \in \mathbf{Z}\}$.* □

Convolution of doubly infinite sequences

Here $G = \mathbf{Z}$, the integers with addition, and the abstract integral reduces to the infinite sum

$$u * v(x) = \sum_{y=-\infty}^{\infty} u(y)v(x-y).$$

Proposition 1.25 *If $u \in \ell^1(\mathbf{Z})$ and $v \in \ell^p(\mathbf{Z})$ for $1 \leq p \leq \infty$, then $u * v \in \ell^p$.* □

We can compute convolutions efficiently by multiplication of Fourier transforms:

Proposition 1.26 *If u and v are infinite sequences such that $\hat{u}$ and $\hat{v}$ exist a.e., then $\widehat{u * v}(\xi) = \hat{u}(\xi)\hat{v}(\xi)$ for almost every $\xi \in \mathbf{T}$.* □

Proposition 1.27 *If $u \in \ell^1(\mathbf{Z})$, then the map $v \mapsto u * v$ has operator norm $\max_{\xi \in \mathbf{T}} |\hat{u}(\xi)|$ as a map from $L^2(\mathbf{T})$ to $L^2(\mathbf{T})$.* □

The special case which will interest us the most is that of "finitely supported" sequences, *i.e.*, those for which $u(x) = 0$ except for finitely many integers x. Such sequences are obviously summable, and it is easy to show that the convolution of finitely supported sequences is also finitely supported. Furthermore, if u is finitely supported, then $\hat{u}$ is a trigonometric polynomial and we may use many powerful tools from classical analysis to study it.

So, let $u = u(x)$ and $v = v(x)$ be finitely supported sequences taking values at integers $x \in \mathbf{Z}$, with $u(x) = 0$ unless $a \leq x \leq b$ and $v(x) = 0$ unless $c \leq x \leq d$. We call $[a, b]$ and $[c, d]$ the *support intervals* $\operatorname{supp} u$ and $\operatorname{supp} v$, respectively, and $b - a$ and $d - c$ the *support widths* for the sequences u and v. Then $u * v(x) = 0$ unless there is some $y \in \mathbf{Z}$ for which $y \in [a, b]$ and $x - y \in [c, d]$, which requires that $c + a \leq x \leq d + b$. Hence $u * v$ is also finitely supported, with the width of its support growing to $(d + b) - (c + a) = (b - a) + (d - c)$, or the sum of the support widths of u and v. The convolution at x is a sum over $y \in [a, b] \cap [x - d, x - c]$, and will take the following values:

$$u * v(x) = \begin{cases} 0, & \text{if } x < c + a, \\ \sum_{y=a}^{x-c} u(y)v(x - y), & \text{if } c + a \leq x < c + b, \\ \sum_{y=a}^{b} u(y)v(x - y), & \text{if } c + b \leq x \leq d + a, \\ \sum_{y=x-d}^{b} u(y)v(x - y), & \text{if } d + a < x \leq d + b, \\ 0, & \text{if } x > d + b. \end{cases} \tag{1.65}$$

Note that the middle term will only be present if $c - a \leq d - b$, *i.e.*, if $|\operatorname{supp} u| = b - a \leq |\operatorname{supp} v| = d - c$. We call this the *long signal* case, since we will be thinking of v as the signal and u as the filter, with Uv being a filtering operation on the signal v. For theoretical purposes we may assume that this holds, without any loss of generality, because u and v are interchangeable by the formula $u * v(x) = v * u(x)$ so we may simply choose u to be the shorter of the two. However, in the software implementation of convolution it may be more convenient to allow arbitrary support lengths for u and v, so we will also derive the formulas for $u * v$ in the *short signal* case $b - a > d - c$:

$$u * v(x) = \begin{cases} 0, & \text{if } x < c + a, \\ \sum_{y=a}^{x-c} u(y)v(x - y), & \text{if } c + a \leq x < d + a, \\ \sum_{y=x-d}^{x-c} u(y)v(x - y), & \text{if } d + a \leq x \leq c + b, \\ \sum_{y=x-d}^{b} u(y)v(x - y), & \text{if } c + b < x \leq d + b, \\ 0, & \text{if } x > d + b. \end{cases} \tag{1.66}$$

Convolution of periodic sequences

Let $G = \mathbf{Z}/q\mathbf{Z}$ be the integers $\{0, 1, \ldots, q-1\}$ with addition modulo q, for which the convolution integral becomes a finite sum:

$$u * v(x) = \sum_{y=0}^{q-1} u(y)v(x - y \bmod q).$$

Since all sequences in this case are finite, there is no question of summability. Convolution becomes multiplication if we use Equation 1.52:

Proposition 1.28 *If u, v are q-periodic sequences, then $\widehat{u * v}(y) = \hat{u}(y)\hat{v}(y)$.* □

Thus we can compute the norm of discrete convolution operators:

Proposition 1.29 *The operator norm of the map $v \mapsto u*v$ from $\ell^2(\mathbf{Z}/q\mathbf{Z})$ to itself is $\max_{0 \le y < q} |\hat{u}(y)|$.*

Proof: The maximum is achieved for the sequence $v(x) = \exp 2\pi i x y_0/q$, where y_0 is the maximum for $|\hat{u}|$, since then $\hat{v}(y) = \sqrt{q}\,\delta(y - y_0)$. □

Periodic convolution is the efficient way to apply a convolution operator to a periodic sequence. Suppose that $v \in \ell^\infty(\mathbf{Z})$ happens to be q-periodic, namely that $v(x+q) = v(x)$ for all $x \in \mathbf{Z}$. Then for $u \in \ell^1(\mathbf{Z})$ we can compute the convolution of u and v by decomposing $y = k + qn$:

$$\begin{aligned} u * v(x) &= \sum_{y=-\infty}^{\infty} u(y)v(x-y) = \sum_{n=-\infty}^{\infty} \sum_{k=0}^{q-1} u(k+qn)v(x-k-qn) \\ &= \sum_{k=0}^{q-1} \left(\sum_{n=-\infty}^{\infty} u(k+qn) \right) v(x-k). \end{aligned}$$

Now let us define the *q-periodization* u_q of $u \in \ell^1(\mathbf{Z})$ to be the q-periodic function

$$u_q(k) \stackrel{\text{def}}{=} \sum_{n=-\infty}^{\infty} u(k+qn). \tag{1.67}$$

Thus starting with a single sequence u, we can get a family of convolution operators, one on $\mathbf{Z}/q\mathbf{Z}$ for each integer $q > 0$:

$$U_q : \ell^2(\mathbf{Z}/q\mathbf{Z}) \to \ell^2(\mathbf{Z}/q\mathbf{Z}); \qquad U_q v(x) = u_q * v(x) = \sum_{k=0}^{q-1} u_q(k)v(x-k). \tag{1.68}$$

In effect, we *preperiodize* the sequence u to any desired period q before applying the convolution operator.

1.3.7 Dilation, decimation, and translation

Dilation by s, for $s > 0$, can be regarded as the process of stretching a function over an s times larger domain. We will denote this by the operator σ_s, which acts on functions of a continuous variable by the following formula:

$$[\sigma_s u](x) \stackrel{\text{def}}{=} \frac{1}{\sqrt{s}} u\left(\frac{x}{s}\right). \tag{1.69}$$

The normalization by $1/\sqrt{s}$ makes this operator unitary on L^2. Observe that σ_1 is the identity. This operation has the following properties:

Proposition 1.30 $\|\sigma_s u\| = \|u\|$ *and* $\int \sigma_s u = \sqrt{s} \int u$, *whenever these make sense. Also, if* $\operatorname{supp} u = E \subset \mathbf{R}$, *then* $\operatorname{supp} \sigma_s u = sE$, *where* $sx \in sE \iff x \in E$. *Hence* $|\operatorname{supp} \sigma_s u| = s\, |\operatorname{supp} u|$ *and* $\operatorname{diam} \operatorname{supp} \sigma_s u = s \operatorname{diam} \operatorname{supp} u$. □

If u is a one-periodic function, then $\sigma_s u$ is s-periodic; we write $u \in L^2(\mathbf{T}) \Rightarrow \sigma_s u \in L^2(s\mathbf{T})$.

Decimation by q, on the other hand, can be regarded as the process of discarding all values of a sampled function except those indexed by a multiple of $q > 0$. We denote it by d_q, and we have

$$[d_q u](n) \stackrel{\text{def}}{=} u\,(qn)\,. \tag{1.70}$$

If $u = \{u(n) : n \in \mathbf{Z}\}$ is an infinite sequence, then the new infinite sequence $d_q u$ is just $\{u(qn) : n \in \mathbf{Z}\}$ or every q^{th} element of the original sequence.

If u is finitely supported and $\operatorname{supp} u = [a, b]$, then $d_q u$ is also finitely supported and $\operatorname{supp} d_q u = [a, b] \bigcap q\mathbf{Z}$. This set contains either $\left\lfloor \frac{|b-a|}{q} \right\rfloor$ or $\left\lfloor \frac{|b-a|}{q} \right\rfloor + 1$ elements.

If u is a periodic sequence of period p, then $d_q u$ has period $q/\gcd(p, q)$. Counting degrees of freedom, the number of q-decimated subsequences of a p-periodic sequence needed to reproduce it is exactly $\gcd(p, q)$. If $\gcd(p, q) = 1$, then decimation is just a permutation of the original sequence and there is no reason to perform it. Thus, in the typical case of $q = 2$ we will always assume that p is even.

The *translation* or *shift* operator τ_y is defined by

$$\tau_y u(x) = u(x - y). \tag{1.71}$$

Whatever properties u has at $x = 0$ the function $\tau_y u$ has at $x = y$. Observe that τ_0 is the identity operator. Translation invariance is a common property of formulas derived from physical models because the choice of "origin" 0 as in $u(0)$ for an infinite sequence is usually arbitrary. Any functional or measurement computed for

u which does not depend on this choice of origin must give the same value for the sequence $\tau_y u$, regardless of y. For example, the energy $\|u\|^2$ in a sequence does not depend on the choice of origin:

$$\text{For all } y, \qquad \|u\|^2 = \|\tau_y u\|^2. \tag{1.72}$$

Such invariance can be used to algebraically simplify formulas for computing the measurement.

Translation and dilation do not commute in general, but there is an "intertwining" relation

$$\text{For all } x, y, p, \qquad \tau_y \sigma_p u(x) = \sigma_p \tau_{y/p} u(x). \tag{1.73}$$

Let us use t_y to denote translation in the discrete case: $t_y u(n) \stackrel{\text{def}}{=} u(n-y)$. The intertwining relation in this case is $t_y d_p u = d_p t_{py} u$, and we have an elementary result:

Lemma 1.31 *If $\|u\|$ is finite, then so are $\|d_p \tau_y u\|$ for $y = 0, 1, \ldots, p-1$, plus we have the relation*

$$\|u\|^2 = \|d_p t_1 u\|^2 + \|d_p t_2 u\|^2 + \ldots + \|d_p t_p u\|^2.$$

□

1.4 Approximation

We can use only finitely many parameters in any computation. A function of a real variable must first be approximated by its projection into a finite-dimensional subspace, the linear span of a finite collection $\Phi_N = \{\phi_{N,k}(t) : k = 0, 1, 2, \ldots, N-1\}$ of *synthesis functions.* So that we can improve the approximation, we must fit Φ_N into a family of finite-dimensional collections $\Phi = \{\Phi_N : N = 1, 2, \ldots\}$. Such functions can be as simple as characteristic functions of translates and dilates of a *sampling interval,* or as complicated as the eigenfunctions of a linear operator. They can also be singular measures. When there is a choice of projections, there can be competition between ease of evaluation and ease of subsequent computation.

1.4.1 Averaging and sampling

The rank-N approximation to $f \in L^2$, using the family Φ, is

$$f_N(t) = \sum_{k=0}^{N-1} a_N(k)\, \phi_{N,k}(t). \tag{1.74}$$

We insist that $\|f - f_N\| \to 0$ as $N \to \infty$. The constants $a_N(k)$ may be chosen, for example, to minimize $\|f-f_N\|$ for each fixed N: this is called *Galerkin's method* and results in an orthogonal projection $f \mapsto f_N$. In that case, $a_N(k)$ can be computed as an inner product $a_N(k) = \langle \psi_{N,k}, f\rangle$, where $\Psi_N = \{\psi_{N,k} : k = 0, 1, \ldots, N-1\}$ is a collection of *sampling* or *analysis functions*, which is a *dual set* of the functions $\phi_{N,k}$: $\langle \phi_{N,k}, \psi_{N,j}\rangle = \delta(k-j)$. The analysis functions are not uniquely determined by the synthesis functions, in general.

If the family Φ_N is orthonormal, then we can use the synthesis functions as analysis functions and the minimizing constants are given by

$$a_N(k) = \langle \phi_{N,k}, f\rangle. \tag{1.75}$$

This fact is a consequence of *Bessel's inequality*; see [3], p.309 for a proof.

The simplest finite-rank projection of a function $f = f(t)$ is a sampling at finitely many points: $f \mapsto \{f(t_0), \ldots, f(t_{N-1})\}$. This is equivalent to using the Dirac masses $\{\delta(t-t_k) : k = 0, 1, \ldots, N-1\}$ as the analysis "functions" Ψ_N. This operation costs very few computations, namely $O(N)$ function evaluations, and we can refine the sampling by increasing N and refining the grid t_k. One choice of synthesis functions Φ_N is then the characteristic functions $\mathbf{1}_k(t)$ supported in the interval $[t_k, t_{k+1}]$ between adjacent sample points. In this case, $f(t)$ is approximated by the constant $f(t_k)$ whenever t_k is the nearest *sample point* to the left of t. Another choice of synthesis functions is the set of *hat functions*:

$$h_k(t) = \begin{cases} 0, & \text{if } t \le t_{k-1} \text{ or } t \ge t_{k+1}; \\ \frac{t-t_{k-1}}{t_k - t_{k-1}}, & \text{if } t_{k-1} < t \le t_k; \\ \frac{t_{k+1}-t}{t_{k+1}-t_k}, & \text{if } t_k < t < t_{k+1}. \end{cases} \tag{1.76}$$

Notice that $h_k(t_k) = 1$ is the maximum. Figure 1.1 shows an example plot of such a function. With hats, the approximating function will be piecewise linear and continuous rather than piecewise constant as in the indicator function case. A whole family of synthesis functions can be built by smoothing up the hat functions. For example, we can use *cubic splines* centered at t_k, with vanishing derivatives at t_{k-1} and t_{k+1}:

$$c_k(t) = \begin{cases} 0, & \text{if } t \le t_{k-1} \text{ or } t \ge t_{k+1}; \\ 3\left(\frac{t-t_{k-1}}{t_k-t_{k-1}}\right)^2 - 2\left(\frac{t-t_{k-1}}{t_k-t_{k-1}}\right)^3, & \text{if } t_{k-1} < t < t_k; \\ 3\left(\frac{t_{k+1}-t}{t_{k+1}-t_k}\right)^2 - 2\left(\frac{t_{k+1}-t}{t_{k+1}-t_k}\right)^3, & \text{if } t_k < t < t_{k+1}. \end{cases} \tag{1.77}$$

Figure 1.2 shows an example of such a cubic spline. Note that here, too, the maximum value is $c_k(t_k) = 1$. If we stick to lone Dirac masses as the analysis functions,

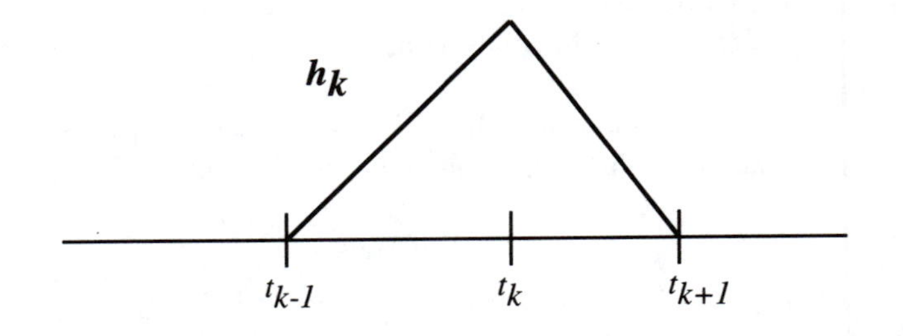

Figure 1.1: Hat function for piecewise linear approximation.

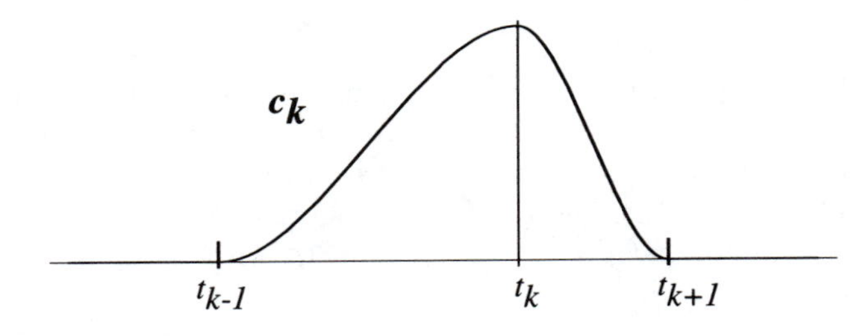

Figure 1.2: Cubic spline function for smoother approximation.

then the synthesis functions will not know about trends and therefore cannot match the derivatives of the approximated function at the sampling points. This problem can be remedied by using *basis splines* (or *B-splines*), which are synthesis functions dual to combinations of Dirac masses.

Regular sampling means that we take equally spaced sample points. In the Dirac mass case just discussed, and for functions supported on the interval $[0,1[$, we can compute the analysis and synthesis functions explicitly: $\psi_{N,k}(t) = N\delta(Nt-k)$, and $\phi_{N,k}(t) = \mathbf{1}(Nt-k)$. Here δ is the Dirac mass supported at zero and $\mathbf{1}$ is the indicator function of the interval $[0,1[$. Thus $\psi_{N,k}$ is the unit Dirac mass at $\frac{k}{N}$, while $\phi_{N,k}$ is the indicator function of the subinterval $[\frac{k}{N}, \frac{k+1}{N}[$. The projection with these ϕ's is

$$f_N(t) = \sum_{k=0}^{N-1} f\left(\frac{k}{N}\right) \mathbf{1}(Nt-k). \tag{1.78}$$

We should remark that f_N is discontinuous at the sample points k/N; this problem can be remedied by sampling at $(k+1/2)/N$, or else using continuous synthesis functions.

Analysis with Dirac masses makes sense only if f is a continuous function. Samples of more general functions in $L^2([0,1])$ must be regarded as averages, and the analysis functions must be at least regular enough to belong to L^2. As a consequence, they must spread out a bit. We will have to control both the analysis and synthesis functions of the dual pair of function families; the analyzing functions influence the coefficient sequence a_N, while the synthesis functions determine the properties of the approximating function f_N.

1.4.2 Band-limited functions

Suppose that $f = f(t)$ is a square-integrable function on all of $\mathbf{R}$. It is called *band-limited* if there is a finite $K > 0$ such that $\hat{f}(\xi) = 0$ whenever $|\xi| > K$. The values of f at any set of sampling points are well-defined because a band-limited function must be continuous. In practice all signals are indistinguishable from band-limited functions, since we are limited to finite sampling rates and finite precision. Shannon's sampling theorem ([99], p.53) asserts that f is determined exactly at all points by its values at discrete points spaced $1/2K$ apart:

$$f(t) = \sum_{n=-\infty}^{\infty} f\left(\frac{n}{2K}\right) \frac{\sin \pi(2Kt-n)}{\pi(2Kt-n)}. \tag{1.79}$$

The *Shannon synthesis function* used here is $\operatorname{sinc}(2Kt-n)$, where $\operatorname{sinc}(t) = \frac{\sin \pi t}{\pi t}$. The function sinc is perfectly band-limited to the interval $[-\frac{1}{2}, \frac{1}{2}]$, but it is not compactly supported in t and thus Equation 1.79 is not a finite sum. In fact, sinc (t) has quite poor decay as $t \to \pm\infty$, but for every square-integrable f the remainder part of the series where $n \notin [-N, N]$ goes to zero as $N \to \infty$. If $|f(t)| \to 0$ rapidly as $t \to \pm\infty$, then we will have rapid convergence of f_N to f.

If the function f is periodic and band-limited to $]-K, K[$, then at most $2K$ samples in a period suffice to determine its exact value everywhere. This follows from the observation that a polynomial of degree $2K-1$ is determined by its values at any $2K$ distinct points.

Averages of a band-limited signal may be computed exactly from samples. Let $\phi = \phi(t)$ be any smooth and integrable (hence square-integrable) real-valued averaging function; this means that ϕ has *unit mass* $\int \phi = 1$. Then

$$\begin{aligned}\langle \phi, f\rangle &= \int \phi(t) f(t)\, dt = \sum_{n=-\infty}^{\infty} \left(\int \phi(t) \frac{\sin \pi(2Kt-n)}{\pi(2Kt-n)} \right) f(n/2K)\, dt \\ &= \sum_{n=-\infty}^{\infty} a(n) f(n/2K), \qquad \text{for } a(n) \stackrel{\text{def}}{=} \int \phi(t) \frac{\sin \pi(2Kt-n)}{\pi(2Kt-n)}\, dt.\end{aligned}$$

The sequence $\{a\}$ is square-summable, because the functions $\{\text{sinc}\,(2Kt-n) : n \in \mathbf{Z}\}$ are orthonormal in $L^2(\mathbf{R})$. We can evaluate $a(n)$ by Plancherel's formula; $\hat{\phi}(\xi) = 0$ if $|\xi| > K$, so we have

$$a(n) = \frac{1}{2K} \int_{-\infty}^{\infty} e^{2\pi i \frac{n}{2K} \xi} \hat{\phi}(\xi)\, d\xi = \frac{1}{2K} \phi(n/2K). \tag{1.80}$$

This proves the following:

Proposition 1.32 *If the functions f and ϕ are band-limited to $]-K, K[$, with ϕ real-valued, then*

$$\langle \phi, f\rangle = \frac{1}{2K} \sum_{n=-\infty}^{\infty} \phi(n/2K) f(n/2K).$$

□

In particular, if we take $\phi(t) = \text{sinc}\,(2Kt)$, then $\phi(n/2K) = 0$ unless $n = 0$ while $\phi(0) = 1$. Thus $\langle \phi, f\rangle = f(0)$; the inner product is just a single sample.

1.4.3 Approximation by polynomials

We can use the functions $\phi_k(t) = t^n$, $n = 0, 1, \ldots$ as our synthesis functions on the interval $[-1, 1]$, since polynomials are dense in $L^2([-1, 1])$. The easiest way to get a dual set is to orthogonalize them via the *Gram–Schmidt* algorithm. The result is called the *Legendre polynomials*. The first few of these are 1, t, $\frac{-1+3t^2}{2}$, $\frac{-3t+5t^3}{2}$.

A *trigonometric polynomial* is a polynomial $P(z)$, $z \in \mathbf{C}$, with the specialization $z = e^{2\pi i\xi}$. Alternatively, it may be defined explicitly as $P(z) = \sum_{k=-M}^{N} a_k z^k = \sum_{k=-M}^{N} a_k e^{2\pi i k\xi}$. We will say that the *degree* of P is $N + M$.

Theorem 1.33 (Weierstrass Approximation) *If $f = f(\xi)$ is continuous on $\mathbf{T}$, then for each $\epsilon > 0$ we can find a trigonometric polynomial $g(\xi) = \sum_{|k|<N} a_k e^{2\pi i k\xi}$ with the property that $|f(\xi) - g(\xi)| < \epsilon$ for all $\xi \in \mathbf{T}$.* □

Note that we may have to use extremely high degrees to get good approximations.

1.4.4 Smooth functions and vanishing moments

If f has $d-1$ continuous derivatives and a finite d^{th} derivative in a neighborhood of a point t_0, then Taylor's theorem ([3],p.113) states that for each t in the neighborhood we can find a $t_1 = t_1(t)$ in the neighborhood such that

$$f(t) = f(t_0) + \sum_{k=1}^{d-1} \frac{f^{(k)}(t_0)}{k!}(t - t_0)^k + \frac{f^{(d)}(t_1)}{d!}(t - t_0)^d. \tag{1.81}$$

If the neighborhood is small and the dth derivative never gets too large, then the unknown *remainder* term $\frac{f^{(d)}(t_1)}{d!}(t - t_0)^d$ will be small.

We will say that a function ϕ has a *vanishing k^{th} moment* at the point t_0 if the following equality holds with the integral converging absolutely:

$$\int (t - t_0)^k \phi(t)\, dt = 0. \tag{1.82}$$

We will say that ϕ has d *vanishing moments* if its first through d^{th} moments vanish. We will discuss examples of such functions in a subsequent chapter. If $\int \phi = 0$, then we will say that the 0th *moment* also vanishes; here it is meaningless to specify t_0. If ϕ has d vanishing moments at zero, then the Fourier transform $\hat{\phi}$ is d-times differentiable at zero with $\hat{\phi}^{(k)}(0) = 0$ for $k = 1, 2, \ldots, d$. Then too, the function $\phi(at - t_0)$ will have d vanishing moments at t_0 for every $a > 0$.

Now suppose that f has d continuous derivatives in a neighborhood of t_0, and that $|f^{(d)}(t)| \le M < \infty$ in this neighborhood. Suppose that ϕ is a real-valued function supported in the interval $[-R, R]$ with d vanishing moments at zero, and $\int \phi = 1$. Define an approximate identity by $\phi_a(t) = a\phi(at - t_0)$. Then we have the following:

$$\begin{aligned}\langle \phi_a, f\rangle &= \int a\phi(at-t_0)f(t)\,dt \\ &= f(t_0) + \frac{1}{d!}\int f^{(d)}\left(t_1(t)\right)(t-t_0)^d a\phi(at-t_0)\,dt.\end{aligned}$$

Here we have used Equations 1.81 and 1.82 to get the second equality. Rewriting and estimating the integral with the triangle inequality gives:

$$|\langle \phi_a, f\rangle - f(t_0)| \leq \frac{2M}{d!}\left(\frac{R}{a}\right)^d. \tag{1.83}$$

Thus the inner product has d^{th} order convergence to the sample as $a \to \infty$, or equivalently, the sample gives a dth order accurate estimate of the inner product.

1.5 Exercises

1. Show that for every N the function $f(t) = e^{-[\log|t|]^2}$ is $O(|t|^{-N})$ as $t \to \infty$, while there is no $\epsilon > 0$ such that $f(t) = O(e^{-\epsilon t})$ as $t \to \infty$.

2. Let T be a tempered distribution, define $\phi_y(x) = \phi(x+y)$ for a Schwartz function ϕ, and define $\psi(y) = T(\phi_y)$. Show that ψ is also a Schwartz function.

3. Show that the set of functions $\{1, \sqrt{2}\cos 2\pi nx, \sqrt{2}\sin 2\pi nx : n = 1, 2, \ldots\}$ is orthonormal with respect to the (Hermitean) inner product in Equation 1.11, that is, show that

$$\begin{aligned}\langle \sqrt{2}\cos 2\pi nx, \sqrt{2}\sin 2\pi mx\rangle &= 0, \qquad \text{for all } n, m \in \mathbf{Z}; \\ \langle \sqrt{2}\cos 2\pi nx, \sqrt{2}\cos 2\pi mx\rangle &= 0, \qquad \text{for all } n, m \in \mathbf{Z},\ n \neq m; \\ \langle \sqrt{2}\sin 2\pi nx, \sqrt{2}\sin 2\pi mx\rangle &= 0, \qquad \text{for all } n, m \in \mathbf{Z},\ n \neq m; \text{ and} \\ \|1\| = \|\sqrt{2}\sin 2\pi nx\| &= \|\sqrt{2}\cos 2\pi nx\| = 1.\end{aligned}$$

4. Show that if $f = g$ in $L^2([0,1])$ and f and g are continuous functions, then $f(x) = g(x)$ for every x.

5. Show that the functions $\phi_k(x) = \mathbf{1}(Nx - k)$, $k = 0, 1, \ldots, N-1$ are an orthogonal collection in $L^2([0,1])$. How can this collection be made orthonormal? Show that it is not a complete system for any N.

6. Show that the vectors $\omega_j \in \mathbf{C}^N$, $j = 0, 1, \ldots, N-1$ defined by $\omega_j(k) = \frac{1}{\sqrt{N}}\exp(2\pi ijk/N)$ form an orthonormal basis.

7. Prove that the infima $\triangle x$ and $\triangle\xi$ are attained by taking x_0 as in Equation 1.60 and ξ_0 as in Equation 1.61.

8. Show that if $\psi = \psi(x)$ belongs to $L^2(\mathbf{R})$ and $\triangle x(\psi)$ is finite, then ψ belongs to $L^1(\mathbf{R})$. (Hint: use the Cauchy–Schwarz inequality on $\int \frac{\sqrt{1+x^2}\,\psi(x)}{\sqrt{1+x^2}}\,dx$.)

9. Compute $\triangle x(u)$ and $\triangle\xi(u)$ for $u(x) = e^{-\pi x^2/2}$. (Hint: $\int_{\mathbf{R}} |u(x)|^2\,dx = 1$.)

10. Show that $\triangle x(u) \cdot \triangle\xi(u)$ is preserved if we replace u with v, where $v = v(x)$ is one of the following:

    (a) $v(x) = s^{-1/2}u(x/s)$ for any $s > 0$;

    (b) $v(x) = e^{2\pi i f x}u(x)$ for any $f \in \mathbf{R}$;

    (c) $v(x) = u(x - p)$ for any $p \in \mathbf{R}$.

11. Set $u = u_0 = \mathbf{1}_{[0,1]}$ and define u_n recursively by

$$u_{n+1}(t) = u_n(t)\;[u_n(3t) + u_n(3t-2)]\,.$$

Show that $\operatorname{diam}\operatorname{supp} u_n = 1$ even though $|\operatorname{supp} u_n| = (2/3)^n$, and prove that $\operatorname{diam}\operatorname{supp} u_n * u_n = |\operatorname{supp} u_n * u_n| = 2$.

Chapter 2

Programming Techniques

Our goal is to convert some of the algorithms of mathematical analysis into computer programs. This requires as much thought and subtlety as devising the algorithms themselves. It is also fraught with hazards, since there is no acceptable procedure for proving that a computer implementation of a nontrivial algorithm is correct. Even if the underlying algorithm is supported by mathematical proof, the peculiarities of computation in practice might still lead to unexpected results. Thus the first part of this chapter is devoted to listing some of the most common problems encountered while transferring mathematical analysis into software.

The second part is devoted to listing some of the data structures used in the analyses which follow. Others will be introduced as the need arises. The purpose is to make the reader familiar with our scheme for presenting algorithms, as well as to review some of the common manipulation techniques used with trees, arrays, and so on.

2.1 Computation in the real world

Our algorithms are intended to be implemented on typical computing machines, which are subject to many practical constraints.

2.1.1 Finiteness

A computer can only represent real numbers to a finite degree of precision. For reasons of efficiency, this precision is normally fixed for all quantities used during the calculation. Thus for each computer there is a small number $\epsilon_f > 0$ such that for all $0 \leq \epsilon \leq \epsilon_f$, we have $1.0 + \epsilon = 1.0$ but $1.0 + \epsilon' \neq 1.0$ for all $\epsilon' > \epsilon_f$. Then

$\log_{10}(1/\epsilon_f)$ is the maximum number of digits of accuracy for any calculation with that machine; for a binary machine, there are no more than $\log_2(1/\epsilon_f)$ bits in the mantissa of a floating-point number. In Standard C [98], ϵ_f is given the logical name `FLT_EPSILON`. On one typical desk top computer, namely the one I am using to write these lines, that quantity is $1.19209290 \times 10^{-7}$.

For reasons of tradition, computers are also equipped to efficiently perform calculations in "double" precision. The corresponding and much smaller ϵ_f is given the logical name `DBL_EPSILON` in Standard C. On my rather typical machine it has the value $2.2204460492503131 \times 10^{-16}$.

Because of the precision limit, computer arithmetic has some peculiar differences from exact arithmetic, such as:

- Addition is not associative: $(1.0 + \epsilon_f) + \epsilon_f = 1.0 \neq 1.0 + (\epsilon_f + \epsilon_f)$.
- An infinite series $\sum_{n=1}^{\infty} a(n)$ converges if and only if $a(n) \to 0$ as $n \to \infty$. If $a(n) \to 0$ but the series diverges in exact arithmetic, then the limit in finite-precision arithmetic depends both upon the value of ϵ_f and upon the order of summation.
- Every matrix is invertible. Every matrix is also diagonalizable. In general, if a property is satisfied by a dense open subset of matrices, then in finite precision arithmetic it is satisfied by all matrices.

These peculiar properties may be avoided by two simple policies:

1. No two of the magnitudes in an arithmetic statement $a = b + c$ should have a ratio close to or less than ϵ_f;
2. We must avoid any algorithm which magnifies the relative error of the parameters by $1/\epsilon_f$ or more.

We must adhere to the second policy because the error in a computed quantity can never be less than the amount by which it varies when the input parameters are changed by a relative error of ϵ_f or less. For operations like applying an $n \times n$ matrix A to a vector, this error magnification is given by the *condition number cond*(A), which might grow rapidly as $n \to \infty$. An error magnification of $1/\epsilon_f$ renders the result completely meaningless, but sometimes even a smaller magnification will result in an unacceptable loss of accuracy.

The second consideration is that the number of memory locations is finite, so that any analysis must of necessity take place in a finite-dimensional space. All the values of a "general" continuous function cannot be stored, since continuous functions even on the interval $[0, 1]$ form an infinite-dimensional space. We can at best store a finite-rank approximation of a continuous function, such as a finite sequence of samples $\{f(k/N) : k = 0, 1, 2, \ldots, N-1\}$. The details of the approximation influence any subsequent computation.

2.1.2 Validity

Between the mathematical derivation of an algorithm and the execution of a particular implementation on a particular machine, there are many opportunities to introduce errors. The formulas of exact arithmetic may become unstable algorithms on a finite-precision machine. Ambiguities in the programming language might result in unexpected results. Other errors might be introduced if there are automatic translation steps such as compilation between the programmer and the machine instructions. Finally, the behavior of a real-world machine may not exactly match its specifications, or those specifications may be ambiguous. We will therefore not attempt to prove that a given computer program is correct. However, we can employ certain sensible techniques to reduce the probability of error.

Rule 1: Keep the implementation simple. We use a well-defined language, we stick to its simplest constructs, we will be explicit rather than rely on defaults, and we will avoid combining calculations into long complex statements.

Rule 2: Keep track of the assumptions. Standard C comes with an `assert()` function which tests a logical expression and exits with an error message if the expression evaluates to 'false.' This expression or "assertion" is the only input parameter: for example,

```
assert( N>0 )
```

will return an error message and terminate the program if the variable N is zero or negative. Each time we make an assumption, we can assert it before further calculation. We will not attempt to recover from assertion failures.

Rule 3: Perform sanity checks on the input and the output. Each function to be computed will have a valid range for each of its input parameters. Also, there may be certain required relationships among the input parameters. We can thus test the input for validity before execution. Likewise, each function will have a valid range of output values and possible relationships among them. We can test them upon exit or just prior. These steps prevent certain types of errors from propagating.

Rule 4: Write short function definitions and test them individually. Our algorithms break down into many small computations, so it is natural to implement them as many small functions. We will take full advantage of the resulting opportunities for asserting our assumptions and testing them. We can also test the individual pieces on inputs with known outputs, prior to assembling them into a large whole which is too huge to test. If a function is invertible, then one simple way of testing it is to implement the inverse and test whether the composition is the identity on a large set of inputs.

2.1.3 Pseudocode

Standard C programming language implementations of all the algorithms in this text can be found on the optional diskette, available as described in the back of the book. My experience convinces me that the syntactic features of a real programming language only obscure the reader's understanding of the main steps in an algorithm. Also, good use of white space for a video display is enormously wasteful of paper so that the electronic form and the printed form of a function definition could look quite different. Thus, rather than provide printouts of those implementations inside the text, I have chosen to present the algorithms in pseudocode. In many cases, the pseudocode of the text can be found in source code comments on the diskette. Any reader objecting to pseudocode is free to reformat and print out any of the Standard C language programs, and to study the algorithms from those instead.

The pseudocode bears some resemblance to Algol-based languages like Pascal, FORTRAN, and BASIC as well as Standard C. I expect that the reader can easily translate what is printed into any of those languages. Data type and variable declarations are omitted. There is no naming convention, but data type can be inferred from context, and variables will be set in upper-case letters to make them easy to spot. A variable will always make its first appearance as the left-hand side of an assignment statement. We will observe a strict taboo against using uninitialized (global) variables.

Function names will always be in lower-case, followed by parentheses enclosing a parameter list. When we define a function for the first time, we will use a colon after the parameter list to signify the beginning of the definition. We will avoid most other punctuation marks except to signify parts of data structures. Keywords and action descriptions will either be capitalized or all lower-case, but can be distinguished from function names by the absence of parentheses.

Array indexing is signified by placing the index inside square brackets. Members of data structures are set apart with periods. Operations will always be evaluated from left to right, so that `Let LIST[J].VAL = X-Y-Z` means "assign the value of $(x - y) - z$ to member `VAL` of the data structure at index j in the array of data structures `LIST[]`." Notice how we emphasize that a variable represents an array by appending an empty set of brackets. An array of arrays can be denoted by two sets of empty brackets: `MATRIX[][]`. Likewise, writing `sqrt()` emphasizes that we are discussing a function which may have a nonempty list of parameters, though at the moment those parameters are not the focus of attention.

We will use the operators `%`, `<<`, `>>`, `&`, `|`, and `^` from Standard C to mean remainder, left-shift, right-shift, bitwise conjunction, bitwise disjunction, and bitwise exclusive-or, respectively. Likewise, we will use the Standard C logical operators `!`, `&&`, `||`, `==` and `!=` to signify NOT, AND, OR, EQUAL, and NOT EQUAL, respectively. We also borrow the preprocessor syntax from Standard C. Consider the following pair of expressions:

```
#define PI (3.141593)
#define SQUARE(X) ((X)*(X))
```

This means that an expression such as `SQUARE(PI*Y)` appearing in a pseudocode fragment should be replaced by `((PI*Y)*(PI*Y))` and then further expanded into `(((3.141593)*Y)*((3.141593)*Y))`.

We set the pseudocode apart from the text in a typewriter font for the amusement of old-time computer users. Indentation shows the structure. Our statements and keywords will be similar to those in the following list of examples:

- Assignments:

```
Let VAR = VALUE
```

- Calls to previously defined functions, including recursive calls:

```
allocate( ARRAY, LENGTH )
```

- Increments and decrements, multipliers and normalizers:

```
VAR1 += INCR          VAR3 *= MULT
VAR2 -= DECR          VAR4 /= NORM
```

- Conditionals:

```
If X is even then
   Let Y = X/2
Else
   Let Y = (X+1)/2
```

- For loops:

```
For J = A to B
   Let ARRAY[J] = J
```

The loop index is incremented by +1 at each pass. If `B<A`, then no statement in the body of the loop will ever be executed.

- While loops:

```
Let K = N
While K > 0
   NFACFAC *= K
   K -= 2
```

- Return statements, which indicate the value to be returned by a function:

```
Return OUTVAL
```

If no value is named it means that the function acts solely by side-effect and returns no explicit output. Since a function returns by default after the last line of its definition, we can omit the return statement in this case.

2.1.4 Common utilities

Certain functions are used so ubiquitously that we will suppose they are available in every context:

- *max(x,y)*: return the larger of x or y. This can be implemented in Standard C as a preprocessor macro which uses the *conditional* expression `Z?X:Y`, which equals `X` if `Z` is true and `Y` if `Z` is false:

```
#define max(x,y)          ((x)>(y)?(x):(y))
```

- *min(x,y)*: return the smaller of x or y. This can also be implemented in Standard C as a preprocessor macro:

```
#define min(x,y)          ((x)<(y)?(x):(y))
```

- *absval(x)*: return the absolute value of x, independent of arithmetic type. In Standard C we can use the following preprocessor macro:

```
#define absval(x)          ((x)<0? -(x):(x))
```

- *sqrt(x)*: return the square root of the floating-point number x. We assume that exceptions such as $x < 0$ cause immediate termination.

- *sin(x),cos(x),exp(x),log(x)* return the values of the sine, cosine, exponential and logarithm functions evaluated at the floating-point number x. We assume that exceptions such as $x \leq 0$ in $\log x$ cause immediate termination.

2.2 Structures

Here we collect the definitions of basic data structures and the functions that manipulate them. Most of these are defined in standard texts on computer programming; they are included here for completeness and to give examples of the pseudocode.

2.2.1 Complex arithmetic

We begin by choosing whether to do floating-point arithmetic in single or double precision. In Standard C we can use the unreserved *REAL* data type to denote floating-point numbers. This can be set using a preprocessor macro, so it can be easily changed:

```
#ifndef REAL
# define REAL float
#endif
```

To switch to double precision, we simply override this definition:

```
#define REAL double
```

Next, we can define a COMPLEX data type to use for input and output to the DFT. It should have two members:

- *COMPLEX.RE*, the real part,

- *COMPLEX.IM*, the imaginary part.

The two members should be of the same REAL data type as ordinary floating-point numbers.

Complex addition is simple enough to do componentwise when needed, but complex multiplication is enough of a bother that we should implement it with preprocessor macros:

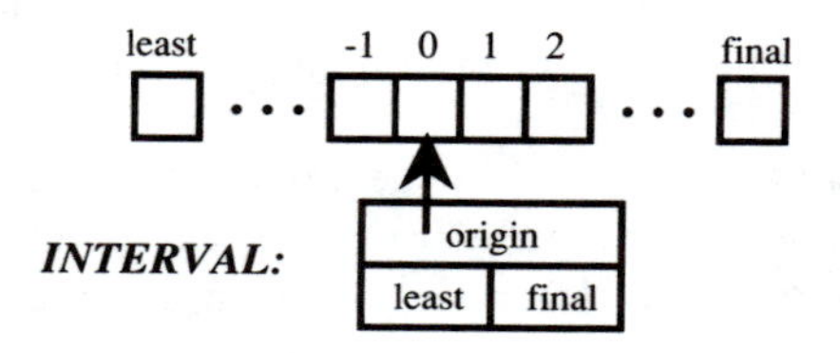

Figure 2.1: INTERVAL data structure and its array.

Multiplication of two COMPLEX data structures

```
#define CCMULRE(Z1, Z2)        ( Z1.RE*Z2.RE - Z1.IM*Z2.IM )
#define CCMULIM(Z1, Z2)        ( Z1.RE*Z2.IM + Z1.IM*Z2.RE )
```

Notice that we compute the real and imaginary parts of the product separately. Also, if we wish to refer explicitly to the real and imaginary parts of the second factor, we can use the following variants:

Multiplication of a COMPLEX data structure by two REALs

```
#define CRRMULRE(Z, YRE, YIM)     ( Z.RE*YRE - Z.IM*YIM )
#define CRRMULIM(Z, YRE, YIM)     ( Z.RE*YIM + Z.IM*YRE )
```

2.2.2 Intervals

To describe a segment of a sampled signal, we need an array of approximate real numbers. For convenience, we would like to index this array using both positive and negative integer indices, but then we must store the *least* and *final* indices as well as a pointer to the *origin* of the array. So, we define an *INTERVAL* data structure, which contains the following:

- *INTERVAL.ORIGIN*, a pointer to the origin of a data array,
- *INTERVAL.LEAST*, the least valid index in the array,
- *INTERVAL.FINAL*, the final valid index in the array.

An INTERVAL `X` contains `1+X.FINAL-X.LEAST` elements in its data array. This arrangement is depicted in Figure 2.1, which also shows the ORIGIN member pointing to the array.

We allocate an interval after giving the least and final values as parameters, and copy a data array into it if one is provided:

Allocate an INTERVAL and assign its data array

```
makeinterval( DATA, LEAST, FINAL ):
   Allocate an INTERVAL at SEG with all members 0
   Let LENGTH = 1+FINAL-LEAST
   If LENGTH>0 then
      Allocate an array of LENGTH REALs at SEG.ORIGIN
      Shift SEG.ORIGIN -= LEAST
      If DATA != NULL then
         For K = LEAST to FINAL
            Let SEG.ORIGIN[K] = DATA[K-LEAST]
   Let SEG.LEAST = LEAST
   Let SEG.FINAL = FINAL
   Return SEG
```

To deallocate an interval's data array, it is necessary to shift its pointer back to its position at allocation time:

Deallocate an INTERVAL and its data array

```
freeinterval( SEG ):
   If SEG != NULL then
      If SEG.ORIGIN != NULL then
         Shift SEG.ORIGIN += LEAST
         Free SEG.ORIGIN
      Free SEG
   Return NULL
```

We can add some elementary range checking so we reference only those array locations which are in bounds:

Check if an offset is within an INTERVAL

```
ininterval( SEGMENT, OFFSET ):
   If SEGMENT.ORIGIN != NULL then
      If OFFSET>=SEGMENT.LEAST && OFFSET<=SEGMENT.FINAL then
         Return TRUE
   Return FALSE
```

Here `TRUE` and `FALSE` are predefined constants which evaluate to "true" and "false" respectively in conditional statements.

Next, we write a short loop to calculate the total length of an array of disjoint intervals:

Total length in a list of nonoverlapping INTERVALs

```
intervalstotal( IN, N):
   Let TOTAL = 0
   For K = 0 to N-1
      TOTAL += 1 + IN[K].FINAL - IN[K].LEAST
   Return TOTAL
```

Finally, we write another utility to enlarge the data array in an interval to accommodate a new offset. If there is already some data in the old interval, we copy that into the newly enlarged array:

Enlarge an INTERVAL, preserving any contents

```
enlargeinterval( OLD, LEAST, FINAL ):
  If OLD.ORIGIN == NULL then
     Let LENGTH = 1+FINAL-LEAST
     If LENGTH>0 then
        Allocate an array of LENGTH REALs at OLD.ORIGIN
        Shift OLD.ORIGIN -= LEAST
        Let OLD.LEAST = LEAST
        Let OLD.FINAL = FINAL
  Else
     If OLD.LEAST<LEAST || OLD.FINAL>FINAL then
        Let LEAST = min( OLD.LEAST, LEAST )
        Let FINAL = max( OLD.FINAL, FINAL )
        Let LENGTH = 1 + FINAL-LEAST
        If LENGTH>0 then
           Allocate an array of LENGTH REALs at NEWDATA
           Shift NEWDATA -= LEAST
           For J = OLD.LEAST to OLD.FINAL
              Let NEWDATA[J] = OLD.ORIGIN[J]
           Shift OLD.ORIGIN += OLD.LEAST
           Deallocate OLD.ORIGIN[]
           Let OLD.ORIGIN = NEWDATA
           Let OLD.LEAST = LEAST
           Let OLD.FINAL = FINAL
  Return OLD
```

We assume that OLD is preallocated, though it may have a null pointer as its data array. Notice that we test if the length of the new data array is positive. If not, then

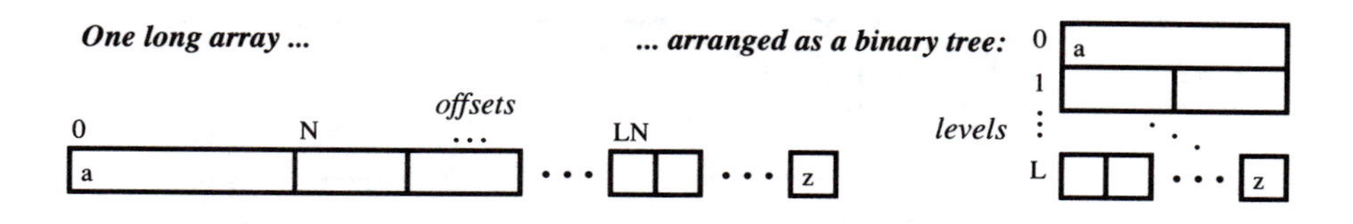

Figure 2.2: Multilevel analyses concatenated into one long array.

we change nothing. This is for proper handling of certain degenerate inputs. Also, it is important for the deallocation routine that `OLD.ORIGIN+OLD.LEAST` always be the first location of the data array allocated to the INTERVAL.

2.2.3 Binary trees

Array binary trees

An adapted analysis contains multiple representations of the signal, divided into successively smaller blocks and expanded in local bases in each block. If these can be conventionally indexed, so that each subinterval is a portion of an array starting at offset zero, then the multiple copies may be concatenated into one long array, as depicted in Figure 2.2.

Using one long array has the advantage that only one allocation of storage is needed. Accessing an element just requires knowing its offset in the array. Also, almost every programming language has an array data type, so that no additional types need to be defined. To make an array binary tree for L decompositions of a signal of length N, we allocate an array of length $(L+1)N$. Thenceforth we keep track of the pointer to its first element until it is time to deallocate the storage.

The first index of a block in the array binary tree may be computed by a function so simple, it is best implemented as a preprocessor macro:

```
#define abtblock(N,L,B) ((L)*(N)+(B)*((N)>>(L)))
```

Likewise, the length of a block can be computed by another preprocessor macro:

```
#define abtblength(N,L) ((N)>>(L))
```

Binary tree nodes

The basic element in a binary tree is a *binary tree node* data structure, which we define as a type *BTN* containing at least the following members:

- *BTN.CONTENT*, the interval or whatnot represented by the node,

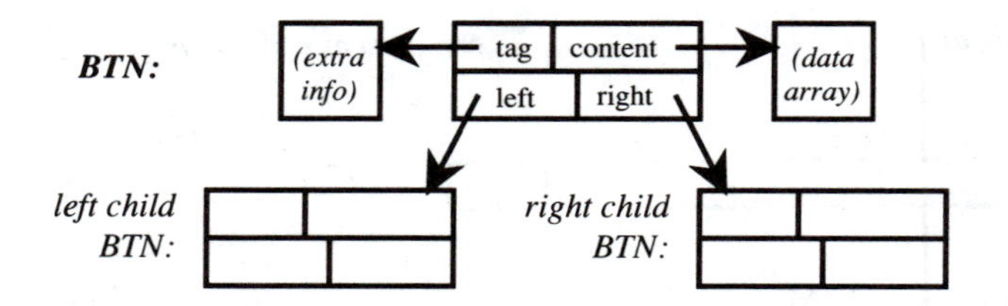

Figure 2.3: Part of a tree of BTN data structures.

- *BTN.LEFT*, a pointer to the left descendent of this node,
- *BTN.RIGHT*, a pointer to the right descendent of this node.

It is often useful to store additional information in such a structure, so we include a member that can be used to point to a block of identifying information:

- *BTN.TAG*, a pointer to a data structure containing further information about this node.

Figure 2.3 gives one illustration of how the nodes are arranged into a tree. As an example, the content member might be an INTERVAL data structure, which might be separately allocated for each node. Alternatively, it may be a pointer into a single long array that is shared by all nodes in the tree with the boundaries computable from the position of the BTN within its tree.

When we allocate a BTN data structure, we can simultaneously assign to it some content and a tag:

Allocate a BTN data structure with given members

```
makebtn( CONTENT, LEFT, RIGHT, TAG ):
   Allocate a BTN data structure at NODE
   Let NODE.TAG = TAG
   Let NODE.CONTENT  = CONTENT
   Let NODE.LEFT = LEFT
   Let NODE.RIGHT = RIGHT
   Return NODE
```

By calling `makebtn(CONTENT,NULL,NULL,TAG)`, we can allocate a node with null pointers in place of its children.

To deallocate a BTN structure, we first check if it has valid content or tag pointers and deallocate their targets first:

Deallocate a BTN structure, its content, and its tag

```
freebtn( NODE, FREECONTENT, FREETAG ):
   If NODE != NULL then
      If NODE.CONTENT != NULL && FREECONTENT != NULL then
         Deallocate NODE.CONTENT with FREECONTENT()
      If NODE.TAG != NULL && FREETAG != NULL then
         Deallocate NODE.TAG with FREETAG()
      Deallocate NODE
   Return NULL
```

We are cheating here by not specifying how to deallocate the various kinds of contents and tags. Those procedures may require separate utilities.

BTN trees

We build a completely empty tree down to a specified level with a recursive function:

Allocate a complete BTN tree of null contents to a specified depth

```
makebtnt( LEVEL ):
   Let ROOT = makebtn(NULL,NULL,NULL,NULL)
   If LEVEL>0 then
      Let ROOT.LEFT = makebtnt( LEVEL-1 )
      Let ROOT.RIGHT = makebtnt( LEVEL-1 )
   Return ROOT
```

We get a pointer to a particular node in a BTN tree as follows:

Get a BTN from a binary tree

```
btnt2btn( ROOT, LEVEL, BLOCK ):
   If LEVEL==0 || ROOT==NULL then
      Let NODE = ROOT
   Else
      If BLOCK is even then
         Let NODE = btnt2btn( ROOT.LEFT, LEVEL-1, BLOCK/2 )
      Else
         Let NODE = btnt2btn( ROOT.RIGHT, LEVEL-1, (BLOCK-1)/2 )
   Return NODE
```

To deallocate an entire tree of BTN structures, we employ a recursive function that deallocates all the descendent nodes as well. We return a null pointer to save

Figure 2.4: One branch of a BTN tree.

a line of code later:

Deallocate a complete binary tree of BTN structures

```
freebtnt( ROOT, FREECONTENT, FREETAG ):
   If ROOT != NULL then
      freebtnt( ROOT.LEFT, FREECONTENT, FREETAG )
      freebtnt( ROOT.RIGHT, FREECONTENT, FREETAG )
      freebtn( ROOT, FREECONTENT, FREETAG )
   Return NULL
```

To preserve the contents while freeing the tree, we call `freebtnt()` with a NULL argument for the FREECONTENT function. Likewise, to preserve the tag members, we call `freebtnt()` with a NULL argument for the FREETAG function.

A *branch* of a BTN tree is a subset of nodes which all lie on a single path to the root. This is illustrated in Figure 2.4.

To link together one branch of a BTN tree, the one between the root and a target specified by its level and block indices, we use the following utility:

Allocate a BTN tree branch out to a target node

```
btn2branch( SELF, LEVEL, BLOCK ):
  If LEVEL>0 then
    If BLOCK is even then
      If SELF.LEFT==NULL then
        Let SELF.LEFT = makebtn( NULL, NULL, NULL, NULL )
      Let SELF = btn2branch( SELF.LEFT, LEVEL-1, BLOCK/2 )
    Else
      If SELF.RIGHT==NULL then
        Let SELF.RIGHT = makebtn( NULL, NULL, NULL, NULL )
      Let SELF = btn2branch( SELF.RIGHT, LEVEL-1, (BLOCK-1)/2 )
  Return SELF
```

This function expects as input a preallocated BTN data structure, regarded as the root of the binary tree, and valid level and block indices. It returns a pointer

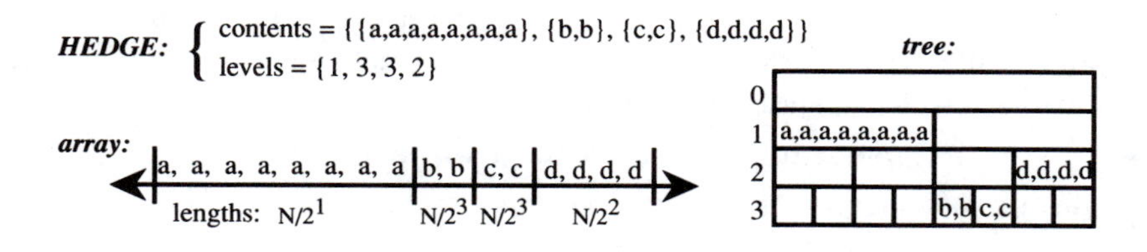

Figure 2.5: Interpretation of coefficients in a hedge.

to the node corresponding to that level and block in the binary tree. It allocates any intermediate nodes along the branch, including the target node itself if that is necessary, and leaves null pointers to any unused children in the allocated nodes.

2.2.4 Hedges

In an adapted analysis we must choose a basis subset from the library tree. Once a subset is chosen, it is necessary to describe it in a convenient manner, so that the subset and the coefficients from inner products with its elements may be used in further computation. For example, they may be used to reconstruct the original signal. One method is to store an array of *coefficients* and an array of *levels*, together with some additional information such as the maximum decomposition level and the number of samples in the original signal.

So let us introduce a structure for all this data. We will call it a *hedge*, to suggest its connection with a tree. The *HEDGE* data structure contains the following elements:

- *HEDGE.CONTENTS*, an array of data structures containing the coefficients of the signal in the adapted basis,
- *HEDGE.LEVELS*, the sequence of levels in encounter order,
- *HEDGE.BLOCKS*, the number of elements in the levels array, also the number of elements in the contents array,
- *HEDGE.TAG*, a pointer to an optional structure containing extra information about the decomposition.

Figure 2.5 shows the relationship between the contents of a HEDGE data structure and the numbers in an array binary tree.

We can simultaneously assign the members of a hedge when we allocate it. If we write the allocation utility as below, then we can create a hedge containing zero arrays of specified length just by using null pointers as parameters:

Allocate a hedge with given members

```
makehedge( BLOCKS, CONTENTS, LEVELS, TAG ):
   Allocate a HEDGE data structure at OUT
   Let OUT.BLOCKS = BLOCKS
   If CONTENTS==NULL then
      Allocate an array of BLOCKS pointers at OUT.CONTENTS
   Else
      Let OUT.CONTENTS = CONTENTS
   If LEVELS==NULL then
      Allocate an array of BLOCKS bytes at OUT.LEVELS
   Else
      Let OUT.LEVELS = LEVELS
   Let OUT.TAG = TAG
   Return OUT
```

Before deallocating the hedge itself, we deallocate the contents and levels arrays and the tag pointer:

Deallocate a hedge and its members

```
freehedge( IN, FREECONTENT, FREETAG ):
   If IN.CONTENTS != NULL then
      For B = 0 to IN.BLOCKS-1
         Deallocate IN.CONTENTS[B] with FREECONTENT()
      Deallocate IN.CONTENTS
   If IN.LEVELS != NULL then
      Deallocate IN.LEVELS
   If IN.TAG != NULL then
      Deallocate IN.TAG with FREETAG()
   Deallocate IN
   Return NULL
```

We are being deliberately vague about the procedure for deallocating the contents, levels and tag arrays; each may require its own separate function. Similar data structures are useful in multidimensions as well, so we prefer not to make any assumptions about the contents or the tag. The interpretation of the contents might depend upon the dimension and the decomposition method, and can be defined elsewhere. Likewise, the tag might store additional data to describe different decomposition methods used at different levels. This could be done by making the tag an array of data structures, one for each level.

2.2.5 Atoms

The coefficients in the output arrays of an adapted wavelet analysis consist of amplitudes tagged with associated scale, frequency, and position indices. To hold this information we can define *time-frequency atom* data structures for one, two, and D dimensions named *TFA1*, *TFA2*, and *TFAD*.

The one-dimensional atom has the following four members:

- *TFA1.AMPLITUDE*, the amplitude of the coefficient,
- *TFA1.LEVEL*, the scale index,
- *TFA1.BLOCK*, the frequency index,
- *TFA1.OFFSET*, the position index.

We will use these parameters to signify the level in the decomposition tree, the block number within that level, and the offset within the data array of the block. These quantities are related to the nominal width of 2^{scale} sampling intervals, the nominal frequency of $\frac{1}{2}(frequency)$ oscillations over the nominal width, and the nominal position of *position* sampling intervals with respect to the origin. How we compute the relationship depends on the underlying waveform.

The two-dimensional *TFA2* structure has seven members: an amplitude and six tags.

- *TFA2.AMPLITUDE*, the amplitude of the coefficient,
- *TFA2.XLEVEL*, the scale index along the rows,
- *TFA2.YLEVEL*, the scale index along the columns,
- *TFA2.XBLOCK*, the number of oscillations along a row,
- *TFA2.YBLOCK*, the number of oscillations along a column,
- *TFA2.XOFFSET*, the position index within a row,
- *TFA2.YOFFSET*, the position index within a column.

Even at two dimensions we see the beginnings of combinatorial explosion. The storage requirements for signals in the form of time-frequency atoms become prohibitive as the dimension increases, and the problem is aggravated by the proliferation of the number of tags needed to describe the atoms. Therefore, for the D-dimensional atom $TFAD$ we will combine the scale, frequency, and position tags for all dimensions into a single coded integer:

- *TFAD.AMPLITUDE*, the amplitude of the coefficient,
- *TFAD.DIMENSION*, the number of dimensions of the signal,
- *TFAD.LEVELS*, a coded form of the D scale indices,
- *TFAD.BLOCKS*, a coded form of the D frequency indices,
- *TFAD.OFFSETS*, a coded form of the D position indices.

For example, if the computer uses 32-bit integers, it is possible to distinguish more than 16×10^{18} different frequency and position combinations using just two integer frequency and position tags. This is a comfortable margin larger than the biggest data sets that present-day computers can easily handle.

2.3 Manipulation

2.3.1 Hedges and trees

Given a long array corresponding to a complete adapted wavelet analysis and an array of levels describing a graph basis subset, we can extract the corresponding hedge of coefficients with the following function:

Complete a levels-specified hedge from an array binary tree

```
abt2hedge( GRAPH, DATA, LENGTH ):
  Let COLUMN = 0
  For I = 0 to GRAPH.BLOCKS-1
    Let GRAPH.CONTENTS[I] = DATA+COLUMN+LENGTH*GRAPH.LEVELS[I]
    COLUMN += LENGTH>>GRAPH.LEVELS[I]
```

Here `LENGTH` is the length of the original signal and we assume that `OUT.CONTENTS[]` has been allocated with at least `OUT.BLOCKS` locations. We assume that the hedge's contents array has the same type as the binary tree; it holds pointers into the tree indicating the origins of all the blocks in the hedge basis set. We also assume that the list of levels is valid for the array binary tree at `DATA[]`.

Given a valid hedge, we can superpose its amplitudes into an array binary tree. The following code assumes that the output array at `DATA[]` has been preallocated to an adequate size, and that the contents of the input hedge are pointers into an array of the same type as `DATA[]`. It superposes the hedge amplitudes into the array, for greater code generality:

Superpose amplitudes from a hedge into an array binary tree

```
hedge2abt( DATA, GRAPH, LENGTH ):
   Let COLUMN = 0
   For I = 0 to GRAPH.BLOCKS-1
      Let BLENGTH = LENGTH>>GRAPH.LEVELS[I]
      Let BLOCK = DATA + LENGTH*GRAPH.LEVELS[I] + COLUMN
      For J = 0 to BLENGTH-1
         BLOCK[J] += GRAPH.CONTENTS[I][J]
      COLUMN += BLENGTH
```

BTN trees

We first define a function to extract the content members from those BTNs specified in the levels list of a hedge:

Fill a levels-specified hedge from a BTN tree

```
btnt2hedge( GRAPH, ROOT ):
   Let MAXLEVEL = 0
   Let FRACTION = 0
   For I = 0 to GRAPH.BLOCKS-1
      Let LEVEL = GRAPH.LEVELS[I]
      If LEVEL>MAXLEVEL then
         FRACTION <<= LEVEL-MAXLEVEL
         Let MAXLEVEL = LEVEL
         Let BLOCK = FRACTION
      Else
         Let BLOCK = FRACTION>>(MAXLEVEL-LEVEL)
      Let NODE = btnt2btn( ROOT, LEVEL, BLOCK )
      Let GRAPH.CONTENTS[I] = NODE.CONTENT
      FRACTION += 1<<LEVEL
   Return MAXLEVEL
```

Given a BTN tree corresponding to a complete adapted wavelet analysis and a partially-assigned HEDGE with its levels array describing a graph basis subset within the tree, `btnt2hedge()` extracts the content members of that graph. We assume that the contents array is preallocated to a sufficient length to hold the entire graph.

We extract the blocks in the hedge from left to right and append each one's content member to the contents array of the hedge. We assume that utilities already exist for copying various kinds of contents from a BTN to the hedge contents array, so we do not have to keep track of the exact type. We will also be vague about what happens to the information in the tag data structure. It might be possible to digest it and store it compactly in the hedge.

Conversely, given a valid hedge with both its levels and contents arrays properly assigned, we can build a partial BTN tree to hold the contents in the levels-specified nodes:

Make a partial BTN tree from a hedge

```
hedge2btnt( ROOT, GRAPH ):
   Let MAXLEVEL = 0
   Let FRACTION = 0
   For I = 0 to GRAPH.BLOCKS-1
      Let LEVEL = GRAPH.LEVELS[I]
      If LEVEL>MAXLEVEL then
         FRACTION <<= LEVEL-MAXLEVEL
         Let MAXLEVEL = LEVEL
         Let BLOCK = FRACTION
      Else
         Let BLOCK = FRACTION>>(MAXLEVEL-LEVEL)
      Let NODE = btn2branch( ROOT, LEVEL, BLOCK )
      Let NODE.CONTENT = GRAPH.CONTENTS[I]
      FRACTION += 1<<LEVEL
   Return MAXLEVEL
```

This function puts the hedge contents into the leaf nodes of a partial binary tree. It also creates contentless nodes between the leaves and the root.

Notice that both `hedge2btnt()` and `btnt2hedge()` return the maximum depth of any node in their tree. In fact, the two functions differ in just two lines.

The hedge is finished when `FRACTION==(1<<MAXLEVEL)`; this can be used as a termination condition if we do not know the number of blocks in the hedge.

It should be noted that there are natural recursive versions of these algorithms as well. The recursive versions are a preferred way to generalize to multidimensions.

2.3.2 Atoms and trees

We assume that an array binary tree has been preallocated at DATA and filled with zeroes. Into it we can superpose amplitudes from a list ATOMS[] of TFA1 data structures using the following loop:

Superpose TFA1s into an array binary tree

```
tfa1s2abt( DATA, N, ATOMS, NUM ):
   For K = 0 to NUM-1
      Let START = abtblock( N, ATOMS[K].LEVEL, ATOMS[K].BLOCK )
      DATA[START + ATOMS[K].OFFSET] += ATOMS[K].AMPLITUDE
```

It is wise to perform range checking before trying to fill the array. For example, the following function tests whether an atom fits into the array binary tree of dimensions N by 1+MAXLEVEL:

Verify that a TFA1 fits into an array binary tree

```
tfa1inabt( ATOM, N, MAXLEVEL ):
  If ATOM.LEVEL>=0 && ATOM.LEVEL<=MAXLEVEL then
    If ATOM.BLOCK>=0 && ATOM.BLOCK<(1<<ATOM.LEVEL) then
      If ATOM.OFFSET>=0 && ATOM.OFFSET<(N>>ATOM.LEVEL) then
         Return TRUE
  Return FALSE
```

To test an array of atoms we use a loop:

Verify that a list of TFAs fits into an array binary tree

```
tfa1sinabt( ATOMS, NUM, LENGTH, MAXLEVEL ):
   For K = 0 to NUM-1
      If !tfa1inabt( ATOMS[K], LENGTH, MAXLEVEL ) then
         Return FALSE
   Return TRUE
```

We now define a function to superpose an amplitude from a BTN tree onto the amplitude member of a TFA1 which specifies the location of the amplitude in an array binary tree built from a signal of LENGTH samples:

Fill a partially defined TFA1 from an array binary tree

```
abt2tfa1( ATOM, DATA, LENGTH ):
  Let BLOCK = abtblock( LENGTH, ATOM.LEVEL, ATOM.BLOCK )
  ATOM.AMPLITUDE += DATA[BLOCK+ATOM.OFFSET]
```

Given a preallocated array ATOMS[] of TFA1 data structures with their level, block, and offset members filled with valid indices, we can superpose the remaining amplitude member for each atom by reading it from an array binary tree:

Fill an array of TFA1s from an array binary tree

```
abt2tfa1s( ATOMS, NUM, DATA, LENGTH ):
   For K = 0 to NUM-1
      abt2tfa1( ATOMS[K], DATA, LENGTH )
```

Atoms and BTN trees

We suppose that the content member of a BTN is an INTERVAL data structure. We can superpose one amplitude in a TFA1 data structure ATOM into a BTN tree at ROOT with the following function:

Superpose one TFA1 into a partial BTN tree

```
tfa12btnt( ROOT, ATOM ):
   Let NODE = btn2branch( ROOT, ATOM.LEVEL, ATOM.BLOCK )
   Let LEAST = min( NODE.CONTENT.LEAST, ATOM.OFFSET )
   Let FINAL = max( NODE.CONTENT.FINAL, ATOM.OFFSET )
   enlargeinterval( NODE.CONTENT, LEAST, FINAL )
   NODE.CONTENT.ORIGIN[ATOM.OFFSET] += ATOM.AMPLITUDE
```

Then, superposing a list of amplitudes into a preallocated tree can be done with a loop:

Superpose a list of TFA1s into a BTN tree

```
tfa1s2btnt( ROOT, ATOMS, NUM ):
   For K = 0 to NUM-1
      tfa12btnt( ROOT, ATOMS[K] )
```

The efficiency of this algorithm depends greatly upon the order of the list of atoms. Best efficiency comes with the fewest reallocations, which means that for every pair (level, block) in the list of atoms, the least and greatest values of the offset members should appear before any intermediate values. To get such an arrangement, we can sort the atoms by their tags into lexicographical order by level and block, then within each stretch of constant level and block we simply swap the second and the last elements.

Given a TFA1 data structure ATOM with valid level, block, and offset indices for the binary tree of BTN structures at ROOT, calling btnt2tfa1(ATOM,ROOT) superposes any nonzero amplitude it finds in the tree onto the amplitude member of that atom:

Complete a TFA1 from a BTN tree of intervals

```
btnt2tfa1( ATOM, ROOT ):
  Let NODE = btnt2btn( ROOT, ATOM.LEVEL, ATOM.BLOCK )
  If NODE != NULL
    If NODE.CONTENT != NULL
      If ininterval( NODE.CONTENT, ATOM.OFFSET ) then
        ATOM.AMPLITUDE += NODE.CONTENT.ORIGIN[ATOM.OFFSET]
```

If we have a valid index subset for the BTN tree at `ROOT`, then we can complete a preallocated array `ATOMS[]` of TFA1 data structures by superposing the amplitudes found in the tree. We will assume that the index subset is already assigned into the array, *i.e.*, `ATOMS[K].LEVEL`, `ATOMS[K].BLOCK` and `ATOMS[K].OFFSET` are valid indices for all `K=0,...,N-1`. Then we superpose an amplitude from the tree onto the remaining member `ATOMS[K].AMPLITUDE` for each `K` as follows:

Complete an array of TFA1s from a BTN tree

```
btnt2tfa1s( ATOMS, NUM, ROOT ):
   For K = 0 to NUM-1
      btnt2tfa1( ATOMS[K], ROOT )
```

2.3.3 Hedges to atoms

Given a hedge, we can produce an array of atoms by calculating the block and offset indices from the sequence of levels, then copying the amplitudes from the data arrays of the hedge contents.

We first write a utility that assigns TFA1s from a list of amplitudes, assuming that they are all from the same level and block and have offsets beginning with zero:

Assign TFA1s from an array of amplitudes, a block, and a level

```
array2tfa1s( ATOMS, NUM, AMPLITUDES, BLOCK, LEVEL ):
   For I = 0 to NUM-1
      Let ATOMS[I].AMPLITUDE = AMPLITUDES[I]
      Let ATOMS[I].BLOCK = BLOCK
      Let ATOMS[I].LEVEL = LEVEL
      Let ATOMS[I].OFFSET = I
```

Suppose we are given a hedge representing a basis in an array binary tree developed from a one-dimensional signal. Suppose that we are also given a preallocated array of TFA1s equal in length to the signal. In addition, we specify the length of the signal and the depth of the decomposition. The following function fills the list of atoms with the ordered quadruplets defined by the hedge:

Convert an array binary tree HEDGE of known depth to a list of TFA1s

```
hedgeabt2tfa1s( ATOMS, GRAPH, LENGTH, MAXLEVEL ):
  Let START = 0
  For J = 0 to GRAPH.BLOCKS-1
    Let LEVEL = GRAPH.LEVELS[J]
    Let BLOCK = START>>(MAXLEVEL-LEVEL)
    Let NUM = abtblength( LENGTH, LEVEL )
    array2tfa1s( ATOMS, NUM, GRAPH.CONTENTS+J, BLOCK, LEVEL )
    ATOMS += NUM
```

We need not specify the maximum level in the hedge or tree at the outset, since that can be computed as we go. Notice the great similarity between the following function and `btnt2hedge()`:

Convert an array binary tree HEDGE to a list of TFA1s

```
abthedge2tfa1s( ATOMS, GRAPH, LENGTH ):
  Let MAXLEVEL = 0
  Let FRACTION = 0
  For J = 0 to GRAPH.BLOCKS-1
    Let LEVEL = GRAPH.LEVELS[J]
    If LEVEL>MAXLEVEL then
      FRACTION <<= LEVEL-MAXLEVEL
      Let MAXLEVEL = LEVEL
      Let BLOCK = FRACTION
    Else
      Let BLOCK = FRACTION>>(MAXLEVEL-LEVEL)
    Let NUM = abtblength( LENGTH, LEVEL )
    array2tfa1s( ATOMS, NUM, GRAPH.CONTENTS+J, BLOCK, LEVEL )
    FRACTION += 1<<LEVEL
    ATOMS += NUM
  Return MAXLEVEL
```

In the more general case, we assume instead that the hedge contains INTERVAL data structures. The first task is to compute the total number of amplitudes to be extracted, or the sum of the measures of all the intervals, and allocate enough TFA1s to hold the lot. This is the return value of the function call `intervalstotal(GRAPH.CONTENTS,GRAPH.BLOCKS)`, where `GRAPH` is the hedge in question.

Next, we need a utility that writes an interval of amplitudes to a list of atoms with a specified level and block:

Assign TFA1s from an array of amplitudes, a block, and a level

```
interval2tfa1s( ATOMS, SEGMENT, BLOCK, LEVEL ):
  For I = 0 to SEGMENT.FINAL-SEGMENT.LEAST
    Let ATOMS[I].AMPLITUDE = SEGMENT.ORIGIN[I+SEGMENT.LEAST]
    Let ATOMS[I].BLOCK = BLOCK
    Let ATOMS[I].LEVEL = LEVEL
    Let ATOMS[I].OFFSET = I
```

Now suppose that we have enough TFA1s allocated to hold all the atoms. The following function assigns them with amplitudes from the hedge with their block, level, and offset indices:

Convert a HEDGE of INTERVALs to a list of TFA1s

```
intervalhedge2tfa1s( ATOMS, GRAPH ):
   Let MAXLEVEL = 0
   Let FRACTION = 0
   For J = 0 to GRAPH.BLOCKS-1
      Let LEVEL = GRAPH.LEVELS[J]
      If LEVEL>MAXLEVEL then
         FRACTION <<= LEVEL-MAXLEVEL
         Let MAXLEVEL = LEVEL
         Let BLOCK = FRACTION
      Else
         Let BLOCK = FRACTION>>(MAXLEVEL-LEVEL)
      Let SEGMENT = GRAPH.CONTENTS + J
      interval2tfa1s( ATOMS, SEGMENT, BLOCK, LEVEL )
      ATOMS += 1 + SEGMENT.FINAL - SEGMENT.LEAST
      FRACTION += 1<<LEVEL
   Return MAXLEVEL
```

Remark. The HEDGE data structure is more efficient at storing the output of an analysis, whereas an array of TFA1s is easier to interpret and manipulate. We can use the former for data compression and the latter for computation and processing.

2.4 Exercises

1. Find out what FLT_EPSILON and DBL_EPSILON are for your computer. Compute their logarithms base two, to find out how many bits there are in the mantissas of single and double-precision numbers on your machine.

2. Write a pseudocode function that returns the square root of a number. Use Newton's method and stop when the error is less than FLT_EPSILON.

3. Write a pseudocode utility to convert an array binary tree into a BTN tree.

4. Rewrite the following recursive function in pseudocode without using recursion:

```
bisect( ARRAY, N, U ):
   If N>0 then
      If U is even then
         Let ARRAY = bisect(  ARRAY,  N/2, U/2 )
      Else
         Let ARRAY = bisect( ARRAY+N/2, N/2, U/2 )
   Return ARRAY
```

Chapter 3

The Discrete Fourier Transform

To implement the Fourier transform directly, it is necessary to evaluate integrals involving highly oscillatory integrands. In rare cases this can be done analytically, such as for certain distributions or some rather simple functions. An exact analytical formula gives considerable insight into a problem, but the Fourier transform is far too useful to be restricted to just those few cases. In the more general case, we can employ a numerical integration method such as the *extended Simpson's rule* ([2], p.605ff), but the rapid oscillation of e^{ikx} for large $|k|$ will necessitate many small subintervals and much work. Also, Simpson's rule gives good approximations only when the function being transformed has four continuous derivatives.

Alternatively, we can approximate a function by samples and approximate the Fourier integral by the discrete Fourier transform or DFT. This approach requires applying a matrix whose order is the number of sample points. Since multiplying an $N \times N$ matrix by a vector costs $O(N^2)$ arithmetic operations, the problem scales rather badly as the number of sample points increases. However, if the samples are uniformly spaced, then the Fourier matrix can be factored into a product of just a few sparse matrices, and the resulting factors can be applied to a vector in a total of $O(N \log N)$ arithmetic operations. This is the so-called fast Fourier transform, or FFT.

FFT plays an enormous role in numerical analysis and signal processing, and as a consequence there exist many specialized and highly-engineered versions. Some of the more exotic variations are described in [10, 108]. We will describe the relatively simple FFT implementation based on the Danielson–Lanczos lemma and popularized in [33]. In addition, we will consider the related discrete sine, cosine, and Hartley transforms. These can be obtained by slightly modifying the FFT implementation, or else by composing it with a small number of additional low complexity transformations.

3.1 The Fourier transform on $\mathbf{C}^N$

Suppose $v \in \mathbf{C}^N$, $v = \{v(n)\}_{n=0}^{N-1}$. The *discrete Fourier transform* of v is the vector $\hat{v} \in \mathbf{C}^N$ defined by

$$\hat{v}(n) = \frac{1}{\sqrt{N}} \sum_{j=0}^{N-1} v(j) \exp\left(-2\pi i \frac{jn}{N}\right),$$

for $n = 0, 1, \ldots, N-1$.

We will also denote the discrete Fourier transform by the matrix application $\hat{v} = \mathcal{F}v$, where $\mathcal{F} : \mathbf{C}^N \to \mathbf{C}^N$ is the matrix defined by

$$\mathcal{F}(n,j) = \frac{1}{\sqrt{N}} \exp\left(-2\pi i \frac{jn}{N}\right). \tag{3.1}$$

We will add a subscript and write $\mathcal{F}_N$ when this is necessary to emphasize the dimension of the range and domain. If we write $\omega_n = \exp\left(-2\pi i \frac{n}{N}\right)$, then $\mathcal{F}_N$ is given by the matrix below:

$$\mathcal{F} = \frac{1}{\sqrt{N}} \begin{pmatrix} 1 & 1 & 1 & \cdots & 1 \\ 1 & \omega_1 & \omega_1^2 & \cdots & \omega_1^{N-1} \\ 1 & \omega_2 & \omega_2^2 & \cdots & \omega_2^{N-1} \\ \vdots & & \ddots & & \vdots \\ 1 & \omega_{N-1} & \omega_{N-1}^2 & \cdots & \omega_{N-1}^{N-1} \end{pmatrix}; \qquad \mathcal{F}(n,j) = \frac{1}{\sqrt{N}} \left(\omega_n^j\right). \tag{3.2}$$

Such a matrix is called a *Vandermonde* matrix: its nth row consists of successive powers $\{1, \omega_n, \omega_n^2, \ldots\}$ of ω_n. We can prove by induction on N that its determinant is given by the following equation:

$$\det \mathcal{F} = \prod_{0 \le n < m < N} (\omega_m - \omega_n).$$

Since $\omega_n \neq \omega_m$ for $n \neq m$, this matrix is nonsingular. Also, $\omega_n^m = \omega_m^n$ (so $\mathcal{F}$ is a symmetric matrix), $\bar{\omega}_n = \omega_n^{-1}$ (since $|\omega_n| = 1$), and $\omega_n^N = 1$ for all $n =$

$0, 1, \ldots, N-1$. Thus we can show that $\mathcal{F}$ is unitary:

$$\begin{aligned}
\mathcal{F}\mathcal{F}^*(n,j) &= \frac{1}{N}\sum_{k=0}^{N-1} \omega_n^k \bar{\omega}_j^k = \frac{1}{N}\sum_{k=0}^{N-1} \omega_1^{(n-j)k} = \begin{cases} 1, & \text{if } n = j, \\ \frac{1}{N}\frac{1-\omega_1^{(n-j)N}}{1-\omega_1^{n-j}}, & \text{if } n \neq j, \end{cases} \\
&= \delta(n-j).
\end{aligned}$$

Since v can be represented by $v = \mathcal{F}^*\hat{v}$, it lies in the column space of the adjoint matrix $\mathcal{F}^*$. These columns are the *discrete Fourier basis functions* $\frac{1}{\sqrt{N}} \exp\left(2\pi i \frac{jn}{N}\right)$. Figure 3.1 shows the real and imaginary parts of an example with $N = 256$ and $j = 3$. Notice that with so many sample points it is difficult to distinguish the graph from that of a smooth function.

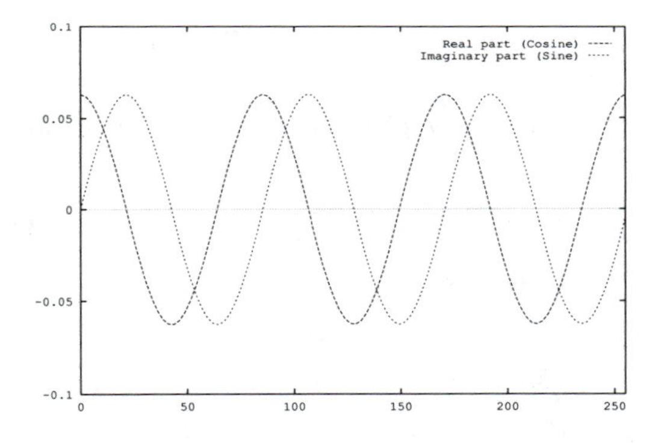

Figure 3.1: Real and imaginary parts of a Fourier basis function $\frac{1}{16} \exp \frac{2\pi 3n}{256}$.

3.1.1 The "fast" Fourier transform

Every coefficient of the matrix $\mathcal{F}(n,j)$ has absolute value one, so it is very far from sparse. However, the remarkable properties of the exponential function allow us to factor $\mathcal{F}$ into a product of just a few sparse matrices. The algorithm for computing the discrete Fourier transform via this factorization is called the "fast" Fourier transform (or *FFT*) and has a long history, although it is safe to say that it became widely used only after the publication of Cooley and Tukey's paper [33] in 1965. Much effort has been devoted to this subject since then [10, 108]. We will develop the basic algorithm below.

Lemma 3.1 (Danielson-Lanczos, 1942) *Suppose that M is a positive integer. Then the matrix $\mathcal{F}_{2M}$ of the $2M$-point Fourier transform can be factored as follows:*

$$\mathcal{F}_{2M} = \frac{1}{\sqrt{2}} E_{2M} \left(\mathcal{F}_M \oplus \mathcal{F}_M\right) P_{2M}$$

where P_M^e and P_M^o are $M \times 2M$ matrices given by $P_M^e(m,n) \stackrel{\text{def}}{=} \delta(2m-n)$ and $P_M^o(m,n) \stackrel{\text{def}}{=} \delta(2m+1-n)$ for $m = 0, 1, \ldots, M-1$ and $n = 0, 1, \ldots, 2M-1$, so that $P_{2M} \stackrel{\text{def}}{=} \begin{pmatrix} P_M^e \\ P_M^o \end{pmatrix}$ is a $2M \times 2M$ permutation matrix. Also,

$$\mathcal{F}_M \oplus \mathcal{F}_M \stackrel{\text{def}}{=} \begin{pmatrix} \mathcal{F}_M & 0 \\ 0 & \mathcal{F}_M \end{pmatrix},$$

and

$$E_{2M} \stackrel{\text{def}}{=} \begin{pmatrix} I_M & \Omega_M \\ I_M & -\Omega_M \end{pmatrix},$$

where I_M is the $M \times M$ identity matrix and Ω_M is a diagonal $M \times M$ matrix defined by

$$\Omega_M = \operatorname{diag}\{1,\ \exp\frac{-\pi i}{M},\ \exp\frac{-2\pi i}{M},\ \ldots,\ \exp\frac{-(M-1)\pi i}{M}\}.$$

Proof: For $0 \le n < M$, we have

$$\begin{aligned}
\sqrt{2M}\mathcal{F}_{2M}v(n) &= \sum_{k=0}^{2M-1} v(k)\exp\frac{-2\pi ikn}{2M} \\
&= \sum_{j=0}^{M-1} v(2j)\exp\frac{-2\pi i(2j)n}{2M} + \sum_{j=0}^{M-1} v(2j{+}1)\exp\frac{-2\pi i(2j{+}1)n}{2M} \\
&= \sum_{j=0}^{M-1} v(2j)\exp\frac{-2\pi ijn}{M} + \exp\frac{-\pi in}{M}\sum_{j=0}^{M-1} v(2j{+}1)\exp\frac{-2\pi ijn}{M} \\
&= \sqrt{M}\mathcal{F}_M v^e(n) + \sqrt{M}\left(\exp\frac{-\pi in}{M}\right)\mathcal{F}_M v^o(n).
\end{aligned}$$

Here $v^e(j) \stackrel{\text{def}}{=} v(2j)$ and $v^o(j) \stackrel{\text{def}}{=} v(2j+1)$ for $j = 0, 1, \ldots, M-1$. In particular, we note that $v^e = P^e v$ and $v^o = P^o v$.

If $M \le n < 2M-1$, then we observe that $\exp\left(\frac{-2\pi i(2j)n}{2M}\right) = \exp\left(\frac{-2\pi ijn}{M}\right) = \exp\left(\frac{-2\pi ij(n-M)}{M}\right)$, and

$$\exp\left(\frac{-2\pi i(2j{+}1)n}{2M}\right) = \exp\left(\frac{-\pi in}{M}\right)\exp\left(\frac{-2\pi ijn}{M}\right)$$

$$= \quad -\exp\left(\frac{-\pi i(n-M)}{M}\right)\exp\left(\frac{-2\pi ij(n-M)}{M}\right),$$

since $\exp\frac{-\pi iM}{M} = -1$. This lets us reduce to the previous case by substituting $n-M$ for n:

$$\begin{aligned}
\sqrt{2M}\mathcal{F}_{2M}v(n) &= \sum_{j=0}^{M-1} v(2j)\exp\frac{-2\pi i(2j)n}{2M} + \sum_{j=0}^{M-1} v(2j{+}1)\exp\frac{-2\pi i(2j{+}1)n}{2M} \\
&= \sum_{j=0}^{M-1} v(2j)\exp\frac{-2\pi ij(n-M)}{M} \\
&\qquad -\exp\frac{-\pi i(n-M)}{M}\sum_{j=0}^{M-1} v(2j{+}1)\exp\frac{-2\pi ij(n-M)}{M} \\
&= \sqrt{M}\mathcal{F}_M v^e(n-M) - \sqrt{M}\left(\exp\frac{-\pi i(n-M)}{M}\right)\mathcal{F}_M v^o(n-M).
\end{aligned}$$

We have thus produced $\mathcal{F}_{2M}v$ in two pieces:

$$\mathcal{F}_{2M}v(n) = \begin{cases} \frac{1}{\sqrt{2}}\left(\mathcal{F}_M P^e + \Omega_M\mathcal{F}_M P^o\right)v(n), & \text{if } 0 \le n < M, \\ \frac{1}{\sqrt{2}}\left(\mathcal{F}_M P^e - \Omega_M\mathcal{F}_M P^o\right)v(n-M), & \text{if } M \le n < 2M-1. \end{cases}$$

This can be now be written as the matrix form statement of the lemma. □

Notice that P_{2M} is a $2M \times 2M$ permutation matrix which may be written as follows:

$$P_{2M} = \begin{pmatrix} P^e_M \\ - \\ P^o_M \end{pmatrix} = \begin{pmatrix}
1 & 0 & 0 & 0 & 0 & \cdots & 0 & 0 & 0 \\
0 & 0 & 1 & 0 & 0 & \cdots & 0 & 0 & 0 \\
\vdots & & \cdots & & \ddots & & \cdots & & \vdots \\
0 & 0 & 0 & 0 & 0 & \cdots & 0 & 1 & 0 \\
- & - & - & - & - & - & - & - & \\
0 & 1 & 0 & 0 & 0 & \cdots & 0 & 0 & 0 \\
0 & 0 & 0 & 1 & 0 & \cdots & 0 & 0 & 0 \\
\vdots & & \cdots & & \ddots & & \cdots & & \vdots \\
0 & 0 & 0 & 0 & 0 & \cdots & 0 & 0 & 1
\end{pmatrix}.$$

Thus Lemma 3.1 implies that we can factor the dense matrix $\mathcal{F}_{2M}$ into sparser matrices: E_{2M} has exactly two elements per row or column, and P_{2M} is a permutation matrix. We now recursively factor the direct summands in the middle to get the "fast" discrete Fourier transform factorization:

Theorem 3.2 (Fast Fourier Transform) *If $N = 2^q$ for a positive integer q, then the discrete Fourier transform on N points can be factored as follows:*

$$\mathcal{F}_N = \frac{1}{\sqrt{N}} F_0 F_1 \cdots F_{q-1} P^N,$$

where P^N is an $N \times N$ permutation matrix, and

$$F_k = F_k^N = \overbrace{E_{(N/2^k)} \oplus \cdots \oplus E_{(N/2^k)}}^{2^k \text{ times}}$$

is an $N \times N$ block diagonal matrix, having 2^k blocks of order $2^{-k}N = 2^{q-k}$ each.

Proof: If $N = 2$, then $F_0 = E_2 = \begin{pmatrix} 1 & 1 \\ 1 & -1 \end{pmatrix}$ and $P^2 = I$ in Lemma 3.1, so that $\mathcal{F}_2 = \frac{1}{\sqrt{2}} F_0 P^2$ as claimed.

Now suppose that the formula holds for $M = 2^q$. We show that it holds for $N = 2M = 2^{q+1}$. We first note that $F_k^M \oplus F_k^M = F_{k+1}^{2M}$, and that $E_N = F_0^N$. Then by Lemma 3.1 we have

$$\begin{aligned}
\mathcal{F}_N &= \frac{1}{\sqrt{2}} E_N (\mathcal{F}_M \oplus \mathcal{F}_M) P_N \\
&= \frac{1}{\sqrt{2}} \frac{1}{\sqrt{M}} E_N \left((F_0^M F_1^M \cdots F_{q-1}^M P^M) \oplus (F_0^M F_1^M \cdots F_{q-1}^M P^M) \right) P_N \\
&= \frac{1}{\sqrt{2M}} E_N (F_0^M \oplus F_0^M) \cdots (F_{q-1}^M \oplus F_{q-1}^M)(P^M \oplus P^M) P_N \\
&= \frac{1}{\sqrt{N}} E_N F_1^{2M} \cdots F_q^{2M} (P^M \oplus P^M) P_N \\
&= \frac{1}{\sqrt{N}} F_0^N F_1^N \cdots F_q^N P^N.
\end{aligned}$$

The product $P^N \stackrel{\text{def}}{=} (P^M \oplus P^M) P_N$ is a permutation matrix. □

Proposition 3.3 *The permutation matrix P^N in Theorem 3.2 gives the N-point* bit-reversal *involution $n \mapsto n'$: if $n = (a_{q-1} a_{q-2} \cdots a_1 a_0)_{base\ 2}$ for $a_k \in \{0, 1\}$, then $n' = (a_0 a_1 \cdots a_{q-2} a_{q-1})_{base\ 2}$.*

Proof: If $N = 2$, then $P^N = Id$ which is the (trivial) two-point bit-reversal involution. Borrowing from the proof of Theorem 3.2, we must show that if P^N is N-point bit-reversal, then $P^{2N} = \left(P^N \oplus P^N\right) P_{2N}$ is $2N$-point bit reversal. But if $x \in \mathbf{C}^{2N}$, then $P_{2N} x = (P_N^e x, P_N^o x)$. If we take $n = (a_q a_{q-1} \cdots a_1 a_0)_{\text{base } 2}$, then P_{2N} exchanges $x(n)$ with $x(n'')$, where $n'' = (a_0 a_q a_{q-1} \cdots a_1)_{\text{base } 2}$. The subsequent application of P^N to the two half-vectors

$$(P_N^e x(0), \ldots, P_N^e x(N-1)) \quad \text{and} \quad (P_N^o x(N), \ldots, P_N^o x(2N-1))$$

performs two N-point bit-reversals on the lowest-order bits $a_q a_{q-1} \cdots a_1$ of n'', resulting in the exchange of $x(n)$ and $x(n')$. □

Each matrix F_k^N has exactly two nonzero elements in each row and column, so it costs only $2N$ operations to apply F_k^N to a vector. The permutation matrix P requires N operations to apply to a vector, and the scalar multiplication by $\frac{1}{\sqrt{N}}$ costs another N operations. Thus the total cost of applying $\mathcal{F}_N$ this way for $N = 2^q$ is $(2q+2)N = O(N \log_2 N)$ as $N \to \infty$. This has low complexity compared with the $O(N^2)$ cost of applying a dense matrix to a vector.

The inverse discrete Fourier transform has virtually the same matrix as the discrete Fourier transform, only with $+i$ rather than $-i$ in the exponential function: $\mathcal{F}^{-1} = \overline{\mathcal{F}}$. But $\mathcal{F}$ is also a symmetric matrix, so another way of saying this is that $\mathcal{F}^{-1} = \mathcal{F}^*$, or that the Fourier transform matrix is unitary.

Corollary 3.4 (Inverse Fast Fourier Transform) *If $N = 2^q$ for a positive integer q, then the inverse discrete Fourier transform on N points can be factored as follows:*

$$\mathcal{F}_N^{-1} = \frac{1}{\sqrt{N}} \bar{F}_0 \bar{F}_1 \cdots \bar{F}_{q-1} P^N,$$

where P^N is an $N \times N$ permutation matrix, and

$$\bar{F}_k = \bar{F}_k^N = \overbrace{\bar{E}_{(N/2^k)} \oplus \cdots \oplus \bar{E}_{(N/2^k)}}^{2^k \text{ times}}$$

is an $N \times N$ block diagonal matrix, having 2^k blocks of order $2^{-k}N = 2^{q-k}$ each. This is equivalent to replacing Ω_M with its complex conjugate $\bar{\Omega}_M$ in each matrix E_{2M}. □.

In fact, P^N is N-point bit-reversal, just like for $\mathcal{F}_N$.

3.1.2 Implementation of DFT

The discrete Fourier transform can be easily implemented in any programming language which supports complex arithmetic. Even when such a facility is absent, as in Standard C which has no built-in complex types, we can substitute the COMPLEX data structure plus the functions and utilities defined in Chapter 2. Using complex arithmetic mainly serves to organize the calculation; it is of course possible to use pairs of real numbers instead.

Factored DFT

We now implement the "fast" discrete Fourier transform described in Section 3.1.1. We first write a function which computes the initial bit-reversal permutation $f \to Pf$ needed by the "fast" discrete Fourier transform. The following function computes and returns the integer u whose bits are the reverse of the input n.

Return the input integer bit-reversed

```
br( N, LOG2LEN ):
   Let U = N&1
   For J = 1 to LOG2LEN
      N >>= 1
      U <<= 1
      U += N&1
   Return U
```

Note that bit-reversal is its own inverse: `br(br(`N`,`L`),`L`)==`N for $0 \leq N < 2^L$.

Second, we construct a bit-reversal permutation between two arrays. We will assume that `IN[]` and `OUT[]` are disjoint arrays of COMPLEX data structures:

Permute to a disjoint array by bit-reversing the indices

```
bitrevd( OUT, IN, Q ):
   Let M = 1<<Q
   For N = 0 to M-1
      Let U = br(N, Q)
      Let OUT[U] = IN[N]
```

We could also perform the bit-reversal in place by exchanging elements:

Permute an array in place via index bit-reversal

```
bitrevi( X, Q ):
   Let M = 1<<Q
   For N = 1 to M-2
      Let U = br(N, Q)
      If U > N then
         Let TEMP = X[N]
         Let X[N] = X[U]
         Let X[U] = TEMP
```

The binary numbers $00\ldots0_2$ and $111\ldots1_2$ are never moved by bit-reversal permutation, so we need only count from index 1 to index $2^q - 2$. Also, since bit-reversal is an involution, we must at most make $N/2$ exchanges; the skipped exchanges would invert the first ones. We arbitrarily choose to exchange index N for index U if and only if $U > N$.

Third, we need a routine which generates the vectors $\Omega_{N/2}, \Omega_{N/4}, \Omega_{N/8}, \ldots$. We observe that

$$\Omega_{M/2}(j) = \Omega_{N/2}(\frac{N}{M} * j),$$

so that we need only compute $\Omega_{N/2}$ and then decimate to get the smaller vectors. At this time it is easy to write a slightly more general routine which will produce $\bar{\Omega}_{N/2}$ as well. The following function fills the vector `W[]` of M COMPLEX's with the values $e^{-\pi i n/M}$, $n = 0, 1, ..., |M|-1$. If $M < 0$, then `W[]` is assigned the complex conjugate of $\Omega_{|M|}$. In this and later routines, we use the tag `PI` for the constant π. It is also useful to define `SQH` $= \sqrt{1/2}$ and `SQ2` $= \sqrt{2}$:

```
#define PI (3.1415926535897932385)
#define SQH (0.70710678118654752440)
#define SQ2 (1.4142135623730950488)
```

Remark. We can place these three constants and any others we might need into a separate file, to be included in any computer program that requires them. That will ensure that each portion of the software uses the same degree of precision for its numerical constants.

Compute sines and cosines for DFT

```
fftomega( W, M ):
   Let FACTOR = -PI/M
   If M < 0 then
      Let M = -M
   For K = 0 to M-1
      Let W[N].RE = cos(K*FACTOR)
      Let W[N].IM = sin(K*FACTOR)
```

This should be compared to the function `dhtcossin()`, defined below.

Next, we define the function which applies the sparse factors $F_{q-1}F_{q-2}\cdots F_1F_0$ made from $\Omega_{N/2}$ to the permuted vector Pf. This can be done in place if we use just one temporary variable. Note that the innermost loop requires just one complex addition, one complex subtraction, and one complex multiplication. The function below successively applies the sparse matrices $\{F_{q-1}, F_{q-2}, ..., F_1\}$, and finally F_0 to the complex input vector F, transforming it in place. Each F_k has 2^k blocks E_{2M}, where $2M = 2^{q-k}$, and the matrix E_{2M} is defined as in Lemma 3.1.

Product of sparse matrices for DFT

```
fftproduct( F, Q, W ):
   Let N = 1<<Q
   Let K = Q
   While K > 0
      K -= 1
      Let N1 = N>>K be the block size
      Let M  = N1/2 be the butterfly size
      B = 0
      While B < N
         Let TMP.RE = F[B+M].RE
         Let TMP.IM = F[B+M].IM
         Let F[B+M].RE = F[B].RE - TMP.RE
         Let F[B+M].IM = F[B].IM - TMP.IM
         F[B].RE  += TMP.RE
         F[B].IM  += TMP.IM
         For J = 1 to M-1
            Let TMP.RE  = CCMULRE( F[B+M+J], W[J*(N/N1)] )
            Let TMP.IM  = CCMULIM( F[B+M+J], W[J*(N/N1)] )
            Let F[B+M+J].RE = F[B+J].RE - TMP.RE
            Let F[B+M+J].IM = F[B+J].IM - TMP.IM
            F[B+J].RE  += TMP.RE
            F[B+J].IM  += TMP.IM
         B += N1
```

Finally, we need a routine which multiplies a vector of length N by $\frac{1}{\sqrt{N}}$ in order to produce a unitary Fourier transform.

Normalization for unitary DFT

```
fftnormal( F, N ):
   Let NORM = sqrt(1.0/N)
   For K = 0 to N-1
      Let F.RE = NORM * F.RE
      Let F.IM = NORM * F.IM
```

Now we put the bit-reversal permutation together with the loops which compute the applications of the matrices F_k (or $\bar{F}_k$) in Theorem 3.2. The function below allocates, computes, and returns a vector of COMPLEX data structuress which is the discrete Fourier transform of the complex input vector `F[]`. If the logarithm-of-length parameter Q is negative, then it computes the inverse discrete Fourier transform.

Unitary DFT and iDFT

```
dft( F, Q ):
   Let N = 1<<(absval(Q))
   Allocate an array FHAT[] of N COMPLEX's
   If N == 1 then
      Let FHAT[0].RE = F[0].RE
      Let FHAT[0].IM = F[0].IM
   Else
      If N == 2 then
         Let FHAT[0].RE = (F[0].RE + F[1].RE)*SQH
         Let FHAT[0].IM = (F[0].IM + F[1].IM)*SQH
         Let FHAT[1].RE = (F[0].RE - F[1].RE)*SQH
         Let FHAT[1].IM = (F[0].IM - F[1].IM)*SQH
      Else
         Allocate an array W[] of N/2 COMPLEX's
         If Q < 0 then
            Let Q = -Q
            fftomega(W, -N/2)
         Else
            fftomega(W, N/2)
         bitrevd( FHAT, F, Q)
         fftproduct(FHAT, Q, W)
         Deallocate the array W[]
         fftnormal(FHAT, N)
   Return FHAT
```

Remark. If we will be using the DFT function over and over, then it makes sense to allocate and assign the array W[] once and reuse it.

3.2 The discrete Hartley transform

The Hartley or CAS transform is a more symmetric and purely real-valued version of the Fourier transformation; its matrix is given by the following formula:

$$H_N : \mathbf{R}^N \to \mathbf{R}^N; \quad H_N(n,m) = \frac{1}{\sqrt{N}} \left[\cos\frac{2\pi nm}{N} + \sin\frac{2\pi nm}{N} \right].$$

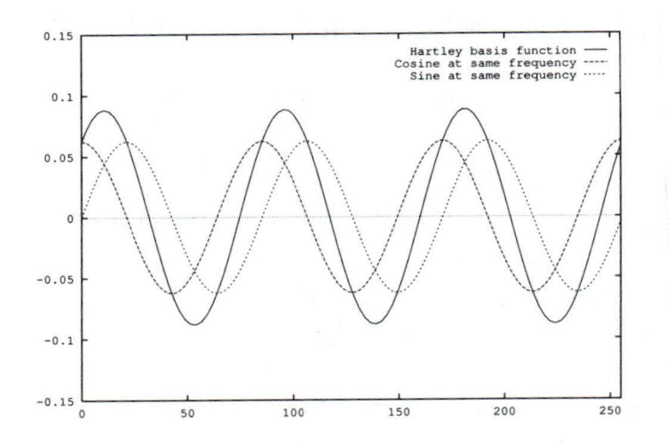

Figure 3.2: Hartley basis function $\frac{1}{16}\left[\cos\frac{2\pi 3n}{256} + \sin\frac{2\pi 3n}{256}\right]$.

An elementary identity shows that the Hartley basis functions are just cosines with amplitude $\sqrt{\frac{2}{N}}$ and with the origin translated by one-eighth of a period. Figure 3.2 compares one example of such a basis function with sine and cosine functions of the same frequency.

3.2.1 The "fast" discrete Hartley transform

The Hartley transform can be recursively factored in virtually the same way as the Fourier transform, using slightly different sparse factors. To obtain the analog of Lemma 3.1, we use the following identity:

$$\operatorname{cas}\frac{2\pi nm}{N} \stackrel{\text{def}}{=} \cos\frac{2\pi nm}{N} + \sin\frac{2\pi nm}{N} = \Re\, e^{-2\pi i\frac{nm}{N}} - \Im\, e^{-2\pi i\frac{nm}{N}}. \tag{3.3}$$

So if $f \in \mathbf{R}^N$ is purely real, then

$$\begin{aligned} H_N f(m) &= \frac{1}{\sqrt{N}}\sum_{m=0}^{N-1} f(m)\Re\exp\frac{-2\pi inm}{N} - \frac{1}{\sqrt{N}}\sum_{m=0}^{N-1} f(m)\Im\exp\frac{-2\pi inm}{N} \\ &= \Re\frac{1}{\sqrt{N}}\sum_{m=0}^{N-1} f(m)\exp\frac{-2\pi inm}{N} - \operatorname{Im}\frac{1}{\sqrt{N}}\sum_{m=0}^{N-1} f(m)\exp\frac{-2\pi inm}{N} \\ &= \operatorname{Re}\mathcal{F}_N f(n) - \Im\,\mathcal{F}_N f(n). \end{aligned}$$

If $f \in \mathbf{C}^N$, then we can use linearity to compute the transform in two pieces:

$$\begin{aligned} H_N f &= \Re\,\mathcal{F}_N(\Re f) - \Im\,\mathcal{F}_N(\Re f) \\ &\quad + i\Re\,\mathcal{F}_N(\Im f) + i\Im\,\mathcal{F}_N(\Im f). \end{aligned}$$

But a "fast" Hartley transform of a real vector can be performed purely in real arithmetic. The Hartley transform is distinguished by its similarity to the Fourier transform: the basis functions are $\cos \frac{2\pi nm}{N} + \sin \frac{2\pi nm}{N}$ rather than $\cos \frac{2\pi nm}{N} - i \sin \frac{2\pi nm}{N}$. This similarity produces an analog to the Danielson–Lanczos lemma:

Lemma 3.5 *If M is a positive even number, then the $2M$-point Hartley transform factors as follows:*

$$H_{2M} = \frac{1}{\sqrt{2}} A_{2M} \left(H_M \oplus H_M \right) P_{2M},$$

where P_M^e and P_M^o are $M \times 2M$ matrices given by $P_M^e(m,n) \stackrel{\text{def}}{=} \delta(2m - n)$ and $P_M^o(m,n) \stackrel{\text{def}}{=} \delta(2m + 1 - n)$ for $m = 0, 1, \ldots, M-1$ and $n = 0, 1, \ldots, 2M-1$, so that $P_{2M} \stackrel{\text{def}}{=} \begin{pmatrix} P_M^e \\ P_M^o \end{pmatrix}$ is a $2M \times 2M$ permutation matrix. Also,

$$H_M \oplus H_M \stackrel{\text{def}}{=} \begin{pmatrix} H_M & 0 \\ 0 & H_M \end{pmatrix},$$

and

$$A_{2M} \stackrel{\text{def}}{=} \begin{pmatrix} I_M & B_M \\ I_M & -B_M \end{pmatrix},$$

where I_M is the $M \times M$ identity matrix, and B_M is an $M \times M$ "butterfly" matrix defined by

$$B_M \stackrel{\text{def}}{=} \begin{pmatrix} 1 & 0 & & & \cdots & & & 0 \\ 0 & c_1 & & & & & & s_1 \\ & & \ddots & & & & \iddots & \\ & & & c_{\frac{M}{2}-1} & 0 & s_{\frac{M}{2}-1} & & \\ \vdots & & & 0 & 1 & 0 & & \\ & & & s_{\frac{M}{2}-1} & 0 & -c_{\frac{M}{2}-1} & & \\ & & \iddots & & & & \ddots & \\ 0 & s_1 & & & & & & -c_1 \end{pmatrix}$$

for $c_n \stackrel{\text{def}}{=} \cos \frac{\pi n}{M}$ and $s_n \stackrel{\text{def}}{=} \sin \frac{\pi n}{M}$.

Proof: Our basic identity is the following:

$$\sqrt{2M}\, H_{2M} f(n) = \sum_{m=0}^{2M-1} f(m) \operatorname{cas} \frac{2\pi nm}{2M}$$

$$
\begin{aligned}
&= \sum_{m=0}^{M-1} f(2m)\mathrm{cas}\,\frac{2\pi n(2m)}{2M} + \sum_{m=0}^{M-1} f(2m+1)\mathrm{cas}\,\frac{2\pi n(2m+1)}{2M} \\
&= \sum_{m=0}^{M-1} f(2m)\mathrm{cas}\,\frac{2\pi nm}{M} + \sum_{m=0}^{M-1} f(2m+1)\mathrm{cas}\left(\frac{2\pi nm}{M} + \frac{\pi n}{M}\right).
\end{aligned}
$$

Now $\mathrm{cas}\,(A+B) = \cos B\ \mathrm{cas}\,A + \sin B\ \mathrm{cas}\,(-A)$, so this can be rewritten as

$$
\sqrt{2}\, H_{2M} f(n) = H_M f^e(n) + \cos\frac{\pi n}{M}\ H_M f^o(n) + \sin\frac{\pi n}{M}\ H_M f^o(2M-n), \qquad (3.4)
$$

where $f^e = P^e_M f$ and $f^o = P^o_M f$. In this formula, we treat $H_M f(n)$ as M-periodic in the index n. Then we use the identities

$$
\begin{aligned}
\cos\frac{\pi n}{M} &= -\cos\frac{\pi(n-M)}{M}, & \sin\frac{\pi(n-M)}{M} &= -\sin\frac{\pi n}{M}, \\
\cos\frac{\pi(\frac{M}{2}+n)}{M} &= -\cos\frac{\pi(\frac{M}{2}-n)}{M}, & \sin\frac{\pi(\frac{M}{2}+n)}{M} &= \sin\frac{\pi(\frac{M}{2}-n)}{M},
\end{aligned}
$$

so that we only need to evaluate the trigonometric functions $c_1, \ldots, c_{\frac{M}{2}-1}$ and $s_1, \ldots, s_{\frac{M}{2}-1}$. This shows that the sparse factor has the stated form. □

Now we apply this lemma recursively to obtain one of the "fast" Hartley transforms (the radix-2 transform, described in [102]):

Theorem 3.6 (Radix-2 Fast Hartley Transform) *If $N = 2^q$ for a positive integer q, then the discrete Hartley transform on N points can be factored as follows:*

$$
H_N = \frac{1}{\sqrt{N}} G_0 G_1 \cdots G_{q-1} P,
$$

where P is the $N \times N$ "bit-reversal" permutation matrix, and

$$
G_k = G_k^N = \overbrace{A_{(N/2^k)} \oplus \cdots \oplus A_{(N/2^k)}}^{2^k \text{ times}}
$$

is an $N \times N$ block diagonal matrix, whose 2^k blocks of order $2^{-k}N = 2^{q-k}$ each are the sparse factors A of Lemma 3.5.

Proof: The proof is virtually identical to that of Theorem 3.2. We note that $G_{q-1} = A_2 \oplus \cdots \oplus A_2$ with $A_2 = \begin{pmatrix} 1 & 1 \\ 1 & -1 \end{pmatrix}$ is the same innermost sparse factor as F_{q-1} in the FFT. □

3.2.2 Implementation of DHT

The fast discrete Hartley transformation (or DHT) has much in common with FFT. We obtain the factorization using a close cousin of the Danielson–Lanczos lemma. The main difference is that we need a somewhat different inner loop to apply the sparse factors.

The initial permutation $f \to P^N f$ is just bit-reversal, so we would use the same integer bit-reversal functions `bitrevi()` and `bitrevd()` as for FFT. We may need to rewrite the permutation and normalization routines to handle arrays of REALs rather than COMPLEX's, but the amount of rewriting depends on the programming language. For example, in Standard C it is possible to assign or exchange stored values knowing only their size, regardless of their type. We will assume that any changes to the initial permuation function are minor.

To illustrate how minor such changes are, we explicitly modify the normalization routine to work with arrays of REAL data types:

Normalize an array

```
dhtnormal( F, N ):
   Let NORM = sqrt(1.0/N)
   For K = 0 to N-1
      F[K] *= NORM
```

We also need to modify the routine which generates the table of sines and cosines, as demanded by the comment before the function `dhtproduct()` below. This function fills the preallocated arrays C and S of REALs, each of length $N/2$, with the respective values $C[K] = \cos(\pi K/N)$ and $S[K] = \sin(\pi K/N)$ for $K = 0, 1, ..., N/2-1$.

Generate tables of sines and cosines

```
dhtcossin( C, S, N ):
   Let FACTOR = PI/N
   For K = 0 to (N/2)-1
      Let C[K] = cos(K*FACTOR)
      Let S[K] = sin(K*FACTOR)
```

The function `dhtcossin()` should be compared with the function `fftomega()` used by `dft()`. Besides writing REAL rather than COMPLEX output arrays, we need only half as many angles.

Finally, we modify `fftproduct()` to obtain a function which applies the sparse factors $G_{q-1}G_{q-2}\cdots G_1G_0$ of Theorem 3.5 to the bit-reversed vector $P^N f$. As for FFT, we use temporary variables so we can apply the factors in place. By dividing up the arithmetic, we reduce the operations in the innermost loop to four real multiplications and six real additions.

The function below assumes that $q \geq 1$:

Sparse matrices for the radix-2 Hartley transform

```
dhtproduct( F, Q, C, S ):
   Let N = 1<<Q
   Let B = 0
   While B < N
      Let TEMP   = F[B] - F[B+1]
      Let F[B]   = F[B] + F[B+1]
      Let F[B+1] = TEMP
      B += 2
   Let K = Q-1
   While K > 0
      K -= 1
      Let N1 = N>>K be the block size
      Let M  = N1/2 be the butterfly size
      Let M2 = M/2 be the butterfly midpoint
      Let B = 0
      While B < N
         Let TEMP = F[B] - F[B+M]
         Let F[B] = F[B] + F[B+M]
         Let F[B+M] = TEMP
         Let TEMP   = F[B+M2] - F[B+M+M2]
         Let F[B+M2] = F[B+M2] + F[B+M+M2]
         Let F[B+M+M2] = TEMP
         For J = 1 to M2-1
            Let TMP1 = F[B+M+J]*C[J*N/N1] + F[B+N1-J]*S[J*N/N1]
            Let TMP2 = F[B+M+J]*S[J*N/N1] - F[B+N1-J]*C[J*N/N1]
            Let F[B+M+J]   = F[B+J]   - TMP1
            Let F[B+N1-J] = F[B+M-J] - TMP2
            Let F[B+J]     = F[B+J]   + TMP1
         B += N1
```

We assemble the complete DHT from the bit-reversal, sparse matrix multiplication, and normalization functions just defined. We also allocate the temporary sine and cosine arrays needed by intermediate stages of the transform, and allocate an output array which will be filled and returned upon termination. We assume only that $q \geq 0$, but we isolate the trivial one-point, two-point, and four-point Hartley transforms as special cases. The DHT implementation given here performs those three by explicit matrix multiplication, without factorization:

Radix-2 discrete Hartley transform

```
dht( F, Q ):
   Let N = 1<<Q
   Allocate an output array FH[] of length N
   If N == 1 then
      Let FH[0] = F[0]
   Else
      If N == 2 then
         Let FH[0] = (F[0] + F[1])*SQH
         Let FH[1] = (F[0] - F[1])*SQH
      Else
         If N == 4 then
            Let FH[0] = (F[0]+F[1]+F[2]+F[3])*0.5
            Let FH[1] = (F[0]+F[1]-F[2]-F[3])*0.5
            Let FH[2] = (F[0]-F[1]+F[2]-F[3])*0.5
            Let FH[3] = (F[0]-F[1]-F[2]+F[3])*0.5
         Else
            If N > 4 then
               bitrevd(FH, F, Q)
               Allocate array C[] to length N/2
               Allocate array S[] to length N/2
               dhtcossin( C, S, N/2 )
               dhtproduct( FH, Q, C, S )
               dhtnormal( FH, N )
               Deallocate C[] and S[]
   Return FH
```

Remark. As with the DFT, if we will be applying this function several times, the arrays of sines and cosines may be allocated and assigned once and retained. We can also modify `dht()` to perform DHT in place; this is left as an exercise.

3.3 Discrete sine and cosine transforms

There are several other real-valued transforms which are "near" the Fourier transform in the sense that computing them requires only an additional sparse matrix application after FFT. We classify these in the manner of [92], by listing the basis vectors in their conventional normalization. The numbers $b(k)$ in the formulas below are weights which are needed to insure orthonormality; we have

$$b(k) = \begin{cases} 0, & \text{if } k < 0 \text{ or } k > N; \\ 1/\sqrt{2}, & \text{if } k = 0 \text{ or } k = N; \\ 1, & \text{if } 0 < k < N. \end{cases} \tag{3.5}$$

In the following list of matrices, the indices always start at zero.

DCT-I.

$$C^I_{N+1} : \mathbf{R}^{N+1} \to \mathbf{R}^{N+1}; \quad C^I_{N+1}(n,m) = b(n)b(m)\sqrt{\frac{2}{N}} \cos \frac{\pi nm}{N}.$$

DCT-II.

$$C^{II}_N : \mathbf{R}^N \to \mathbf{R}^N; \quad C^{II}_N(n,m) = b(n)\sqrt{\frac{2}{N}} \cos \frac{\pi n(m+\frac{1}{2})}{N}.$$

DCT-III.

$$C^{III}_N : \mathbf{R}^N \to \mathbf{R}^N; \quad C^{III}_N(n,m) = b(m)\sqrt{\frac{2}{N}} \cos \frac{\pi (n+\frac{1}{2})m}{N}.$$

DCT-IV.

$$C^{IV}_N : \mathbf{R}^N \to \mathbf{R}^N; \quad C^{IV}_N(n,m) = \sqrt{\frac{2}{N}} \cos \frac{\pi (n+\frac{1}{2})(m+\frac{1}{2})}{N}.$$

DST-I.

$$S^I_{N-1} : \mathbf{R}^{N-1} \to \mathbf{R}^{N-1}; \quad S^I_{N-1}(n,m) = \sqrt{\frac{2}{N}} \sin \frac{\pi nm}{N}.$$

DST-II.

$$S^{II}_N : \mathbf{R}^N \to \mathbf{R}^N; \quad S^{II}_N(n,m) = b(n+1)\sqrt{\frac{2}{N}} \sin \frac{\pi (n+1)(m+\frac{1}{2})}{N}.$$

DST-III.

$$S^{III}_N : \mathbf{R}^N \to \mathbf{R}^N; \quad S^{III}_N(n,m) = b(m+1)\sqrt{\frac{2}{N}} \sin \frac{\pi (n+\frac{1}{2})(m+1)}{N}.$$

DST-IV.

$$S^{IV}_N : \mathbf{R}^N \to \mathbf{R}^N; \quad S^{IV}_N(n,m) = \sqrt{\frac{2}{N}} \sin \frac{\pi (n+\frac{1}{2})(m+\frac{1}{2})}{N}.$$

The inner products with these vectors can be rapidly computed using the FFT algorithm with simple modifications. The trigonometric transforms are just conjugates of the FFT by an appropriate sparse matrix; we will compute these sparse matrices below.

3.3.1 DCT-I and DST-I

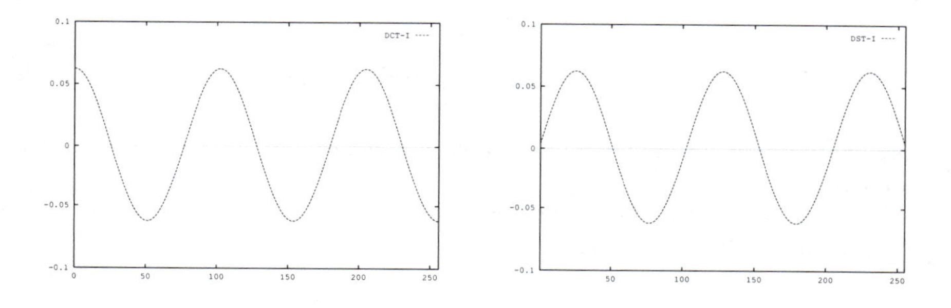

Figure 3.3: DCT-I and DST-I basis functions.

We can relate S^I_{N-1} to $\mathcal{F}_{2N}$ by using the following substitutions:

$$\cos\frac{\pi nm}{N} = \frac{1}{2}\left[\exp\frac{-2\pi i(2N-n)m}{2N} + \exp\frac{-2\pi inm}{2N}\right]; \tag{3.6}$$

$$\sin\frac{\pi nm}{N} = \frac{1}{2i}\left[\exp\frac{-2\pi i(2N-n)m}{2N} - \exp\frac{-2\pi inm}{2N}\right]. \tag{3.7}$$

We deduce that, for an arbitrary function $g \in \mathbf{C}^{2N}$, we have the following identity:

$$\begin{aligned}
\mathcal{F}_{2N}g(n) \quad - \quad \mathcal{F}_{2N}g(2N-n) \quad &= \\
&= \frac{1}{\sqrt{2N}}\sum_{m=0}^{2N-1} g(m)\left[\exp\frac{-2\pi inm}{2N} - \exp\frac{-2\pi i(2N-n)m}{2N}\right] \\
&= -i\sqrt{\frac{2}{N}}\sum_{m=0}^{2N-1} g(m)\sin\frac{\pi nm}{N} \\
&= -i\sqrt{\frac{2}{N}}\sum_{m=1}^{N-1}\left[g(m)\sin\frac{\pi nm}{N} + g(2N-m)\sin\frac{\pi n(2N-m)}{N}\right]
\end{aligned}$$

$$= -i\sqrt{\frac{2}{N}}\sum_{m=1}^{N-1}[g(m)-g(2N-m)]\sin\frac{\pi nm}{N}. \qquad (3.8)$$

Now if we suppose that $y \in \mathbf{R}^{N-1}$ and

$$-g(2N-m) = g(m) = iy(m)/\sqrt{2} \text{ for } m = 1,2,\ldots,N-1, \text{ with } g(0) = g(N) = 0,$$

then we get the following identity for $n = 1,2,\ldots,N-1$:

$$\mathcal{F}_{2N}g(n) - \mathcal{F}_{2N}g(2N-n) = \sqrt{2}\sqrt{\frac{2}{N}}\sum_{m=1}^{N-1} y(m)\sin\frac{\pi nm}{N} = \sqrt{2}\, S^I_{N-1}y(n). \qquad (3.9)$$

Note that if $y = y(n)$ is purely real, then the right-hand side of Equation 3.9 is purely real.

The map $y \mapsto g$ can also be written in matrix form $g = iV_N y$, where V_N is the following $2N \times (N-1)$ matrix:

$$V_N = \frac{1}{\sqrt{2}}\begin{pmatrix} 0 & 0 & \cdots & 0 \\ 1 & & & \\ & 1 & & \\ & & \ddots & \\ & & & 1 \\ 0 & 0 & \cdots & 0 \\ & & & -1 \\ & & -1 & \\ & \cdot^{\cdot^{\cdot}} & & \\ -1 & & & \end{pmatrix}. \qquad (3.10)$$

We observe that this is very similar to the matrix which produces $\mathcal{F}_{2N}g(n) - \mathcal{F}_{2N}g(2N-n)$ from $\mathcal{F}_{2N}g$; in fact, we observe that S^I_{N-1} is just $\mathcal{F}_{2N}$ conjugated by V_N:

Proposition 3.7 $\quad S^I_{N-1} = i\,V_N^*\mathcal{F}_{2N}V_N$. □

Likewise, we can relate C^I_{N+1} to $\mathcal{F}_{2N}$. Let $f \in \mathbf{C}^{2N}$ be an arbitrary vector. We have the following identity for $n = 1,2,\ldots,N$:

$$\begin{aligned} \mathcal{F}_{2N}f(n) \quad + \quad \mathcal{F}_{2N}f(2N-n) \quad &= \\ &= \frac{1}{\sqrt{2N}}\sum_{m=0}^{2N-1} f(m)\left[\exp\frac{-2\pi inm}{2N} + \exp\frac{-2\pi i(2N-n)m}{2N}\right] \end{aligned}$$

$$
\begin{aligned}
&= \sqrt{\frac{2}{N}} \sum_{m=0}^{2N-1} f(m) \cos \frac{\pi n m}{N} \\
&= \sqrt{\frac{2}{N}} \sum_{m=1}^{N-1} \left[f(m) \cos \frac{\pi n m}{N} + f(2N-m) \cos \frac{\pi n (2N-m)}{N} \right] \\
&\qquad + \frac{2}{\sqrt{2N}} \left[f(0) \cos 0 + f(N) \cos \pi n \right] \\
&= \sqrt{\frac{2}{N}} \sum_{m=1}^{N-1} \left[f(m) + f(2N-m) \right] \cos \frac{\pi n m}{N} \\
&\qquad + \sqrt{\frac{2}{N}} \left[f(0) \cos 0 + f(N) \cos \pi n \right]. \qquad (3.11)
\end{aligned}
$$

Now if we let $x \in \mathbf{R}^{N+1}$ be arbitrary and put $f(2N-m) = f(m) = x(m)/\sqrt{2}$ for $m = 1, \ldots, N-1$, $f(N) = x(N)$ and $f(0) = x(0)$, then for $n = 1, 2, \ldots, N-1$ we have the following identity:

$$
\begin{aligned}
\mathcal{F}_{2N} f(n) + \mathcal{F}_{2N} f(2N-n) &= \sqrt{2} \sqrt{\frac{2}{N}} b(n) \sum_{m=0}^{N} b(m) x(m) \cos \frac{\pi n m}{N} \\
&= \sqrt{2}\, C_{N+1}^{I} x(n). \qquad (3.12)
\end{aligned}
$$

For $n = N$ we have $\mathcal{F}_{2N} f(N) + \mathcal{F}_{2N} f(2N-N) = 2\mathcal{F}_{2N} f(N)$. Recalling that $b(N) = \frac{1}{\sqrt{2}}$, we can write the following:

$$
\mathcal{F}_{2N} f(N) = \sqrt{\frac{2}{N}} b(N) \sum_{m=0}^{N} b(m) x(m) \cos \frac{\pi m N}{N} = C_{N+1}^{I} x(N). \qquad (3.13)
$$

For $n = 0$ we have the identity

$$
\mathcal{F}_{2N} f(0) = \sqrt{\frac{2}{N}} b(0) \sum_{m=0}^{N} b(m) x(m) = C_{N+1}^{I} x(0). \qquad (3.14)
$$

Note that if $x = x(n)$ is purely real, then both sides of Equations 3.12 to 3.14 are purely real.

The map $x \mapsto f$ can also be written in matrix form $f = U_N x$, where U_N is the following $2N \times (N+1)$ matrix:

$$U_N = \frac{1}{\sqrt{2}} \begin{pmatrix} \sqrt{2} & 0 & & \cdots & & 0 \\ & 1 & & & & \\ & & 1 & & & \\ & & & \ddots & & \\ & & & & 1 & \\ 0 & & \cdots & & 0 & \sqrt{2} \\ & & & & 1 & \\ & & & \iddots & & \\ & & 1 & & & \\ 0 & 1 & 0 & \cdots & & 0 \end{pmatrix}. \tag{3.15}$$

We observe, as in the case of DST-I, that this is very similar to the matrix which produces $\mathcal{F}_{2N}f(n) + \mathcal{F}_{2N}f(2N-n)$ from $\mathcal{F}_{2N}f$. In this case, we see that C^I_{N+1} is just $\mathcal{F}_{2N}$ conjugated by U_N:

Proposition 3.8 $\quad C^I_{N+1} = U^*_N \mathcal{F}_{2N} U_N.$ □

We obtain the following technical result:

Lemma 3.9 *The matrices V_N defined as in Equation 3.10 and U_N defined as in Equation 3.15 satisfy the following relations:*

1. $V^*_N V_N = I_{N-1}$,
2. $U^*_N U_N = I_{N+1}$,
3. $U^*_N V_N = 0_{(N+1)\times(N-1)}$,
4. $V^*_N U_N = 0_{(N-1)\times(N+1)}$,
5. $V_N V^*_N + U_N U^*_N = I_{2N}$, *with*

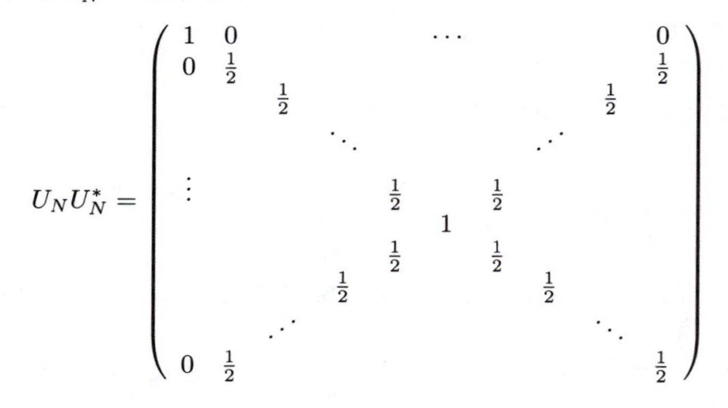

and

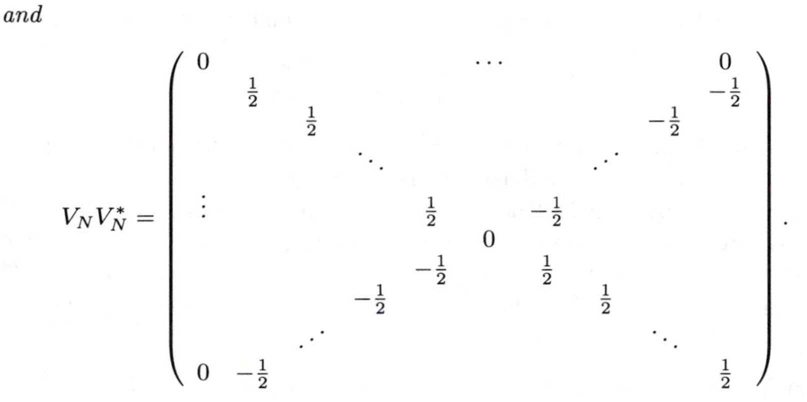

Thus $V_N V_N^*$ and $U_N U_N^*$ are orthogonal projections onto orthogonal subspaces of $\mathbf{C}^{2N}$.

Proof: These calculations are left to the reader (Exercise 1). □

We can form the $2N \times 2N$ matrix

$$T = \left(\begin{array}{c|c} U_N & iV_N \end{array} \right). \tag{3.16}$$

Then $T^*T = I_{2N}$, so $TT^* = I_{2N}$. Also, T is sparse—it has at most two nonzero elements in each row or column.

The DST-I and DCT-I transforms can be performed simultaneously by applying $T^*\mathcal{F}T$ to the two vectors $x \in \mathbf{R}^{N+1}$ and $y \in \mathbf{R}^{N-1}$ concatenated as follows:

$$X = \begin{bmatrix} x(0) \\ x(1) \\ \vdots \\ x(N-1) \\ x(N) \\ y(1) \\ \vdots \\ y(N-1) \end{bmatrix} \in \mathbf{R}^{2N}; \qquad TX = \begin{bmatrix} \sqrt{2}\,x(0) \\ x(1) + iy(1) \\ \vdots \\ x(N-1) + iy(N-1) \\ \sqrt{2}\,x(N) \\ x(1) - iy(1) \\ \vdots \\ x(N-1) - iy(N-1) \end{bmatrix} \in \mathbf{C}^{2N}.$$

Theorem 3.10 $\quad T^*\mathcal{F}_{2N}T = C^I_{N+1} \oplus (-i)S^I_{N-1}.$

Proof: We will drop all subscripts for clarity. The matrix splits as follows:

$$T^*\mathcal{F}T = \begin{pmatrix} U^*\mathcal{F}U & iU^*\mathcal{F}V \\ -iV^*\mathcal{F}U & V^*\mathcal{F}V \end{pmatrix}.$$

The diagonal blocks are C^I and $(-i)S^I$ by Propositions 3.8 and 3.7, respectively. We need to show that the off-diagonal blocks are zero. But if we put $g = Uy$ rather than $g = Vy$ in Equation 3.8 (that is, put $g(m) = g(2N-m)$ for $n = 1, 2, \dots, N-1$), then the right-hand side vanishes identically, proving that $V^*\mathcal{F}U = 0$. Similarly, if we put $f = Vx$ in Equation 3.11 (or $f(m) = -f(2N-m)$ for $n = 1, 2, \dots, N-1$, and $f(0) = f(N) = 0$), then the right-hand side vanishes identically and we have shown that $U^*\mathcal{F}V = 0$. □

Remark. Since $\mathcal{F}_{2N}$ has a sparse factorization requiring only $4N + 4N\log_2 2N$ operations to apply, and each of T and T^* requires $4N$ operations, the total cost for computing both $S^I_{N-1}y$ and $C^I_{N+1}x$ is $4N(4 + \log_2 N)$ operations.

3.3.2 DCT-II, DCT-III, DST-II, and DST-III

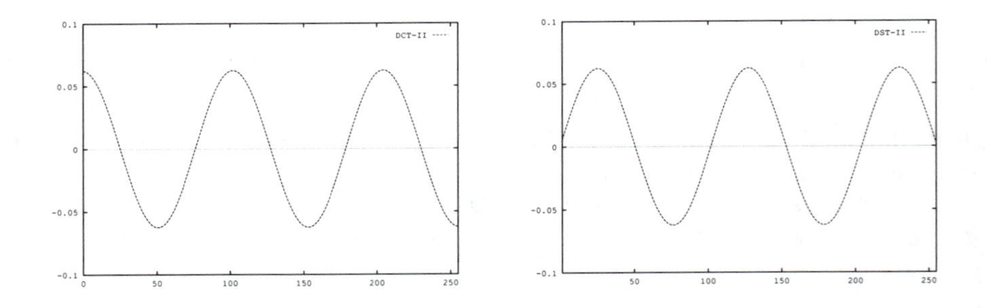

Figure 3.4: DCT-II and DST-II basis functions.

We first observe that $C^{II}_N(n,m) = C^{III}_N(m,n)$ and $S^{II}_N(n,m) = S^{III}_N(m,n)$ for all $0 \le n, m < N$. Thus it suffices to consider just the -II case.

Note that for $f \in \mathbf{C}^{2N}$ we have the following:

$$\begin{aligned} \mathcal{F}_{2N}f(n) &= \frac{1}{\sqrt{2N}} \sum_{m=0}^{2N-1} f(m)e^{\frac{-2\pi inm}{2N}} \qquad (3.17) \\ &= \frac{1}{\sqrt{2N}} \sum_{m=0}^{N-1} \left[f(m)e^{\frac{-2\pi inm}{2N}} + f(2N-m-1)e^{\frac{-2\pi in(2N-m-1)}{2N}} \right] \end{aligned}$$

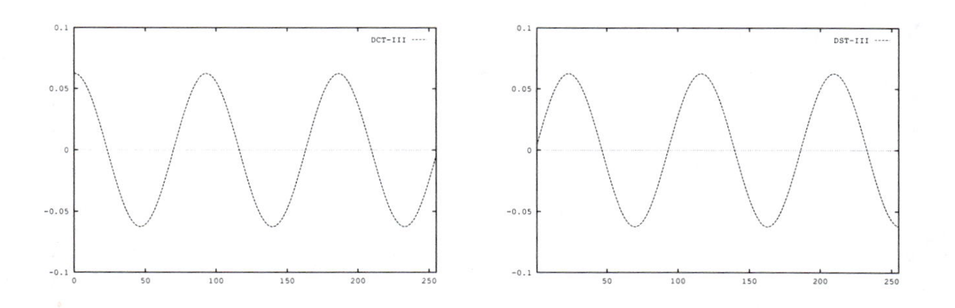

Figure 3.5: DCT-III and DST-III basis functions.

$$= \frac{e^{\frac{\pi i n}{2N}}}{\sqrt{2N}} \sum_{m=0}^{N-1} \left[f(m) e^{\frac{-2\pi i n (m+\frac{1}{2})}{2N}} + f(2N-m-1) e^{\frac{2\pi i n(m+\frac{1}{2})}{2N}} \right].$$

Thus if we let $f(m) = f(2N-m-1) = x(m)/\sqrt{2}$ for $x \in \mathbf{C}^N$ and $n = 0, 1, \ldots, N-1$, we get

$$b(n)\mathcal{F}_{2N} f(n) = b(n) \frac{e^{\frac{\pi i n}{2N}}}{\sqrt{N}} \sum_{m=0}^{N-1} x(m) \cos \frac{\pi n(m+\frac{1}{2})}{N} = \frac{e^{\frac{\pi i n}{2N}}}{\sqrt{2}} C_N^{II} x(n). \tag{3.18}$$

Then using the identities

$$\cos \pi \frac{(2N-n)(m+\frac{1}{2})}{N} = -\cos \pi \frac{n(m+\frac{1}{2})}{N} \tag{3.19}$$

and

$$\exp \frac{\pi i(2N-n)}{2N} = -\exp \frac{-\pi i n}{2N}, \tag{3.20}$$

we obtain the formula below for $n = 1, \ldots, N-1$:

$$\mathcal{F}_{2N} f(2N-n) = e^{\frac{-\pi i n}{2N}} \sqrt{\frac{2}{N}} \sum_{m=0}^{N-1} x(m) \cos \frac{\pi n(m+\frac{1}{2})}{N} = e^{\frac{-\pi i n}{2N}} C_N^{II} x(n). \tag{3.21}$$

Thus letting $w = e^{\frac{\pi i n}{2N}}$, so $\bar{w} = e^{\frac{-\pi i n}{2N}}$, gives the following formula:

$$\sqrt{2}\, C_N^{II} x(n) = \begin{cases} \sqrt{2}\, \mathcal{F}_{2N} f(0), & \text{if } n = 0; \\ \bar{w}^n \mathcal{F}_{2N} f(n) + w^n \mathcal{F}_{2N} f(2N-n), & \text{if } 0 < n < N. \end{cases} \tag{3.22}$$

Now we define the following $2N \times N$ matrices:

$$P_N = \frac{1}{\sqrt{2}} \begin{pmatrix} \sqrt{2} & & & \\ & w & & \\ & & \ddots & \\ & & & w^{N-1} \\ 0 & \cdots & & 0 \\ & & & \bar{w}^{N-1} \\ & & \iddots & \\ 0 & \bar{w} & & \end{pmatrix}; \qquad Q_N = \frac{1}{\sqrt{2}} \begin{pmatrix} 1 & & & \\ & 1 & & \\ & & \ddots & \\ & & & 1 \\ & & & 1 \\ & & 1 & \\ & \iddots & & \\ 1 & & & \end{pmatrix}.$$

Then the following proposition follows from our calculations:

Proposition 3.11 $\quad C_N^{II} = P_N^* \mathcal{F}_{2N} Q_N.$ □

We can also let $f(m) = -f(2N-m-1) = iy(m)/\sqrt{2}$ in Equation 3.17 for $y \in \mathbf{C}^N$ and $n = 0, 1, \ldots, N-1$, to get

$$\begin{aligned} b(n+1)\mathcal{F}_{2N} f(n+1) &= b(n+1)\frac{e^{\frac{\pi i(n+1)}{2N}}}{\sqrt{N}} \sum_{m=0}^{N-1} y(m) \sin \frac{\pi(n+1)(m+\frac{1}{2})}{N} \\ &= \frac{e^{\frac{\pi i(n+1)}{2N}}}{\sqrt{2}} S_N^{II} y(n). \end{aligned} \tag{3.23}$$

Using Equation 3.20 at $n+1$ and the identity

$$\sin \pi \frac{(2N-n-1)(m+\frac{1}{2})}{N} = \sin \pi \frac{(n+1)(m+\frac{1}{2})}{N} \tag{3.24}$$

we obtain the formula below for $n = 0, 1, \ldots, N-1$:

$$\begin{aligned} b(n+1)\mathcal{F}_{2N} f(2N-n-1) &= -b(n+1)\frac{e^{\frac{-\pi i(n+1)}{2N}}}{\sqrt{N}} \sum_{m=0}^{N-1} y(m) \sin \frac{\pi(n+1)(m+\frac{1}{2})}{N} \\ &= -\frac{e^{\frac{-\pi i(n+1)}{2N}}}{\sqrt{2}} S_N^{II} y(n). \end{aligned} \tag{3.25}$$

Note that Equation 3.23 and Equation 3.25 agree at $n = N-1$; in both cases the coefficient before $S_N^{II} y(N-1)$ is $i/\sqrt{2}$. Thus for $w = e^{\frac{\pi i}{2N}}$ we obtain the following formulas:

$$\sqrt{2}\, S_N^{II} y(n) = \begin{cases} \bar{w}^{n+1}\mathcal{F}_{2N} f(n+1) - w^{n+1}\mathcal{F}_{2N} f(2N-n-1), & \text{if } 0 \le n < N-1; \\ -i\sqrt{2}\,\mathcal{F}_{2N} f(N), & \text{if } n = N-1. \end{cases}$$

So we define the $2N \times N$ matrices $\tilde{P}_N$ and $\tilde{Q}_N$ by the following formulas:

$$\tilde{P}_N = \frac{1}{\sqrt{2}} \left(\begin{array}{cccc} 0 & \cdots & 0 & 0 \\ w & & & \\ & \ddots & & \\ & & w^{N-1} & \\ 0 & \cdots & 0 & i\sqrt{2} \\ & & -\bar{w}^{N-1} & \\ & \iddots & & \\ -\bar{w} & & & \end{array} \right); \qquad \tilde{Q}_N = \frac{1}{\sqrt{2}} \left(\begin{array}{ccccc} 1 & & & & \\ & 1 & & & \\ & & \ddots & & \\ & & & & 1 \\ & & & & -1 \\ & & & -1 & \\ & & \iddots & & \\ -1 & & & & \end{array} \right).$$

Then similarly to the C_N^{II} case we obtain the following proposition:

Proposition 3.12 $\qquad S_N^{II} = i\tilde{P}_N^* \mathcal{F}_{2N} \tilde{Q}_N.$ □

By forming two $2N \times 2N$ combinations of the above matrices, $\mathcal{P}_N = (P_N | i\tilde{P}_N)$ and $\mathcal{Q}_N = (Q_N | \tilde{Q}_N)$, we can compute the DCT-II and DST-II transforms simultaneously:

Theorem 3.13 $\qquad \mathcal{P}_N^* \mathcal{F}_{2N} \mathcal{Q}_N = C_N^{II} \oplus (-i) S_N^{II}.$

Proof: Dropping subscripts, we have the following matrix:

$$\mathcal{P}^* \mathcal{F} \mathcal{Q} = \begin{pmatrix} P^* \mathcal{F} Q & iP^* \mathcal{F} \tilde{Q} \\ -i\tilde{P}^* \mathcal{F} Q & \tilde{P}^* \mathcal{F} \tilde{Q} \end{pmatrix}.$$

From Equation 3.17, we see that $\tilde{P}^* \mathcal{F} Q = P^* \mathcal{F} \tilde{Q} = 0$. Referring to Propositions 3.11 and 3.12 completes the proof. □

Since C_N^{III} is the transpose of C_N^{II} and S_N^{III} is the transpose of S_N^{II}, we have also proved similar results for DCT-III and DST-III. Note that $\mathcal{F}_{2N}$ is its own transpose, and that $(P^*)^t = \overline{P}$ and $Q^t = \bar{Q}^t = Q^*$, so that for example $(P^* \mathcal{F} Q)^t = Q^t \mathcal{F}^t (P^*)^t = Q^* \mathcal{F} \overline{P}$:

Proposition 3.14 $\qquad C_N^{III} = Q_N^* \mathcal{F}_{2N} \overline{P_N}.$ □

Proposition 3.15 $\qquad S_N^{III} = i\tilde{Q}_N^* \mathcal{F}_{2N} \overline{\tilde{P}_N}.$ □

Theorem 3.16 $\qquad \bar{\mathcal{Q}}_N^* \mathcal{F}_{2N} \bar{\mathcal{P}}_N = C_N^{III} \oplus (-i) S_N^{III}.$ □

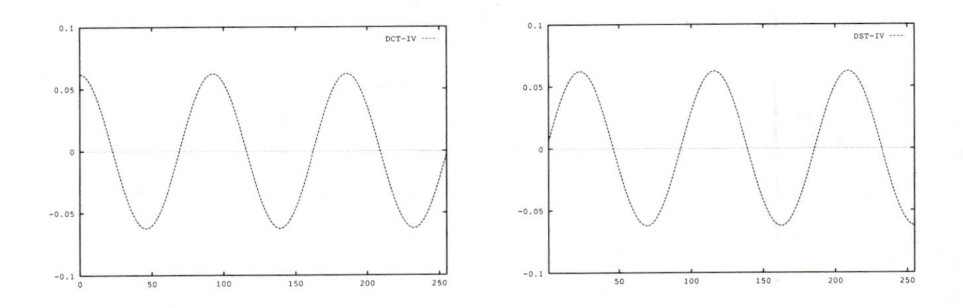

Figure 3.6: DCT-IV and DST-IV basis functions.

3.3.3 DCT-IV and DST-IV

For any two constants α, β we have

$$
\begin{aligned}
\alpha\mathcal{F}_{2N}f(n) \quad + \quad & \beta\mathcal{F}_{2N}f(2N-n-1) = \\
= \quad & \frac{1}{\sqrt{2N}}\sum_{m=0}^{2N-1} f(m)\left[\alpha e^{\frac{-2\pi i n m}{2N}} + \beta e^{\frac{-2\pi i(2N-n-1)m}{2N}}\right] \\
= \quad & \frac{1}{\sqrt{2N}}\sum_{m=0}^{2N-1} f(m)e^{\frac{\pi i m}{2N}}\left[\alpha e^{\frac{-\pi i(n+\frac{1}{2})m}{N}} + \beta e^{\frac{\pi i(n+\frac{1}{2})m}{N}}\right] \\
= \quad & \frac{1}{\sqrt{2N}}\sum_{m=0}^{N-1}\Bigg(f(m)e^{\frac{\pi i m}{2N}}\left[\alpha e^{\frac{-\pi i(n+\frac{1}{2})m}{N}} + \beta e^{\frac{\pi i(n+\frac{1}{2})m}{N}}\right] \\
& \qquad + f(2N-m-1)e^{\frac{\pi i(2N-m-1)}{2N}} \\
& \qquad\qquad \times \left[\alpha e^{\frac{-\pi i(n+\frac{1}{2})(2N-m-1)}{N}} + \beta e^{\frac{\pi i(n+\frac{1}{2})(2N-m-1)}{N}}\right]\Bigg) \\
= \quad & \frac{1}{\sqrt{2N}}\sum_{m=0}^{N-1}\Bigg(f(m)e^{\frac{\pi i m}{2N}}\left[\alpha e^{\frac{-\pi i(n+\frac{1}{2})m}{N}} + \beta e^{\frac{\pi i(n+\frac{1}{2})m}{N}}\right] \\
& \qquad + f(2N-m-1)e^{\frac{-\pi i(m+1)}{2N}} \\
& \qquad\qquad \times \left[\alpha e^{\frac{\pi i(n+\frac{1}{2})(m+1)}{N}} + \beta e^{\frac{-\pi i(n+\frac{1}{2})(m+1)}{N}}\right]\Bigg).
\end{aligned}
\tag{3.26}
$$

Now suppose that $x = x(n)$ is an arbitrary vector in $\mathbf{C}^N$. For DCT-IV, we let $\alpha = \frac{1}{\sqrt{2}}\exp\frac{-\pi i(n+\frac{1}{2})}{2N}$ and $\beta = \bar{\alpha} = \frac{1}{\sqrt{2}}\exp\frac{\pi i(n+\frac{1}{2})}{2N}$, and put $f(m) = \frac{x(m)}{\sqrt{2}}\exp\frac{-\pi i m}{2N}$ and $f(2N-m-1) = \frac{x(m)}{\sqrt{2}}\exp\frac{\pi i(m+1)}{2N}$ for $m = 0, 1, \ldots, N-1$. This gives the following equation for $n = 0, 1, \ldots, N-1$:

$$\begin{aligned}\frac{1}{\sqrt{2}}\exp\frac{-\pi i(n+\frac{1}{2})}{2N}\mathcal{F}_{2N}f(n) \quad &+ \quad \frac{1}{\sqrt{2}}\exp\frac{\pi i(n+\frac{1}{2})}{2N}\mathcal{F}_{2N}f(2N-n-1)\\ &= \quad \sqrt{\frac{2}{N}}\sum_{m=0}^{N-1}x(m)\cos\frac{\pi(n+\frac{1}{2})(m+\frac{1}{2})}{N}\\ &= \quad C_N^{IV}x(n).\end{aligned}$$

Thus, if we put $w = \exp\frac{\pi i}{2N}$ and define the matrix

$$R_N = \frac{1}{\sqrt{2}}\begin{pmatrix} 1 & & & & & & \\ & \bar{w} & & & & & \\ & & \ddots & & & & \\ & & & \bar{w}^{N-1} & & & \\ & & & i & & & \\ & & w^{N-1} & & & & \\ & \iddots & & & & & \\ w & & & & & & \end{pmatrix},$$

then we have proved the following proposition:

Proposition 3.17 $\quad C_N^{IV} = e^{\frac{-\pi i}{4N}}R_N^t\mathcal{F}_{2N}R_N.$ □

For DST-IV, we let $\alpha = \frac{i}{\sqrt{2}}\exp\frac{-\pi i(n+\frac{1}{2})}{2N}$ and $\beta = -\bar{\alpha} = -\frac{-i}{\sqrt{2}}\exp\frac{\pi i(n+\frac{1}{2})}{2N}$, and keep $f(m) = \frac{x(m)}{\sqrt{2}}\exp\frac{-\pi i m}{2N}$ as before but use

$$f(2N-m-1) = \frac{-x(m)}{\sqrt{2}}\exp\frac{\pi i(m+1)}{2N}.$$

Then we get the following equation for $n = 0, 1, \ldots, N-1$:

$$\begin{aligned}\frac{i}{\sqrt{2}}\exp\frac{-\pi i(n+\frac{1}{2})}{2N}\mathcal{F}_{2N}f(n) \quad &- \quad \frac{i}{\sqrt{2}}\exp\frac{\pi i(n+\frac{1}{2})}{2N}\mathcal{F}_{2N}f(2N-n-1)\\ &= \quad \sqrt{\frac{2}{N}}\sum_{m=0}^{N-1}x(m)\sin\frac{\pi(n+\frac{1}{2})(m+\frac{1}{2})}{N}\\ &= \quad S_N^{IV}x(n).\end{aligned}$$

Thus, if we keep $w = \exp\frac{\pi i}{2N}$ and define the matrix

$$\tilde{R}_N = \frac{1}{\sqrt{2}} \begin{pmatrix} 1 & & & & & & \\ & \bar{w} & & & & & \\ & & \ddots & & & & \\ & & & & \bar{w}^{N-1} & & \\ & & & & -i & & \\ & & & -w^{N-1} & & & \\ & & \cdot^{\cdot^{\cdot}} & & & & \\ -w & & & & & & \end{pmatrix},$$

then we have proved the following proposition:

Proposition 3.18 $S_N^{IV} = i\, e^{\frac{-\pi i}{4N}} \tilde{R}_N^t \mathcal{F}_{2N} \tilde{R}_N$. □

As before, we can define $\mathcal{R}_N = \left(R_N | i\tilde{R}_N\right)$ and we get the following theorem:

Theorem 3.19 $e^{\frac{-\pi i}{4N}} \mathcal{R}_N^t \mathcal{F}_{2N} \mathcal{R}_N = C_N^{IV} \oplus i\, S_N^{IV}$.

Proof: Dropping subscripts, we have the following matrix:

$$\mathcal{R}^t \mathcal{F} \mathcal{R} = \begin{pmatrix} R^t \mathcal{F} R & i R^t \mathcal{F} \tilde{R} \\ i\tilde{R}^t \mathcal{F} R & -\tilde{R}^t \mathcal{F} \tilde{R} \end{pmatrix}.$$

From Equation 3.26, we see that $\tilde{R}^t \mathcal{F} R = R^t \mathcal{F} \tilde{R} = 0$. Referring to Propositions 3.17 and 3.18 completes the proof. □

3.3.4 Implementations of DCT and DST

The central ingredient of each of these algorithms is a $2N$-point fast Fourier transform. This must be done with the aid of a temporary array of $2N$ COMPLEX data structures.

$2N$-point DFT fragment inside N-point DST or DCT

```
...
Let N = 1<<Q
Allocate and zero an array F[] of 2*N COMPLEX's
Allocate an array W[] of N COMPLEX's
bitrevi(F, Q+1)
fftomega(W, N)
fftproduct(F, Q+1, W)
Deallocate the array W[]
...
```

The normalization can be performed at the end, if at all. Also, the temporary array F[] will be deallocated at the very end.

For each of the variations, there is a different routine which performs the conjugation. Each of these "pre-matrices" and "post-matrices" requires its own code fragment; some involve just addition or subtraction:

U_N: Inject the input vector X[] into F[] using U:

```
Let F[0].RE = X[0]*SQ2
Let F[N].RE = X[N]*SQ2
For K = 1 to N-1
   Let F[2*N-K].RE = X[K]
   Let F[K].RE     = X[K]
```

U_N^*: Project back to the output array X[] using U^*:

```
Let X[0]  = F[0].RE * SQ2
Let X[N]  = F[N].RE * SQ2
For K = 1 to N-1
   Let X[K]  = F[2*N-K].RE + F[K].RE
```

V_N: Inject the input vector X[] into F[] using V:

```
For K = 1 to N-1
   Let F[K].RE      =  X[K]
   Let F[2*N-K].RE = -X[K];
```

V_N^*: Project back to the output array X[] using iV^*:

```
For K = 1 to N-1
   Let X[K]  = F[2*N-K].IM - F[K].IM
```

Q_N: Inject the input vector X[] into F[] using Q:

```
For K = 0 to N-1
   Let F[K].RE      = X[K]
   Let F[2*N-K].RE = X[K]
```

$\tilde{Q}_N$: Inject the input vector X[] into f[] using $i\tilde{Q}$:

```
For K = 0 to N-1
   Let F[K].IM       =  X[K]
   Let F[2*N-K-1].IM = -X[K]
```

Q_N^*: Project back to the output array X[] using Q^*:

```
For K = 0 to N-1
   Let X[K] = F[2*N-K-1].RE + F[K].RE
```

$\tilde{Q}_N^*$: Project back to the output array x[] using $i\tilde{Q}^*$:

```
For K = 0 to N-1
   Let X[K] =  F[2*N-K-1].IM - F[K].IM
```

Some of the conjugations require additional tables of cosines and sines, so we must allocate and assign these. This may be done by a utility function, or we may elect to include the following code fragment where necessary. It allocates arrays C[] and S[], each of length N, and assigns them with the respective values C[K] $= \cos(\pi K/2N)$, S[n] $= \sin(\pi K/2N)$, for $K = 0, 1, ..., N-1$. Note how this differs from dftomega() and dhtcossin().

Sine/cosine table fragment inside N-point DCT and DST

```
...
Allocate arrays C[] and S[] of length N
Let FACTOR = PI/(2*N)
For K = 0 to N-1
   C[K] = cos(K*FACTOR)
   S[K] = sin(K*FACTOR)
...
```

We continue with the code fragments giving the pre- and post-matrices for various DCTs and DSTs:

P_N^*: Project back to the output array X[] using P^*:

```
Let X[0] = F[0].RE * SQ2
For K = 1 to N-1
   Let X[K] = CRRMULRE( F[K], C[K], -S[K] )
            + CRRMULRE( F[2*N-K], C[K], S[K] )
```

$\tilde{P}_N^*$: Project back to the output array `X[]` using $\tilde{P}^*$:

```
For K = 1 to N-1
   Let X[K-1] = CRRMULRE(F[K], C[K], -S[K])
                  + CRRMULRE(F[2*N-K], C[K], S[K])
Let X[N-1] = F[N].IM * SQ2
```

$\overline{P_N}$: Inject the input vector `X[]` into `F[]` using $\bar{P}$:

```
Let F[0].RE = X[0] * SQ2
For K = 1 to N-1
   Let F[2*N-K].RE = X[K] * C[K]
   Let F[2*N-K].IM = X[K] * S[K]
   Let F[K].RE =  F[2*N-K].RE
   Let F[K].IM = -F[2*N-K].IM
```

$\overline{\tilde{P}_N}$: Inject the input array `X[]` into `F[]` using $\overline{\tilde{P}}$:

```
For K = 1 to N-1
    Let F[K].RE =  X[K-1] * C[K]
    Let F[K].IM = -X[K-1] * S[K]
    Let   F[2*N-K].RE = -F[K].RE
    Let   F[2*N-K].IM =  F[K].IM
Let F[N].RE = -X[N-1] * SQ2
```

R_N: Inject the input vector `X[]` into `F[]` using R:

```
Let F[0].RE = X[0]
Let F[N].IM = X[N-1]
For K = 1 to N-1
   Let F[K].RE =  X[K] * C[K]
   Let F[K].IM = -X[K] * S[K]
   Let F[2*N-K].RE = X[K-1]
   Let F[2*N-K].IM = X[K-1]
```

R_N^t: Project back to the output array `X[]` using ωR^t:

```
Let W.RE = cos(-PI/(4*N))
Let W.IM = sin(-PI/(4*N))
Let TMP.RE = F[0].RE + CRRMULRE(F[2*N-1], C[1], S[1])
```

```
Let TMP.IM = F[0].IM + CRRMULIM(F[2*N-1], C[1], S[1])
Let X[0]   = CCMULRE(TMP, W);
Let TMP.RE = CRRMULRE(F[N-1], C[N-1], -S[N-1]) -F[N].IM
Let TMP.IM = CRRMULIM(F[N-1], C[N-1], -S[N-1]) +F[N].RE
Let X[N-1] = CCMULRE(TMP, W)
For K = 1 to N-2
   Let TMP.RE = CRRMULRE( F[K], C[K], -S[K] )
                + CRRMULRE( F[2*N-K-1], C[K+1], S[K+1])
   Let TMP.IM = CRRMULIM(F[K], C[K], -S[K])
                + CRRMULIM(F[2*N-K-1], C[K+1], S[K+1])
   Let X[K] = CCMULRE(TMP, W)
```

$\tilde{R}_N$: Inject the input vector `X[]` into `F[]` using $\tilde{R}$:

```
Let F[0].RE = X[0]
Let F[N].IM = -X[N-1]
For K = 1 to N-1
   Let F[K].RE =  X[K] * C[K]
   Let F[K].IM = -X[K] * S[K]
   Let F[2*N-K].RE = -X[K-1] * C[K]
   Let F[2*N-K].IM = -X[K-1] * S[K]
```

$\tilde{R}_N^t$: Project back to the array `X[]` using $i\omega\tilde{R}^t$:

```
Let W.RE = cos(-PI/(N/4))
Let W.IM = sin(-PI/(N/4))
Let TMP.RE = F[0].RE - CRRMULRE(F[2*N-1], C[1], S[1])
Let TMP.IM = F[0].IM - CRRMULIM(F[2*N-1], C[1], S[1])
Let X[0]   = -CCMULIM( TMP, W )
Let TMP.RE =  CRRMULRE(F[N-1], C[N-1], -S[N-1]) +F[N].IM
Let TMP.IM =  CRRMULIM(F[N-1], C[N-1], -S[N-1]) -F[N].RE
Let X[N-1] = -CCMULIM( TMP, W )
For K = 1 to N-2
   Let TMP.RE = CRRMULRE(F[K], C[K], -S[K])
                - CRRMULRE(F[2*N-K-1], C[K+1], S[K+1])
   Let TMP.IM = CRRMULIM(F[K], C[K], -S[K])
                - CRRMULIM(F[2*N-K-1], C[K+1], S[K+1])
   Let X[K] = -CCMULIM(TMP, W)
```

The normalization is performed all at once at the end, combining the constants in the conjugating matrices plus the factor $1/\sqrt{2N}$ from the $2N$-point FFT. Notice that the normalizing factor will be a rational number if q is odd in $N = 2^q$.

Normalization for the N-point DCT or DST

```
If Q is odd then
    Let NORM = 0.5/(1<<((Q+1)/2))
Else
    Let NORM = 0.5/sqrt(1<<(Q+1))
For K = 0 to N-1
    Let X[K] *= NORM
```

At the end, we can deallocate the temporary vectors `C[]`, `S[]`, and `F[]`. The DCT or DCT will be returned by side-effect in the input array `X[]`.

3.4 Exercises

1. Prove Lemma 3.9.
2. Write out explicitly the matrices for the 2×2 and 4×4 DFT and DHT.
3. Write out explicitly the matrices for the 2×2 and 4×4 DCT-II, DCT-III, and DCT-IV.
4. Implement DCT-I in pseudocode.
5. Implement DST-I in pseudocode.
6. Implement DCT-II in pseudocode.
7. Implement DST-II in pseudocode.
8. Implement DCT-III in pseudocode.
9. Implement DST-III in pseudocode.
10. Implement DCT-IV in pseudocode.
11. Implement DST-IV in pseudocode.
12. Find the "butterfly" matrix for the fast Hartley transform factorization in the case of *odd* M.
13. Implement `dft()` as a transform in place.
14. Implement `dht()` as a transform in place.

Chapter 4

Local Trigonometric Transforms

A remarkable observation made independently by several individuals [91, 40, 64, 74, 4, 24] allows us to construct smooth orthogonal bases subordinate to arbitrary partitions of the line. The bases consist of sines or cosines multiplied by smooth, compactly supported cutoff functions, or more generally they are arbitrary periodic functions smoothly restricted to adjacent overlapping intervals. These *localized trigonometric functions* remain orthogonal despite the overlap, and the decomposition maps smooth functions to smooth compactly supported functions. We will describe both the bases and transformations into those bases in this chapter.

A generalization introduced in [117] of the local trigonometric transform provides an orthogonal projection onto periodic functions which also preserves smoothness. This *smooth local periodization* permits us to study smooth functions restricted to intervals with arbitrary smooth periodic bases, without creating any discontinuities at the endpoints or introducing any redundancy. The inverse of periodization takes a smooth periodic function and produces a smooth, compactly supported function on the line. The injection is also orthogonal, so we can generate smooth, compactly supported orthonormal bases on the line from arbitrary smooth periodic bases. In particular, we get *localized exponential functions* by applying the injection to complex exponentials.

The basis functions produced in this way are "windowed" in the sense that they are just the original functions multiplied by a compactly supported bump function. Hence we can control their analytic properties by choosing the bump, which we can do in an almost arbitrary manner. In particular, we can arrange for the resulting orthonormal bases to have a small Heisenberg product.

The method evades the Balian–Low obstruction, which prevents windowed exponential functions from simultaneously being a frame and having finite Heisenberg product, by a modification of the definition of "window." It also permits us to construct smooth orthonormal wavelet and wavelet packet bases on the line, which we will use later in Chapters 10 and 11.

Localized trigonometric functions can also be combined into a library of orthonormal bases. Elements of the library are windowed bases and we adapt the window sizes by choosing an appropriate window-determining partition of the line. Such a library can be made into a tree: two adjacent intervals I and J may be regarded as branches of the interval $I \cup J$, because the basis functions over I and J span the space of windowed functions over $I \cup J$. This fact will be used in Chapter 8 to construct search algorithms for the best-adapted local trigonometric transform.

4.1 Ingredients and examples

The smooth projections, smooth periodizations, and the library of adapted windowed trigonometric transforms are all constructed by unitary conjugation of trivial operations like restriction to an interval. We will first lay out these various constructions, then we will study their analytic properties and see how the free parameters affect them.

4.1.1 Unitary folding and unfolding

The main ingredient in the recipe for these transforms is a unitary operator which conjugates a sharp cutoff into a smooth orthogonal projection. The unitary operator depends on a function of one real variable, which must satisfy a simple algebraic symmetry condition. The degree of smoothness preserved by the projection depends on this parameter function's smoothness.

Rising cutoff functions

Let $r = r(t)$ be a function in the class $C^d(\mathbf{R})$ for some $0 \leq d \leq \infty$, satisfying the following conditions:

$$|r(t)|^2 + |r(-t)|^2 = 1 \quad \text{for all } t \in \mathbf{R}; \qquad r(t) = \begin{cases} 0, & \text{if } t \leq -1, \\ 1, & \text{if } t \geq 1. \end{cases} \tag{4.1}$$

The function r need not be increasing or even real-valued. We will call it a *rising cutoff function* because $r(t)$ rises from being identically zero to being identically one as t goes from $-\infty$ to $+\infty$.

A general construction for such functions is given in [5], but we will repeat the salient details here to clarify the implementations at the end of this chapter. We notice that any function satisfying Equation 4.1 must be of the form

$$r(t) \stackrel{\text{def}}{=} \exp[i\rho(t)] \sin[\theta(t)], \tag{4.2}$$

where ρ and θ are real-valued functions satisfying

$$\rho(t) = \begin{cases} 2n\pi, & \text{if } t < -1, \\ 2m\pi, & \text{if } t > 1; \end{cases} \quad \theta(t) = \begin{cases} 0, & \text{if } t < -1, \\ \frac{\pi}{2}, & \text{if } t > 1, \end{cases} \quad \theta(t)+\theta(-t) = \frac{\pi}{2}. \tag{4.3}$$

Since $|r(-t)| = |\sin[\frac{\pi}{2} - \theta(t)]| = |\cos\theta(t)|$, we have $|r(t)|^2 + |r(-t)|^2 = 1$.

Let $r = r(t)$ be any differentiable rising cutoff functions. Then $\frac{d}{dt}|r(t)|^2$ is a symmetric continuous bump function supported on $[-1, 1]$. This observation provides a mechanism for parametrizing the cutoffs: one way to obtain θ is to start with an integrable bump function $\phi = \phi(t)$ satisfying the following conditions:

$$\phi(t) = 0 \text{ if } |t| > 1; \quad \phi(t) = \phi(-t) \text{ for all } t; \quad \int_{-\infty}^{\infty} \phi(s)\, ds = \frac{\pi}{2}. \tag{4.4}$$

Then we take $\theta(t) = \int_{-1}^{t} \phi(s)\, ds$. An example of this construction is to use a quadratic polynomial for the bump, which gives a cubic spline for the angle function:

$$\phi(t) = \frac{3\pi}{8}(1-t)(1+t); \qquad \theta(t) = \frac{\pi}{8}\left[2 + 3t - t^3\right]; \qquad -1 \le t \le 1. \tag{4.5}$$

Smooth cutoffs by sine iteration

We obtain real-valued cutoff functions by choosing $\rho \equiv 0$. An example real-valued continuous function which satisfies the conditions is the following:

$$r_{\sin}(t) = \begin{cases} 0, & \text{if } t \le -1, \\ \sin\left[\frac{\pi}{4}(1+t)\right], & \text{if } -1 < t < 1, \\ 1, & \text{if } t \ge 1. \end{cases} \tag{4.6}$$

We can also obtain real-valued d-times continuously differentiable functions ($r \in C^d$) for arbitrarily large fixed d by repeatedly replacing t with $\sin(\pi t/2)$:

$$r_{[0]}(t) \stackrel{\text{def}}{=} r_{\sin}(t); \qquad r_{[n+1]}(t) \stackrel{\text{def}}{=} r_{[n]}(\sin \frac{\pi}{2} t). \tag{4.7}$$

Using induction we can show that $r_{[n]}(t)$ has $2^n - 1$ vanishing derivatives at $t = +1$ and $t = -1$; this is left as an exercise. It means that $r_{[n]} \in C^{2^n-1}$. The case $r_{[1]} \in C^1$ is depicted in Figure 4.1.

Notice that the example cutoff of Equation 4.6 comes from $\phi(t) = \frac{\pi}{4}\mathbf{1}_{[-1,1]}$, which results in $\theta(t) = \frac{\pi}{4}(1+t)$ for $t \in [-1, 1]$. Likewise, taking $\phi(t) = \frac{\pi^2}{8} \cos \frac{\pi}{2} t$ for $t \in [-1, 1]$ yields $r_{[1]}$ from $\theta(t) = \frac{\pi}{4}(1 + \sin \frac{\pi}{2} t)$.

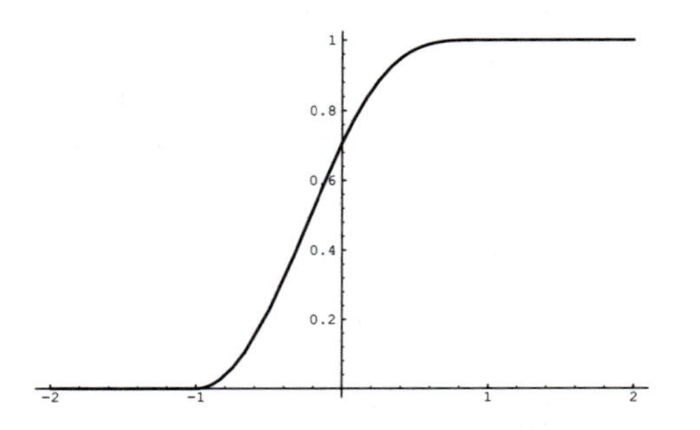

Figure 4.1: Example of C^1 cutoff function.

Folding and unfolding

The *folding* operator U and its adjoint *unfolding* operator U^* are defined as follows:

$$Uf(t) = \begin{cases} r(t)f(t) + r(-t)f(-t), & \text{if } t > 0, \\ \bar{r}(-t)f(t) - \bar{r}(t)f(-t), & \text{if } t < 0; \end{cases} \tag{4.8}$$

$$U^*f(t) = \begin{cases} \bar{r}(t)f(t) - r(-t)f(-t), & \text{if } t > 0, \\ r(-t)f(t) + \bar{r}(t)f(-t), & \text{if } t < 0. \end{cases} \tag{4.9}$$

When it is necessary to emphasize that both of these operators depend on a rising cutoff function r we will write $U(r)$ and $U^*(r)$, but in most cases this excess notation can be omitted without ambiguity.

Observe that $Uf(t) = f(t)$ and $U^*f(t) = f(t)$ if $t \geq 1$ or $t \leq -1$. Also, $U^*Uf(t) = UU^*f(t) = \left(|r(t)|^2 + |r(-t)|^2\right)f(t) = f(t)$ for all $t \neq 0$, so that U and U^* are unitary isomorphisms of $L^2(\mathbf{R})$. As an example we compute Uf for the particularly simple case of the constant function $f \equiv 1$. The cutoff function is the one defined in Equation 4.6. The result is plotted in Figure 4.2. Notice how folding introduces a particular discontinuity at 0: $U1(t)$ smoothly approaches the normalized even part ($\sqrt{2}$) of 1 as $t \to 0+$ and the normalized odd part (which is 0 for every continuous function) as $t \to 0-$.

Arbitrary action regions

We now shift and dilate the folding operator so its action takes place on an interval $(\alpha - \epsilon, \alpha + \epsilon)$ rather than $[-1, 1]$. We do this with the translation and rescaling operators:

$$\tau_\alpha f(t) = f(t-\alpha); \qquad \tau_\alpha^* f(t) = f(t+\alpha); \tag{4.10}$$

$$\sigma_\epsilon f(t) = \epsilon^{-1/2} f(t/\epsilon); \qquad \sigma_\epsilon^* f(t) = \epsilon^{1/2} f(\epsilon t). \tag{4.11}$$

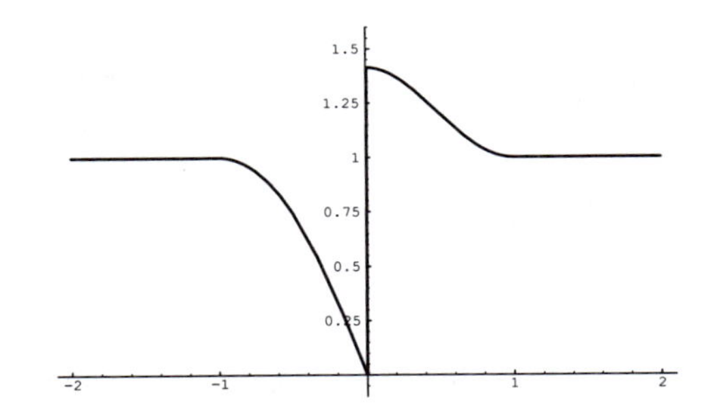

Figure 4.2: Action of U on the constant function $f(t) = 1$.

Here α and $\epsilon > 0$ are real numbers. We dilate U and U^* by conjugation with σ_ϵ and then translate by conjugation with τ_α. This gives folding and unfolding operators indexed by the triple (r, α, ϵ):

$$U(r, \alpha, \epsilon) = \tau_\alpha \sigma_\epsilon U(r) \sigma_\epsilon^* \tau_\alpha^*; \qquad U^*(r, \alpha, \epsilon) = \tau_\alpha \sigma_\epsilon U^*(r) \sigma_\epsilon^* \tau_\alpha^*. \tag{4.12}$$

For future reference, we expand the formulas for $U(r, \alpha, \epsilon)f$ and $U^*(r, \alpha, \epsilon)f$ and write them explicitly:

$$U(r, \alpha, \epsilon) f(t) = \begin{cases} r(\frac{t-\alpha}{\epsilon}) f(t) + r(\frac{\alpha-t}{\epsilon}) f(2\alpha - t), & \text{if } \alpha < t < \alpha + \epsilon, \\ \bar{r}(\frac{\alpha-t}{\epsilon}) f(t) - \bar{r}(\frac{t-\alpha}{\epsilon}) f(2\alpha - t), & \text{if } \alpha - \epsilon < t < \alpha, \\ f(t), & \text{otherwise;} \end{cases} \tag{4.13}$$

$$U^*(r, \alpha, \epsilon) f(t) = \begin{cases} \bar{r}(\frac{t-\alpha}{\epsilon}) f(t) - r(\frac{\alpha-t}{\epsilon}) f(2\alpha - t), & \text{if } \alpha < t < \alpha + \epsilon, \\ r(\frac{\alpha-t}{\epsilon}) f(t) + \bar{r}(\frac{t-\alpha}{\epsilon}) f(2\alpha - t), & \text{if } \alpha - \epsilon < t < \alpha, \\ f(t), & \text{otherwise.} \end{cases} \tag{4.14}$$

Where convenient we will write U_0 for $U(r_0, \alpha_0, \epsilon_0)$, and so on.

Let us denote the interval $(\alpha-\epsilon, \alpha+\epsilon)$ by $B_\epsilon(\alpha)$, the "ball" of radius ϵ centered at α. We can call this the *action region* of the folding operator $U = U(r, \alpha, \epsilon)$, since outside this ball it acts like the identity. Our later constructions will use families of folding and unfolding operators with various action regions. We can abuse the notation and say that a family of folding and unfolding operators is *disjoint* if the following dichotomy applies to their action regions and rising cutoff functions:

- Every two distinct action regions are disjoint;
- Each action region has its own unique rising cutoff function.

Every pair of operators in a disjoint family will commute.

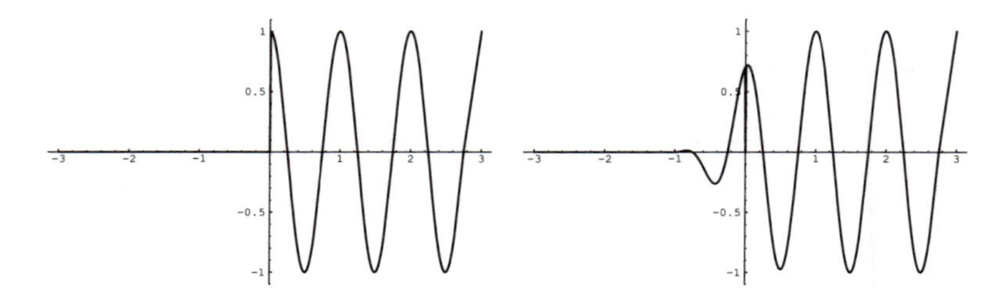

Figure 4.3: Cosine restricted to the right half-line, and unfolded.

Relation to windowing

If the function $f = f(t)$ is purely even on $[-1, 1]$, *i.e.*, if it satisfies the relation $f(t) = f(-t)$ for all $-1 \leq t \leq 1$, then the folding operator becomes just multiplication by a function:

$$Uf(t) = \begin{cases} [r(t) + r(-t)]\, f(t), & \text{if } t > 0, \\ [\bar{r}(-t) - \bar{r}(t)]\, f(t), & \text{if } t < 0. \end{cases} \tag{4.15}$$

On the other hand, if $f(t) = -f(-t)$ is purely odd on $[-1, 1]$, then

$$Uf(t) = \begin{cases} [r(t) - r(-t)]\, f(t), & \text{if } t > 0, \\ [\bar{r}(-t) + \bar{r}(t)]\, f(t), & \text{if } t < 0. \end{cases} \tag{4.16}$$

If r is real-valued, then these two multipliers are just reflections of each other. Figure 4.2 may also be viewed as the plot of the first multiplier, since the constant function $f(t) = 1$ is even.

If f is supported on the right half-line, then the unfolding operator likewise simplifies to multiplication by the (complex conjugate of the) rising cutoff function:

$$U^* f(t) = \bar{r}(t) f^{even}(t) \stackrel{\text{def}}{=} \begin{cases} \bar{r}(t) f(t), & \text{if } t > 0, \\ \bar{r}(t) f(-t), & \text{if } t < 0. \end{cases} \tag{4.17}$$

Here f^{even} is the even extension of f to the left half-line. As an example, we have $U^* \mathbf{1}_{\mathbf{R}_+} \cos(2\pi t) = r(t) \cos(2\pi t)$, which is as smooth as r but supported in the interval $[-1, \infty)$. Figure 4.3 shows the effect of this unfolding.

Similarly, if f is supported on the left half-line, then we just multiply the odd extension f^{odd} of f to the right half-line by the reflection of the rising cutoff function:

$$U^* f(t) = r(-t) f^{odd}(t) \stackrel{\text{def}}{=} \begin{cases} -r(-t) f(-t), & \text{if } t > 0, \\ r(-t) f(t), & \text{if } t < 0. \end{cases} \tag{4.18}$$

This time the example is $U^* \mathbf{1}_{\mathbf{R}_+} \sin(2\pi t) = r(-t) \sin(2\pi t)$, which has a smooth odd extension. Figure 4.4 shows how this unfolding results in a smooth function supported in $(-\infty, 1]$.

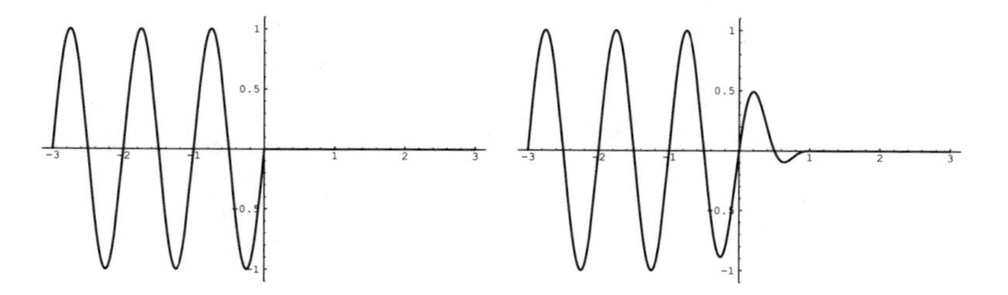

Figure 4.4: Sine restricted to the left half-line, and unfolded.

Local cosines and local sines

Let n be an integer and define $C_n(t) = \cos\left[\pi(n+\frac{1}{2})t\right]$, a cosine function at half-integer frequency. Consider the function $\mathbf{1}(t)C_n(t)$, which is this cosine restricted to the interval $[0,1]$; we may call this a *block cosine.* At zero, it looks like the chopped-off cosine in the left part of Figure 4.3, but at one it looks like the chopped-off sine of Figure 4.4. This is the main effect of the half-integer frequency.

Now suppose we apply two unfolding operators $U_0^* = U^*(r,0,1/3)$ and $U_1^* = U^*(r,1,1/2)$, whose action regions are disjoint and contain only one of the two ends of the support interval each. Disjointness allows us to apply these in any order. An example of the resulting function $U_0^*U_1^*\mathbf{1}\,C_4$ is plotted in the right half of Figure 4.5, superposed under its envelope. It is an even extension of the block cosine function to the left of the interval $[0,1]$ and an odd extension to the right, multiplied by $r(3t)r(2-2t)$.

This multiplier is a *bell* or *window* function which is supported in the interval $[-\frac{1}{3},\frac{3}{2}]$. The left half of Figure 4.5 shows the bell function obtained by using the rising cutoff $r_{[1]}$ defined by Equation 4.7. Since $r_{[1]}$ is smooth on $(-1,1)$ with vanishing derivatives at the boundary points, the bell has a continuous derivative on $\mathbf{R}$. We can use $r_{[m]}$ with $m>1$ to obtain additional derivatives.

The block cosine functions may be dilated, normalized, and translated to the interval $I_k = [\alpha_k, \alpha_{k+1}]$ by the formulas $C_{n,k}(t) = \sqrt{\frac{2}{|I_k|}}C_n\left(\frac{t-\alpha_j}{|I_k|}\right)$ and $\mathbf{1}_{I_k}(t) = \mathbf{1}\left(\frac{t-\alpha_j}{|I_k|}\right)$. We can then unfold $\mathbf{1}_{I_k}\,C_{n,k}$ with action regions of radii ϵ_k and ϵ_{k+1} respectively, possibly using different rising cutoffs r_k and r_{k+1} in each action region. This is equivalent to multiplying $C_{n,k}$ by the window function b_k, supported on $[\alpha_k - \epsilon_k, \alpha_{k+1}+\epsilon_{k+1}]$ and defined by the following formula:

$$b_k(t) \stackrel{\text{def}}{=} r_k\left(\frac{t-\alpha_k}{\epsilon_k}\right) r_{k+1}\left(\frac{\alpha_{k+1}-t}{\epsilon_{k+1}}\right). \tag{4.19}$$

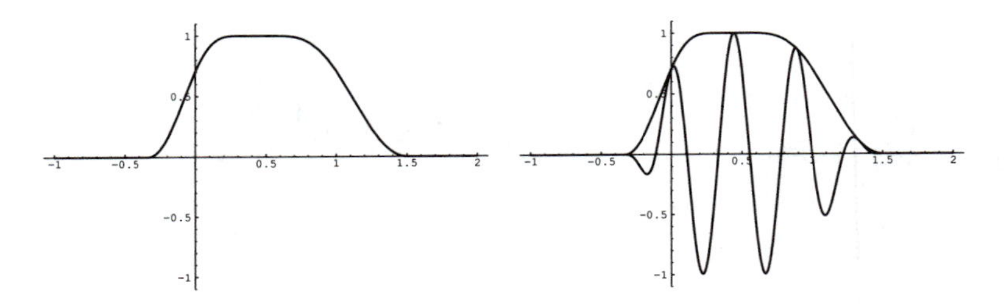

Figure 4.5: The bell and its local cosine function.

We will call such windowed or unfolded block cosine functions the *local cosines*. For integers $n \geq 0$ and k, they have the following formulas:

$$\begin{aligned} \psi_{nk}(t) &= U^*(r_k, \alpha_k, \epsilon_k)\, U^*(r_{k+1}, \alpha_{k+1}, \epsilon_{k+1})\, \mathbf{1}_{I_k}(t)\, C_{n,k}(t) \\ &= b_k(t) C_{n,k}(t) \\ &= \sqrt{\frac{2}{\alpha_{k+1}-\alpha_k}}\, r_k\left(\frac{t-\alpha_k}{\epsilon_k}\right) r_{k+1}\left(\frac{\alpha_{k+1}-t}{\epsilon_{k+1}}\right) \cos\left[\frac{\pi(n+\frac{1}{2})(t-\alpha_k)}{\alpha_{k+1}-\alpha_k}\right]. \end{aligned} \tag{4.20}$$

We note that cosine may be replaced by sine to get *local sine* functions. Other modifications are possible as well; for example, we may use combinations of sines and cosines on different intervals if we adjust the bell functions appropriately.

Boundary conditions at the point of folding

It does not matter how we define $Uf(0)$ or $U^*f(0)$ for functions $f \in L^2$; for smooth f we may just as well define $Uf(0) \stackrel{\text{def}}{=} f(0)$, and for f satisfying certain smoothness and boundary limit conditions we will show that there is a unique smooth extension of U^*f across $t = 0$.

Lemma 4.1 *Suppose $r \in C^d(\mathbf{R})$ for some $0 \leq d \leq \infty$. If $f \in C^d(\mathbf{R})$, then Uf has d continuous derivatives in $\mathbf{R} \setminus \{0\}$, and for all $0 \leq n \leq d$ there exist limits $[Uf]^{(n)}(0+)$ and $[Uf]^{(n)}(0-)$ which satisfy the following conditions:*

$$\lim_{t\to 0+} [Uf]^{(n)}(t) = 0 \qquad \text{if } n \text{ is odd,} \tag{4.21}$$

$$\lim_{t\to 0-} [Uf]^{(n)}(t) = 0 \qquad \text{if } n \text{ is even.} \tag{4.22}$$

Conversely, if f belongs to $C^d(\mathbf{R} \setminus \{0\})$ and has limits $f^{(n)}(0+)$ and $f^{(n)}(0-)$ for all $0 \leq n \leq d$ which satisfy the equations

$$\lim_{t\to 0+} f^{(n)}(t) = 0 \qquad \text{if } n \text{ is odd,} \tag{4.23}$$

$$\lim_{t\to 0-} f^{(n)}(t) \quad = \quad 0 \qquad \textit{if } n \textit{ is even,} \tag{4.24}$$

then U^*f has a unique continuous extension (across $t = 0$) which belongs to $C^d(\mathbf{R})$.

Proof: The smoothness of Uf and U^*f on $(0,\infty)$ and $(-\infty, 0)$ follows from elementary calculus. We can calculate the one-sided limits of the derivatives as follows:

$$\lim_{t\to 0+} [Uf]^{(n)}(t) = \sum_{k=0}^{n} \binom{n}{k} \Big[r^{(n-k)}(0) f^{(k)}(0+) + (-1)^n r^{(n-k)}(0) f^{(k)}(0-) \Big] \tag{4.25}$$

$$\lim_{t\to 0-} [Uf]^{(n)}(t) = \sum_{k=0}^{n} \binom{n}{k} (-1)^k \Big[(-1)^n \overline{r^{(n-k)}(0)} f^{(k)}(0-) - \overline{r^{(n-k)}(0)} f^{(k)}(0+) \Big]. \tag{4.26}$$

If n is odd, with $0 \le n \le d$, then the summands in the right-hand side of Equation 4.25 are $r^{(n-k)}(0)[f^{(k)}(0+) - f^{(k)}(0-)] = 0$, since $f^{(k)}$ is continuous at zero for all $0 \le k \le d$. If n is even, then the summands in the right-hand side of Equation 4.26 are $r^{(n-k)}(0)[f^{(k)}(0-) - f^{(k)}(0+)] = 0$ for the same reason.

The converse requires showing the equality of two one-sided limits:

$$\begin{aligned}
[U^*f]^{(n)}(0+) \quad - \quad & [U^*f]^{(n)}(0-) \\
= \quad & \sum_{k=0}^{n} \binom{n}{k} \Big[\overline{r^{(n-k)}(0)} f^{(k)}(0+) - (-1)^n r^{(n-k)}(0) f^{(k)}(0-) \\
& - (-1)^{n-k} r^{(n-k)}(0) f^{(k)}(0-) - (-1)^k \overline{r^{(n-k)}(0)} f^{(k)}(0+) \Big] \\
= \quad & \sum_{k=0}^{n} \binom{n}{k} \Big[\left\{1 - (-1)^k\right\} \overline{r^{(n-k)}(0)} f^{(k)}(0+) \\
& - (-1)^n \left\{1 + (-1)^k\right\} r^{(n-k)}(0) f^{(k)}(0-) \Big].
\end{aligned} \tag{4.27}$$

The right-hand side is zero, since $\left\{1 - (-1)^k\right\} f^{(k)}(0+)$ and $\left\{1 + (-1)^k\right\} f^{(k)}(0-)$ both vanish for all k. Thus $\lim_{t\to 0} [U^*f]^{(n)}(t)$ exists for $0 \le n \le d$, since the one-sided limits agree. Now the function U^*f has a unique continuous extension across $t = 0$. By the mean value theorem, for each $t \ne 0$ there is some t_0 between 0 and t such that

$$\frac{[U^*f]^{(k)}(t) - [U^*f]^{(k)}(0)}{t} = [U^*f]^{(k+1)}(t_0). \tag{4.28}$$

By letting $t \to 0$ in this equation, we show that $[U^*f]^{(k)}(0) = \lim_{t\to 0}[U^*f]^{(k)}(t) \Rightarrow [U^*f]^{(k+1)}(0) = \lim_{t\to 0}[U^*f]^{(k+1)}(t)$ for $0 \le k < d$. Induction on k then shows that the unique continuous extension of U^*f belongs to $C^d(\mathbf{R})$. □

This lemma shows that just a trivial boundary condition is needed to obtain smoothness. In particular, the constant function $f(t) = 0$ satisfies the condition at every point. This fact gives another perspective on Figures 4.3 and 4.4.

The boundary conditions at α after folding with action region $B_\epsilon(\alpha)$ will be the same as the boundary conditions at zero described in Lemma 4.1. Likewise, unfolding with this action region undoes the boundary conditions at α. Every $\epsilon > 0$ will yield the same boundary conditions.

4.1.2 Smooth orthogonal projections

In [5], two orthogonal projections were defined directly in terms of the rising cutoff function:

$$P_0 f(t) = |r(t)|^2 f(t) + \bar{r}(t) r(-t) f(-t); \tag{4.29}$$

$$P^0 f(t) = |r(-t)|^2 f(t) - \bar{r}(t) r(-t) f(-t). \tag{4.30}$$

We can relate P_0, P^0 to the trivial orthogonal projections given by restriction to intervals. Define first the restriction operator:

$$\mathbf{1}_I f(t) = \begin{cases} f(t), & \text{if } t \in I, \\ 0, & \text{otherwise.} \end{cases} \tag{4.31}$$

A straightforward calculation shows that P^0 and P_0 are just conjugates of restriction to the half-line:

$$P_0 = U^* \mathbf{1}_{\mathbf{R}^+} U; \qquad P^0 = U^* \mathbf{1}_{\mathbf{R}^-} U. \tag{4.32}$$

It is evident from both of the formulas that $P_0 + P^0 = I$ and that both P_0 and P^0 are selfadjoint. Also, we have $P^0 P_0 = P_0 P^0 = 0$.

Since the operators P_0, P^0 are obtained from the trivial orthogonal projections by unitary conjugation, they are themselves orthogonal projections on $L^2(\mathbf{R})$. We call them *smooth projections* onto half-lines because of the following property:

Corollary 4.2 *If $f \in C^d(\mathbf{R})$, then the unique continuous extensions of $P_0 f$ and $P^0 f$ belong to $C^d(\mathbf{R})$, and* $\operatorname{supp} P^0 f \subset (-\infty, 1]$ *and* $\operatorname{supp} P_0 f \subset [-1, \infty)$.

Proof: The result follows from two applications of Lemma 4.1, since $\mathbf{1}_{\mathbf{R}^+} Uf(t)$ and $\mathbf{1}_{\mathbf{R}^-} Uf(t)$ satisfy Equation 4.23 at $t = 0$. □

We remark that these definitions follow the "local cosine" polarity, and that we can just as well exchange the + and − in the equations for P^0 and P_0 (likewise for U and U^*) to obtain the "local sine" polarity.

Historical note The projection operators P^0 and P_0 were first described to me by R. R. Coifman in 1985, in a discussion about some work of Y. Meyer. They were originally used to approximate the Hilbert transform with H_ϵ defined by $(\widehat{H_\epsilon f}) = P_{0\epsilon}\hat{f} - P^{0\epsilon}\hat{f}$, a better behaved operator which retains the algebraic properties of the original.

Projection over intervals

The smooth projections P^0 and P_0 may also be shifted and rescaled to move their action regions to $(\alpha-\epsilon, \alpha+\epsilon)$ rather than $[-1, 1]$. This is done with the translation and rescaling operators τ_α and σ_ϵ defined in Equations 4.10 and 4.11:

$$P_{\alpha\epsilon} = \tau_\alpha \sigma_\epsilon P_0 \sigma_\epsilon^* \tau_\alpha^*; \qquad P^{\alpha\epsilon} = \tau_\alpha \sigma_\epsilon P^0 \sigma_\epsilon^* \tau_\alpha^*. \tag{4.33}$$

If $\epsilon_0+\epsilon_1 < \alpha_1-\alpha_0$, then the action regions $B_{\epsilon_0}(\alpha_0)$ and $B_{\epsilon_1}(\alpha_1)$ are disjoint and the operators $P^{\alpha_0\epsilon_0}$ and $P_{\alpha_1\epsilon_1}$ will commute. In that case the following operator is an orthogonal projection:

$$P_{(\alpha_0,\alpha_1)} = P_{\alpha_0\epsilon_0} P^{\alpha_1\epsilon_1}. \tag{4.34}$$

This projection maps smooth functions on the line into smooth functions supported in $[\alpha_0-\epsilon_0,\ \alpha_1+\epsilon_1]$. P may also be written in terms of translated and dilated folding and unfolding operators, which themselves commute because of the disjointness:

$$P_{(\alpha_0,\alpha_1)} = U_0^* U_1^* \mathbf{1}_{(\alpha_0,\alpha_1)} U_1 U_0. \tag{4.35}$$

Adjacent compatible intervals are defined in [5] to be the intervals $I = (\alpha_0, \alpha_1)$ and $J = (\alpha_1, \alpha_2)$ corresponding to mutually disjoint action regions $B_{\epsilon_i}(\alpha_i)$, $i = 0, 1, 2$, and with associated smooth rising cutoffs r_i, $i = 0, 1, 2$ satisfying Equation 4.1. Using the factorization of P, we can give a simple proof of one of the lemmas which appears in that paper:

Lemma 4.3 *If I and J are adjacent compatible intervals, then $P_I + P_J = P_{I\cup J}$ and $P_I P_J = P_J P_I = 0$.*

Proof: The two sets of operators $\{U_0, U_1, U_2\}$ and $\{U_0^*, U_1^*, U_2^*\}$ form commuting families because of the disjointness condition. Furthermore, U_0 and U_0^* commute with $\mathbf{1}_J$ and U_2 and U_2^* commute with $\mathbf{1}_I$. Thus:

$$\begin{aligned} P_I + P_J &= U_0^* U_1^* \mathbf{1}_I U_1 U_0 + U_1^* U_2^* \mathbf{1}_J U_2 U_1 \\ &= U_1^* \left[U_0^* \mathbf{1}_I U_0 + U_2^* \mathbf{1}_J U_2 \right] U_1 \\ &= U_0^* U_2^* U_1^* \left[\mathbf{1}_I + \mathbf{1}_J \right] U_1 U_2 U_0. \end{aligned}$$

We note that U_1 and U_1^* commute past $[\mathbf{1}_I + \mathbf{1}_J] = \mathbf{1}_{I \cup J}$ and cancel. This shows that $P_I + P_J = P_{I \cup J}$.

Similarly, after interchanging various commuting operators we obtain $P_J P_I = P_I P_J = U_0^* U_1^* \mathbf{1}_I U_1 U_0 U_1^* U_2^* \mathbf{1}_J U_2 U_1 = U_0^* U_1^* U_2^* \mathbf{1}_I \mathbf{1}_J U_0 U_2 U_1 = 0$. □

Smooth orthogonal maps

The factored construction for P_I not only simplifies the proof of Lemma 4.3, it also points the way to a natural generalization: we can use different rising cutoffs for folding and unfolding. We make the following assumptions:

- $I = (\alpha_0, \alpha_1)$ as before;
- r_0, r_1, r_2, r_3 are rising cutoff functions belonging to $C^d(\mathbf{R})$;
- $\epsilon_0 > 0$, $\epsilon_1 > 0$ are chosen so that $U_0 = U(r_0, \alpha_0, \epsilon_0)$ and $U_1 = U(r_1, \alpha_1, \epsilon_1)$ are disjoint folding operators;
- $\epsilon_2 > 0$, $\epsilon_3 > 0$ are chosen so that $V_0^* = U^*(r_2, \alpha_0, \epsilon_2)$ and $V_1^* = U^*(r_3, \alpha_1, \epsilon_3)$ are disjoint unfolding operators.

We define $Q_I f \stackrel{\text{def}}{=} V_0^* V_1^* \mathbf{1}_I U_1 U_0 f$.

Lemma 4.4 *Q_I is a unitary isomorphism of Hilbert spaces between $Q_I^* L^2(\mathbf{R})$ and $Q_I L^2(\mathbf{R})$, where $Q_I^* \stackrel{\text{def}}{=} U_0^* U_1^* \mathbf{1}_I V_0 V_1$. Furthermore, if $f \in C^d(\mathbf{R})$, then $Q_I f$ has a unique continuous extension in $C^d(\mathbf{R})$ which is supported in the interval $[\alpha_0 - \epsilon_2,\ \alpha_1 + \epsilon_3]$.*

Proof: The smoothness and support properties of $Q_I f$ follow from Lemma 4.1, since $\mathbf{1}_I U_1 U_0 f$ satisfies Equation 4.23 at $t = \alpha_0$ and $t = \alpha_1$ and has support in I.

Since U_0 and U_1 are isomorphisms we may write $Q_I L^2(\mathbf{R}) \cong V_0^* V_1^* \mathbf{1}_I L^2(\mathbf{R})$. Similarly we may write $Q_I^* L^2(\mathbf{R}) \cong U_0^* U_1^* \mathbf{1}_I L^2(\mathbf{R})$, which shows that Q_I is an isomorphism between $Q_I^* L^2(\mathbf{R})$ and $Q_I L^2(\mathbf{R})$, with the inverse map being Q_I^*. □

Notice that $Q_I L^2(\mathbf{R})$ is also unitarily isomorphic as a Hilbert space to $L^2(I)$, with the isomorphism given by $V_0 V_1$. Now we restrict our attention to I-periodic functions and show how they may be injected into $Q_I L^2(\mathbf{R})$:

Corollary 4.5 *Suppose that Q_I is the operator defined in Lemma 4.4. Then if f and h are two orthogonal I-periodic functions, we have*

$$\langle Q_I f, Q_I h\rangle = \langle \mathbf{1}_I f, \mathbf{1}_I h\rangle.$$

Proof: To avoid repeating a tedious calculation, we defer part of the proof until the next section. There we introduce a unitary operator $W = W(r, I, \epsilon)$, and in Lemma 4.7 and Equation 4.46 show that for I-periodic functions f it satisfies

$$\mathbf{1}_I U_0 U_1 f = W \mathbf{1}_I f.$$

But then, since V_0^* and V_1^* are also unitary, we have

$$\begin{aligned}\langle Q_I f, Q_I h\rangle &= \langle V_0^* V_1^* \mathbf{1}_I U_0 U_1 f, V_0^* V_1^* \mathbf{1}_I U_0 U_1 h\rangle \\ &= \langle \mathbf{1}_I U_0 U_1 f, \mathbf{1}_I U_0 U_1 h\rangle = \langle W \mathbf{1}_I f, W \mathbf{1}_I h\rangle = \langle \mathbf{1}_I f, \mathbf{1}_I h\rangle.\end{aligned}$$

□

In particular, if f is I-periodic, then $\|Q_I f\| = \|\mathbf{1}_I f\|$, and if f and h are I-periodic and orthogonal then $Q_I f$ and $Q_I h$ are orthogonal.

This isomorphism can be used in place of the traditional windowing of periodic functions as a method of smoothly restricting them to intervals. We remark that Q_I will be a projection if and only if $V_0^* U_0 = U_0 V_0^* = Id$ and $V_1^* U_1 = U_1 V_1^* = Id$. Since these are all unitary, it is equivalent that $U_0 = V_1$ and $U_1 = V_2$.

If we take $U_0 \neq V_1$ and $U_1 \neq V_2$, then the operator Q_I will *warp* the function near the endpoints of the interval. This may be used for certain signal processing applications, particularly in the two-dimensional case.

Corollary 4.6 *If I and J are adjacent compatible intervals, then $Q_I + Q_J = Q_{I\cup J}$ and $Q_I Q_J = Q_J Q_I = 0$.*

Proof: Replace U_i^* with V_i^*, $i = 0, 1, 2$, in the proof of Lemma 4.3. □

Relation with windowing

If $f(t) = f(-t)$ for all $-1 \leq t \leq 1$, then the two basic projection operators simplify to multiplication:

$$P_0 f(t) = \left[|r(t)|^2 + \bar{r}(t) r(-t)\right] f(t); \tag{4.36}$$

$$P^0 f(t) = \left[|r(-t)|^2 - \bar{r}(t) r(-t)\right] f(t). \tag{4.37}$$

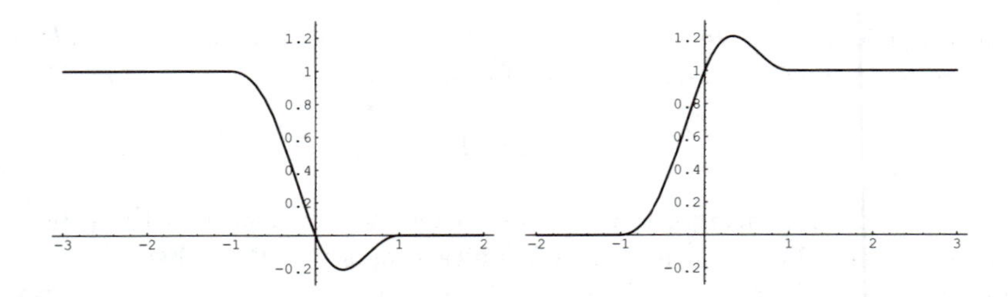

Figure 4.6: Multipliers for the smooth projections on $\mathbf{R}_-$ and $\mathbf{R}_+$, restricted to purely even functions.

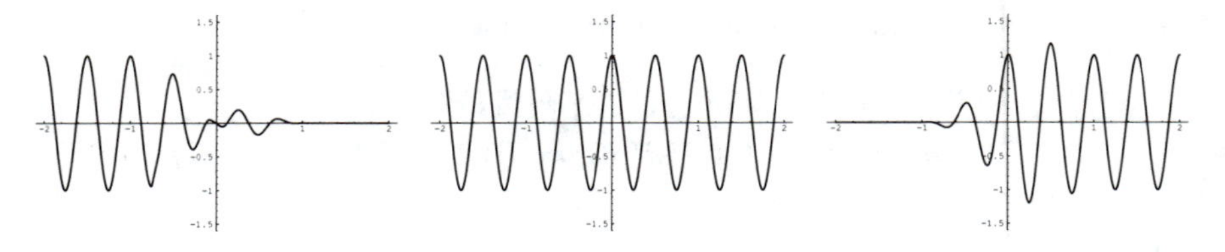

Figure 4.7: The projection of $\cos(4\pi t)$ on $\mathbf{R}_-$ and $\mathbf{R}_+$.

On the other hand, if $f(t) = -f(-t)$ is purely odd on $[-1, 1]$, then

$$P_0 f(t) \quad = \quad \left[|r(t)|^2 - \bar{r}(t)r(-t)\right] f(t); \tag{4.38}$$

$$P^0 f(t) \quad = \quad \left[|r(-t)|^2 + \bar{r}(t)r(-t)\right] f(t). \tag{4.39}$$

The two multiplier functions sum to 1 for all t, and they have as many continuous derivatives as the rising cutoff r. If we start with an even function such as $\cos(t)$, then the smooth orthogonal projections are equivalent to multiplication by a smooth partition of unity subordinate to the decomposition $\mathbf{R} = \mathbf{R}_- \cup \mathbf{R}_+$. We have plotted an example of this case in Figure 4.6, again using the rising cutoff function $r_{[1]}$.

Applying the projections to $\cos(4\pi t)$ yields the plots in Figure 4.7.

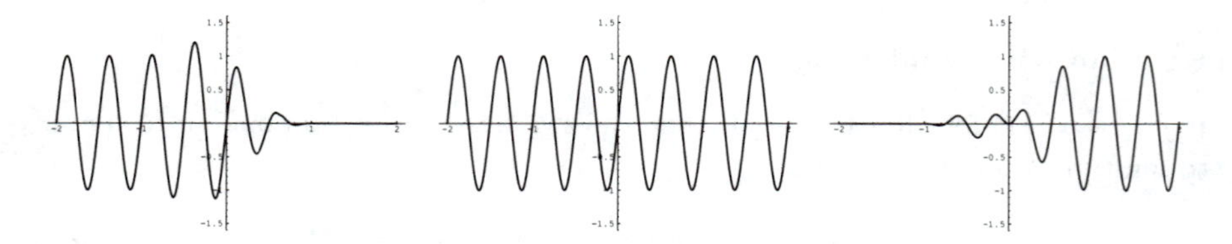

Figure 4.8: The projection of $\sin(4\pi t)$ on $\mathbf{R}_-$ and $\mathbf{R}_+$.

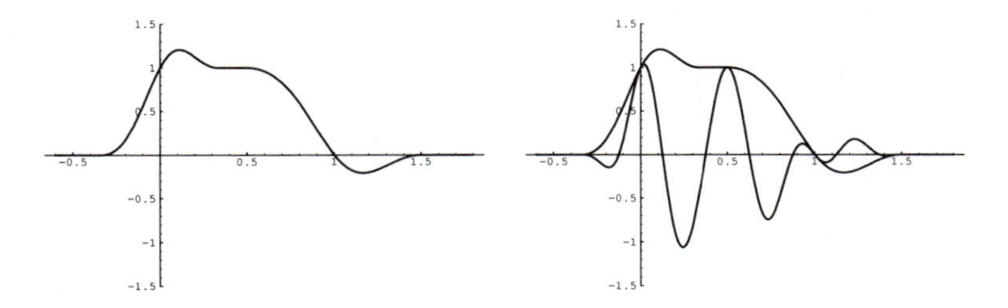

Figure 4.9: The window and the smooth projection of a cosine function.

A similar result holds for an odd function like $\sin(t)$, but the multipliers will be reflected as if we sent $t \to -t$ and took complex conjugates. The example $\sin(4\pi t)$ and its projections are plotted in Figure 4.8. Note the subtle difference at zero between these projections and the ones for $\cos(4\pi t)$.

If we combine these two smooth projection operators but move them to disjoint action regions, then they will orthogonally project a smooth function to a smooth, compactly supported function. Figure 4.9 shows the example function $\cos(4\pi t)$ projected using operators with action regions $B_{1/3}(0)$ and $B_{1/2}(1)$.

The plan for getting a smooth windowed orthonormal basis is to start with a collection of functions which are either "locally even" or "locally odd" at the endpoints of the interval I, then applying the shifted and rescaled smooth orthogonal projection P_I. The orthogonality follows from the special properties of folding and unfolding, but the result looks like multiplication by a smooth, compactly supported window function.

Local exponential functions

Any smooth periodic function may be orthogonally mapped to a smooth, compactly supported function. This is a convenient method for building test functions.

The complex exponential function $E_n(t) = e^{2\pi int}$, for integer n, has a real part which is even about zero and one and an imaginary part which is odd at each of those points. If we apply a smooth projection operator P_I to E_n, the effect will be to multiply the real and imaginary parts of E_n by different window functions supported mainly on I.

The example $P_I E_1$ in Figure 4.10, where $I = (0, 4)$ and the action regions for folding and unfolding are $(-1, 1)$ and $(3, 5)$, uses the rising cutoff function $r = r_{[1]}$ defined in Equation 4.7.

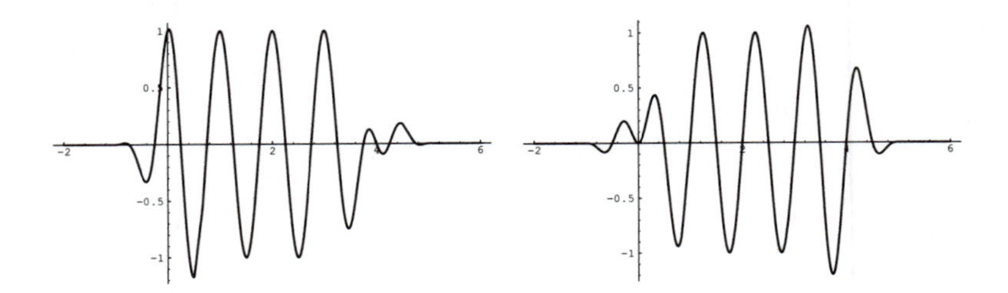

Figure 4.10: Real and imaginary parts of a localized exponential.

4.1.3 Periodization

If we use a single ϵ and a single r throughout, then the interval $I = (\alpha_0, \alpha_1)$ and its adjacent translate by $|I| = \alpha_1 - \alpha_0$ are compatible. The main consequence of this observation is that P_I composed with I-periodization is still a unitary isomorphism.

The *q-periodization* f_q of a function $f = f(t)$ is defined by the following formula:

$$f_q(t) \stackrel{\text{def}}{=} \sum_{k\in\mathbf{Z}} f(t + kq) = \sum_{k\in\mathbf{Z}} \tau_{kq} f(t). \tag{4.40}$$

If f belongs to $L^2(\mathbf{R})$ and is compactly supported, then f_q belongs to $L^2_{loc}(\mathbf{R})$ and is periodic of period q. If in addition f belongs to $C^d(\mathbf{R})$, then f_q also belongs to $C^d(\mathbf{R})$. We will abuse notation in the interest of simplicity and write f_I for f_q with $q = |I|$, just as we have written I-periodic rather than $|I|$-periodic.

Periodized folding and unfolding

We can now define versions of the folding and unfolding operators periodized to the interval $I = (\alpha_0, \alpha_1)$. The periodization is hidden in the following formulas:

$$\begin{aligned} Wf(t) &= W(r, I, \epsilon) f(t) \qquad (4.41) \\ &= \begin{cases} r(\frac{t-\alpha_0}{\epsilon}) f(t) + r(\frac{\alpha_0-t}{\epsilon}) f(\alpha_0 + \alpha_1 - t), & \text{if } \alpha_0 < t \le \alpha_0 + \epsilon, \\ \bar{r}(\frac{\alpha_1-t}{\epsilon}) f(t) - \bar{r}(\frac{t-\alpha_1}{\epsilon}) f(\alpha_0 + \alpha_1 - t), & \text{if } \alpha_1 - \epsilon \le t < \alpha_1, \\ f(t), & \text{otherwise;} \end{cases} \\ W^*f(t) &= W^*(r, I, \epsilon) f(t) \qquad (4.42) \\ &= \begin{cases} \bar{r}(\frac{t-\alpha_0}{\epsilon}) f(t) - r(\frac{\alpha_0-t}{\epsilon}) f(\alpha_0 + \alpha_1 - t), & \text{if } \alpha_0 < t \le \alpha_0 + \epsilon, \\ r(\frac{\alpha_1-t}{\epsilon}) f(t) + \bar{r}(\frac{t-\alpha_1}{\epsilon}) f(\alpha_0 + \alpha_1 - t), & \text{if } \alpha_1 - \epsilon \le t < \alpha_1, \\ f(t), & \text{otherwise.} \end{cases} \end{aligned}$$

These operators fold and unfold the right end of the segment with the left end, thereby avoiding the need to step out of the interval. We will assume that $B_\epsilon(\alpha_0)$ and $B_\epsilon(\alpha_1)$ are disjoint, so that the operators are well-defined. We will also write W_I for $W(r, I, \epsilon)$, suppressing the r and ϵ when there is no possibility of confusion.

A direct calculation shows that W_I is unitary: $W_I^* W_I = W_I W_I^* = Id$. This follows as before from the identity $|r(t)|^2 + |r(-t)|^2 = 1$, all $t \in \mathbf{R}$.

The action region for W_I and W_I^* is the union of two half-balls $[\alpha_0, \alpha_0 + \epsilon] \cup [\alpha_1 - \epsilon, \alpha_1]$. If we restrict to I and periodize, then these two half-balls merge into $B_\epsilon(\alpha_0)$ and its translates by $k|I|$, $k \in \mathbf{Z}$. Thus by using I-periodization, we can write a relation between W and U:

Lemma 4.7 *Suppose that $B_{\epsilon_0}(\alpha_0)$ and $B_{\epsilon_1}(\alpha_1)$ are disjoint and r is a smooth rising cutoff. Then for any function f we have:*

$$W(r, I, \epsilon)\, \mathbf{1}_I f = (\mathbf{1}_I\, U(r, \alpha_0, \epsilon) U(r, \alpha_1, \epsilon)\, \mathbf{1}_I f)_I\,; \tag{4.43}$$

$$W^*(r, I, \epsilon)\, \mathbf{1}_I f = (\mathbf{1}_I\, U^*(r, \alpha_0, \epsilon) U^*(r, \alpha_1, \epsilon)\, \mathbf{1}_I f)_I\,. \tag{4.44}$$

Proof: We observe that the following identity holds for the periodic function $\tilde{f} = (\mathbf{1}_I f)_I$ of period $|I| = \alpha_1 - \alpha_0$:

$$\tilde{f}(\alpha_0 + \alpha_1 - t) = \tilde{f}(2\alpha_0 - t) = \tilde{f}(2\alpha_1 - t). \tag{4.45}$$

Also, $f(t) = \tilde{f}(t)$ for all $t \in I$. Using these facts and Equation 4.13, the formula for $W_I f$ inside I becomes the following:

$$W(r, I, \epsilon)\, f(t) = \begin{cases} r(\frac{t-\alpha_0}{\epsilon})\tilde{f}(t) + r(\frac{\alpha_0 - t}{\epsilon})\tilde{f}(2\alpha_0 - t), & \text{if } \alpha_0 < t \leq \alpha_0 + \epsilon, \\ \bar{r}(\frac{\alpha_1 - t}{\epsilon})\tilde{f}(t) - \bar{r}(\frac{t-\alpha_1}{\epsilon})\tilde{f}(2\alpha_1 - t), & \text{if } \alpha_1 - \epsilon \leq t < \alpha_1, \\ f(t), & \text{otherwise;} \end{cases}$$

$$= \begin{cases} U(r, \alpha_0, \epsilon)\tilde{f}(t), & \text{if } \alpha_0 < t \leq \alpha_0 + \epsilon, \\ U(r, \alpha_1, \epsilon)\tilde{f}(t), & \text{if } \alpha_1 - \epsilon \leq t < \alpha_1. \\ f(t), & \text{otherwise.} \end{cases}$$

Because the two disjoint folding operators commute, the last formula can be rewritten as $U(r, \alpha_0, \epsilon) U(r, \alpha_1, \epsilon)\tilde{f}(t)$.

Finally, we note that for $t \neq I$, both sides of the equation are 0. □

Notice that if f is I-periodic, then $f = (\mathbf{1}_I f)_I$, so that

$$W(r, I, \epsilon)\, \mathbf{1}_I\, f = \mathbf{1}_I\, U(r, \alpha_0, \epsilon) U(r, \alpha_1, \epsilon)\, f. \tag{4.46}$$

This fact was used in the proof of Corollary 4.5.

We illustrate the action of W_I on cosines and sines restricted to an interval. Figure 4.11 shows a plot of the real and imaginary parts of $W_I E_1$, where $I = (0, 4)$ and the action regions for periodized folding are $(0, 1)$ and $(3, 4)$. It uses the rising cutoff function $r = r_{[1]}$ defined in Equation 4.7.

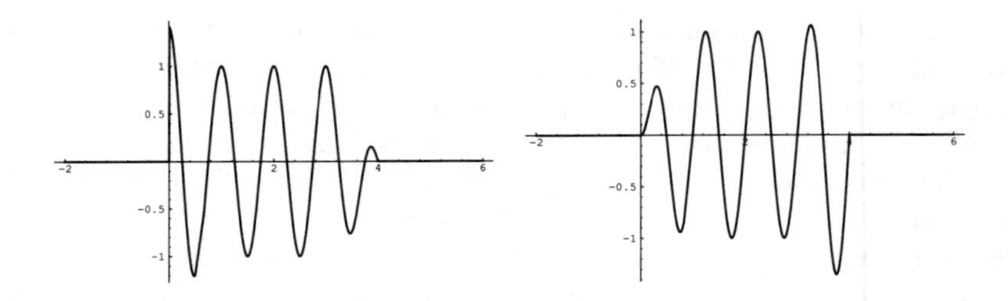

Figure 4.11: Block sine and cosine after periodic folding.

Boundary conditions and periodic extensions

We observe that W_I and W_I^* are unitary isomorphisms of $L^2(I)$ (*i.e.*, $W_I^* W_I = W_I W_I^* = Id$) because $|r(t)|^2 + |r(-t)|^2 = 1$ for all t. Also, if $t < \alpha_0$ or $t > \alpha_1$, then $W_I f(t) = f(t)$ and $W_I^* f(t) = f(t)$, so they are also unitary isomorphisms of $L^2(\mathbf{R})$. Another consequence is that if I and J are disjoint intervals, the operators W_I, W_I^*, W_J, and W_J^* all commute.

Now fix $I = (\alpha_0, \alpha_1)$ and suppose that

- $B_\epsilon(\alpha_0)$ and $B_\epsilon(\alpha_1)$ are disjoint for $\epsilon > 0$;
- r is a rising cutoff function in C^d;
- $f \in C^d(\mathbf{R})$ is I-periodic.

We take $W_i = W(r, I, \epsilon)$. Note that if f belongs to $C^d(I)$, then $W_I f \in C^d(I)$. We now show that $W_I f$ satisfies the same boundary conditions at α_0+ and α_1- as $U_0 U_1 f$ did:

Lemma 4.8 *If f belongs to $C^d(\mathbf{R})$ then $W_I f$ belongs to $C^d(\mathbf{R} \setminus \{\alpha_0, \alpha_1\})$, has one-sided limits $[W_I f]^{(n)}(\alpha_0+)$ and $[W_I f]^{(n)}(\alpha_1-)$ for all $0 \le n \le d$, and satisfies the following conditions:*

$$\lim_{t \to \alpha_0+} [W_I f]^{(n)}(t) = 0, \qquad \text{if } n \text{ is odd;} \tag{4.47}$$

$$\lim_{t \to \alpha_1-} [W_I f]^{(n)}(t) = 0, \qquad \text{if } n \text{ is even.} \tag{4.48}$$

Conversely, if f belongs to $C^d(I)$ with one-sided limits $f^{(n)}(\alpha_0+)$ and $f^{(n)}(\alpha_1-)$ for all $0 \le n \le d$ which satisfy

$$\lim_{t \to \alpha_0+} f^{(n)}(t) = 0, \qquad \text{if } n \text{ is odd;} \tag{4.49}$$

$$\lim_{t \to \alpha_1-} f^{(n)}(t) = 0, \qquad \text{if } n \text{ is even,} \tag{4.50}$$

then $W_I^* f$ satisfies the equation

$$\lim_{t\to\alpha_0+} [W_I^* f]^{(n)}(t) = \lim_{t\to\alpha_1-} [W_I^* f]^{(n)}(t), \qquad \textit{for all } 0 \le n \le d. \tag{4.51}$$

Thus $W_I^* \mathbf{1}_I f$ has a continuous periodic extension in $C^d(\mathbf{R})$.

Proof: If f is I-periodic, then $(\mathbf{1}_I f)_I = f$. Thus Equation 4.47 follows from Lemma 4.7 and an application of Lemma 4.1 at α_0+ and α_1-.

Given Equation 4.8 we deduce that $\tilde{f} = (\mathbf{1}_I f)_I$ satisfies the conditions

$$\tilde{f}^{(n)}(\alpha_0+) = \tilde{f}^{(n)}(\alpha_1+) = 0, \qquad \text{if } n \text{ is odd}; \tag{4.52}$$

$$\tilde{f}^{(n)}(\alpha_0-) = \tilde{f}^{(n)}(\alpha_1-) = 0, \qquad \text{if } n \text{ is even}. \tag{4.53}$$

We can evaluate the one-sided limits of Equation 4.51 by using Lemma 4.7:

$$[W^*(r, I, \epsilon) f]^{(n)}(\alpha_0+) = \left[U_0^* \tilde{f}\right]^{(n)}(\alpha_0+), \qquad \text{for all } 0 \le n \le d; \tag{4.54}$$

$$[W^*(r, I, \epsilon) f]^{(n)}(\alpha_1-) = \left[U_1^* \tilde{f}\right]^{(n)}(\alpha_1-), \qquad \text{for all } 0 \le n \le d. \tag{4.55}$$

But $U_1^* \tilde{f} = \tau_{|I|} U_0^* \tau_{|I|}^* \tilde{f} = \tau_{|I|} U_0^* \tilde{f}$, since $\tilde{f}$ is I-periodic, so that $\left[U_0^* \tilde{f}\right]^{(n)}(\alpha_0+) = \left[U_1^* \tilde{f}\right]^{(n)}(\alpha_1+)$ and $\left[U_0^* \tilde{f}\right]^{(n)}(\alpha_0-) = \left[U_1^* \tilde{f}\right]^{(n)}(\alpha_1-)$. Finally, the converse of Lemma 4.1 applied at α_0 (or just as well at α_1) implies that $\left[U_0^* \tilde{f}\right]^{(n)}(\alpha_0+) = \left[U_0^* \tilde{f}\right]^{(n)}(\alpha_0-)$ for all $0 \le n \le d$, from which follows Equation 4.51. □

Smooth orthogonal periodization

The main result of this section is the following construction. We use folding and unfolding to build an orthogonal transformation which restricts functions on the line to intervals and then periodizes them while preserving smoothness. In practice it is often useful to treat functions on intervals as periodic, for example for short-time discrete Fourier analysis.

The ingredients are:

- Three rising cutoff functions r_0, r_1, r in $C^d(\mathbf{R})$;
- An interval $I = (\alpha_0, \alpha_1)$;
- Folding operators $U_0 = U(r_0, \alpha_0, \epsilon_0)$ and $U_1 = U(r_1, \alpha_1, \epsilon_1)$ at the endpoints of I, with disjoint action regions;

- A periodized unfolding operator $W_I^* = W^*(r, I, \epsilon)$ whose two action regions are disjoint.

The conditions $2\epsilon \leq \alpha_1 - \alpha_0$ and $\epsilon_0 + \epsilon_1 \leq \alpha_1 - \alpha_0$ are enough to guarantee disjointness.

We define a *smooth periodic interval restriction* operator with the formula

$$T_I f \stackrel{\text{def}}{=} W_I^* \mathbf{1}_I U_0 U_1 f. \tag{4.56}$$

Then we have the main result of this section:

Theorem 4.9 *If f belongs to $C^d(\mathbf{R})$, then $T_I f$ has an I-periodic extension which belongs to $C^d(\mathbf{R})$. In addition, T_I is a unitary isomorphism of Hilbert spaces mapping $U_0^* U_1^* \mathbf{1}_I L^2(\mathbf{R})$ to $L^2(I)$.*

Proof: Since $\mathbf{1}_I U_0 U_1 f$ satisfies Equation 4.8, the converse part of Lemma 4.8 implies that for all $0 \leq n \leq d$,

$$\lim_{t \to \alpha_0 +} [W_I^* \mathbf{1}_I U_0 U_1 f]^{(n)}(t) = \lim_{t \to \alpha_1 -} [W_I^* \mathbf{1}_I U_0 U_1 f]^{(n)}(t). \tag{4.57}$$

Hence $W_I^* \mathbf{1}_I U_0 U_1 f$ has a unique continuous periodic extension in $C^d(\mathbf{R})$.

To prove the second part, we note that $U_0 U_1$ provides a unitary isomorphism from $U_0^* U_1^* \mathbf{1}_I L^2(\mathbf{R})$ to $\mathbf{1}_I L^2(\mathbf{R}) \cong L^2(I)$, and $W_I^* \mathbf{1}_I$ is a unitary automorphism on $\mathbf{1}_I L^2(\mathbf{R})$. □

4.1.4 Some analytic properties

Time-frequency localization

Observe that each local cosine function ψ_{nk} defined in Equation 4.20 is well-localized in both time and frequency. In time, it is supported on $[\alpha_k - \epsilon_k, \alpha_{k+1} + \epsilon_{k+1}]$ and thus has position uncertainty at most equal to the width of that compact interval. In frequency, $\hat{\psi}_{nk}$ consists of two modulated bumps centered at $n + \frac{1}{2}$ and $-n - \frac{1}{2}$, respectively, with uncertainty equal to that of the Fourier transform $\hat{b}_k$ of the bell.

Let us drop the subscript and note that this frequency uncertainty depends upon the relative steepness of the sides of the bell; in fact, it is exactly $\|b'\|/\|b\|$. Now suppose that the bell b is real-valued, symmetric about $t = 0$, and supported in the interval $[-a, a]$. Then $b(t) = r(2t/a - 1)r(1 - 2t/a)$ for some rising cutoff function r. In this case, b is subordinate to the interval $[-a/2, a/2]$ and has action regions

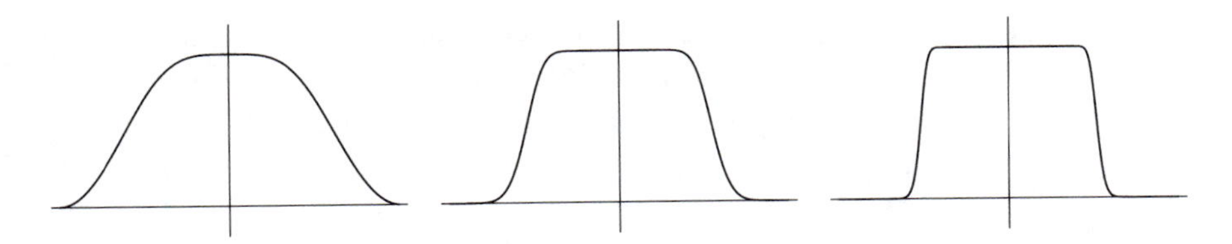

Figure 4.12: Bells associated (from left to right) with $r_{[1]}$, $r_{[3]}$, and $r_{[5]}$.

$B_{a/2}(\pm a/2)$. Thus

$$\begin{aligned}\|b\|^2 &= \int_{-a}^{a} |r(2t/a+1)r(1-2t/a)|^2\,dt = \frac{a}{2}\int_{-2}^{2} |r(t+1)r(1-t)|^2\,dt \\ &= \frac{a}{2}\int_{-1}^{1} |r(t)\cdot 1|^2 + |1\cdot r(-t)|^2\,dt = a.\end{aligned}$$

Also, a may also be called the characteristic width of b, since by computing $\int t^2|b(t)|^2\,dt$ with the slowest and fastest possible decays of b we get the inequalities

$$\frac{a}{\sqrt{12}} \le \frac{1}{\|b\|}\left[\int_{-a/2}^{a/2} t^2\,dt\right]^{1/2} \le \Delta x(b) \le \frac{1}{\|b\|}\left[\int_{-a}^{-a/2} + \int_{a/2}^{a} t^2\,dt\right]^{1/2} \le a\sqrt{\frac{7}{12}}.$$

If b has finite frequency uncertainty, then b' must be in L^2. Since $b'(t) = 0$ for all $|t| > a$, it must also be integrable. Because $b(-a) = b(a) = 0$ while $b(0) = 1$, the total variation of b on $[-a, a]$ must be at least 2. Thus by the Cauchy–Schwarz inequality,

$$2 \le \int_{-a}^{a} |b'(t)|\,dt \le \left(\int_{-a}^{a} |b'(t)|^2\,dt\right)^{\frac{1}{2}} \left(\int_{-a}^{a} 1\,dt\right)^{\frac{1}{2}} = \|b\| \bigtriangleup \xi(b)\sqrt{2a}.$$

Combining the two inequalities allows us to estimate the Heisenberg product for this bell:

$$\bigtriangleup x(b) \cdot \Delta\xi(b) \ge \frac{1}{\sqrt{6}} \approx 0.41. \tag{4.58}$$

This is appreciably larger than the minimum $\frac{1}{4\pi} \approx 0.08$, but is still reasonably small. There is no upper bound on this product; we can make it as large as we like by steepening r. This steepening is visible in Figure 4.12, which plots, from left to right, the 256-point approximations to the bell functions associated to $r_{[1]}$, $r_{[3]}$, and $r_{[5]}$.

Minimizing this variance is an exercise in the calculus of variations. We can compute it for the iterated sine bells. First note that the product in Equation 4.58 is independent of a, so we may use the most convenient value $a = 2$. Then

$$\begin{aligned} \|b\| &= \sqrt{2}; \\ \triangle\xi(b) &= \frac{1}{\|b\|}\left(\int_{-2}^{2} |b'(t)|^2\,dt\right)^{1/2} = \left(\int_{-1}^{1} |r'(t)|^2\,dt\right)^{1/2}; \\ \triangle x(b) &= \frac{1}{\|b\|}\left(\int_{-2}^{2} t^2|b(t)|^2\,dt\right)^{1/2} = \left(\frac{4}{3} - 2\int_{-1}^{1} t|r(t)|^2\,dt\right)^{1/2}. \end{aligned}$$

Some of these integrals can be evaluated analytically, the others must be done by numerical quadrature. We will only be needing crude approximations; the first six iterates are in Table 4.1. Notice that the $\triangle x$ column is decreasing to $1/\sqrt{3} \approx 0.577$ even as the $\triangle\xi$ column is increasing up to $+\infty$.

n	$\triangle x(r_{[n]})$	$\triangle\xi(r_{[n]})$	$\triangle x \cdot \triangle\xi$
0	$\left(\frac{4}{3} - \frac{8}{\pi^2}\right)^{1/2} \approx 0.723$	$\frac{\pi}{4} \approx 0.785$	0.568
1	0.653	$\frac{\pi^2}{\sqrt{128}} \approx 0.872$	0.570
2	0.612	1.053	0.645
3	0.592	1.302	0.771
4	0.584	1.622	0.947
5	0.580	2.029	1.177

Table 4.1: Heisenberg products for the iterated sine bells.

Alternatively, we can set up the finite-dimensional optimization problem for the discrete bell and its Fourier transform. We will not take this idea further than to plot the 256-point discrete Fourier transforms of the bell functions associated to $r_{[1]}$, $r_{[3]}$, and $r_{[5]}$. Notice that since the bells are even functions, their Fourier transforms are real-valued. In Figure 4.13 we have plotted the most interesting of these values, the ones very close to the zero frequency. Notice how the shoulders off the main peak become larger as the sides of the bell become steeper.

Evading the Balian–Low obstruction

Let $G = \{g(t-n)e^{2\pi imt} : n, m \in \mathbf{Z}\}$ be a collection of functions in $L^2(\mathbf{R})$, where g is some fixed square-integrable bump. The Balian–Low theorem (see [37], p.108) states that if G is an orthogonal basis, then the Heisenberg product of g must be infinite. This obstruction inhibits our strong legitimate urge to build orthonormal

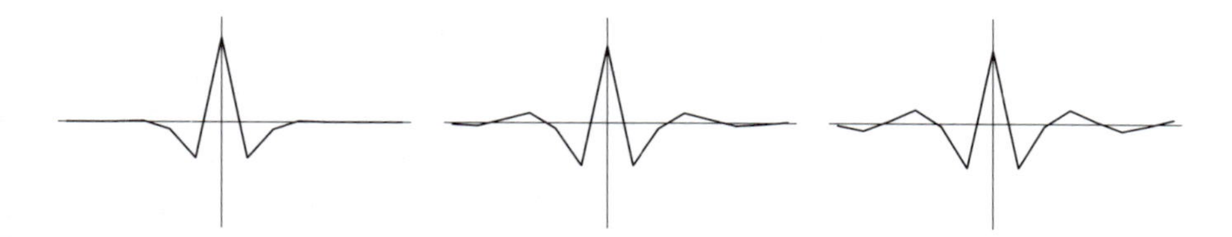

Figure 4.13: Fourier transforms of (from left to right) the $r_{[1]}$, $r_{[3]}$ and $r_{[5]}$ bells.

bases from windowed exponentials. The functions in such bases have a desirable time-frequency localization property: in time, they consist of single bumps g, while in frequency they are likewise single bumps $\hat{g}$. The Gaussian function $g(t) = e^{-t^2}$, or anything sufficiently like it, has the property that both g and $\hat{g}$ are virtually indistinguishable from zero not far from their maxima. This means that we can say unambiguously that the function $g(t-n)e^{2\pi imt}$ has *position* n and *frequency* m.

Various "Wilson bases" have been constructed which evade the Balian–Low obstruction: see [64, 40, 4, 117] for a few of the many examples. All are windowed, modulated functions, and all have in common the use of sines and cosines rather than exponentials. The local cosine and lapped orthogonal functions of [74, 24] share this property. Such a mechanism results in basis functions that contain equal energy at both positive and negative frequencies.

With smooth localized orthonormal bases, we can avoid the Balian–Low phenomenon while using functions that have all but an arbitrarily small amount of energy localized in just the positive part of the frequency spectrum. To see this, consider the effect on the exponential function of the smooth projection operator over $I = [-1, 1]$ defined in Equation 4.35:

$$P_I f(t) = |r(1-t)|^2|r(t+1)|^2 f(t) + \bar{r}(t+1)r(-t-1)f(-2-t) - \bar{r}(t-1)r(1-t)f(2-t).$$

Here we have used $\epsilon = 1$ to get the maximal action regions $B_1(\pm 1)$. Applied to the exponential, this gives us two terms:

$$\begin{aligned} P_I e^{2\pi imt} &= |r(1-t)|^2|r(t+1)|^2 e^{2\pi imt} \\ &\quad + \left[\bar{r}(t+1)r(-t-1)e^{-4\pi im} - \bar{r}(t-1)r(1-t)e^{4\pi im}\right] e^{2\pi i(-m)t} \\ &\stackrel{\text{def}}{=} b_+(t)e^{2\pi imt} + b_-(t)e^{2\pi i(-m)t}. \end{aligned}$$

The coefficient functions b_+ and b_- indicate the relative strengths of the positive and negative frequency components of $P_I e^{2\pi imt}$. Figure 4.14 shows the graphs of the absolute values of b_+ and b_- for the rising cutoff $r_{[2]}$. Table 4.2 shows the relative

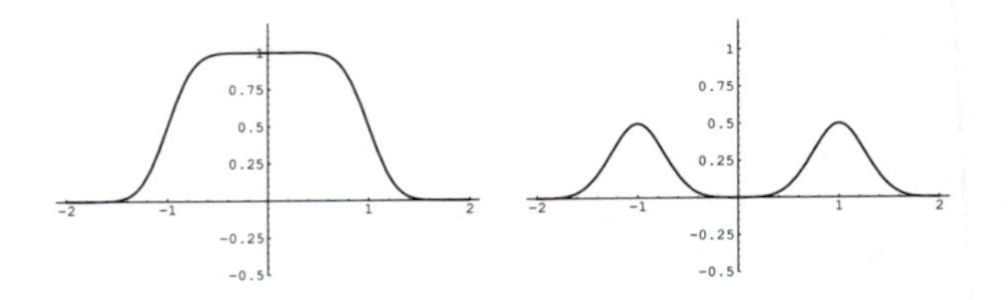

Figure 4.14: The functions b_+ and b_- for $r_{[2]}$.

amounts of energy in the b_+ and b_- associated to the six rising cutoff functions $r_{[0]}$–$r_{[5]}$.

n	$\|b_+\|^2$ for $r_{[n]}$	$\|b_-\|^2$ for $r_{[n]}$	$\|b_-\|^2/(\|b_+\|^2 + \|b_-\|^2)$
0	1.5	0.5	0.25
1	1.65212	0.347879	0.173939
2	1.76898	0.231016	0.115508
3	1.85019	0.149811	0.0749057
4	1.90389	0.0961145	0.0480572
5	1.93862	0.0613835	0.0306918

Table 4.2: Relative energies of positive and negative frequency components.

In fact $\|b_+\|^2 \to 2$ and $\|b_-\|^2 \to 0$ as $n \to \infty$, which leads us to the following conclusion:

Theorem 4.10 *For every $\epsilon > 0$ there is a smooth orthogonal projection $P_{[-1,1]}$ such that $\|b_-\|^2/(\|b_+\|^2 + \|b_-\|^2) < \epsilon$.* □

4.2 Orthogonal bases

We can use the projections and unitary operators to produce orthonormal bases of smooth, compactly supported functions on the line which have many properties in common with sines and cosines. The line may be replaced by an interval or by the circle: we still benefit from the added time localization.

Such smooth localized bases have uses in mathematical analysis, for example to build special function spaces. In addition, they are useful in practice because we can get discrete versions of the bases just by regularly sampling the continuous basis functions. The discrete bases have fast transforms related to FFT; the folding and unfolding operators do not increase the complexity of the transform.

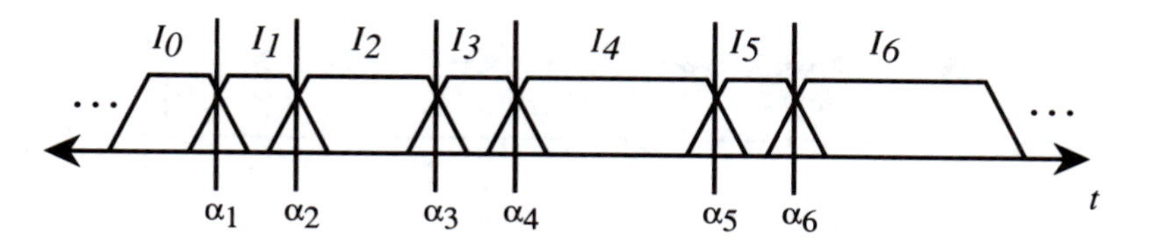

Figure 4.15: A few of the windows for a smooth localized orthonormal basis subordinate to a partition of the line.

4.2.1 Compatible partitions

We start with a partition $\mathbf{R} = \bigcup_k I_k$ and construct an orthonormal basis subordinate to it by using the projections P_{I_k}. We may imagine putting a smooth bell function or window over each interval and modulating the bells to get an orthonormal set over each. Each bell overlaps one additional interval on either side. This arrangement is depicted schematically in Figure 4.15. By taking sufficiently many modulations we get an orthonormal basis for each interval; by taking all the intervals we get an orthonormal basis for $L^2(\mathbf{R})$.

Construction from middle points

We first need to construct a partition $\mathbf{R} = \bigcup_k I_k$ of disjoint intervals $\{I_k\}$, and put disjoint action regions at their endpoints. To do this with the simplest possible notation consistent with generality, it is more convenient to treat the intervals as derived from a list of points within the intervals. Adjacent listed points will bound the action region of the folding operators used to merge adjacent windows. The interval endpoints will then be the midpoints between adjacent list points, the centers of the action regions.

So let $\{c_k : k \in \mathbf{Z}\}$ be any sequence which satisfies the following conditions:

increase: $k < j \Rightarrow c_k < c_j$;

unboundedness: $c_k \to \infty$ and $c_{-k} \to -\infty$ as $k \to \infty$.

For each integer k, define $\alpha_k = \frac{1}{2}(c_k + c_{k+1})$. The intervals will then be $I_k = [\alpha_k, \alpha_{k+1}]$. We also define the half-lengths $\epsilon_k = \frac{1}{2}(c_{k+1} - c_k) = c_{k+1} - \alpha_k = \alpha_k - c_k$, so that $|I_k| = \alpha_{k+1} - \alpha_k = \epsilon_k + \epsilon_{k+1}$. This indexing scheme is depicted in Figure 4.16.

Remark. It is also permissible that the points $\{c_k\}$ accumulate somewhere, with

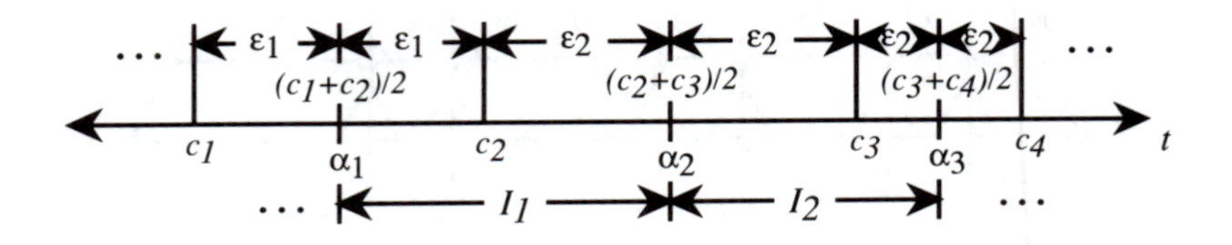

Figure 4.16: Adjacent compatible intervals defined by middle points.

the windows there shrinking to zero width. This happens at zero in the local sine and cosine construction of the Meyer wavelets [5].

The balls $B_k \stackrel{\text{def}}{=} B_{\epsilon_k}(\alpha_k)$ and $B_{k+1} \stackrel{\text{def}}{=} B_{\epsilon_{k+1}}(\alpha_{k+1})$ define the action regions at the left and right endpoints of I_k. They are disjoint by construction, since $\alpha_k + \epsilon_k = c_{k+1} = \alpha_{k+1} - \epsilon_{k+1}$. Thus we have produced a partition of $\mathbf{R}$ into adjacent compatible intervals. The chosen points $\{c_k\}$ serve as the nominal centers of the intervals. In the local sine and cosine case, they may also be regarded as the "peaks" of the window functions we will use in our basis constructions.

We let $r_k = r_k(t)$ be a family of smooth, real-valued rising cutoff functions with action regions $[-1, 1]$. We will move them to the action regions $B_{\epsilon_k}(\alpha_k) = [c_k, c_{k+1}]$ by translation and dilation. Notice that we can use different functions at each point of folding, although for simplicity we will normally keep $r_k \equiv r$ constant on all the intervals.

If we partition the integers instead of the line, and use discrete orthonormal bases on the finite subsets of integers, then we will get an orthonormal basis of ℓ^2. It is likewise possible to partition a finite interval of integers, or the "periodic" integers, to obtain discrete versions of local bases for approximations on an interval or on the circle.

Successive refinements

We may also consider several partitions at once. If these are a succession of refinements, then there is a natural tree structure to the windows. The subintervals of an interval after one refinement step may be regarded as its children, and the totality of the intervals will form a forest of family trees with each interval at the coarsest refinement being a root. This arrangement is depicted schematically in Figure 4.17, which shows one tree with dyadic refinement at each level.

Suppose that we begin with the partition of the line into unit intervals with

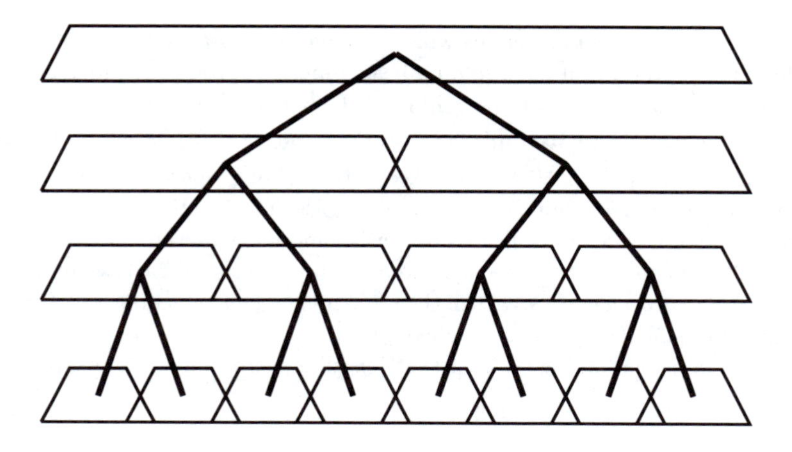

Figure 4.17: Tree of successive refinements to adjacent orthogonal windows.

integer endpoints:

$$\mathbf{R} = \bigcup_{k\in\mathbf{Z}} [k, k+1) \stackrel{\text{def}}{=} \bigcup_{k\in\mathbf{Z}} I_{0,k}. \tag{4.59}$$

We then refine the partition by breaking the intervals at their midpoints. At the jth level, the intervals will have width 2^{-j}:

$$I_{j,k} \stackrel{\text{def}}{=} [\frac{k}{2^j}, \frac{k+1}{2^j}); \qquad \mathbf{R} = \bigcup_{k\in\mathbf{Z}} I_{j,k}. \tag{4.60}$$

The genealogy of this refinement is expressed by the equation $I_{j,k} = I_{j+1,2k} \cup I_{j+1,2k+1}$, where the union is disjoint. By induction, we see that at level L we have

$$I_{0,0} = \bigcup_{k=0}^{2^L-1} I_{L,k}. \tag{4.61}$$

After a refinement of $I_{j,k}$ into its two descendents $I_{j+1,2k}$ and $I_{j+1,2k+1}$, we will perform folding at the midpoint $(k+\frac{1}{2})/2^j$. There are two procedures we can employ. In *fixed folding*, we build action regions whose radii are a fixed $\epsilon > 0$, then the folding operators used in the refinement will remain disjoint only as long as $2^{-j-1} \geq \epsilon$. Thus fixed folding can only be used to produce a limited number of refinements. However, it has the advantage that the support of the basis function built over $I_{j,k}$ is $[(k-\frac{1}{2})/2^j, (k+\frac{3}{2})/2^j]$, which is always just twice the diameter of $I_{j,k}$. Thus basis functions from small intervals will have small support. A disadvantage is that the Heisenberg product of fixed folding basis functions grows with increasing j.

In *multiple folding*, we set the radius of the action region depending upon the level of refinement. For example, at the midpoint $(k+\frac{1}{2})/2^j$ we could use the maximum radius $\epsilon_j \stackrel{\text{def}}{=} 2^{-j-1}$ which keeps the folding operators disjoint. Then there will be no *a priori* limit on the number of levels of refinement. However, the support diameter of a multiple folding basis function at level j might be almost one regardless of the depth j, because even small intervals might have quite distant points folded in during early stages of the refinement. This phenomenon of spreading is discussed in [45]. It may be controlled by judicious choice of the function $j \mapsto \epsilon_j$. We can use multiple folding to keep the Heisenberg product of the basis functions uniformly small.

4.2.2 Orthonormal bases on the line

We start with the construction of local cosine bases and other smooth localized orthonormal bases on the line, since the formulas are simplest in that case. All the constructions are "local" in the sense that we apply operators which act like the identity outside the tiny region of interest. Thus the same constructions may be applied to the circle or the interval case, since these differ from the line only in what happens "far away" from the interior points.

Continuous local cosine bases

The functions $\{\sqrt{2}\cos\left[\pi(n+\frac{1}{2})t\right] : n = 0, 1, \ldots\}$ form an orthonormal basis for $L^2([0,1])$ when restricted to the interval $[0,1]$, since they are the complete set of eigenfunctions for a Sturm–Liouville eigenvalue problem:

$$-y'' = \lambda y; \qquad y'(0) = 0, y(1) = 0. \tag{4.62}$$

The eigenvalues are $\{\pi^2\left(n+\frac{1}{2}\right)^2 : n = 0, 1, 2, \ldots\}$. The restriction may also be viewed as multiplication by the characteristic function $\mathbf{1}$ of $[0,1]$.

If $\mathbf{R} = \bigcup_k I_k$ is a compatible partition into intervals, then the collection $\{\psi_{nk} : n \in \mathbf{N}, k \in \mathbf{Z}\}$ defined in Equation 4.20 forms an orthonormal basis for $L^2(\mathbf{R})$. The proof of this fact is left as an exercise.

Since $\psi_{nk} = U_k^* U_{k+1}^* \mathbf{1}_{I_k} C_{n,k}$, we can compute the inner product of a function f with a local cosine as follows:

$$\langle f, \psi_{nk}\rangle = \langle U_k U_{k+1} f, \mathbf{1}_{I_k} C_{nk}\rangle. \tag{4.63}$$

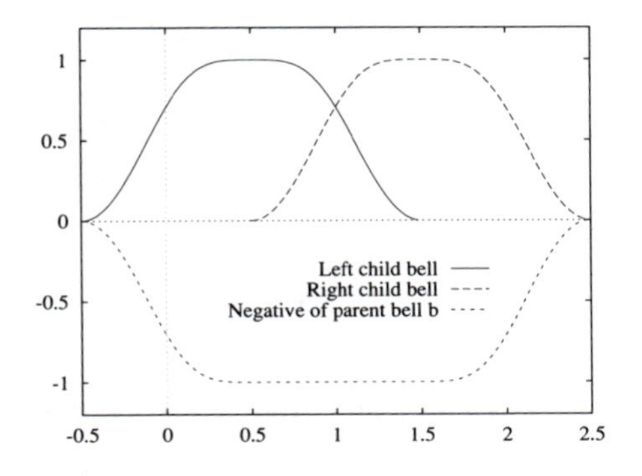

Figure 4.18: Combining adjacent windows.

This simple observation has great importance in practice, since it means that we can preprocess f by folding and then use an ordinary cosine transform to compute the local cosine transform.

Multidimensions

If we define the functions $\Psi_{nk}(x) = \psi_{n_1k_1}(x_1)\ldots\psi_{n_dk_d}(x_d)$ for multi-indices n and k, we will obtain an orthonormal basis for $\mathbf{R}^d$ made of tensor products. Of course, it is possible to use a different partition in each dimension, as well as different windows. The multidimensional case will be examined in detail in Chapter 9.

Adapted local cosine bases

The subspaces of $L^2(\mathbf{R})$ spanned by local cosines in adjacent windows are orthogonal, and their direct sum equals the space of local cosines in the parent window. If $b_{j,k}$ is the bell function subordinate to the interval $I_{j,k}$, then we have:

$$|b_{j,k}(t)|^2 = |b_{j+1,2k}(t)|^2 + |b_{j+1,2k+1}(t)|^2. \tag{4.64}$$

The overlap of these bells is displayed in Figure 4.18. Adjacent parent windows span orthogonal subspaces, which may be combined into grandparent subspaces, and so on. We will call the collection of local cosines built under these different-sized bells the *adapted local cosines.*

We may draw the tree of the local trigonometric function library in the fixed folding case schematically as in Figure 4.17, where we restrict ourselves to just a

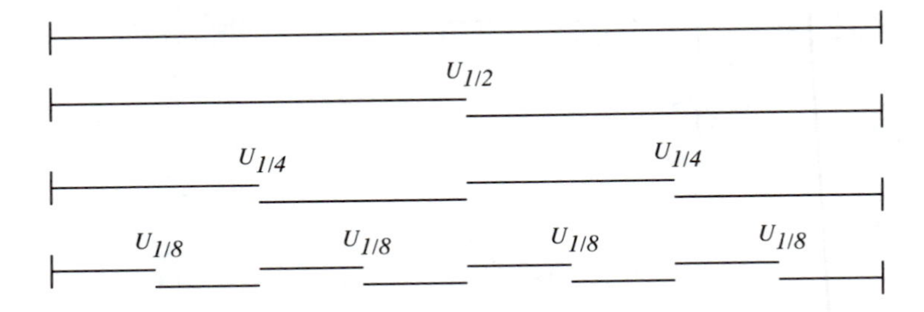

Figure 4.19: Folding at midpoints produces successive refinements.

few levels. Since the local trigonometric transform imposes no restrictions on the support intervals of the bells, the subspaces need not be of equal size. Nor is it necessary to combine windows in pairs: the tree can be inhomogeneous if desired.

We use Equation 4.63 to compute inner products with adapted local cosines. This requires successive pretreatment of the function to be expanded: we successively fold the function inside the intervals at level j to get the proper boundary conditions at the endpoints of the intervals at level $j+1$. The procedure is depicted, for the dyadic refinement case, in Figure 4.19.

Note that in the multiple folding algorithm, the action regions of the folding at the left and right endpoints of an interval will not in general have the same diameters. The multiple folding algorithm produces asymmetrical bells.

Smooth localized orthonormal bases

The local cosine basis exploits two miracles. First, the cosines at half-integer frequencies form an orthonormal basis for $L^2([0,1])$. Second, such cosines produce smooth, compactly supported functions when multiplied by the sharp cutoff $\mathbf{1}$ and then unfolded at zero and one. The result is a remarkably simple windowed orthonormal basis, but the technique cannot be applied to arbitrary bases because they may not satisfy the needed boundary conditions.

The remedy is to start with periodic functions and use the folding operator to impose the right boundary conditions upon them. This is equivalent to applying the orthogonal transformation P_I defined in Lemma 4.4. When applied to a periodic basis, the result is a local orthonormal basis.

Suppose that

- $I = (\alpha_0, \alpha_1)$;

- r, r_0, and r_1 are rising cutoff functions in $C^d(\mathbf{R})$;
- $W_I = W(r, I, \epsilon)$ is a periodized folding operator with disjoint action regions $B_\epsilon(\alpha_1)$ and $B_\epsilon(\alpha_0)$;
- $U_0^* = U^*(r_0, \alpha_0, \epsilon_0)$ and $U_1^* = U^*(r_1, \alpha_1, \epsilon_1)$ are disjoint unfolding operators;
- $\{e_j : j \in \mathbf{Z}\}$ is a collection of I-periodic functions which form an orthonormal basis for $L^2(I)$ when restricted to I.

Write $T_I^* = U_0^* U_1^* \mathbf{1}_I W_I$. We can use this operator to generate a smooth, compactly supported orthogonal set of functions.

Corollary 4.11 *The set $\{T_I^* e_j : j \in \mathbf{Z}\}$ is an orthonormal basis of $U_0^* U_1^* \mathbf{1}_I\, L^2(\mathbf{R})$. In addition, if each $e_j \in C^d(\mathbf{R})$, then each function $T_I^* e_j$ belongs to $C_0^d(\mathbf{R})$.*

Proof: The functions $\{\mathbf{1}_I\, W_I\, e_j : j \in \mathbf{Z}\}$ form an orthonormal basis of $L^2(I)$, since W_I is unitary. Thus $\{T_I^* e_j : j \in \mathbf{Z}\}$ is an orthonormal basis of $U_0^* U_1^* \mathbf{1}_I L^2(\mathbf{R})$ since $U_0^* U_1^*$ is unitary on $L^2(\mathbf{R})$.

Lemma 4.8 implies that $\mathbf{1}_I\, W_I\, e_j$ satisfies Equation 4.23 at α_0 and α_1. Then the converse of Lemma 4.1 implies that each function in the basis belongs to $C_0^d(\mathbf{R})$, in fact with support in the interval $[\alpha_0 - \epsilon_0,\ \alpha_1 + \epsilon_1]$. □

Notice that T_I^* agrees with the operator Q_I defined for Lemma 4.4, if we write $U(r, \alpha_0, \epsilon) U_1(r, \alpha_1, \epsilon)$ instead of W_I. We use the new notation to emphasize that this basis construction is the inverse of smooth periodic interval restriction.

In practice it is often better to transform a smooth function into a smooth periodic function and then expand it in a periodic basis, rather than expand a smooth function in the basis described here. This is because well-tested computer programs exist for the first algorithm but not the second. But we may once again compute inner products by taking adjoints:

$$\langle f, T_I^* e_j \rangle = \langle T_I f, e_j \rangle. \tag{4.65}$$

Hence we may compute the transform by preapplying smooth periodic interval restriction, then using preexisting software to expand in the basis $\{e_j\}$.

If we take an arbitrary segmentation $\mathbf{R} = \bigcup_{k \in \mathbf{Z}} I_k$ into adjacent compatible intervals, we can build a smooth, *compactly supported* orthonormal basis for $L^2(\mathbf{R})$. We suppose that for each $k \in \mathbf{Z}$, $\{e_{kj} : j \in \mathbf{Z}\}$ is a family of I_k-periodic functions with the property that when restricted to I_k it forms an orthonormal basis of $L^2(I_k)$. Let T_k^* be the operator of Corollary 4.11 over the interval I_k.

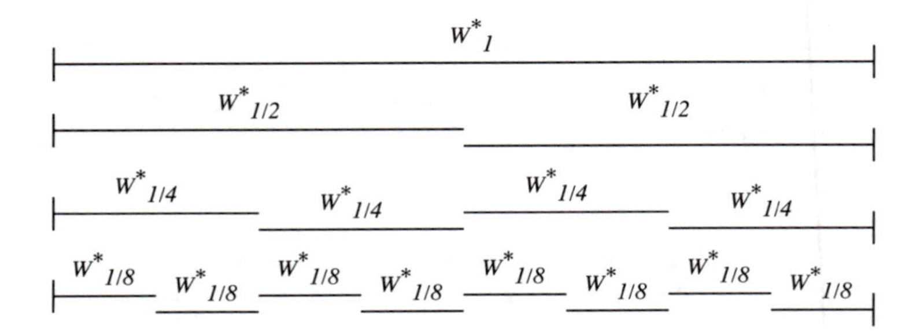

Figure 4.20: Local periodization after folding to successive refinements.

Theorem 4.12 *The collection $\{T_k^* e_{kj} : j,k \in \mathbf{Z}\}$ is an orthonormal basis for $L^2(\mathbf{R})$ consisting of functions of compact support. If in addition all the functions e_{kj} and r_k, $k,j \in \mathbf{Z}$ belong to $C^d(\mathbf{R})$, then the basis functions belong to $C_0^d(\mathbf{R})$.*

Proof: Since adjacent intervals I_k, I_{k+1} are compatible for all $k \in \mathbf{Z}$, Lemma 4.6 gives us the decomposition $L^2(\mathbf{R}) = \bigoplus_{k \in \mathbf{Z}} T_k^* L^2(\mathbf{R})$. By Corollary 4.11, each of the spaces $U_0^* U_1^* \mathbf{1}_{I_k} L^2(\mathbf{R})$ has an orthonormal basis $\{T_k^* e_{kj} : j \in \mathbf{Z}\}$. Putting these bases together yields the result. □

Adapted localized orthonormal bases

To build a sequence of transforms over successive refinements of a partition, we perform either fixed or multiple folding at the new midpoints introduced at each level of refinement. This is done just as in the local cosine case depicted in Figure 4.19. But then rather than perform the half-integer cosine transform on each interval, we preapply the periodic unfolding operator to each interval. This is illustrated in Figure 4.20.

The two steps depicted in Figures 4.19 and 4.20 apply an operator $T_{I_{j,k}}$ on each interval $I_{j,k}$. Afterwards, we may perform any periodic transform within the subintervals. By Equation 4.65, this gives us the expansion of a function in the smooth localized periodic basis.

4.2.3 Discrete orthonormal bases

We obtain discrete versions of the local cosine and other bases by replacing the line with the integers and using the discrete cosine transform.

Discrete local cosines

We can use nearly the same formulas in the discrete case that we used in the continuous case, only the variables will take integer values rather than real values. We assume that:

- $\alpha_k < \alpha_{k+1}$ are integers;
- the signal is sampled at integer points t, $\alpha_k \leq t < \alpha_{k+1}$, which gives us $N \stackrel{\text{def}}{=} \alpha_{k+1} - \alpha_k$ samples;
- r_k and r_{k+1} are rising cutoff functions;
- $\epsilon_k > 0$ and $\epsilon_{k+1} > 0$, with $\epsilon_k + \epsilon_{k+1} \leq N$ to insure that the action regions are disjoint.

Since there is a whole family of discrete cosine transforms, we have to make a choice. We begin by exploring what happens if we simply take the continuous formula and replace the real variables with integers. Thus, we define the following discrete local cosine basis functions:

$$\begin{aligned}\psi_{nk}^{\text{DCT-III}}(t) &= r_k\left(\frac{t-\alpha_k}{\epsilon_k}\right) r_{k+1}\left(\frac{\alpha_{k+1}-t}{\epsilon_{k+1}}\right) \\ &\quad \times \sqrt{\frac{2}{\alpha_{k+1}-\alpha_k}} \cos\left[\frac{\pi(n+\frac{1}{2})(t-\alpha_k)}{\alpha_{k+1}-\alpha_k}\right].\end{aligned} \tag{4.66}$$

If r is a rising cutoff function on the line, then the samples $r(n/N)$ satisfy Equation 4.1:

$$|r(n/N)|^2 + |r(-n/N)|^2 = 1. \tag{4.67}$$

By taking r so sharp that $r(n/N) = 1$ for all samples $n > 0$ and $r(n/N) = 0$ for all samples $n < 0$, we will get the ordinary discrete cosine transform. Notice that $|r(0)| = 1/\sqrt{2}$ supplies the correction term at 0; the resulting discrete cosine function is evidently the one used in DCT-III.

The discrete folding and unfolding operators in this case are defined using the same formulas as Equations 4.13 and 4.14. We notice that the effect of folding on the sample at the left endpoint α_k is simply to leave it alone. This version of the local cosine transform thus requires applying $U(r_k, \alpha_k, \epsilon_k)$ and $U(r_{k+1}, \alpha_{k+1}, \epsilon_{k+1})$ in either order, then performing DCT-III on the N points $\alpha_k, \ldots, \alpha_{k+1} - 1$. Notice that the right-hand folding operator is applied one point past the end of the interval. To compute the inverse transform we apply DCT-II, which is the inverse of DCT-III, followed by $U^*(r_k, \alpha_k, \epsilon_k)$ and $U^*(r_{k+1}, \alpha_{k+1}, \epsilon_{k+1})$ in either order.

These formulas make a distinction between the left and right endpoints, because we are sampling at the left endpoint of each sampling interval. It is more symmetric to sample in the middle of the intervals, which can be accomplished by taking the function in Equation 4.66 and replacing every instance of t by $t+\frac{1}{2}$. We may regard the basis functions for this transform as being cosines sampled between grid points. The rising cutoff function is also sampled between grid points. The result is the following collection of discrete local cosines:

$$\begin{aligned} \psi_{nk}^{\text{DCT-IV}}(t) &= r_k\left(\frac{t+\frac{1}{2}-\alpha_k}{\epsilon_k}\right) r_{k+1}\left(\frac{\alpha_{k+1}-t-\frac{1}{2}}{\epsilon_{k+1}}\right) \\ &\quad \times \sqrt{\frac{2}{\alpha_{k+1}-\alpha_k}}\cos\left[\frac{\pi(n+\frac{1}{2})(t+\frac{1}{2}-\alpha_k)}{\alpha_{k+1}-\alpha_k}\right]. \end{aligned} \tag{4.68}$$

It is evident that $\psi_{nk}^{\text{DCT-IV}}(t) = \psi_{nk}^{\text{DCT-III}}(t+\frac{1}{2})$. The cosines are just those of DCT-IV. The folding and unfolding operators in this case also need to be shifted by 1/2.

To apply this DCT-IV version of the local cosine transform, we first fold at α_k and at α_{k+1}, then apply DCT-IV. The inverse transform is to apply DCT-IV, which is its own inverse, then unfold at α_k and at α_{k+1}. The only subtlety is the indexing of the samples. It is traditional in this shifted-by-1/2 case to treat the signal indexed by k as being the sample from gridpoint $k+\frac{1}{2}$. This means for instance that the reflection of $r(t)$ will be $r(-t-1)$ rather than $r(-t)$. Thus there is no unaffected zero-point for folding and unfolding; every sample of the signal is changed by the discrete operator U. We will return to these technical points in the section on implementation.

We can replace cosine with sine in all these formulas to get the local sine transform. We can also use DCT-I on the odd intervals and DST-I on the even intervals, and switch polarities at the points of folding depending on whether there are sines or cosines in the left interval, to the get the mixed local cosine/sine basis. Some of these variations are explored in the exercises.

Discrete adapted local cosine functions

Since the factored DCT algorithms work best when the number of sample points is a power of two, it makes sense to start with a root interval $N = 2^L$ samples and use dyadic refinement to produce the subintervals for an adapted local cosine analysis. The resulting tree will have at most L levels of decomposition.

If we use fixed folding with an action region radius of 2^R samples, $0 \le R < L$, then we can descend at most $L - R$ levels. If we use multiple folding and reduce the folding radius as we descend, then we can always descend the L levels allowed by the number of samples.

Discrete versions of smooth localized orthonormal bases

The choice in this case is whether to sample at the integers or at the half-integers. This affects the indexing of the folding and unfolding operators. In computing T_I, the same indexing convention must be used by both the two folding operators U_0, U_1 and the periodized unfolding operator W_I^*.

The adapted version of this algorithm also comes in two variants: fixed folding and multiple folding. It is a fruitful exercises to generate examples of the basis functions in both cases, for various underlying periodic bases.

Likewise, the discrete version of smooth periodic interval restriction must use the same indexing convention for all folding and unfolding operators.

4.3 Basic implementation

It is possible to compute numerical approximations to the continuous local trigonometric transform, using symbolic integration when possible or numerical quadrature in general. The folding and unfolding operators can be applied prior to integration, and the result will be either analytic formulas or numerical approximations to the continuous transforms. Such an approach leads to the same difficulties with rapidly oscillating integrands that we faced when considering numerical approximations to the Fourier integral transform.

Instead, we shall only implement the discrete local sine and cosine transforms and the discrete versions of the local periodization algorithm. This allows us to use the FFT and its DCT and DST variants (see Chapter 3) rather than symbolic or numerical quadrature. It places the burden of approximation on the user who provides the sampled signal, but for band-limited signals we can obtain rapid convergence to samples of the continuous transform by increasing the sampling rate.

4.3.1 Rising cutoff functions

We begin by writing some utility functions to fill an array with samples of a smoothly rising cutoff function satisfying Equation 4.1.

Iterated sines

Given a desired number of iterations n and a point t, this function returns $r_{[n]}(t)$ using the formulas in Equations 4.6 and 4.7.

Iterated sine rising cutoff function

```
rcfis( N, T ):
   If T > -1.0 then
      If T < 1.0 then
         For I = 0 to N-1
            Let T = sin( 0.5 * PI * T )
         Let T = sin( 0.25 * PI * (1.0 + T) )
      Else
         Let T = 1.0
   Else
      Let T = 0.0
   Return T
```

Recall that `PI` has been defined to be the constant π.

Integrals of smooth bumps

To avoid issues of numerical integration, we will define the general real-valued rising cutoff from the function θ, rather than from an integrable bump which is the derivative of θ. We first implement this function, for example with a preprocessor macro that takes one argument, a real number between -1 and 1, and computes a value between 0 and $\frac{\pi}{2}$. For example, we can take $\theta(t) \stackrel{\text{def}}{=} \frac{\pi}{4}(1+t)$ to be the angle function for $r = r_{[0]}$, or we could use the cubic polynomial of Equation 4.5. Then we choose one of these with a third macro:

```
#define theta1(T)  (0.25*PI * (1.0 + (T)))
#define theta3(T)  (0.125*PI * (2.0 +(T)*(3.0 - (T)*(T))))
#define th  theta1
```

We must check that θ is defined at the value of the sample point before sampling the rising cutoff. This can be done with a (rather complicated) macro using the Standard C conditional operator. For θ defined on $[-1, 1]$, this gives:

```
#define rcfth0(T) ( T>-1 ? ( T<1 ? sin(th(T)) :1 ) : 0 )
```

But it suffices to define θ on $[0, 1]$, since $\theta(-t) = \frac{\pi}{2} - \theta(t)$. This observation is especially helpful if we use a precomputed table of values for θ. We implement it by checking whether $t \in [-1, 0[$ or $t \in [0, 1]$ before evaluating, then using both sine and cosine:

```
#define rcfth(T) (T>-1?(T<0?cos(th(-T)):(T<1?sin(th(T)):1)):0)
```

We can choose among these with yet another macro:

Two ways to define the rising cutoff function

```
#define rcf(T)  rcfis( 1, T )
       OR
#define th theta3
#define rcf  rcfth
```

In a fixed implementation, it is reasonable to merge computing θ and r into one function, which can also sample the result at the gridpoints used in the discrete algorithm.

Gridpoints or midpoints?

We will precompute the samples of the rising cutoff function r to be used in folding and unfolding, and place them into an INTERVAL data structure. We must decide in advance, though, whether to fold at gridpoints or between gridpoints. The two possibilities give different interpretations for the sampled r and result in different indexing conventions. In either case, we will assume that an array `RISE.ORIGIN[]` has been preallocated with enough memory locations to hold all the samples of r that we will need.

Let E be the radius of the folding: this means that the gridpoints labeled $\pm E$ are the first ones outside the action region at zero.

If folding takes place at gridpoints, then `R.ORIGIN[N]`$= r(N/E)$ and for every valid N we have the identity

```
R.ORIGIN[N] * R.ORIGIN[N] + R.ORIGIN[-N] * R.ORIGIN[-N] == 1.
```

Then `R.LEAST`$= -E+1$ and `R.FINAL`$= E-1$ and we must fill the array with $2*E-1$ values. This is done by the following function:

Rising cutoff function, sampled at gridpoints

```
rcfgrid( R ):
   Let X  = 0.0
   Let DX = 1.0/(R.FINAL+1.0)
   Let R.ORIGIN[0] = sqrt(2.0)
   For J = 1 to R.FINAL
      X += DX
      Let R.ORIGIN[J]  = rcf(X)
      Let R.ORIGIN[-J] = rcf(-X)
```

On the other hand, if our sampled function is to be folded between gridpoints, then R.ORIGIN[N]$= r\left(\frac{N+1/2}{E}\right)$ and the identity derived from Equation 4.1 is

```
R.ORIGIN[N]*R.ORIGIN[N]  + R.ORIGIN[-N-1]*R.ORIGIN[-N-1]  == 1.
```

This must hold for every N that indexes a valid sample. In this case, we must fill the array with $2E$ values, R.LEAST$= -E$ and R.FINAL$= E-1$.

Rising cutoff function, sampled between gridpoints

```
rcfmidp( R ):
   Let X =  0.5/(R.FINAL+1.0)
   Let DX = 1.0/(R.FINAL+1.0)
   For J = 0 to R.FINAL
      Let R.ORIGIN[J]     = rcf(X)
      Let R.ORIGIN[-J-1] = rcf(-X)
      X += DX
```

Notice how in each case we start at the middle gridpoints and then move out to either side. The main advantage of doing this, rather than starting at one end and advancing to the other, is that we preserve the identity $|r(t)|^2 + |r(-t)|^2 = 1$ to high order despite the truncation errors which accumulate as we compute gridpoints.

4.3.2 Midpoint folding and unfolding functions

We must distinguish the gridpoint and midpoint indexing conventions. To be specific, we will show the complete implementations for the midpoint folding case only. This can be done by allocating an INTERVAL and preparing it with `rcfmidp()`. Also, we will use two input arrays for the signal to the right and to the left of the point of folding.

It is also useful to define both an in place and a disjoint version of the algorithm. Folding in place can be used to save memory; folding while copying can be used in multidimensions and in the adapted local cosine analysis to avoid some steps.

Disjoint input and output

We first implement the disjoint, or compute-and-copy version of folding and unfolding at midpoints. Suppose that the input arrays contain N elements each. Then N elements will be assigned to the output array, even if the radius of folding E is less than N. The part of the input array which is outside the radius is copied to the output array. The part inside the action region is computed.

Allowing for positive and negative halves of the output arrays and for the sine and cosine polarities, there are four functions in all.

fdcn(): [F]old a given array into a [D]isjoint output array, with [C]osine polarity (odd to the left and even to the right), into the [N]egative part of the output array.

Fold disjoint cosine negative

```
fdcn( ONEG, STEP, INEG, IPOS, N, RISE ):
  For K = -N to RISE.LEAST-1
    Let ONEG[K*STEP] = INEG[K]
  For K = RISE.LEAST to -1
    Let ONEG[K*STEP] = RISE.ORIGIN[-1-K] * INEG[K]
                          - RISE.ORIGIN[K] * IPOS[-1-K]
```

fdcp(): [F]old a given array into a [D]isjoint output array, with [C]osine polarity (odd to the left and even to the right), into the [P]ositive part of the output array.

Fold disjoint cosine positive

```
fdcp( OPOS, STEP, INEG, IPOS, N, RISE ):
  For K = 0 to RISE.FINAL
    Let OPOS[K*STEP] = RISE.ORIGIN[K] * IPOS[K]
                          + RISE.ORIGIN[-1-K] * INEG[-1-K]
  For K = RISE.FINAL+1 to N-1
    OPOS[K*STEP]=IPOS[K]
```

fdsn(): [F]old a given array into a [D]isjoint output array, with [S]ine polarity (even to the left and odd to the right), into the [N]egative part of the output array. The conditions are the same as for `fdcn()`:

Fold disjoint sine negative

```
fdsn( ONEG, STEP, INEG, IPOS, N, RISE ):
  For K = -N to RISE.LEAST-1
    Let ONEG[K*STEP] = INEG[K]
  For K = RISE.LEAST to -1
    Let ONEG[K*STEP] = RISE.ORIGIN[-1-K] * INEG[K]
                          + RISE.ORIGIN[K] * IPOS[-1-K]
```

fdsp(): [F]old a given array into a [D]isjoint output array, with [S]ine polarity (even to the left and odd to the right), into the [P]ositive part of the output array. The conditions are the same as for `fdcp()`:

Fold disjoint sine positive

```
fdsp( OPOS, STEP, INEG, IPOS, N, RISE ):
  For K = 0 to RISE.FINAL
    Let OPOS[K*STEP] = RISE.ORIGIN[K] * IPOS[K]
                       - RISE.ORIGIN[-1-K] * INEG[-1-K]
  For K = RISE.FINAL+1 to N-1
    OPOS[K*STEP]=IPOS[K]
```

Since the input and output arrays are disjoint, we can perform multidimensional array transposition at the same time as the folding and unfolding. For this extra utility we need to allow output increments other than one. This is given as one of the input parameters, STEP. The N output array locations will be separated by offsets of STEP. In the disjoint case, these locations must not overlap with any of the input arrays. We also assume that $N \geq E > 0$ and that STEP> 0.

We observe that the unfolding functions are the same as the folding functions, only with the opposite polarities. Thus, they can be implemented using preprocessor macros which just rename:

```
#define udcn fdsn  /* [U]nfold [D]isj. [C]os. [N]egative. */
#define udcp fdsp  /* [U]nfold [D]isj. [C]os. [P]ositive. */
#define udsn fdcn  /* [U]nfold [D]isj. [S]in. [N]egative. */
#define udsp fdcp  /* [U]nfold [D]isj. [S]in. [P]ositive. */
```

Folding and unfolding in place

With but a single temporary variable, we can perform the folding and unfolding transformations on the same array. This is implemented as a transformation in place of two arrays:

```
       --the left half--      |    --the right half--
     ONEG[-E] ... ONEG[-1]    |   OPOS[0] ... OPOS[E-1]
```

Here E is a positive integer. This indexing is chosen so that OPOS and ONEG are typically identical pointers to the first element of a block of the given array. The function then folds the leading edge of the block into the trailing edge of the previous block. The array locations ONEG[-E],...,ONEG[-1], OPOS[0],...,OPOS[E-1] must not overlap.

There are two polarities, sine and cosine:

fipc(): Given an array, *cosine polarity* folding smoothly maps the odd part into the left half and the even part into the right half.

Fold in place, cosine polarity

```
fipc( ONEG, OPOS, RISE ):
  For K = 0 to RISE.FINAL
    Let TEMP = RISE.ORIGIN[K] * OPOS[K]
               + RISE.ORIGIN[-K-1] * ONEG[-K-1]
    Let ONEG[-K-1] = RISE.ORIGIN[K] * ONEG[-K-1]
                     - RISE.ORIGIN[-K-1] * OPOS[K]
    Let OPOS[K] = TEMP
```

fips(): Alternatively, we can smoothly fold the even part into the left half and the odd part into the right half. This is the *sine polarity*:

Fold in place, sine polarity

```
fips( ONEG, OPOS, RISE ):
  For K = 0 to RISE.FINAL
    Let TEMP = RISE.ORIGIN[K] * OPOS[K]
               - RISE.ORIGIN[-K-1] * ONEG[K]
    Let ONEG[-K-1] = RISE.ORIGIN[K] * ONEG[-K-1]
                     + RISE.ORIGIN[-K-1] * OPOS[K]
    Let OPOS[K] = TEMP
```

Unfolding in one polarity is identical with folding in the opposite polarity, so we can use preprocessor macros to define the corresponding unfolding functions:

```
#define uipc fips  /* [U]nfold [I]n [P]lace, [C]os polarity */
#define uips fipc  /* [U]nfold [I]n [P]lace, [S]in polarity */
```

4.3.3 Midpoint local trigonometric transforms

Suppose that we have an array `SIG[]` representing midpoint samples of a signal, and we are interested in computing its local sine or cosine transform near the interval $[0, N-1]$. We need to include a few extra samples: if the sampled rising cutoff function `RISE[]` is defined for indices $-E, \ldots, E-1$, where E is a positive number, then we will have to involve `SIG[-E]`, ..., `SIG[N+E-1]` in the calculation. We must also assume that $N \geq 2E$. The local cosine transform on the array has the following generic form:

In place local cosine transform on N midpoints

```
lct( SIG, N, RISE ):
   fipc( SIG, SIG, RISE )
   fipc( SIG+N, SIG+N, RISE )
   dctiv( SIG, N )
```

In place local sine transform on N midpoints

```
lst( SIG, N, RISE ):
   fips( SIG, SIG, RISE )
   fips( SIG+N, SIG+N, RISE )
   dstiv( SIG, N )
```

Folding the input array with the appropriate parity at the endpoints 0 and N imposes the boundary conditions which allow a smooth reconstruction later. We then apply the discrete cosine or sine transforms which are adapted to these boundary conditions.

We invert these transforms by first inverting the cosine or sine transform and then unfolding:

In place inverse local cosine transform on N midpoints

```
ilct( SIG, N, RISE ):
   dctiv( SIG, N )
   uipc( SIG, SIG, RISE )
   uipc( SIG+N, SIG+N, RISE )
```

In place inverse local sine transform on N midpoints

```
ilst( SIG, N, RISE ):
   dstiv( SIG, N )
   uips( SIG, SIG, RISE )
   uips( SIG+N, SIG+N, RISE )
```

All four local trigonometric transform functions require certain predefined auxiliary arrays, such as the sampled rising cutoff function and a table of sines or cosines needed by DST or DCT. These may be prepared in advance by an initialization function. If we are to perform several LCTs or LSTs, the auxiliary arrays can be retained until the calculations are complete.

Remark. The `dctiv()` and `dstiv()` transforms are their own inverses. Had we sampled the signal at gridpoints and used `dstiii()` or `dstiii()` to perform the cosine or sine transforms in the "forward" direction, we would need to use `dctii()` (respectively `dstii()`) in the inverse direction.

4.3.4 Midpoint local periodization

With the same conditions on `SIG[]`, namely that it stretches beyond the interval $[0, N-1]$ to include enough extra samples for the folding functions, we can perform an arbitrary locally periodic transform in place on the array:

In place local periodization to N points, cosine polarity

```
lpic( SIG, N, RISE ):
   fipc( SIG, SIG, RISE )
   fipc( SIG+N, SIG+N, RISE )
   uipc( SIG+N, SIG, RISE )
```

The samples `SIG[0]`,...,`SIG[N-1]` are transformed in place so that this portion of the array can be treated as one period of a periodic signal without introducing any new singularities.

We invert the local periodization on a single interval by first folding the right end with the left and then unfolding the the left and right ends individually:

Inverse local periodization from N points, cosine polarity

```
ilpic( SIG, N, RISE ):
   fipc( SIG+N, SIG, RISE )
   uipc( SIG, SIG, RISE )
   uipc( SIG+N, SIG+N, RISE )
```

Remark. We get an equivalent though different function `lpis()` by using the local sine polarity. Likewise, sine polarity works as well for inverse local periodization, though it gives different a different function `ilpis()`. Inverting a sine polarity periodization with cosine polarity does not return a smooth signal, since the affected samples outside the main interval $[0, N-1]$ have the wrong polarity.

The extra samples `SIG[N]` through `SIG[N+E-1]` and `SIG[-E]` through `SIG[-1]` are affected by folding and unfolding in place, but in a way that allows telescoping the local periodization for a sequence of adjacent compatible intervals. The result will be a segmentation of the input into a sequence of adjacent subarrays, each representing one period of different smooth periodic function.

To define such a segmentation, we must specify the lengths of the subarrays. Suppose that `LENGTHS[]` is an array of `NUM` lengths which are all at least twice as long as `E` but which add up to no more than the total length of the array `SIG[]`. We can use these to decompose the signal into a succession of segments and perform local periodization within each segment:

In place local periodizations of adjacent intervals

```
lpica( SIG, LENGTHS, NUM, RISE ):
   fipc( SIG, SIG, RISE )
   For I = 0 to NUM-1
      fipc( SIG+LENGTHS[I], SIG+LENGTHS[I], RISE )
      uipc( SIG+LENGTHS[I], SIG, RISE )
      SIG += LENGTHS[I]
```

Similarly, we can implement the local sine version `lpisa()` just by replacing `fipc()` with `fips()` and `uipc()` with `uips()`.

This implementations of adjacent-interval inverse local periodization functions `ilpica()` and `ilpisa()` are left as exercises.

If we need greater generality, we may for example specify an array of different action regions and different rising cutoff functions for each region.

4.4 Implementation of adapted transforms

The adapted transform produces more data than a simple local cosine transform or local periodic transform on an interval. We will use the various data structures and functions defined in Chapter 2 to organize and manipulate this data.

To be specific, we will limit ourselves to the midpoint or DCT-IV based transformations so we may use the already-defined rising cutoff and folding functions.

4.4.1 Adapted local cosine analysis

An *adapted local cosine analysis* has plenty of parameters in its full generality, making it a bit too complex to use as the first example. We will instead concentrate on the simpler dyadic case down to level L, which contains $1+L$ copies of the signal arranged as the levels of a binary tree. With an initial periodic signal of $N \geq 2^L$ points, this means we will have a total of $N(L+1)$ coefficients to compute.

We will treat the fixed folding and multiple folding cases separately. In the fixed folding case we will use an array binary tree to hold the coefficients; in the multiple folding case we will arrange the output into a binary tree of BTN data structures.

Fixed folding

The intervals at the bottom level L should have length $N/2^L > 1$. We will assume that this is a power of two so that we can use the fast factored DCT algorithm at every level. *Fixed folding* means using at every level the same sampled rising cutoff function with an action radius of half this minimal interval length.

We first need to allocate enough space to hold the $L + 1$ output arrays of length N each, which are respectively the local cosine transforms at window widths $N, N/2, N/4, \ldots, N/2^L$. This can be done with one single array binary tree of length $(L+1)N$, in the manner depicted in Figure 2.2 of Chapter 2.

We begin by copying the input array to the first N locations. If the original signal is periodic, then it is necessary to fold the right end of the input with the left end so that the proper boundary conditions are satisfied at the endpoints. Thus we will use the disjoint folding routines to combine copying and preparing the endpoint boundary conditions. The input parameter `PARENT` should be the array binary tree with the initialized signal in its first row.

To descend the binary tree of successively smaller intervals, we loop over the levels, folding and copying each *parent* interval into two *child* intervals before performing the DCT-IV on the parent:

Complete dyadic adapted local cosine analysis with fixed folding

```
lcadf( PARENT, N, L, RISE ):
   Let NP = N
   For LEVEL = 0 to L-1
      Let NC = NP/2
      For PBLOCK = 1 to 1<<LEVEL
         Let MIDP = PARENT + NC
         Let CHILD = MIDP + N
         fdcn( CHILD, 1, MIDP, MIDP, NC, RISE )
         fdcp( CHILD, 1, MIDP, MIDP, NC, RISE )
         dctiv( PARENT, NP )
         PARENT += NP
      Let NP = NC
   For PBLOCK = 1 to 1<<L
      dctiv( PARENT, NP )
      PARENT += NP
```

We must also allocate and assign the auxiliary arrays used for DCT and folding. For example, allocating and assigning the sampled rising cutoff array can be done with an initialization routine:

Allocate sampled rising cutoff function

```
initrcf( E ):
   Let RISE = makeinterval(NULL, -E, E-1 )
   rcfmidp( RISE )
   Return RISE
```

We assume that the angle and rising cutoff functions have already been defined as macros. We can deallocate the storage for this array later with `freeinterval()`.

Remark. The function `lcadf()` assumes that E is less than half the minimal interval length $n/2^L$. This should be tested with an `assert()` statement.

Multiple folding

We can use a different action region radius at each level:

Different rising cutoffs for each level

```
initrcfs( N, L ):
   Allocate an array of L INTERVALs at RS
   For I = 0 to L-1
      Let RS[I] = initrcf( (N/2)>>I )
   Return RS
```

The action radius function for different levels can be adjusted to obtain variants of the algorithm. Taking a constant $N/2^{L+1}$ will give us the fixed folding transform. If we wish to mix fixed folding and multiple folding we take constant radii to a certain depth and then let them decrease beyond that. But each radius can be no larger than half the corresponding interval width $N/2^K$.

We put the input array into the INTERVAL content of a BTN data structure, after folding one end against the other with suitable polarity:

Prepare input in a BTN node for local cosine analysis

```
initlcabtn( IN, N, RISE )
   Let ROOT = makebtn( NULL, NULL, NULL, NULL )
   Let ROOT.CONTENT = makeinterval( IN, 0, N-1 )
   fipc( ROOT.CONTENT.ORIGIN+N, ROOT.CONTENT.ORIGIN, RISE )
   Return ROOT
```

Subsequent nodes are allocated and assigned using a post-order recursive development of the binary tree:

Complete dyadic adapted local cosine analysis with multiple folding

```
lcadm( ROOT, S, L, RS ):
  Let LENGTH = 1 + ROOT.CONTENT.FINAL
  If S < L then
    Let NC = LENGTH/2
    Let MIDP = ROOT.CONTENT.ORIGIN + NC
    Let LCHILD = makeinterval( NULL, 0, NC-1 )
    Let ROOT.LEFT = makebtn( LCHILD, NULL, NULL, NULL )
    fdcn( LCHILD.ORIGIN+NC, 1, MIDP, MIDP, NC, RS[S] )
    lcadm( ROOT.LEFT, S+1, L, RS )
    Let RCHILD = makeinterval( NULL, 0, NC-1 )
    Let ROOT.RIGHT = makebtn( RCHILD, NULL, NULL, NULL )
    fdcp( RCHILD.ORIGIN, 1, MIDP, MIDP, NC, RS[S] )
    lcadm( ROOT.RIGHT, S+1, L, RS )
  dctiv( ROOT.CONTENT.ORIGIN, LENGTH )
  Return
```

We assume that each INTERVAL content of each BTN begins at initial index 0.

4.4.2 Extraction of coefficients

Most of the functions needed to extract the coefficients produced by a local trigonometric analysis have already been defined in Chapter 2.

Getting a hedge from a fixed folding analysis

Once we have filled an array with the local cosine transform coefficients for all dyadic window sizes, we can extract a particular basis subset by specifying the sizes of the intervals that make up the partition. Namely, we specify the levels array of a HEDGE data structure and use the utility function `abt2hedge()` to fill the coefficients portion of the structure.

Suppose that N is the length of the signal in the input array `IN[]`, and L is the deepest level of decomposition. The hedge must have a preallocated and assigned levels array, and a preallocated contents array long enough to hold pointers to the output blocks of coefficients. The return value is the start of the array binary tree holding the entire analysis, which we must have in order to deallocate the storage.

The following pseudocode shows an example of a complete dyadic fixed folding local cosine analysis, beginning with preparation of the periodic input by folding one end against the other, and followed by the extraction of a basis subset to a hedge:

Fixed folding dyadic adapted local cosine transform to a graph basis

```
lcadf2hedge( GRAPH, IN, N, L )
  Let RISE = initrcf( (N>>L)/2 )
  Allocate an array binary tree of length (1+L)*N at OUT
  fdcp(  OUT,  1, IN+N, IN, N/2, RISE )
  fdcn( OUT+N, 1, IN+N, IN, N/2, RISE )
  lcadf( OUT, N, L, RISE )
  abt2hedge( GRAPH, OUT, N )
  freeinterval( RISE )
  Return OUT
```

It is useful to verify that the input levels list is valid for the given analysis. The main thing to check is that the sum of the lengths of the intervals required by the levels list is N. In addition, we should insure that the maximum level in the list does not exceed the maximum depth of the analysis, and that no insane data such as a negative level is present. All of this may be done with a short loop and a few `assert()` statements.

Getting a hedge from a multiple folding analysis

Suppose we have tree of BTN data structures produced by a complete dyadic local cosine analysis with multiple folding on an N-point periodic signal. We can extract coefficients to fill a partially filled hedge as follows:

Multiple folding dyadic adapted local cosine transform to a graph basis

```
lcadm2hedge( GRAPH, IN, N, L )
   Let RS = initrcfs( N, L )
   Let ROOT = initlcabtn( IN, N, RS[0] )
   lcadm( ROOT, 0, L, RS )
   btnt2hedge( GRAPH, ROOT )
   For I = 0 to L-1
     freeinterval( RS[I] )
   Deallocate the array at RS
   Return ROOT
```

Again, N is the length of the original signal in the input array `IN[]`, while L is the desired number of levels of decomposition.

4.4.3 Adapted local cosine synthesis

The inverse local cosine transform from a single interval has already been described. Now we consider the problem of reconstructing a signal from an adapted local cosine analysis. We may call this an *adapted local cosine synthesis*, since it will involve the superposition of local cosine basis functions in a manner akin to additive synthesis in musical composition.

Let us suppose that we have a HEDGE data structure for a periodic signal of length $N \geq 2^L$, pointing into a joint input and output array `DATA[]` whose N coefficients will be replaced with N signal samples. The following function then reconstructs the periodic signal from its local cosine pieces:

Dyadic iLCT from a hedge, fixed midpoint folding

```
lcsdf( GRAPH, N, RISE ):
   Let DATA = GRAPH.CONTENTS[0]
   Let NSEG = N>>GRAPH.LEVELS[0]
   dctiv( DATA, NSEG )
   For BLOCK = 1 to GRAPH.BLOCKS-1
      Let SEG = GRAPH.CONTENTS[BLOCK]
      Let NSEG = N>>GRAPH.LEVELS[BLOCK]
      dctiv( SEG, NSEG )
      uipc( SEG, SEG, RISE )
   uipc( DATA+N, DATA, RISE )
```

Calling `uipc()` in the last statement unfolds the left end of the signal interval with the right end, undoing the preparation made by `lcadf()`. This assumes that the original signal was periodic, and avoids the question of what happens outside the reconstructed interval. It is not much more difficult to perform an adapted local cosine analysis on an "aperiodic" root interval that was smoothly extracted from a longer signal. The modifications to the algorithm chiefly involve some additional bookkeeping, and replacing the calls to `fdcn()` and `fdcp()` with simple copying.

4.5 Exercises

1. Prove that the function r_n defined by Equation 4.7 has $2^n - 1$ continuous derivatives at $+1$ and -1.

2. Prove that $UU^* = U^*U = I$ for the operators U and U^* defined by Equations 4.8 and 4.9

3. Prove that the operators P^0 and P_0, defined by Equations 4.30 and 4.29, are indeed orthogonal projections. Namely, verify that $P_0P_0 = P_0$, $P^0P^0 = P^0$, and that P^0 and P_0 are selfadjoint. This can also be done by showing that Equation 4.32 implies Equations 4.30 and 4.29.

4. Prove that, if $f = f(t)$ satisfies the symmetry property $f(t) = f(-t)$ for all $-1 \leq t \leq 1$, then with any rising cutoff function $r = r(t)$ satisfying Equation 4.1 we have
$$\int_{-\infty}^{\infty} |r(t)|^2 f(t)\, dt = \int_0^{\infty} f(t)\, dt.$$

5. Prove that the functions $c_k(n) \stackrel{\text{def}}{=} \cos \frac{\pi}{N}(k + \frac{1}{2})(n + \frac{1}{2})$, defined for each $k = 0, 1, \ldots, N-1$ on the integers $n = 0, 1, 2, \ldots, N-1$, form an orthogonal basis for $\mathbf{R}^N$. Note that the same result applies if cosine is replaced with sine.

6. Prove that the collection $\{\psi_{nk} : n \geq 0; k \in \mathbf{Z}\}$ defined in Equation 4.20 forms an orthonormal basis for $L^2(\mathbf{R})$.

7. Generate a few DCT-III local cosines and a few DCT-IV local cosines and plot them. Can you see a difference? Are the differences greatest at low frequencies or at high frequencies?

8. Consider the dyadic adapted local cosine transform on $[0, 1]$, descending three levels, using the rising cutoff $r_{[1]}$.

   (a) Using fixed folding, what is the maximum radius of the action regions?

   (b) Using multiple folding, what is the radius of the action region at each level?

   (c) Using the maximum radius in each case, what is the support of an adapted local cosine function from each level?

9. Write a pseudocode implementation of the local sine versions `lpis()` and `ilpis()`, respectively, of in place local periodization and its inverse.

10. Write a pseudocode implementation of the inverse of `lpica()`, namely inverse local periodization of successive adjacent intervals `ilpica()`. Do the same for `ilpisa()`, the inverse of `lpisa()`.

Chapter 5

Quadrature Filters

We shall use the term *quadrature filter* or just *filter* to denote an operator which convolves and then decimates. We will define filter operators both on sequences and on functions of one real variable. We can also project such actions onto periodic sequences and periodic functions, and define *periodized* filters.

In all cases where f is summable, the operation of convolution by f can be regarded as multiplication by the bounded, continuous Fourier transform of f. The Weierstrass approximation theorem assures us that we can uniformly approximate an arbitrary continuous one-periodic function by a trigonometric polynomial, in other words by the Fourier transform $\hat{f}$ of a finitely supported sequence f. Thus we can arrange for an operator F which involves only finitely many operations per output value to multiply "on the Fourier transform side" by a function that attenuates certain values (*i.e.*, multiplies them by zero or a small number) while it preserves or amplifies certain others (multiplies them by one or a large number). Since the Fourier transform of the input is a decomposition of the input into monochromatic waves $e^{2\pi i x \xi}$, the operator F modifies the frequency spectrum of the input to produce the output. Hence the name "filter" which is inherited from electrical engineering.

If the filter sequence is finitely supported, we have a *finite impulse response* or *FIR* filter; otherwise we have an *IIR* or *infinite impulse response* filter.

An individual quadrature filter is not generally invertible; it loses information during the decimation step. However, it is possible to find two complementary filters with each preserving the information lost by the other; then the pair can be combined into an invertible operator. Each of these complementary operators has an *adjoint* operator: when we use filters in pairs to decompose functions and sequences into pieces, it is the adjoint operators which put these pieces back together. The operation is reversible and restores the original signal if we have so-called *exact reconstruction* filters. The pieces will be orthogonal if we have *orthogonal filters* for which the decomposition gives a pair of orthogonal projections which we will define below. Such pairs must satisfy certain algebraic conditions which are completely derived in [36], pp.156–166.

One way to guarantee exact reconstruction is to have "mirror symmetry" of the Fourier transform of each filter about $\xi = \frac{1}{2}$; this leads to what Esteban and Galand [43] first called *quadrature mirror filters* or *QMFs*. Unfortunately, there are no orthogonal exact reconstruction FIR QMFs.

Mintzer [86], Smith and Barnwell [101], and Vetterli [107] found a different symmetry assumption which does allow orthogonal exact reconstruction FIR filters. Smith and Barnwell called these *conjugate quadrature filters* or *CQFs*.

By relaxing the orthogonality condition, Cohen, Daubechies, and Feauveau [19] obtained a large family of *biorthogonal* exact reconstruction filters. Such filters come in two pairs: the analyzing filters which split the signal into two pieces, and the synthesizing filters whose adjoints reassemble it. All of these can be FIRs, and the extra degrees of freedom are very useful to the filter designer.

5.1 Definitions and basic properties

A convolution-decimation operator has at least three incarnations, depending upon the domain of the functions upon which it is defined. We have three different formulas for functions of one real variable, for doubly infinite sequences, and for $2q$-periodic sequences. We will use the term quadrature filter or *QF* to refer to all three, since the domain will usually be obvious from the context.

5.1.1 Action on sequences

Here $u = u(n)$ for $n \in \mathbf{Z}$, the *aperiodic* case, or $n \in \mathbf{Z}/q\mathbf{Z} = \{0, 1, \ldots, q-1\}$, the *periodic* case. Convolution in this case is a sum.

Aperiodic filters

Suppose that $f = \{f(n) : n \in \mathbf{Z}\}$ is an absolutely summable sequence. We define a *convolution-decimation* operator F and its *adjoint* F^* to be operators acting on doubly infinite sequences, given respectively by the following formulas:

$$Fu(i) = \sum_{j=-\infty}^{\infty} f(2i-j)u(j) = \sum_{j=-\infty}^{\infty} f(j)u(2i-j), \qquad i \in \mathbf{Z}; \tag{5.1}$$

$$F^*u(j) = \sum_{i=-\infty}^{\infty} \bar{f}(2i-j)u(i) = \begin{cases} \sum_{i=-\infty}^{\infty} \bar{f}(2i)u(i+\frac{j}{2}), & j \in \mathbf{Z} \text{ even}, \\ \sum_{i=-\infty}^{\infty} \bar{f}(2i+1)u(i+\frac{j+1}{2}), & j \in \mathbf{Z} \text{ odd}. \end{cases} \tag{5.2}$$

Periodic filters

If f_{2q} is a $2q$-periodic sequence (*i.e.*, with even period), then it can be used to define a *periodic convolution-decimation* F_{2q} from $2q$-periodic to q-periodic sequences and its *periodic adjoint* F_{2q}^* from q-periodic to $2q$-periodic sequences. These are, respectively, the operators

$$F_{2q}u(i) = \sum_{j=0}^{2q-1} f_{2q}(2i-j)u(j) = \sum_{j=0}^{2q-1} f_{2q}(j)u(2i-j), \qquad 0 \leq i < q; \tag{5.3}$$

and

$$F_{2q}^*u(j) = \sum_{i=0}^{q-1} \bar{f}_{2q}(2i-j)u(i) \tag{5.4}$$

$$= \begin{cases} \sum_{i=0}^{q-1} \bar{f}_{2q}(2i)u(i+\frac{j}{2}), & \text{if } j \in [0, 2q-2] \text{ is even}, \\ \sum_{i=0}^{q-1} \bar{f}_{2q}(2i+1)u(i+\frac{j+1}{2}), & \text{if } j \in [1, 2q-1] \text{ is odd}. \end{cases} \tag{5.5}$$

Periodization commutes with convolution-decimation: we get the same periodic sequence whether we first convolve and decimate an infinite sequence and then periodize the result, or first periodize both the sequence and the filter and then perform a periodic convolution-decimation. The following proposition makes this precise:

Proposition 5.1 $\quad (Fu)_q = F_{2q}u_{2q}$ *and* $(F^*u)_{2q} = F_{2q}^*u_q$.

Proof: We have

$$\begin{aligned} (Fu)_q(i) &= \sum_{k=-\infty}^{\infty} Fu(i+qk) = \sum_{k=-\infty}^{\infty} \sum_{j=-\infty}^{\infty} f(2[i+qk]-j)u(j) \\ &= \sum_{j=-\infty}^{\infty} \left(\sum_{k=-\infty}^{\infty} f(2i+2qk-j) \right) u(j) = \sum_{j=-\infty}^{\infty} f_{2q}(2i-j)u(j) \end{aligned}$$

$$
\begin{aligned}
&= \sum_{j=0}^{2q-1} \sum_{k=-\infty}^{\infty} f_{2q}(2i-j-2qk)u(j+2qk) \\
&= \sum_{j=0}^{2q-1} f_{2q}(2i-j) \sum_{k=-\infty}^{\infty} u(j+2qk) = \sum_{j=0}^{2q-1} f_{2q}(2i-j)u_{2q}(j).
\end{aligned}
$$

Also,

$$
\begin{aligned}
(F^*u)_{2q}(j) &= \sum_{k=-\infty}^{\infty} F^*u(j+2qk) = \sum_{k=-\infty}^{\infty} \sum_{i=-\infty}^{\infty} \bar{f}(2i-[j+2qk])u(i) \\
&= \sum_{i=-\infty}^{\infty} \left(\sum_{k=-\infty}^{\infty} \bar{f}(2i-j-2qk) \right) u(i) = \sum_{i=-\infty}^{\infty} \bar{f}_{2q}(2i-j)u(i) \\
&= \sum_{i=0}^{q-1} \sum_{k=-\infty}^{\infty} \bar{f}_{2q}(2i+2qk-j)u(i+qk) \\
&= \sum_{i=0}^{q-1} \bar{f}_{2q}(2i-j) \sum_{k=-\infty}^{\infty} u(i+qk) = \sum_{i=0}^{q-1} \bar{f}_{2q}(2i-j)u_q(i).
\end{aligned}
$$

□

5.1.2 Biorthogonal QFs

A quadruplet H, H', G, G' of convolution-decimation operators or filters is said to form a set of *biorthogonal quadrature filters* or *BQFs* if the filters satisfy the following conditions:

Duality: $H'H^* = G'G^* = I = HH'^* = GG'^*$;

Independence: $G'H^* = H'G^* = 0 = GH'^* = HG'^*$;

Exact reconstruction: $H^*H' + G^*G' = I = H'^*H + G'^*G$;

Normalization: $H\mathbf{1} = H'\mathbf{1} = \sqrt{2}\,\mathbf{1}$ and $G\mathbf{1} = G'\mathbf{1} = \mathbf{0}$, where $\mathbf{1} = \{\ldots,1,1,1,\ldots\}$ is all ones and $\mathbf{0} = \{\ldots,0,0,0,\ldots\}$ is all zeroes.

The first two conditions may be expressed in terms of the filter sequences h, h', g, g' which respectively define H, H', G, G':

$$
\begin{aligned}
\sum_k h'(k)\bar{h}(k+2n) &= \delta(n) = \sum_k g'(k)\bar{g}(k+2n); \\
\sum_k g'(k)\bar{h}(k+2n) &= 0 = \sum_k h'(k)\bar{g}(k+2n).
\end{aligned}
\tag{5.6}
$$

The normalization condition allows us to say that H and H' are the *low-pass* filters while G and G' are the *high-pass* filters. It may be restated as

$$\begin{aligned} \sum_k h(k) &= \sqrt{2}; & \sum_k g(2k) &= -\sum_k g(2k{+}1); \\ \sum_k h'(k) &= \sqrt{2}; & \sum_k g'(2k) &= -\sum_k g'(2k{+}1). \end{aligned} \tag{5.7}$$

Having four operators provides plenty of freedom to construct filters with special properties, but there is also a regular method for constructing the G, G' filters from H, H'. If we have two sequences $\{h(k)\}$ and $\{h'(k)\}$ which satisfy Equation 5.6, then we can obtain two *conjugate quadrature filter sequences* $\{g(k)\}$ and $\{g'(k)\}$ via the formulas below, using any integer M:

$$g(k) = (-1)^k \bar{h}'(2M+1-k); \qquad g'(k) = (-1)^k \bar{h}(2M+1-k). \tag{5.8}$$

We also have the following result, which is related to Lemma 12 in [52] and a similar result in [65]:

Lemma 5.2 *The biorthogonal QF conditions imply* $H^*\mathbf{1} = H'^*\mathbf{1} = \frac{1}{\sqrt{2}}\mathbf{1}$.

Proof: With exact reconstruction, $\mathbf{1} = \left(H'^*H + G'^*G\right)\mathbf{1} = \sqrt{2}\,H'^*\mathbf{1}$, since $H\mathbf{1} = \sqrt{2}\,\mathbf{1}$ and $G\mathbf{1} = \mathbf{0}$. Likewise, $\mathbf{1} = (H^*H' + G^*G')\,\mathbf{1} = \sqrt{2}\,H^*\mathbf{1}$, since $H'\mathbf{1} = \sqrt{2}\,\mathbf{1}$ and $G'\mathbf{1} = \mathbf{0}$. □

Remark. The conclusion of Lemma 5.2 may be rewritten as follows:

$$\sum_k h(2k) = \sum_k h(2k+1) = \frac{1}{\sqrt{2}} = \sum_k h'(2k) = \sum_k h'(2k+1). \tag{5.9}$$

If we have the duality, independence, and exact reconstruction conditions, together with $H\mathbf{1} = H'\mathbf{1} = \sqrt{2}\,\mathbf{1}$ but no normalization on G or G', then at least one of the following must be true:

$$G'\mathbf{1} = \mathbf{0} \text{ and } H^*\mathbf{1} = \frac{1}{\sqrt{2}}\mathbf{1}, \quad \text{or} \quad G\mathbf{1} = \mathbf{0} \text{ and } H'^*\mathbf{1} = \frac{1}{\sqrt{2}}\mathbf{1}.$$

However, the BQF conditions as stated insure that the pairs H, G and H', G' are interchangeable in our analyses.

If H, H', G, G' is a set of biorthogonal QFs, and ρ is any nonzero constant, then $H, H', \bar{\rho}G, \rho^{-1}G'$ is another biorthogonal set. We can use this to normalize the G

and G' filters so that

$$\sum_k g(2k) = -\sum_k g(2k+1) = \frac{1}{\sqrt{2}} = \sum_k g'(2k) = -\sum_k g'(2k+1). \qquad (5.10)$$

This will be called the *conventional normalization* for the high-pass filters.

Since $H^*H'H^*H' = H^*H'$ and $G^*G'G^*G' = G^*G'$, the combinations H^*H' and G^*G' are projections although they will not in general be orthogonal projections. That is because they need not be equal to their adjoint projections H'^*H and G'^*G.

An argument similar to the one in Proposition 5.1 shows that periodization of biorthogonal QFs to an even period $2q$ preserves the biorthogonality conditions. Writing h_{2q}, h'_{2q}, g_{2q}, and g'_{2q} for the $2q$-periodizations of h, h', g, and g', respectively, we have

$$\begin{aligned} \sum_k h'_{2q}(k)\bar{h}_{2q}(k+2n) &= \delta(n \bmod q) = \sum_k g'_{2q}(k)\bar{g}_{2q}(k+2n); \\ \sum_k g'_{2q}(k)\bar{h}_{2q}(k+2n) &= 0 = \sum_k h'_{2q}(k)\bar{g}_{2q}(k+2n). \end{aligned} \qquad (5.11)$$

Here we define the periodized Kronecker delta as follows:

$$\delta(n \bmod q) \stackrel{\text{def}}{=} \sum_{k=-\infty}^{\infty} \delta(n+qk) = \begin{cases} 1, & \text{if } n \equiv 0 \pmod q, \\ 0, & \text{otherwise.} \end{cases} \qquad (5.12)$$

Periodization to an even period also preserves the sums over the even and odd indices, and thus Lemma 5.2 remains true if we replace h, h', g, and g' with h_{2q}, h'_{2q}, g_{2q}, and g'_{2q}.

5.1.3 Orthogonal QFs

If $H = H'$ and $G = G'$ in a biorthogonal set of QFs, then the pair H, G is called an *orthogonal quadrature filter* pair. In that case the following conditions hold:

Self-duality: $HH^* = GG^* = I$;

Independence: $GH^* = HG^* = 0$;

Exact reconstruction: $H^*H + G^*G = I$;

Normalization: $H\mathbf{1} = \sqrt{2}\,\mathbf{1}$, where $\mathbf{1} = \{\ldots,1,1,1,\ldots\}$.

We will use the abbreviation *OQF* to refer to one or both elements of such a pair. In this normalization, H is the low-pass filter while G is the high-pass filter.

If H and G are formed respectively from the sequences h and g, the duality and independence conditions satisfied by an OQF pair are equivalent to the following equations:

$$\begin{aligned} \sum_k h(k)\bar{h}(k+2n) &= \delta(n) = \sum_k g(k)\bar{g}(k+2n); \\ \sum_k g(k)\bar{h}(k+2n) &= 0 = \sum_k h(k)\bar{g}(k+2n). \end{aligned} \tag{5.13}$$

For orthogonal QFs, we have a stronger result than Lemma 5.2:

Lemma 5.3 *The orthogonal QF conditions imply that* $G\mathbf{1} = 0$, $H^*\mathbf{1} = \frac{1}{\sqrt{2}}\mathbf{1}$ *and* $|G^*\mathbf{1}| = \frac{1}{\sqrt{2}}\mathbf{1}$.

Proof: These conditions may be rewritten as follows:

$$\sum_k g(2k) = -\sum_k g(2k+1); \tag{5.14}$$

$$\sum_k h(2k) = \sum_k h(2k+1) = \frac{1}{\sqrt{2}}; \tag{5.15}$$

$$\left|\sum_k g(2k)\right| = \left|\sum_k g(2k+1)\right| = \frac{1}{\sqrt{2}}. \tag{5.16}$$

Using $H\mathbf{1} = \sqrt{2}\,\mathbf{1}$ we calculate as follows:

$$2 = \left|\sum_k h(k)\right|^2 = \sum_j \sum_k h(k)\bar{h}(k+2j) + \sum_j \sum_k h(k)\bar{h}(k+2j+1).$$

The first sum is 1, and the second sum may itself be decomposed into even and odd parts to yield

$$1 = 2\Re\left(\sum_k h(2k)\right)\left(\sum_j \bar{h}(2j+1)\right). \tag{5.17}$$

Putting $X = \sum_k h(2k)$ and $Y = \sum_k h(2k+1)$, we observe that $|X-Y|^2 = |X+Y|^2 - 4\Re X\bar{Y} = 0$, so that $X = Y = \frac{1}{\sqrt{2}}$ or equivalently $H^*\mathbf{1} = \frac{1}{\sqrt{2}}\mathbf{1}$. Then by independence we have $0 = GH^*\mathbf{1} = \frac{1}{\sqrt{2}}G\mathbf{1}$.

Writing $U = \sum_k g(2k)$ and $V = \sum_k g(2k+1)$, we have $|U+V|^2 = 0$ from $G\mathbf{1} = 0$. Self-duality for G implies $|U-V|^2 = 1$; solving gives $|U| = |V| = \frac{1}{\sqrt{2}}$. □

If H, G are a pair of orthogonal QFs and ρ is any constant with $|\rho| = 1$, then $H, \rho G$ are also orthogonal QFs. Hence by taking $\rho = \sqrt{2}\sum_k \bar{g}(2k)$ we can arrange that

$$\sum_k g(2k) = -\sum_k g(2k+1) = \frac{1}{\sqrt{2}}. \tag{5.18}$$

As in Equation 5.10, this will be called the *conventional normalization* of an orthogonal high-pass filter.

Given h satisfying Equation 5.13, we can generate a *conjugate* g to satisfy the rest of the orthogonal QF conditions by choosing its coefficients as follows [36], using any integer M:

$$g(n) = (-1)^n \bar{h}(2M+1-n), \qquad n \in \mathbf{Z}. \tag{5.19}$$

Notice that this sequence g is conventionally normalized.

Proposition 5.1 shows that periodization of an orthogonal QF pair to an even period $2q$ preserves the orthogonality conditions, and also preserves the sums over the even and odd indices, and thus Lemma 5.3 remains true if we replace h and g with h_{2q} and g_{2q}.

Self-duality gives $H^*HH^*H = H^*H$ and $G^*GG^*G = G^*G$. Notice that H^*H and G^*G are selfadjoint, so H^*H and G^*G are orthogonal projections.

5.1.4 Action on functions

We can also define the action of a filter sequence f on a function of one real variable. A *convolution-decimation* F and its *adjoint* F^*, acting as operators on functions of one real variable, are given respectively by the following formulas:

$$Fu(t) = \sqrt{2}\sum_{j\in\mathbf{Z}} f(j)u(2t-j), \qquad t \in \mathbf{R}, \tag{5.20}$$

$$F^*u(t) = \frac{1}{\sqrt{2}}\sum_{k\in\mathbf{Z}} \bar{f}(k)u\left(\frac{t+k}{2}\right), \qquad t \in \mathbf{R}. \tag{5.21}$$

Here $u \in L^2(\mathbf{R})$ and $t \in \mathbf{R}$. We are abusing notation in the interest of simplicity. The normalizing factors $\sqrt{2}$ and $1/\sqrt{2}$ appear because we are using dilation by 2 rather than decimation by 2 so as to preserve $\|\cdot\|$ for functions on the line.

We must take care, though, since the action on L^2 does not satisfy the orthogonal QF conditions even if the filter satisfies Equation 5.13. We will see why using Corollary 5.12 below. However, we still have the following lemma:

Lemma 5.4 *If h, g, h', and g' are a biorthogonal set of conjugate QFs related by Equation 5.8, and H, G, H', and G' are the associated operators on $L^2(\mathbf{R})$, then*

$$H'H^* + G'G^* = 2I; \qquad H'G^* + G'H^* = 0;$$
$$H^*H' + G^*G' = 2I; \qquad H^*G' + G^*H' = 0.$$

Proof: We begin by calculating the action of $H'H^*$:

$$H'H^*u(t) = \sum_{j,k\in\mathbf{Z}} h'(j)\bar{h}(k)u\left(\frac{2t-j+k}{2}\right) = \sum_{j,k\in\mathbf{Z}} h'(j)\bar{h}(k+j)u\left(t+\frac{k}{2}\right)$$

$$= \sum_{k\in\mathbf{Z}}\left(\sum_{j\in\mathbf{Z}} h'(j)\bar{h}(2k+j)\right)u(t+k) + \sum_{k\in\mathbf{Z}}\left(\sum_{j\in\mathbf{Z}} h'(j)\bar{h}(2k+j+1)\right)u(t+k+\frac{1}{2})$$

$$= u(t) + \sum_{k\in\mathbf{Z}} c_{h'h}(k)u(t+k+\frac{1}{2}).$$

Here $c_{h'h}(k) = \sum_{j\in\mathbf{Z}} h'(j)\bar{h}(2k+j+1)$. Likewise,

$$\begin{aligned}
G'G^*u(t) &= u(t) + \sum_{k\in\mathbf{Z}} c_{g'g}(k)u(t+k+\frac{1}{2});\\
H^*H'u(t) &= u(t) + \sum_{k\in\mathbf{Z}} c_{h'h}(k)u(t+2k+1);\\
G^*G'u(t) &= u(t) + \sum_{k\in\mathbf{Z}} c_{g'g}(k)u(t+2k+1).
\end{aligned}$$

Now Equation 5.8 allows us to compute $c_{g'g}$ in terms of $c_{h'h}$:

$$\begin{aligned}
c_{g'g}(k) &= \sum_{j\in\mathbf{Z}} g'(j)\bar{g}(j+2k+1)\\
&= \sum_{j\in\mathbf{Z}}(-1)^j\bar{h}(2M+1-j)(-1)^{j+2k+1}h'(2M-j-2k)\\
&= -\sum_{j\in\mathbf{Z}} \bar{h}(2M+1-j)h'(2M-j-2k)\\
&= -\sum_{j'\in\mathbf{Z}} \bar{h}(j'+2k+1)h'(j') = -c_{h'h}(k).
\end{aligned}$$

In the last step we substituted $j = 2M - 2k - j'$ and summed over j'. But then $c_{h'h}(k) + c_{g'g}(k) = 0$ for all k and thus the "twice identity" equations are proved.

For the "vanishing" equations, we note that

$$H'G^*u(t) = \sum_{k\in\mathbf{Z}} c_{h'g}(k)u(t+k+\frac{1}{2}); \qquad G'H^*u(t) = \sum_{k\in\mathbf{Z}} c_{g'h}(k)u(t+k+\frac{1}{2});$$

$$H^*G'u(t) = \sum_{k\in\mathbf{Z}} c_{g'h}(k)u(t+2k+1); \qquad G^*H'u(t) = \sum_{k\in\mathbf{Z}} c_{h'g}(k)u(t+2k+1).$$

Combining these with the identities $c_{h'g} + c_{g'h} \equiv 0$ and $c_{hg'} + c_{gh'} \equiv 0$, which again follow from Equation 5.8, finishes the proof. □

Remark. The sequences c_{gh} and c_{hg} are related by $c_{gh}(k) = \bar{c}_{hg}(-1-k)$. Thus, if H and G are orthogonal conjugate QFs, then $c_{hh}(k) = \bar{c}_{hh}(-1-k)$ and $c_{gg}(k) = \bar{c}_{gg}(-1-k)$; these sequences are Hermitean symmetric about the point $-1/2$.

In the function case $H'H^*$ and $G'G^*$ do not in general equal the identity operator I, even for an orthogonal pair of QFs. Furthermore, $H'G^*$ and $G'H^*$ do not vanish in general (neither do G^*H' and H^*G', for that matter), as we see from the following example. Let H, G be the Haar–Walsh filters; then

$$c_{hh}(k) = \begin{cases} \frac{1}{2}, & \text{if } k = -1, 0; \\ 0, & \text{otherwise;} \end{cases} \qquad c_{gg}(k) = \begin{cases} -\frac{1}{2}, & \text{if } k = -1, 0; \\ 0, & \text{otherwise;} \end{cases}$$

$$c_{hg}(k) \;=\; -c_{gh}(k) \;=\; \begin{cases} \frac{1}{2}, & \text{if } k = -1; \\ -\frac{1}{2}, & \text{if } k = 0; \\ 0, & \text{otherwise.} \end{cases}$$

Thus $HH^*u(t) = \frac{1}{2}u(t-\frac{1}{2}) + u(t) + \frac{1}{2}u(t+\frac{1}{2})$, while $GG^*u(t) = -\frac{1}{2}u(t-\frac{1}{2}) + u(t) - \frac{1}{2}u(t+\frac{1}{2})$; neither of these is the identity operator. Likewise, $HG^*u(t) = \frac{1}{2}u(t-\frac{1}{2}) - \frac{1}{2}u(t+\frac{1}{2})$ and $GH^*u(t) = -\frac{1}{2}u(t-\frac{1}{2}) + \frac{1}{2}u(t+\frac{1}{2})$, neither of which vanishes in general. The remaining combinations are left as an exercise.

Effect on the support

Suppose that F is an FIR quadrature filter whose filter sequence f is supported on the index interval $[a, b]$; *i.e.*, $f = \{\ldots, 0, f(a), f(a+1), \ldots, f(b-1), f(b), 0, \ldots\}$. Because of the dilation $t \mapsto 2t$, F shrinks the support of functions. This has the beneficial side-effect of concentrating the energy of a square-integrable function into the support interval of the filter. We make these notions precise with two lemmas:

Lemma 5.5 *Suppose that F is an FIR QF supported on the index interval $[a, b]$. Then* $\operatorname{supp} u \subset [c, d]$ *implies*

$$\operatorname{supp} Fu \;\subset\; \left[\frac{a+c}{2}, \frac{b+d}{2}\right], \tag{5.22}$$

$$\operatorname{supp} F^* u \quad \subset \quad [2c - b, 2d - a] . \tag{5.23}$$

Proof: For F, note that $t \in \operatorname{supp} Fu$ implies $2t - j \in [c, d]$ for some $j = a, \ldots, b$. Thus $2t - a \geq c$ and $2t - b \leq d$. F^* is left as an easy exercise. □

Note that after N applications,

$$\operatorname{supp} F^N u \subset \left[2^{-N} c + (1 - 2^{-N}) a, 2^{-N} d + (1 - 2^{-N}) b\right] . \tag{5.24}$$

This settles down to $[a, b]$ as $N \to \infty$.

If u is square-integrable, then an FIR filter F supported on $[a, b]$ reduces the energy of u lying outside that interval. To prove this, let $\epsilon > 0$ be given and suppose that we choose M sufficiently large so that

$$\int_{-\infty}^{-M} |u(t)|^2 \, dt + \int_{M}^{\infty} |u(t)|^2 \, dt < \epsilon \|u\|^2.$$

Write $u_0 = u \mathbf{1}_{[-M,M]}$ and $u_\infty = u - u_0$. Since u_0 is supported in the compact interval $[-M, M]$, the previous lemma implies that for sufficiently large N we have $\operatorname{supp} F^N u_0 \subset [a - \epsilon, b + \epsilon]$. The energy remaining outside $[a, b]$ for all N is bounded by $\sup\{\|F^N u_\infty\|^2 : N \geq 0\}$, but since F has unit operator norm we can estimate $\|F^N u_\infty\|^2 \leq \|u_\infty\|^2 = \epsilon \|u\|^2$ for all $N \geq 0$. We have proved the following lemma:

Lemma 5.6 *Suppose that F is an FIR QF supported on the index interval $[a, b]$. If u is square-integrable, then for every $\epsilon > 0$ we can decompose u into two pieces $u = u_0 + u_\infty$ such that for each sufficiently large N we have $\operatorname{supp} F^N u_0 \subset [a - \epsilon, b + \epsilon]$ and $\|F^N u_\infty\|^2 < \epsilon \|u\|^2$.* □

5.2 Phase response

We wish to recognize features of the original signal from the coefficients produced by transformations involving QFs, so it is necessary to keep track of which portion of the sequence contributes energy to the filtered sequence.

Suppose that F is a finitely supported filter with filter sequence $f(n)$. For any sequence $u \in \ell^2$, if $Fu(n)$ is large at some index $n \in \mathbf{Z}$, then we can conclude that $u(k)$ is large near the index $k = 2n$. Likewise, if $F^* u(n)$ is large, then there must be significant energy in $u(k)$ near $k = n/2$. We can quantify this assertion of nearness using the support of f, or more generally by computing the position of f and its uncertainty computed with Equations 1.57 and 1.60. When the support of f is large, the position method gives a more precise notion of where the analyzed function is concentrated.

Consider what happens when $f(n)$ is concentrated near $n = 2T$:

$$Fu(n) = \sum_{j\in\mathbf{Z}} f(j)u(2n-j) = \sum_{j\in\mathbf{Z}} f(j+2T))u(2n-j-2T). \tag{5.25}$$

Since $f(j+2T)$ is concentrated about $j = 0$, we can conclude by our previous reasoning that if $Fu(n)$ is large, then $u(k)$ is large when $k \approx 2n - 2T$. Similarly,

$$F^*u(n) = \sum_{j\in\mathbf{Z}} \bar{f}(2j-n)u(j) = \sum_{j\in\mathbf{Z}} \bar{f}(2j-n+2T))u(j+T). \tag{5.26}$$

Since $\bar{f}(2j-n+2T)$ is concentrated about $2j - n = 0$, we conclude that if $F^*u(n)$ is big then $u(j+T)$ must be big where $j \approx n/2$, which implies that $u(k)$ is big when $k \approx \frac{n}{2} + T$.

Decimation by 2 and its adjoint respectively cause the doubling and halving of the indices n to get the locations where u must be large. The translation by T or $-2T$ can be considered a "shift" induced by the filter convolution. We can precisely quantify the location of portions of a signal, measure the shift, and correct for it when interpreting the coefficients produced by applications of F and F^*. We will see that nonsymmetric filters might shift different signals by different amounts, with a variation that can be estimated by a simple expression in the filter coefficients. The details of the shift will be called the *phase response* of the filter.

5.2.1 Shifts for sequences

The notion of position for a sequence is the same as the one for functions defined in Equation 1.60, only using sums instead of integrals:

$$c[u] \stackrel{\text{def}}{=} \frac{1}{\|u\|^2} \sum_{k\in\mathbf{Z}} k|u(k)|^2. \tag{5.27}$$

This quantity, whenever it is finite, may also be called the *center of energy* of the sequence $u \in \ell^2$ to distinguish it from the function case.

The center of energy is the first moment of the probability distribution function (or *pdf*) defined by $|u(n)|^2/\|u\|^2$. We will say that the sequence u is *well-localized* if the second moment of that pdf also exists, namely if

$$\sum_{k\in\mathbf{Z}} k^2|u(k)|^2 = \|ku\|^2 < \infty. \tag{5.28}$$

A finite second moment insures that the first moment is also finite, by the Cauchy–Schwarz inequality (Equation 1.9):

$$\sum_{k\in\mathbf{Z}} k|u(k)|^2 = \langle ku, u\rangle \le \|ku\|\,\|u\| < \infty.$$

If $u \in \ell^2$ is a finitely supported sequence (say in the interval $[a,b]$) then $a \le c[u] \le b$.

Another way of writing $c[u]$ is in Dirac's *bra and ket* notation:

$$\|u\|^2 c[u] = \langle u|X|u\rangle \stackrel{\text{def}}{=} \langle u, Xu\rangle = \sum_{i\in\mathbf{Z}} \bar{u}(i)X(i,j)u(j), \tag{5.29}$$

where

$$\begin{aligned} X(i,j) &\stackrel{\text{def}}{=} i\delta(i-j) = \begin{cases} i, & \text{if } i=j,\\ 0, & \text{if } i\ne j,\end{cases} \\ &= \operatorname{diag}[\ldots,-2,-1,0,1,2,3,\ldots]. \end{aligned}$$

To simplify the formulas, we will always suppose that $\|u\|=1$. We can also suppose that f is an orthogonal QF, so $\sum_k \bar{f}(k)f(k+2j) = \delta(j)$. Then $FF^* = I$, F^* is an isometry and F^*F is an orthogonal projection. Since $\|F^*u\| = \|u\| = 1$, we can compute the center of energy of F^*u as $c[F^*u] = \langle F^*u|X|F^*u\rangle = \langle u|FXF^*|u\rangle$. We will call the the double sequence FXF^* between the bra and the ket the *phase response* of the adjoint convolution-decimation operator F^* defined by the filter sequence f. Namely,

$$FXF^*(i,j) = \sum_k kf(2i-k)\bar{f}(2j-k). \tag{5.30}$$

Now

$$FXF^*(i,j) = \sum_k ([i{+}j]+k)f([i{-}j]-k)\bar{f}([j{-}i]-k) \stackrel{\text{def}}{=} 2X(i,j) - C_f(i,j).$$

Here $2X(i,j) = (i+j)\sum_k f([i{-}j]-k)\bar{f}([j{-}i]-k) = 2i\delta(i-j)$ as above, since f is an orthogonal QF, while

$$C_f(i,j) \stackrel{\text{def}}{=} \sum_k kf(k-[i{-}j])\bar{f}(k-[j{-}i]). \tag{5.31}$$

Thus $c[F^*u] = 2c[u] - \langle u|C_f|u\rangle$. C_f is evidently a convolution matrix: $C_f(i,j) = \gamma(i-j)$ so that $C_f u = \gamma * u$. The function γ is defined by the following formula:

$$\gamma(n) \stackrel{\text{def}}{=} \sum_k kf(k-n)\bar{f}(k+n). \tag{5.32}$$

From this formula it is easy to see that $\gamma(n) = \bar{\gamma}(-n)$, thus $\hat{\gamma}(\xi) = \hat{\bar{\gamma}}(-\xi) = \overline{\hat{\gamma}}(\xi) \Rightarrow \hat{\gamma} \in \mathbf{R}$. This symmetry of γ makes the matrix C_f selfadjoint. Along its main diagonal, $C_f(i,i) = \gamma(0) = c[f]$. Other diagonals of C_f are constant, and if f is supported in the finite interval $[a,b]$, then $C_f(i,j) = \gamma(i-j) = 0$ for $|i-j| > |b-a|$.

We can subtract the diagonal from C_f by writing $C_f = C_f^0 + c[f]I$, which is the same as the decomposition $\gamma(n) = \gamma^0(n) + c[f]\delta(n)$. This gives a decomposition of the phase response matrix:

$$FXF^* = 2X - c[f]I - C_f^0.$$

Thus FXF^* is multiplication by the linear function $2x - c[f]$ minus convolution with γ^0. We will say that f has a *linear phase response* if $\gamma^0 \equiv 0$.

Proposition 5.7 *Suppose that $f = \{f(n) : n \in \mathbf{Z}\}$ satisfies $\sum_k \bar{f}(k-n)f(k+n) = \delta(n)$ for $n \in \mathbf{Z}$. If f is Hermitean symmetric or antisymmetric about some integer or half integer T, then the phase response of f is linear.*

Proof: We have $f(n) = \pm\bar{f}(2T-n)$ for all $n \in \mathbf{Z}$, taking $+$ in the symmetric case and $-$ in the antisymmetric case. Now $\gamma^0(0) = 0$ for all filters. For $n \neq 0$ we have

$$\begin{aligned}\gamma^0(n) &= \sum_k kf(k-n)\bar{f}(k+n) = \sum_k k\bar{f}(2T-k+n)f(2T-k-n)\\ &= 2T\sum_k \bar{f}(k+n)f(k-n) - \sum_k k\bar{f}(k+n)f(k-n) = \quad 0 - \gamma^0(n).\end{aligned}$$

Thus we have $\gamma^0(n) = 0$ for all $n \in \mathbf{Z}$. □

The linear function shifts the center of energy x to $2x - c[f]$, and the convolution operator γ^0 perturbs this by a "deviation" $\langle u, \gamma^0 * u\rangle/\|u\|^2$. We can denote the

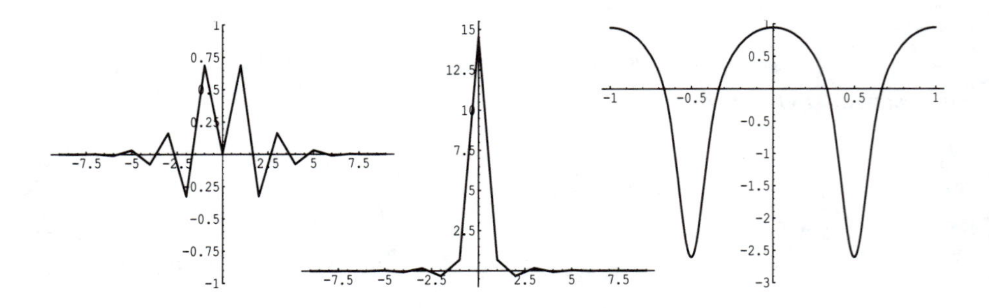

Figure 5.1: γ^0, γ, and $\hat{\gamma}^0$ for "Beylkin 18" high-pass OQF.

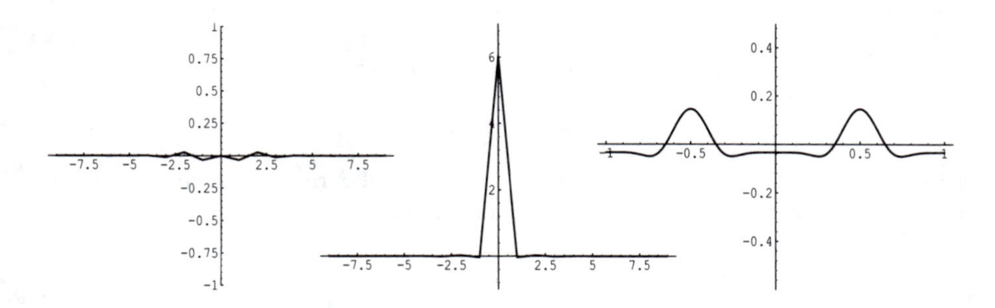

Figure 5.2: γ^0, γ, and $\hat{\gamma}^0$ for "Coiflet 18" low-pass OQF.

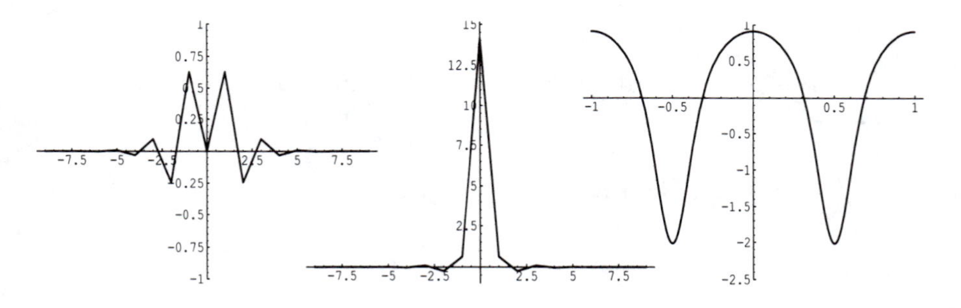

Figure 5.3: γ^0, γ, and $\hat{\gamma}^0$ for "Daubechies 18" high-pass OQF.

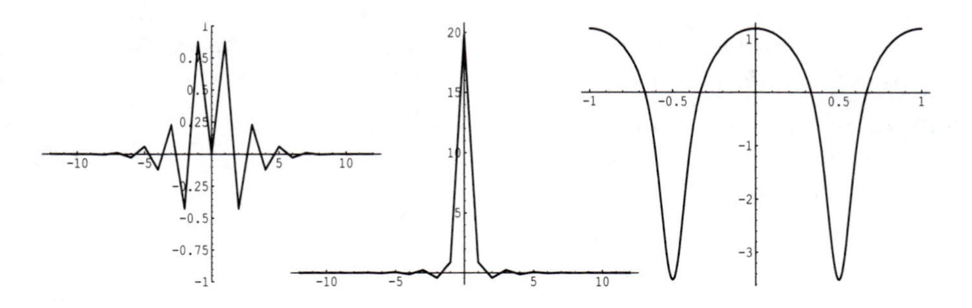

Figure 5.4: γ^0, γ, and $\hat{\gamma}^0$ for "Vaidyanathan 24" low-pass OQF.

maximum value of this perturbation by $d[f]$. By Plancherel's theorem and the convolution theorem, the deviation is $\langle \hat{u}, \hat{\gamma}^0 \hat{u} \rangle / \|u\|^2$ and its maximum value is given (using Proposition 1.19) by the maximum absolute value of $\hat{\gamma}^0(\xi)$:

$$d[f] = \sup\{|\hat{\gamma}^0(\xi)| : \xi \in [0,1]\}. \tag{5.33}$$

Now $\gamma^0(n) = \bar{\gamma}^0(-n)$ is symmetric just like γ, so its Fourier transform $\hat{\gamma}^0$ is purely real and can be computed using only cosines as follows:

$$\hat{\gamma}^0(\xi) = 2\sum_{n=1}^{\infty} \gamma(n) \cos 2\pi n \xi. \tag{5.34}$$

The critical points of $\hat{\gamma}^0$ are found by differentiating Equation 5.34:

$$\hat{\gamma}_0'(\xi) = -4\pi \sum_{n=1}^{\infty} n\gamma(n) \sin 2\pi n \xi. \tag{5.35}$$

It is evident that $\xi = 0$ and $\xi = \frac{1}{2}$ are critical points. For the QFs listed in the appendix, we can show that $|\hat{\gamma}^0(\xi)|$ achieves its maximum at $\xi = \frac{1}{2}$, where

$$\hat{\gamma}^0\left(\frac{1}{2}\right) = 2\sum_{n=1}^{\infty}(-1)^n \gamma(n) = 2\sum_{k=-\infty}^{\infty}\sum_{n=1}^{\infty}(-1)^n k f(k-n)\bar{f}(k+n). \tag{5.36}$$

Graphs of $\hat{\gamma}^0$ for some examples of orthogonal OQFs can be seen in Figures 5.1 through 5.4.

Values of the quantities $c[f]$ and $d[f]$ for example OQFs are listed in Table 5.1. Notice that if $g(n) = (-1)^n \bar{h}(2M+1-n)$, so that h and g are a conjugate pair of filters, and $|\operatorname{supp} g| = |\operatorname{supp} h| = 2M$ is the length of the filters, then $d[g] = d[h]$ and $c[g] + c[h] = 2M - 1$. This also implies that $C_h(i,j) = -C_g(i,j)$, so that the function $\hat{\gamma}^0$ corresponding to the filter h is just the negative of the one corresponding to g.

We can put the preceding formulas together into a single theorem:

Theorem 5.8 (OQF Phase Shifts) *Suppose that $u \in \ell^2$ and that $F : \ell^2 \to \ell^2$ is convolution and decimation by two with an orthogonal QF $f \in \ell^1$. Suppose that $c[u]$ and $c[f]$ both exist. Then*

$$c[F^* u] = 2c[u] - c[f] - \langle u, \gamma^0 * u \rangle / \|u\|^2,$$

where $\gamma^0 \in \ell^2$ is the sequence

$$\gamma^0(n) = \begin{cases} 0, & \text{if } n = 0, \\ \sum_k k f(k-n)\bar{f}(k+n), & \text{if } n \neq 0. \end{cases}$$

f	$\lvert\text{supp}\, f\rvert$	H or G	$c[f]$	$d[f]$
B	18	H	2.4439712920	2.6048841893
		G	14.5560287079	2.6048841893
C	6	H	3.6160691415	0.4990076823
		G	1.3839308584	0.4990076823
	12	H	4.0342243997	0.0868935216
		G	6.9657756002	0.0868935217
	18	H	6.0336041704	0.1453284669
		G	10.9663958295	0.1453284670
	24	H	8.0333521640	0.1953517707
		G	14.9666478359	0.1953517692
	30	H	10.0333426139	0.2400335062
		G	18.9666573864	0.2400330874
D	2	H	0.5000000000	0.0000000000
		G	0.5000000000	0.0000000000
	4	H	0.8504809471	0.2165063509
		G	2.1495190528	0.2165063509
	6	H	1.1641377716	0.4604317871
		G	3.8358622283	0.4604317871
	8	H	1.4613339067	0.7136488576
		G	5.5386660932	0.7136488576
	10	H	1.7491114972	0.9711171403
		G	7.2508885027	0.9711171403
	12	H	2.0307505738	1.2308332718
		G	8.9692494261	1.2308332718
	14	H	2.3080529576	1.4918354676
		G	10.6919470423	1.4918354676
	16	H	2.5821186257	1.7536045071
		G	12.4178813742	1.7536045071
	18	H	2.8536703515	2.0158368941
		G	14.1463296483	2.0158368941
	20	H	3.1232095535	2.2783448731
		G	15.8767904464	2.2783448731
V	24	H	19.8624838621	3.5116226595
		G	3.1375161379	3.5116226595

Table 5.1: Center-of-energy shifts and errors for some example OQFs.

The last term satisfies the sharp inequality

$$|\langle u, \gamma^0 * u\rangle| \leq d[f] \, \|u\|^2,$$

where

$$d[f] = 2\left|\sum_{k=-\infty}^{\infty}\sum_{n=1}^{\infty}(-1)^n k f(k-n)\bar{f}(k+n)\right|.$$

□

If $d[f]$ is small, then we can safely ignore the deviation of F^*u from a pure shift of u by $c[f]$. In that case, we will say that $c[F^*u] \approx 2c[u]-c[f]$ and $c[Fu] \approx \frac{1}{2}c[u]+\frac{1}{2}c[f]$. We note that the "C" filters have the smallest errors $d[f]$; these are the filters to use if we wish to extract reasonably accurate position information.

If we apply a succession of filters $F_1^*F_2^* \cdots F_L^*$, then by induction on L we can compute the shifts as follows:

$$c[F_1^*F_2^* \cdots F_L^*u] = 2^L c[u] - 2^{L-1}c[f_L] - \cdots - 2^1 c[f_2] - c[f_1] - \epsilon^*, \tag{5.37}$$

where

$$|\epsilon^*| \leq 2^{L-1}d[f_L] + \cdots + 2^1 d[f_2] + d[f_1]. \tag{5.38}$$

Similarly, if $v = F_1^*F_2^* \cdots F_L^*u$, so that $F_L \cdots F_2F_1v = u$, then the following holds:

$$c[F_L \cdots F_2F_1v] = 2^{-L}c[v] + 2^{-L}c[f_1] + 2^{-L+1}c[f_2] + \cdots + 2^{-1}c[f_L] + \epsilon, \tag{5.39}$$

where

$$|\epsilon| \leq 2^{-1}d[f_L] + \cdots + 2^{-L+1}d[f_2] + 2^{-L}d[f_1]. \tag{5.40}$$

Now suppose that (h, g) is a conjugate pair of OQFs, so that $f_i \in \{h, g\}$ for each $i = 1, 2, \ldots, L$. Then $d[f_i]$ is constantly $d[h]$ and we have the simpler estimates for the deviation from a pure shift:

$$|\epsilon^*| \leq (2^L - 1)\, d[h] \approx 2^L d[h] \qquad \text{and} \qquad |\epsilon| \leq (1 - 2^{-L})\, d[h] \approx d[h]. \tag{5.41}$$

Suppose that we encode the sequence of filters $F_1^*F_2^* \cdots F_L^*$ as the integer $b = b_1 2^{L-1} + b_2 2^{L-2} + \cdots + b_L 2^0$, where

$$b_k = \begin{cases} 0, & \text{if } F_k = H; \\ 1, & \text{if } F_k = G. \end{cases} \tag{5.42}$$

Then we can write $c[f_k] = b_k c[g] + (1 - b_k)c[h] = c[h] + b_k(c[g] - c[h])$. Notice that the bit-reversal of b, considered as an s-bit binary integer, is the integer $b' = b_1 2^0 + b_2 2^1 + \cdots + b_L 2^{L-1}$. This simplifies the formula for the phase shift as follows:

Corollary 5.9 *If h and g are a conjugate pair of OQFs with centers of energy $c[h]$ and $c[g]$, respectively, then*

$$c[F_1^* F_2^* \cdots F_L^* u] = 2^L c[u] - \left(2^L - 1\right) c[h] - (c[g] - c[h])\, b' - \epsilon^*, \tag{5.43}$$

where $|\epsilon^*| \leq (2^L - 1)\, d[h]$ and $b = b_1 2^{L-1} + b_2 2^{L-2} + \cdots + b_L$ encodes the sequence of filters as in Equation 5.42, and b' is the bit-reversal of b considered as an L-bit binary integer.

Proof: We observe that

$$\begin{aligned} c[F_1^* F_2^* \cdots F_L^* u] &= 2^L c[u] - \sum_{k=1}^{L} 2^{L-k} \Big[c[h] + b_{L-k+1}(c[g] - c[h]) \Big] - \epsilon^* \\ &= 2^L c[u] - c[h] \sum_{s=0}^{L-1} 2^s - (c[g] - c[h]) \sum_{s=0}^{L-1} b_{s+1} 2^s - \epsilon^* \\ &= 2^L c[u] - \left(2^L - 1\right) c[h] - (c[g] - c[h]) b' - \epsilon^*. \end{aligned}$$

The estimate on ϵ^* follows from Equation 5.41. □

5.2.2 Shifts in the periodic case

Defining a center of energy for a periodic signal is problematic. However, if a periodic signal contains a component with a distinguishable scale much shorter than the period, then it may be desirable to locate this component within the period. If the component is characterized by a large amplitude found by filtering, then we can locate it by interpreting the "position" information of the filter output. We must adjust this position information by the center-of-energy shift caused by filtering, and allow for the deviation due to phase nonlinearity. In the periodic case, the shift can be approximated by a cyclic permutation of the output coefficients.

We can compute the center of energy of a nonzero q-periodic sequence u_q as follows:

$$c[u_q] = \frac{1}{\|u_q\|^2} \sum_{k=0}^{q-1} k |u_q(k)|^2.$$

Since $c[u_q]$ is a convex combination of $0, 1, \ldots, q-1$, we have $0 \leq c[u_q] \leq q - 1$. Now suppose that u_q is the q-periodization of u and that all but ϵ of the energy in the sequence u comes from coefficients in one period interval $J_0 \stackrel{\text{def}}{=} [j_0 q, j_0 q + q - 1]$, for some integer j_0 and some positive $\epsilon \ll 1$. We must also suppose that u has a finite position uncertainty which is less than q. These conditions may be succinctly combined into the following:

$$\left(\sum_{j\notin J_0}\left[j-(j_0+\frac{1}{2})q\right]^2|u(j)|^2\right)^{\frac{1}{2}} < q\epsilon\|u\|. \tag{5.44}$$

Equation 5.44 has two immediate consequences. Since $|j-(j_0+\frac{1}{2})q| \geq \frac{q}{2}$ for all $j \notin J_0$, we have

$$\left(\sum_{j\notin J_0}|u(j)|^2\right)^{\frac{1}{2}} < 2\epsilon\|u\|. \tag{5.45}$$

Also, for all $k = 0, 1, \ldots, q-1$ we have

$$\left|\frac{k+jq-(j_0+\frac{1}{2})q}{(j-j_0)q}\right| = \left|1-\frac{\frac{k}{q}-\frac{1}{2}}{j-j_0}\right| \geq \frac{1}{2},$$

so we can use the Cauchy–Schwarz inequality to get the following estimate:

$$\begin{aligned}
\sum_{k=0}^{q-1}\left|\sum_{j\neq j_0}u(k+jq)\right|^2 &\leq \sum_{k=0}^{q-1}\left|\sum_{j\neq j_0}\frac{1}{(j-j_0)q}\left[k+jq-(j_0+\frac{1}{2})q\right]u(k+jq)\right|^2 \\
&\leq 4\sum_{k=0}^{q-1}\left(\sum_{j\neq j_0}\frac{1}{(j-j_0)^2q^2}\right)\left(\sum_{j\neq j_0}\left[k+jq-(j_0+\frac{1}{2})q\right]^2|u(k+jq)|^2\right) \\
&\leq \frac{4\pi^2}{3q^2}\sum_{k=0}^{q-1}\sum_{j\neq j_0}\left[k+jq-(j_0+\frac{1}{2})q\right]^2|u(k+jq)|^2 \\
&= \frac{4\pi^2}{3q^2}\sum_{j\notin J_0}\left[j+(j_0-\frac{1}{2})q\right]^2|u(j)|^2.
\end{aligned}$$

Using Equation 5.44 and a cheap estimate for the constant, this reduces to the following inequality:

$$\left(\sum_{k=0}^{q-1}\left|\sum_{j\neq j_0}u(k+jq)\right|^2\right)^{\frac{1}{2}} < 4\epsilon\|u\|. \tag{5.46}$$

We claim that for sufficiently small ϵ, we can compute the center of energy of u_q in terms of the center of energy of u. We do this through a number of decompositions and estimates. First, we decompose:

$$
\begin{aligned}
\|u_q\|^2 \left[c[u_q] - \frac{q}{2}\right] &= \sum_{k=0}^{q-1} \left[k - \frac{q}{2}\right] |u_q(k)|^2 = \sum_{k=0}^{q-1} \left[k - \frac{q}{2}\right] \left| \sum_{j=-\infty}^{\infty} u(k+jq) \right|^2 \\
&= \sum_{k=0}^{q-1} \left[k - \frac{q}{2}\right] \left| u(k+j_0 q) + \sum_{j \neq j_0} u(k+jq) \right|^2 \\
&= \sum_{k=0}^{q-1} \left[k - \frac{q}{2}\right] |u(k+j_0 q)|^2 \qquad (5.47) \\
&\quad + 2\Re \sum_{k=0}^{q-1} \sum_{j \neq j_0} \left[k - \frac{q}{2}\right] u(k+j_0 q)\, \overline{u(k+jq)} \qquad (5.48) \\
&\quad + \sum_{k=0}^{q-1} \left[k - \frac{q}{2}\right] \left| \sum_{j \neq j_0} u(k+jq) \right|^2 . \qquad (5.49)
\end{aligned}
$$

We use the identity $\sum_{k=0}^{q-1} \left[k - \frac{q}{2}\right] |u(k+j_0 q)|^2 = \sum_{j \in J_0} \left[j - q(j_0 + \frac{1}{2})\right] |u(j)|^2$ and temporarily denote $\left[j - q(j_0 + \frac{1}{2})\right]$ by j' to rewrite 5.47 as follows:

$$
\begin{aligned}
\sum_{j \in J_0} j' |u(j)|^2 &= \sum_{j=-\infty}^{\infty} j' |u(j)|^2 - \sum_{j \notin J_0} j' |u(j)|^2 \\
&= \|u\|^2 (c[u] - j_0 q - \frac{q}{2}) - \sum_{j \notin J_0} j' |u(j)|^2 .
\end{aligned}
$$

We estimate the second term using the Cauchy–Schwarz inequality and Equations 5.45 and 5.46:

$$
\left| \sum_{j \notin J_0} j' |u(j)|^2 \right| \leq \left(\sum_{j \notin J_0} (j')^2 |u(j)|^2 \right)^{\frac{1}{2}} \left(\sum_{j \notin J_0} |u(j)|^2 \right)^{\frac{1}{2}} < 2q\epsilon^2 \|u\|^2 .
$$

In a similar way, we obtain an estimate for 5.48:

$$
\begin{aligned}
\left| \sum_{k=0}^{q-1} \left[k - \frac{q}{2}\right] u(k+j_0 q) \sum_{j \neq j_0} \bar{u}(k+jq) \right| &\leq \frac{q}{2} \|u\| \left(\sum_{k=0}^{q-1} \left| \sum_{j \neq j_0} \bar{u}(k+jq) \right|^2 \right)^{\frac{1}{2}} \\
&< \frac{q}{2} \|u\| \times 4\epsilon \|u\| = 2q\epsilon \|u\|^2 .
\end{aligned}
$$

Finally, we estimate 5.49 as follows:

$$\left|\sum_{k=0}^{q-1}\left[k-\frac{q}{2}\right]\left|\sum_{j\neq j_0} u(k+jq)\right|^2\right| < \frac{q}{2}\sum_{k=0}^{q-1}\left|\sum_{j\neq j_0} u(k+jq)\right|^2 < 8q\epsilon^2\|u\|^2.$$

Putting these estimates together, we get the following:

$$\left|\|u_q\|^2\left[c[u_q]-\frac{q}{2}\right]-\|u\|^2\left[c[u]-j_0q-\frac{q}{2}\right]\right| < 2q\epsilon\,(1+5\epsilon)\,\|u\|^2. \tag{5.50}$$

Now if we were to perform a similar decomposition and estimate for $\|u_q\|^2$ rather than $\|u_q\|^2\left[c[u_q]-\frac{q}{2}\right]$, we would obtain the following inequality:

$$\left|\|u_q\|^2-\|u\|^2\right| < 4\epsilon\,(1+5\epsilon)\,\|u\|^2. \tag{5.51}$$

We can thus replace $\|u_q\|^2$ with $\|u\|^2$ in Equation 5.50:

$$|c[u_q]-c[u]+j_0q| < 4q\epsilon\,(1+5\epsilon)\,. \tag{5.52}$$

We can summarize the calculation in Equations 5.44 to 5.52 as follows: if almost all of the energy of u is concentrated on an interval of length q, then transient features of u have a scale smaller than q and will become transient features of u_q upon q-periodization. Assuming that u satisfies Equation 5.44 with $\epsilon < \frac{1}{8q}$, we can use the following approximation to locate the center of energy of a periodized sequence to within one index:

$$c[u_q] \stackrel{\text{def}}{=} c[u] \bmod q. \tag{5.53}$$

We interpret the expression "$x \bmod q$" to mean the unique real number x' in the interval $[0,q[$ such that $x = x' + nq$ for some integer n.

We can use Proposition 5.1 to compute the following approximation:

$$\begin{aligned} c[F_{2q}^*u_q] &= c[(F^*u)_{2q}] = c[F^*u] \bmod 2q \\ &= 2c[u]-c[f]-\langle u,\gamma^0 * u\rangle/\|u\|^2 \bmod 2q. \end{aligned}$$

Now $\langle u,\gamma^0 * u\rangle/\|u\|^2$ is bounded by $d[f]$ so we plan to ignore it as before, though we must still verify that the OQFs satisfy Equation 5.44 with sufficiently small ϵ. Table 5.2 shows the value of ϵ for a few example OQFs and a few example periodizations. In all cases $\epsilon < 1$, so the table lists only the digits after the decimal point.

Since there is no unique way to *deperiodize* u_q to an infinite sequence u, it is necessary to adopt a convention. The simplest would be the following:

$$u(n) = \begin{cases} u_q(n), & \text{if } 0 \le n < q, \\ 0, & \text{otherwise.} \end{cases} \tag{5.54}$$

$\lvert \text{supp } f \rvert$	H or G	$q=2$ $q=16$	$q=4$ $q=18$	$q=6$ $q=20$	$q=8$ $q=22$	$q=10$ $q=24$	$q=12$ $q=26$	$q=14$ $q=28$
18	H	.703612 .001415	.279300	.142238	.074249	.033688	.014072	.005406
	G	.734120 .001590	.324821	.163452	.087139	.038976	.016137	.006156
6	H	.247013	.102745					
	G	.268885	.069768					
12	H	.263115	.072831	.033281	.010694	.001009		
	G	.251051	.070544	.028711	.009039	.001205		
18	H	.299435 .000040	.100032	.052849	.018963	.007231	.002661	.000621
	G	.291211 .000045	.098243	.046702	.017889	.007332	.002556	.000708
24	H	.329096 .000890	.120402 .000331	.065564 .000036	.027330 .000002	.014121	.005809	.002328
	G	.322880 .000936	.119051 .000367	.060292 .000039	.027004 .000002	.013983	.005754	.002531
30	H	.354113 .002035	.136558 .000958	.075916 .000291	.035107 .000138	.020482 .000026	.009303 .000002	.004743 .000000
	G	.349093 .002111	.135636 .001051	.071338 .000285	.035330 .000134	.020121 .000024	.009401 .000002	.005009 .000000
4	H	.171193						
	G	.273971						
6	H	.304120	.050230					
	G	.259392	.073125					
8	H	.308900	.102651	.017895				
	G	.323009	.122720	.023634				
10	H	.342554	.135552	.040530	.006627			
	G	.449328	.116023	.053618	.008251			
12	H	.422494	.137647	.058646	.016224	.002475		
	G	.463486	.160047	.064599	.020210	.002964		
14	H	.524235	.169394	.072909	.023686	.006412	.000924	
	G	.508880	.223013	.076843	.029062	.007680	.001077	
16	H	.524480	.210433	.085366	.032061	.009408	.002489	.000344
	G	.587024	.220427	.103528	.038321	.011119	.002899	.000393
18	H	.564454 .000128	.243878	.102607	.045068	.014338	.003662	.000948
	G	.636888 .000144	.238832	.128066	.050826	.016666	.004213	.001082
20	H	.634131 .000354	.248979 .000047	.120135	.051443	.024453	.006775	.001411
	G	.672192 .000398	.282813 .000053	.138670	.060597	.025714	.007739	.001591
24	H	.872011 .006270	.390176 .001937	.217686 .000629	.116186 .000191	.062451	.036782	.017151
	G	.829783 .005653	.355441 .001764	.190529 .000574	.101064 .000175	.057180	.034695	.015266

Table 5.2: Concentration of energy for some example orthogonal QFs.

5.3 Frequency response

Applying a convolution-decimation operator to a signal u is equivalent to multiplying the Fourier transform of u by a bounded periodic function. The bounded function is determined by the filter coefficients: This bounded function will be called the *filter multiplier* corresponding to a filter sequence f, and is defined by the following formula:

$$m(\xi) \stackrel{\text{def}}{=} \sum_{k\in\mathbf{Z}} f(k)e^{-2\pi ik\xi} = \hat{f}(\xi). \tag{5.55}$$

This function is also called the *frequency response* of the filter.

With a summable filter sequences, the filter multiplier is a uniformly convergent series of bounded continuous functions, hence is continuous by Proposition 1.1. If the filter sequence decreases rapidly at $\pm\infty$, then the frequency response function is smooth since the Fourier transform converts differentiation (of m) into multiplication (of $f(k)$ by $-2\pi ik$). A little bit of smoothness is very useful, so we will assume that there is some $\alpha > 0$ such that

$$\sum_{k\in\mathbf{Z}} |f(k)|\,|k|^\alpha < \infty. \tag{5.56}$$

This will guarantee that m satisfies a *Hölder condition* at every point, in particular at 0. Namely, we have:

Lemma 5.10 *If f satisfies Equation 5.56 for some $0 < \alpha < 1$, then for all ξ we have $|m(\xi) - m(0)| < C|\xi|^\alpha$.*

Proof: We will show that $|m(\xi) - m(0)|\,|\xi|^{-\alpha}$ is a bounded function. This is true wherever $|\xi| \geq 1$ since f is summable and $|m(\xi) - m(0)|\,|\xi|^{-\alpha} \leq 2|m(\xi)|$ is therefore bounded. For the domain $|\xi| < 1$, we have the following:

$$(m(\xi) - m(0))\,|\xi|^{-\alpha} = \sum_{k\in\mathbf{Z}} f(k)\left(e^{-2\pi ik\xi} - 1\right)|\xi|^{-\alpha}.$$

But if we write $\eta = k\xi$, then we have

$$\frac{|e^{-2\pi ik\xi} - 1|}{|\xi|^\alpha} = \frac{|e^{-2\pi i\eta} - 1|}{|\eta|^\alpha}|k|^\alpha < 2\pi|k|^\alpha.$$

The last inequality holds for all $\alpha < 1$. The result now follows from the triangle inequality and the boundedness of the sum in Equation 5.56. □

If h and g are a conjugate QF pair related by Equation 5.19, then their filter multipliers are related by the formula

$$m_g(\xi) = e^{-2\pi i(2M+1)(\xi+\frac{1}{2})}\bar{m}_h(\xi+\frac{1}{2}). \tag{5.57}$$

5.3.1 Effect of a single filter application

In each of the contexts we are considering—sequences and functions, periodic and aperiodic—it is possible to compute the action of the convolution-decimation operator on the signal in terms of m and the Fourier transform of the signal.

Lemma 5.11 *If $u \in L^2(\mathbf{R})$ then both $\widehat{Fu}(\xi) \in L^2(\mathbf{R})$ and $\widehat{F^*u}(\xi) \in L^2(\mathbf{R})$, with $\widehat{Fu}(\xi) = \frac{1}{\sqrt{2}}m(\frac{\xi}{2})\hat{u}(\frac{\xi}{2})$ and $\widehat{F^*u}(\xi) = \sqrt{2}\,\bar{m}(\xi)\hat{u}(2\xi)$.*

Proof:

$$\begin{aligned}
\widehat{Fu}(\xi) &= \int_{\mathbf{R}} Fu(t)e^{-2\pi it\xi}\,dt \;=\; \int_{\mathbf{R}} \sqrt{2}\sum_k f(k)u(2t-k)e^{-2\pi it\xi}\,dt \\
&= \frac{1}{\sqrt{2}}\sum_k f(k)\int_{\mathbf{R}} u(t)e^{-2\pi i(\frac{t+k}{2})\xi}\,dt \\
&= \frac{1}{\sqrt{2}}\sum_k f(k)e^{-2\pi ik\frac{\xi}{2}}\int_{\mathbf{R}} u(t)e^{-2\pi it(\frac{\xi}{2})}\,dt \;=\; \frac{1}{\sqrt{2}}m\left(\frac{\xi}{2}\right)\hat{u}\left(\frac{\xi}{2}\right); \\
\widehat{F^*u}(\xi) &= \int_{\mathbf{R}} F^*u(t)e^{-2\pi it\xi}\,dt \;=\; \int_{\mathbf{R}} \frac{1}{\sqrt{2}}\sum_k \bar{f}(k)u\left(\frac{t+k}{2}\right)e^{-2\pi it\xi}\,dt \\
&= \sqrt{2}\sum_k \bar{f}(k)\int_{\mathbf{R}} u(t)e^{-2\pi i(2t-k)\xi}\,dt \\
&= \sqrt{2}\sum_k \overline{f(k)e^{-2\pi ik\xi}}\int_{\mathbf{R}} u(t)e^{-2\pi it(2\xi)}\,dt \;=\; \sqrt{2}\,\bar{m}\,(\xi)\,\hat{u}\,(2\xi)\,.
\end{aligned}$$

□

Corollary 5.12 *If $u \in L^2(\mathbf{R})$ then both $F_1^*F_2u \in L^2(\mathbf{R})$ and $F_1F_2^*u \in L^2(\mathbf{R})$, with $\widehat{F_1^*F_2u}(\xi) = \bar{m}_1(\xi)m_2(\xi)\hat{u}(\xi)$ and $\widehat{F_1F_2^*u}(\xi) = m_1(\xi/2)\bar{m}_2(\xi/2)\hat{u}(\xi)$.* □

In particular, $\widehat{F^*Fu}(\xi) = |m(\xi)|^2\hat{u}(\xi)$ and $\widehat{FF^*u}(\xi) = |m(\xi/2)|^2\hat{u}(\xi)$ for $u \in L^2(\mathbf{R})$. As we noted before, neither $H'H^* = I$ or $G'G^* = I$ as operators on $L^2(\mathbf{R})$. In the orthogonal case, the operators H^*H and G^*G will be projections only if $|m|^2$ takes just the values zero or one. Such filter multipliers are either discontinuous or constant and trivial. The discontinuous but nontrivial case $m(\xi) =$

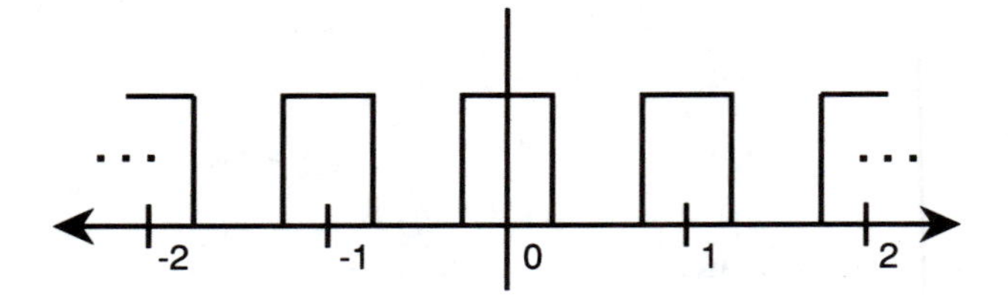

Figure 5.5: A limiting case for the filter multiplier m.

$\sqrt{2}\sum_k \mathbf{1}_{[-\frac{1}{4},\frac{1}{4}]}(\xi - k)$, plotted in Figure 5.5, can never arise from a QF, since absolutely summable filter sequences yield continuous filter multipliers. But that m will nonetheless be useful for calculation because it is a limiting case for which certain formulas become drastically simpler.

The action of the sequence operator is a bit more complicated:

Lemma 5.13 *If $u \in \ell^2$ then both $\widehat{Fu}(\xi) \in L^2(\mathbf{T})$ and $\widehat{F^*u}(\xi) \in L^2(\mathbf{T})$, with*

$$\begin{aligned} \widehat{Fu}(\xi) &= \frac{1}{2}m\left(\frac{\xi}{2}\right)\hat{u}\left(\frac{\xi}{2}\right) + \frac{1}{2}m\left(\frac{\xi}{2}+\frac{1}{2}\right)\hat{u}\left(\frac{\xi}{2}+\frac{1}{2}\right), \\ \widehat{F^*u}(\xi) &= \bar{m}(\xi)\hat{u}(2\xi). \end{aligned}$$

Proof: For Fu:

$$\begin{aligned} \widehat{Fu}(\xi) &= \sum_n Fu(n)e^{-2\pi in\xi} = \sum_n\sum_k f(k)u(2n-k)e^{-2\pi in\xi} \\ &= \sum_n\left(\sum_k f(2k)u(2n-2k)e^{-2\pi in\xi} + \sum_k f(2k+1)u(2n-2k-1)e^{-2\pi in\xi}\right) \\ &= \sum_k f(2k)e^{-2\pi ik\xi}\sum_n u(2n)e^{-2\pi in\xi} \\ &\quad + \sum_k f(2k+1)e^{-2\pi ik\xi}\sum_n u(2n+1)e^{-2\pi in\xi}. \end{aligned}$$

Thus

$$\widehat{Fu}(\xi) = \hat{f}_e(\xi)\hat{u}_e(\xi) + \hat{f}_o(\xi)\hat{u}_o(\xi), \tag{5.58}$$

where we have denoted $f_e(n) = f(2n)$ and $f_o(n) = f(2n+1)$ for all $n \in \mathbf{Z}$. Now, $\hat{f}_e(\xi) + \hat{f}_o(\xi)e^{-\pi i\xi'} = \hat{f}(\frac{\xi}{2})$, and since both $\hat{f}_e$ and $\hat{f}_o$ are one-periodic functions we have $\hat{f}_e(\xi) - \hat{f}_o(\xi)e^{-\pi i\xi} = \hat{f}(\frac{\xi+1}{2})$. Thus

$$\hat{f}_e(\xi) = \frac{1}{2}\left[\hat{f}\left(\frac{\xi}{2}\right) + \hat{f}\left(\frac{\xi}{2}+\frac{1}{2}\right)\right], \tag{5.59}$$

$$\hat{f}_o(\xi) = \frac{1}{2}\left[\hat{f}\left(\frac{\xi}{2}\right) - \hat{f}\left(\frac{\xi}{2}+\frac{1}{2}\right)\right] e^{\pi i \xi}. \tag{5.60}$$

These formulas also hold if we replace f with u. Then substituting for f_e, f_o, u_e, u_o in Equation 5.58 gives $\widehat{Fu}(\xi) = \frac{1}{2}\hat{f}(\frac{\xi}{2})\hat{u}(\frac{\xi}{2}) + \frac{1}{2}\hat{f}(\frac{\xi}{2}+\frac{1}{2})\hat{u}(\frac{\xi}{2}+\frac{1}{2})$, and the result follows.

For F^*u:

$$\begin{aligned}\widehat{F^*u}(\xi) &= \sum_n F^*u(n)e^{-2\pi i n\xi} = \sum_n\sum_k \bar{f}(2k-n)u(n)e^{-2\pi i n\xi} \\ &= \sum_n \overline{f(n)e^{-2\pi i n\xi}} \sum_k u(k)e^{-2\pi i(2k)\xi} = \bar{m}(\xi)\hat{u}(2\xi).\end{aligned}$$

□

Corollary 5.14 *If $u \in \ell^2$, then*

$$\begin{aligned}\widehat{F_1^*F_2u}(\xi) &= \frac{1}{2}\bar{m}_1(\xi)m_2(\xi)\hat{u}(\xi) + \frac{1}{2}\bar{m}_1(\xi)m_2(\xi+\frac{1}{2})\hat{u}(\xi+\frac{1}{2}); \\ \widehat{F_1F_2^*u}(\xi) &= \frac{1}{2}\left(m_1(\frac{\xi}{2})\bar{m}_2(\frac{\xi}{2}) + m_1(\frac{\xi}{2}+\frac{1}{2})\bar{m}_2(\frac{\xi}{2}+\frac{1}{2})\right)\hat{u}(\xi).\end{aligned}$$

Here m_i is the frequency response for the filter operator F_i, where $i = 1, 2$. □

We can now reduce the BQF conditions to functional equations for the filter multipliers.

Theorem 5.15 *The (biorthogonal) quadrature filter conditions for H, H', G, and G' are equivalent to the following equations for their filter multipliers m_h, $m_{h'}$, m_g, and $m_{g'}$:*

1. $m_{h'}(\xi)\bar{m}_h(\xi) + m_{h'}(\xi+\frac{1}{2})\bar{m}_h(\xi+\frac{1}{2}) = 2$;

2. $m_{g'}(\xi)\bar{m}_g(\xi) + m_{g'}(\xi+\frac{1}{2})\bar{m}_g(\xi+\frac{1}{2}) = 2$;

3. $m_{h'}(\xi)\bar{m}_g(\xi) + m_{h'}(\xi+\frac{1}{2})\bar{m}_g(\xi+\frac{1}{2}) = 0$;

4. $m_{g'}(\xi)\bar{m}_h(\xi) + m_{g'}(\xi+\frac{1}{2})\bar{m}_h(\xi+\frac{1}{2}) = 0$;

5. $m_{h'}(\xi)\bar{m}_h(\xi) + m_{g'}(\xi)\bar{m}_g(\xi) = 2$, *for all* $\xi \in \mathbf{R}$;

6. $m_{h'}(0) = m_h(0) = \sqrt{2}$ *and* $m_{g'}(0) = m_g(0) = 0$.

Proof: We can use Corollary 5.14 to rewrite the duality condition as the first pair of equations and independence as the second pair. The fifth equation is equivalent to exact reconstruction by Corollary 5.4. The last pair of equations is equivalent to the normalization conditions. □

The BQF conditions with the conventional normalization are equivalent to the matrix equation $\mathbf{M}^*\mathbf{M}' = 2I$ with the "initial condition" $\mathbf{M}(0) = \mathbf{M}'(0) = \sqrt{2}\, I$, where

$$\mathbf{M} = \mathbf{M}(\xi) \stackrel{\text{def}}{=} \begin{pmatrix} m_h(\xi) & m_g(\xi) \\ m_h(\xi+\frac{1}{2}) & m_g(\xi+\frac{1}{2}) \end{pmatrix}. \tag{5.61}$$

To get $\mathbf{M}' = \mathbf{M}'(\xi)$, we simply replace every h with h' and every g with g' in Equation 5.61. If h, g' and h', g are conjugates, then we can use Equation 5.57 to rewrite $\mathbf{M}$ and $\mathbf{M}'$ in terms of just m_h and $m_{h'}$:

$$\mathbf{M} = \begin{pmatrix} m_h(\xi) & -\bar{m}_{h'}(\xi+\frac{1}{2}) \\ m_h(\xi+\frac{1}{2}) & \bar{m}_{h'}(\xi) \end{pmatrix} \mathbf{D} \stackrel{\text{def}}{=} \mathbf{M}_0\mathbf{D}; \tag{5.62}$$

$$\mathbf{M}' = \begin{pmatrix} m_{h'}(\xi) & -\bar{m}_{h}(\xi+\frac{1}{2}) \\ m_{h'}(\xi+\frac{1}{2}) & \bar{m}_{h}(\xi) \end{pmatrix} \mathbf{D} = 2(\mathbf{M}_0^*)^{-1}\mathbf{D}. \tag{5.63}$$

Here $\mathbf{D} = \operatorname{diag}(1, e^{-2\pi i(2M+1)\xi})$ is a unitary diagonal matrix; notice that it is cancelled out in the matrix equation $\mathbf{M}^*\mathbf{M}' = 2I$, since $\mathbf{D}^*\mathbf{D} = I$. Thus the BQF conditions, for conjugate filters, are equivalent to the the matrix equation $\mathbf{M}_0^*\mathbf{M}_0' = 2I$.

If $H = H'$ and $G = G'$, then $\mathbf{M} = \mathbf{M}'$ and the orthogonal QF conditions are equivalent to the matrix equation $\mathbf{M}^*\mathbf{M} = 2I$. If the associated filter multipliers are m_h and m_g, then for all $\xi \in \mathbf{R}$ we have:

$$\begin{gathered} |m_h(\xi)|^2 + |m_h(\xi+\tfrac{1}{2})|^2 = 2; \qquad |m_g(\xi)|^2 + |m_g(\xi+\tfrac{1}{2})|^2 = 2; \\ m_h(\xi)\bar{m}_g(\xi) + m_h(\xi+\tfrac{1}{2})\bar{m}_g(\xi+\tfrac{1}{2}) = 0. \end{gathered} \tag{5.64}$$

This is equivalent to the following matrix being unitary for all ξ:

$$\frac{1}{\sqrt{2}} \begin{pmatrix} m_h(\xi) & m_g(\xi) \\ m_h(\xi+\frac{1}{2}) & m_g(\xi+\frac{1}{2}) \end{pmatrix}. \tag{5.65}$$

Corollary 5.16 *If H and G are a pair of orthogonal QFs, then their multipliers m_h and m_g satisfy the equation*

$$|m_h(\xi)|^2 + |m_g(\xi)|^2 = 2$$

for all $\xi \in \mathbf{R}$. In particular, $|m_h(\xi)| \le \sqrt{2}$ and $|m_g(\xi)| \le \sqrt{2}$. □

We can also write m_g in terms of m_h in the conjugate QFs case, as in Equations 5.62 and 5.63. Since $\mathbf{D}$ is unitary, it is enough that $\frac{1}{\sqrt{2}}\mathbf{M}_0$ be unitary:

Corollary 5.17 *H and G are conjugate orthogonal QFs if and only if the following matrix is unitary for all $\xi \in \mathbf{R}$:*

$$\frac{1}{\sqrt{2}}\begin{pmatrix} m_h(\xi) & -\bar{m}_h(\xi+\frac{1}{2}) \\ m_h(\xi+\frac{1}{2}) & \bar{m}_h(\xi) \end{pmatrix}.$$

H and G are conventionally normalized if and only if the above matrix takes the value I at $\xi = 0$. □

Remark. Corollary 5.17 and Equations 5.59 and 5.60 show that any sequence f satisfying Equation 5.13 also satisfies:

$$|\hat{f}_e(\xi)|^2 + |\hat{f}_o(\xi)|^2 = 1. \tag{5.66}$$

Remark. We can also use the filter multiplier of an orthogonal QF sequence f to estimate the phase response of the filter operator F. Recalling the definition of γ from Equation 5.32, we compute its Fourier transform:

$$\begin{aligned}
\hat{\gamma}(\xi) &= \sum_n \sum_k k f(k-n)\bar{f}(k+n)e^{-2\pi i n\xi} \\
&= \sum_k 2kf(2k)\sum_n \bar{f}(2k+2n)e^{-2\pi i n\xi} \\
&\qquad + \sum_k (2k+1)f(2k+1)\sum_n \bar{f}(2k+2n+1)e^{-2\pi i n\xi} \\
&= \sum_k 2kf(2k)e^{2\pi i k\xi}\sum_n \overline{f(2n)e^{2\pi i n\xi}} \\
&\qquad + \sum_k (2k+1)f(2k+1)e^{2\pi i k\xi}\sum_n \overline{f(2n+1)e^{2\pi i n\xi}} \\
&= \overline{\check{f}_e(\xi)}\sum_k 2kf(2k)e^{2\pi i k\xi} + \overline{\check{f}_o(\xi)}\sum_k (2k+1)f(2k+1)e^{2\pi i k\xi} \\
&= \frac{1}{i\pi}\overline{\check{f}_e(\xi)}\check{f}_e'(\xi) + \frac{1}{i\pi}\overline{\check{f}_o(\xi)e^{\pi i\xi}}\left(\check{f}_o(\xi)e^{\pi i\xi}\right)'.
\end{aligned}$$

We can use Equations 5.59 and 5.60 plus the observation that $\check{f}(\xi) = \hat{f}(-\xi)$ to reduce $\hat{\gamma}$ to an expression involving only $\check{f}$ and $\check{f}'$:

$$8\pi i\hat{\gamma}(\xi) = \overline{\check{f}\left(\frac{\xi}{2}\right)}\check{f}'\left(\frac{\xi}{2}\right) + \overline{\check{f}\left(\frac{\xi}{2}+\frac{1}{2}\right)}\check{f}'\left(\frac{\xi}{2}+\frac{1}{2}\right). \tag{5.67}$$

The deviation $d[f]$ from linear phase response is just the essential maximum value of $|\hat{\gamma}(\xi) - 1|$.

5.3.2 Effect of iterated filter applications

The seminal work of Daubechies [36] launched an encyclopedic investigation into the properties of limits of iterated convolution-decimation operators. See, for example, [19, 28, 85, 65, 47].

Limits of iterated low-pass QFs

A thorough treatment of this subject may be found in [37]. Here we will consider just two small aspects: the QF interpolation algorithm, and the formula for the scaling function of one real variable. The first can be used to generate smooth curves and sampling functions, as well as graphical approximations of the basis functions we will be using later. The second is central to Daubechies' method for estimating the regularity of wavelets and the analytic properties of the basis functions.

Quadrature filter interpolation may be stated as follows: given samples $\{u_0(i) : i \in \mathbf{Z}\}$, we interpolate new values at half-integers and adjust the old values at the integers, using the adjoint H^* of a low-pass QF:

$$u_1(i/2) = H^*u_0(i) = \sum_{j=-\infty}^{\infty} \bar{h}(2j-i)u_0(j).$$

In general, we must account for the index shift caused by QFs when comparing the values of u_1 with those of u_0. This provides strong motivation to use symmetric QFs centered at zero. Also, nothing prevents us from iterating this operation to produce $u_2, u_3, \ldots, u_L$, the last of which is sampled on a grid with spacing 2^{-L}.

A *scaling function* for a multiresolution analysis of $L^2(\mathbf{R})$ is any normalized fixed point for the low-pass QF operator H. In other words, it is a function $\phi \in L^2 \cap L^1$ with $\int_{\mathbf{R}} \phi = 1$ satisfying $\phi = H\phi$, or

$$\phi(t) = \sqrt{2} \sum_{j\in\mathbf{Z}} h(j)\phi(2t-j), \quad t \in \mathbf{R}. \tag{5.68}$$

By repeated application of Lemma 5.11, this becomes the equation

$$\hat{\phi}(\xi) = \frac{1}{\sqrt{2}} m(\frac{\xi}{2})\hat{\phi}(\frac{\xi}{2}) = 2^{-L/2} m(\frac{\xi}{2}) \cdots m(\frac{\xi}{2^L})\hat{\phi}(\frac{\xi}{2^L}), \qquad L > 0.$$

Integrability and the normalization condition $\int_{\mathbf{R}} \phi = 1$ implies that $\hat{\phi}(0) = 1$ and $\hat{\phi}$ is continuous at 0, so $\hat{\phi}(\xi/2^L) \to 1$ as $L \to \infty$. Also, the filter multiplier $m = m(\xi)$ of a low-pass QF must take the value $\sqrt{2}$ at $\xi = 0$, because of Equation 5.9, so that $\left|\frac{1}{\sqrt{2}} m(\frac{\xi}{2^L}) - 1\right| < 2\pi 2^{-\alpha L}|\xi|^\alpha$ as $L \to \infty$ by Lemma 5.10. Since the series

$\{a(n) = 2\pi 2^{-\alpha n}|\xi|^\alpha : n = 0, 1, \ldots\}$ is absolutely summable, the Weierstrass test implies that the following infinite product converges at each point ξ:

$$\hat{\phi}(\xi) = \prod_{k=1}^{\infty} \frac{1}{\sqrt{2}} m(\frac{\xi}{2^k}). \tag{5.69}$$

Daubechies' original method for insuring that ϕ is regular starts with Equation 5.69 and shows that, for appropriate choices of finitely supported filters h, the function $\hat{\phi}$ so defined will decrease to 0 as $|\xi| \to \infty$. To explain how a smooth ϕ will develop from iteration of H, we examine two example filter multipliers m that need none of the subtle analysis in [36].

Example 1: *The Shannon scaling function.* Take m to be the following step function:

$$m(\xi) = \sqrt{2} e^{-\pi i \xi} \sum_{k=-\infty}^{\infty} \mathbf{1}_{[-\frac{1}{4}, \frac{1}{4}]}(\xi - k). \tag{5.70}$$

This function is one-periodic on $\mathbf{R}$ and Hölder continuous (in fact infinitely differentiable) at 0. Figure 5.5 shows the graph of a few periods of the absolute value of this function. Let H be the "low-pass" convolution-decimation operator whose frequency response is the special function m. We can exactly compute the Fourier transform of the scaling function ϕ satisfying $\phi = H\phi$. In Equation 5.69, we notice that if $2^j < |\xi| \leq 2^{j+1}$ with $j \geq -1$, then for $k = 2 + j \geq 1$ we have $1/4 < |\xi|/2^k \leq 1/2$. But this implies that $m(\xi/2^k) = 0$, so $\hat{\phi}(\xi) = 0$. On the other hand, if $|\xi| \leq 1/2$, then for all $k = 1, 2, \ldots$ we have $|\xi|/2^k \leq 1/4$ and so $\frac{1}{\sqrt{2}} m(\xi/2^k) = e^{-\pi i \xi/2^k}$. But then $\hat{\phi}(\xi) = e^{-\pi i \xi}$; we have shown:

$$\hat{\phi}(\xi) = e^{-\pi i \xi} \mathbf{1}_{[-\frac{1}{2}, \frac{1}{2}]}(\xi). \tag{5.71}$$

Thus for the special m of Equation 5.70, the normalized fixed-point function ϕ satisfying $\phi = H\phi$ has compactly supported Fourier transform and is consequently real analytic. In fact, it is just a translation of the *Shannon scaling function* or *sinc function*:

$$\phi(t) = \frac{\sin \pi(t - \frac{1}{2})}{\pi(t - \frac{1}{2})}. \tag{5.72}$$

The real-valued filter defined by $h(n) = \check{m}(n) = \int_{\mathbf{T}} m(\xi) e^{2\pi i n \xi} \, d\xi$ will satisfy Equation 5.13, one of the orthogonal QF conditions. The associated operator H will produce an orthogonal projection H^*H and will satisfy the other orthogonal QF conditions on functions as well as on sequences. However, h does not have finite support:

$$h(n) = \int_{-1/4}^{1/4} \sqrt{2}\, e^{-\pi i\xi} e^{2\pi i n\xi}\, d\xi = \frac{\sqrt{2}\sin\frac{\pi}{2}(n-\frac{1}{2})}{\pi(n-\frac{1}{2})}. \tag{5.73}$$

Thus $h(n) = O(1/|n|)$ as $n \to \pm\infty$, which is not quite fast enough decay for absolute summability. We can nonetheless approximate m as closely as we like in the L^2 norm (Plancherel's theorem) or in the uniform norm off an arbitrarily small set (Weierstrass' theorem) using finitely supported orthogonal QFs.

Example 2: *The Haar scaling function.* The shortest sequence h which satisfies Equation 5.13 is

$$h(n) = \begin{cases} 1/\sqrt{2}, & \text{if } n = 0 \text{ or } n = 1, \\ 0, & \text{if } n \notin \{0,1\}. \end{cases} \tag{5.74}$$

The filter multiplier associated to h is the following:

$$m(\xi) = \frac{1}{\sqrt{2}}\left(1 + e^{-2\pi i\xi}\right) = \sqrt{2}\, e^{-\pi i\xi} \cos \pi\xi. \tag{5.75}$$

We can compute $\hat{\phi}$ exactly:

$$\prod_{k=1}^{\infty} \frac{1}{\sqrt{2}} m\left(\frac{\xi}{2^k}\right) = e^{-\pi i \sum_{k=1}^{\infty} \frac{\xi}{2^k}} \prod_{k=1}^{\infty} \cos\left(\frac{\pi\xi}{2^k}\right). \tag{5.76}$$

Since $\sum_{k=1}^{\infty} \frac{\xi}{2^k} = \xi$, the first factor is $e^{-\pi i\xi}$. We show that the second factor is $\sin(\pi\xi)/\pi\xi$ by the following classical computation:

$$\begin{aligned} \sin(x) &= 2\sin(x/2)\cos(x/2) = 4\sin(x/4)\cos(x/4)\cos(x/2) = \cdots \\ &= 2^L \sin(x/2^L) \prod_{k=1}^{L} \cos(x/2^k) \\ \Rightarrow \prod_{k=1}^{L} \cos(x/2^k) &= \frac{\sin x}{x} \frac{x/2^L}{\sin(x/2^L)}. \end{aligned} \tag{5.77}$$

The right hand side tends to the limit $\frac{\sin x}{x}$ as $L \to \infty$. Therefore,

$$\hat{\phi}(\xi) = e^{-\pi i\xi} \frac{\sin \pi\xi}{\pi\xi} \qquad \Rightarrow \qquad \phi(t) = \mathbf{1}_{[0,1]}(t). \tag{5.78}$$

This is called the *Haar scaling function.* Notice that the Haar and Shannon scaling functions are almost each other's Fourier transforms.

Support of the scaling function

If H is a low-pass FIR quadrature filter and part of an orthogonal conjugate pair, then there is a compactly supported solution to the fixed-point problem $\phi = H\phi$. This solution may be constructed by iterating the filter and proving a slight generalization of a basic result of Mallat [70]:

Theorem 5.18 *Suppose that H, G' and H', G are conjugate pairs of biorthogonal FIR QFs, whose filter sequences h and h' are supported in the index intervals $[a, b]$ and $[a', b']$, respectively. Suppose there are nonzero solutions $\phi, \phi' \in L^2$ to the fixed-point equations $\phi = H\phi$ and $\phi' = H'\phi'$. Then both ϕ and ϕ' are integrable, $\operatorname{supp}\phi = [a, b]$, $\operatorname{supp}\phi' = [a', b']$, and if we normalize $\int \phi = \int \phi' = 1$ then the solutions ϕ and ϕ' are unique.*

Proof: We first note that the dual filters H and H' are interchangeable, so that all of the results which apply to ϕ apply to ϕ' as well.

If ϕ is a square-integrable solution, then $\phi = H\phi = H^N\phi$ for all $N > 0$. Lemma 5.6 implies that for every $\epsilon > 0$, the total energy of ϕ outside the interval $[a-\epsilon, b+\epsilon]$ must be less than ϵ, so we must have $\operatorname{supp}\phi = [a, b]$.

By the Cauchy–Schwarz inequality, a compactly supported square-integrable function is also integrable: $\|\phi\|_1 = \int_a^b |\phi| \leq \sqrt{b-a}\,\|\phi\| < \infty$. Hence ϕ is also integrable.

To show uniqueness, we note that the normalization $\int \phi = 1$ is preserved through applications of H:

$$\int H\phi(t)\,dt = \sqrt{2}\sum_{k=a}^{b} h(k) \int \phi(2t-k)\,dt = \frac{1}{\sqrt{2}}\sum_{k} h(k) \int \phi(t)\,dt = 1. \qquad (5.79)$$

Now any normalized integrable solution to $\phi = H\phi$ must have a continuous Fourier transform $\hat{\phi} = \hat{\phi}(\xi)$ which by Equation 5.69 is equal to $\prod_{k=1}^{\infty} 2^{-1/2} m(\xi/2^k)$ at each point ξ. Thus the difference between any two normalized solutions must have identically zero Fourier transform, which means that any two such solutions must be equal. □

Frequency response of the scaling function

The step function m of Example 1 is infinitely flat at the origin: $|m(0)|^2 = 2$ and all derivatives of $|m|^2$ vanish at zero. The filter multiplier in Example 2 is only flat to first order at the origin: in that case $|m(\xi)|^2 = 1 + \cos 2\pi\xi$. With finite filters we can only achieve a finite degree of flatness at the origin. However, this is sufficient to provide any desired degree of decay in $|\hat{\phi}(\xi)|$ as $\xi \to \pm\infty$. Daubechies has shown ([37],p.226) that for every sufficiently large $d > 0$, there is a finitely supported low-pass orthogonal QF of support diameter less than $d/10$ such that $|\hat{\phi}(\xi)| \leq C(1+|\xi|)^{-d}$. This decay in turn guarantees that ϕ will have at least $d-2$ continuous derivatives.

To get the kind of sharp frequency localization present in the limiting case of Example 1, it is necessary to use a multiplier similar to that special m. The families of orthogonal QFs which we will be considering give rather good approximations, as can be seen in Figure 5.6. Hence the results which are easy to compute for the step function m can be used to compute at least roughly the properties of our example QFs.

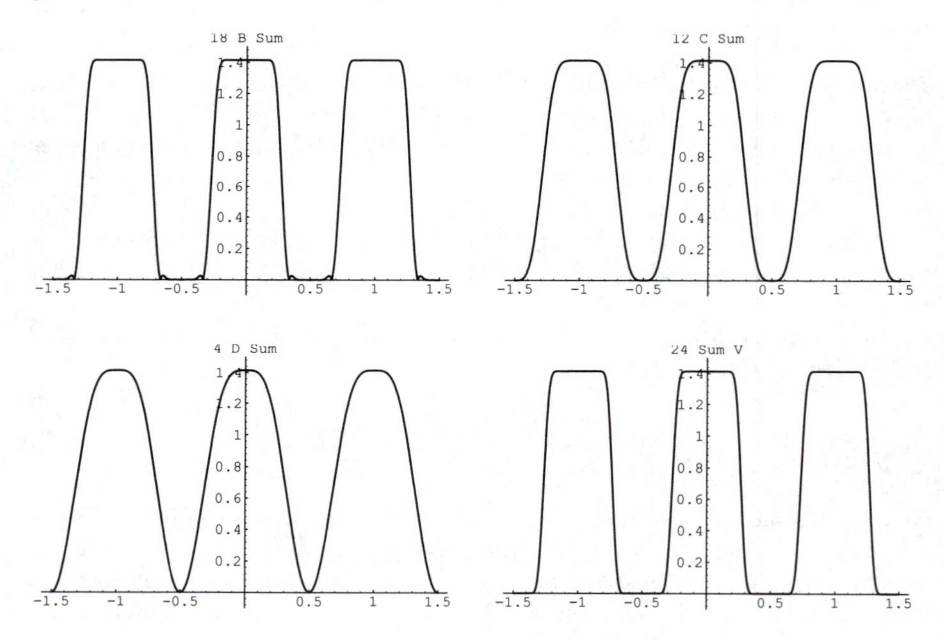

Figure 5.6: Power spectra of some example low-pass orthogonal QFs.

Gray code permutation

We define a map GC of the nonnegative integers by the formula $GC(n)_i = n_i + n_{i+1}$ (mod 2), where n_i is the ith binary digit of n. Then $GC(n)$ is called the *Gray code* of n. It is a permutation since it is invertible with inverse $GC^{-1}(n)_i = n_i + n_{i+1} +$

$n_{i+2} + \cdots \pmod 2$, where the sum is finite because $n_j = 0$ if $n < 2^j$. This permutation preserves all subsets of the form $\{0, 1, \ldots, 2^L - 1\}$, since $GC(n)_i = 0$ if $n_j = 0$ for all $j \geq i$. There is also a recursive definition of the inverse Gray code permutation:

$$GC^{-1}(2n) = \begin{cases} 2GC^{-1}(n), & \text{if } GC^{-1}(n) \text{ is even,} \\ 2GC^{-1}(n) + 1, & \text{if } GC^{-1}(n) \text{ is odd;} \end{cases} \tag{5.80}$$

$$GC^{-1}(2n+1) = \begin{cases} 2GC^{-1}(n) + 1, & \text{if } GC^{-1}(n) \text{ is even,} \\ 2GC^{-1}(n), & \text{if } GC^{-1}(n) \text{ is odd.} \end{cases} \tag{5.81}$$

The Gray code permutation relates Paley order to sequency order for Walsh functions, and thus appears naturally in the frequency localization of the wavelet packets which generalize them.

Gray coding was originally used in the manufacture of electrical counters. Since $GC(n)$ and $GC(n+1)$ differ in exactly one bit, a binary counter which increments itself in Gray code order is never forced to change two bits simultaneously and is therefore immune to certain large transient errors.

In the C programming language, we can use the bitwise exclusive-or operator (^) to implement bit addition modulo 2. Then $GC(n)$ can be defined by a macro:

Gray code permutation

```
#define graycode(N)  ((N)^((N)>>1))
```

The inverse $GC^{-1}(n)$ is computed with a loop:

Inverse Gray code permutation

```
igraycode( N ):
    Let M = N>>1
    While M > 0
        N ^= M
        M >>= 1
    Return N
```

Iteration of high-pass and low-pass QFs

If H and G are orthogonal QFs, they may be iterated to extract specific ranges of frequencies from a signal. We use the Gray code permutation to find the sequence of operators, such as $HGHHG$, which is needed to extract a particular frequency. We will show the reason for this by using the previous two examples.

Let n be a nonnegative integer with binary representation

$$n = (n_{j-1} \cdots n_1 n_0)_2 \stackrel{\text{def}}{=} 2^{j-1} n_{j-1} + \cdots + 2n_1 + n_0,$$

where $2^j > n$. Consider the composition

$$\psi = \psi_n = F_0 F_1 \cdots F_{j-1} \phi, \tag{5.82}$$

where $\phi = H\phi$ as before and we choose filters according to n:

$$F_k = \begin{cases} H, & \text{if } n_k = 0, \\ G, & \text{if } n_k = 1. \end{cases} \qquad \text{for } k = 0, 1, \ldots, j.$$

We determine G from H by the conventional normalization $g(k) = (-1)^k \bar{h}(1-k)$. Notice that ψ_n does not depend upon j: if we write $n = (0 \cdots 0 n_{j-1} n_{j-2} \cdots n_1 n_0)_2$ with leading zeroes, this just gives $F_0 \cdots F_{j-1} H \cdots H\phi = F_0 \cdots F_{j-1}\phi = \psi_n$.

Example 1: *Shannon functions.* Let m_h be the filter multiplier of Equation 5.70, so that ϕ is defined by $\hat{\phi} = e^{-\pi i \xi} \mathbf{1}_{[-\frac{1}{2}, \frac{1}{2}]}$ as in Equation 5.71. The filter sequence g and multiplier m_g associated to G are given by

$$g(n) = (-1)^n \frac{\sqrt{2} \sin \frac{\pi}{2}(n - \frac{1}{2})}{\pi(n - \frac{1}{2})} \tag{5.83}$$

$$m_g(\xi) = \sqrt{2} e^{-\pi i (\xi + \frac{1}{2})} \sum_{k=-\infty}^{\infty} \mathbf{1}_{[\frac{1}{4}, \frac{3}{4}]}(\xi - k). \tag{5.84}$$

Theorem 5.19 *ψ_n is given by $\hat{\psi}_n(\xi) = e^{-\pi i (\xi + \xi_n)} \mathbf{1}_{[0,1]}(2|\xi| - n')$, where $2\xi_n = \#\{k : n_k = 1\}$ is the number of 1s in the dyadic expansion of n, and $n' = GC^{-1}(n)$.*

Proof: Since this is true for $n' = 0 = GC^{-1}(0)$ because $\psi_0 = \phi$, let us make the inductive step and calculate $\hat{\psi}_{2n}$ and $\hat{\psi}_{2n+1}$ from $\hat{\psi}_n$. But if $n = (n_{j-1} n_{j-2} \cdots n_1 n_0)_2$, then $2n = (n_{j-1} \cdots n_1 n_0 0)_2$, so

$$\begin{aligned}
\hat{\psi}_{2n}(\xi) &= \frac{1}{\sqrt{2}} \hat{\psi}_n(\xi/2) m_h(\xi/2) \\
&= e^{-\pi i (\frac{\xi}{2} + \xi_n)} \mathbf{1}_{[0,1]}(|\xi| - n') e^{-\pi i \frac{\xi}{2}} \sum_{k=-\infty}^{\infty} \mathbf{1}_{[-\frac{1}{4}, \frac{1}{4}]}(\frac{\xi}{2} - k) \\
&= \begin{cases} e^{-\pi i (\xi + \xi_n)} \mathbf{1}_{[0,1]}(2|\xi| - 2n'), & \text{if } n' \text{ is even}, \\ e^{-\pi i (\xi + \xi_n)} \mathbf{1}_{[0,1]}(2|\xi| - (2n' + 1)), & \text{if } n' \text{ is odd}; \end{cases} \\
&= e^{-\pi i (\xi + \xi_{2n})} \mathbf{1}_{[0,1]}(2|\xi| - (2n)').
\end{aligned}$$

In the last step we use Equation 5.80. Notice that $\xi_{2n} = \xi_n$.

Likewise, $2n+1 = (n_{j-1}\cdots n_1 n_0 1)_2$, so

$$\begin{aligned}
\hat{\psi}_{2n+1}(\xi) &= \frac{1}{\sqrt{2}}\hat{\psi}_n(\xi/2)m_g(\xi/2) \\
&= e^{-\pi i(\frac{\xi}{2}+\xi_n)}\mathbf{1}_{[0,1]}(|\xi|-n')e^{-\pi i(\frac{\xi}{2}+\frac{1}{2})}\sum_{k=-\infty}^{\infty}\mathbf{1}_{[\frac{1}{4},\frac{3}{4}]}(\frac{\xi}{2}-k) \\
&= \begin{cases} e^{-\pi i(\xi+[\xi_n+\frac{1}{2}])}\mathbf{1}_{[0,1]}\left(2|\xi|-(2n'+1)\right), & \text{if } n' \text{ is even,} \\ e^{-\pi i(\xi+[\xi_n+\frac{1}{2}])}\mathbf{1}_{[0,1]}\left(2|\xi|-2n'\right), & \text{if } n' \text{ is odd;} \end{cases} \\
&= e^{-\pi i(\xi+\xi_{2n+1})}\mathbf{1}_{[0,1]}\left(2|\xi|-(2n+1)'\right).
\end{aligned}$$

We use Equation 5.81 in the last step, and we compute $\xi_{2n+1} = \xi_n + \frac{1}{2}$. To prove that $2\xi_n = \#\{k : n_k = 1\}$, observe that $\xi_0 = 0$ and that if the relation holds for n it also holds for $2n$ and $2n+1$. □

Because of this theorem, we can say that the function ψ_n has "frequency" $n'/2$, since its Fourier transform is supported on the intervals $[\frac{n'}{2}, \frac{n'+1}{2}]$ and $[\frac{-n'-1}{2}, \frac{-n'}{2}]$.

Example 2: *Haar–Walsh functions.* Let m_h be the filter multiplier of Equation 5.75, so that ϕ is defined by $\phi = \mathbf{1}_{[0,1]}$ as in Equation 5.78. The filter sequence g and multiplier m_g associated to G are given by

$$g(k) = \begin{cases} 1/\sqrt{2}, & \text{if } k=0, \\ -1/\sqrt{2}, & \text{if } k=1, \\ 0, & \text{if } k \notin \{0,1\}; \end{cases} \tag{5.85}$$

$$m_g(\xi) = \frac{1}{\sqrt{2}}\left(1-e^{-2\pi i\xi}\right) = \sqrt{2}e^{-\pi i(\xi-\frac{1}{2})}\sin \pi\xi. \tag{5.86}$$

Iterating this H and G produces the *Walsh functions on the interval* $[0,1]$, which can be defined by various formulas including the following recursion:

$$\psi_{2n}(t) = \psi_n(2t)+\psi_n(2t-1); \tag{5.87}$$

$$\psi_{2n+1}(t) = \psi_n(2t)-\psi_n(2t-1). \tag{5.88}$$

Note that $\operatorname{supp}\psi_n = [0,1]$ for all n, and that the function ψ_n takes only the values 1 and -1 on $[0,1]$. In fact, ψ_{2n} has squeezed versions of ψ_n as its left and right halves, while ψ_{2n+1} has a squeezed version of ψ_n as its left half and a squeezed version of $-\psi_n$ as its right half.

Theorem 5.20 *The number of times $\psi_n(t)$ changes sign as t moves from 0 to 1 is given by $GC^{-1}(n)$.*

Proof: Let n' be the number of zero-crossings of ψ_n. Since $\psi_0 = \phi = \mathbf{1}_{[0,1]}$ has no zero-crossings, we have $0' = GC^{-1}(0)$. Now suppose that ψ_n has $n' = GC^{-1}(n)$ zero-crossings. If n' is even, then $\psi_n(0)$ and $\psi_n(1)$ have the same sign; then ψ_{2n} has $2n'$ zero-crossings while ψ_{2n+1} has $2n'+1$ zero-crossings. If n' is odd, then $\psi_n(0)$ and $\psi_n(1)$ have opposite sign; then ψ_{2n} has $2n'+1$ zero-crossings while ψ_{2n+1} has $2n'$ zero-crossings. Hence

$$(2n)' = \begin{cases} 2n', & \text{if } n' \text{ is even,} \\ 2n'+1, & \text{if } n' \text{ is odd;} \end{cases} \qquad (2n+1)' = \begin{cases} 2n'+1, & \text{if } n' \text{ is even,} \\ 2n', & \text{if } n' \text{ is odd.} \end{cases}$$

Comparing with Equations 5.80 and 5.81 completes the proof. □

The number of sign changes or zero-crossings gives another notion of frequency. Notice that the function $\cos kt$, which has a "traditional" frequency k, has $2k$ zero-crossings on a period interval $[0, 2\pi]$. Thus we will say that the frequency of ψ_n in the Haar–Walsh case is $n'/2$.

The two examples use two notions of frequency: localization of the Fourier transform, and number of zero-crossings or "oscillations." The first makes sense for the Shannon scaling function and its descendents because the Fourier transforms of those functions are compactly supported. The second makes sense for the Haar–Walsh case because Walsh functions oscillate between the values 1 and -1. The Shannon and Haar–Walsh examples are extremes, but the relationship between n and the frequency is the same in both cases.

We would like to generalize this notion to other finitely supported filters, since those will provide intermediate cases between Haar–Walsh and Shannon functions. This can be done, but we will avoid the technicalities of that construction. We will simply say that the frequency of the function ψ_n generated by iterating conjugate QFs H and G is $\frac{1}{2}GC^{-1}(n)$.

An interesting difference in all the finitely supported cases except Haar–Walsh is that the ℓ^∞ norm of the ψ_n is not bounded. It grows no faster than n^δ, for some $0 < \delta \leq 1/4$ [28]. This implies that the support of $\hat{\psi}_n$ spreads like n^δ as n increases, making the frequency localization less and less precise. The practical consequences of such spreading can be severe, but can be controlled in one of two ways. We can limit the size of n by performing a windowed analysis, to prevent the number of sample points within a window from exceeding a fixed limit determined by the required frequency precision. We discussed techniques for extracting smooth periodic segments from a long sampled signal in Chapter 4. Alternatively, we can use longer filters with better frequency resolution or even increase the filter length as we iterate. As described in [56], we can construct a family of filters whose length is $O(L^2)$ at iteration number L, such that the functions ψ_n will be uniformly bounded.

5.4 Implementing convolution-decimation

We will restrict our attention to convolution of sequences. We distinguish two cases: *aperiodic* convolution-decimation of two finitely supported infinite sequences, and *periodic* convolution-decimation of two periodic sequence.

5.4.1 General assumptions

It is simplest, safest, and most flexible to write a generic convolution-decimation function which can be reused in various transformations. This function will be last on the execution stack and first off, again and again, because it performs the most basic arithmetic and because the transforms we are considering are recursive. Hence we can gain much speed and reliability by a careful design at this step.

It is good programming practice to confine memory manipulation tasks to the least-called functions. It is also good practice to avoid restrictive assumptions about which input values are read before an output value is written. Thus, we will always assume that the input and output are distinct arrays (or streams) and have been preallocated. We will also assume that any necessary scratch arrays are distinct and have been preallocated.

Since convolution-decimation is a linear operation, it is most general to superpose the computed values onto the output array. But this requires that we initialize the output array, either with zeroes or with some previously computed signal. We will also assume that the scratch array contains all zeroes, as well.

If the convolution-decimation is part of a separable multidimensional discrete wavelet transform (see Chapter 9), we may wish to write the output array coefficients noncontiguously in order to eliminate the explicit transposition step. We will assume that the input array is contiguous, however, since that is the typical arrangement for line-scanned multidimensional signals.

These assumptions, plus the *variable lengths of the filter arrays*, force us to use a rather large number of parameters in the generic convolution-decimation routine:

- A pointer to the output array;
- The increment to use between output values;
- A pointer to the input array;
- The integers describing the support of the input array;
- The quadrature filter specification.

From this data we can compute the length of the output array. Indeed, we must compute this length before applying the convolution-decimation, since it is necessary to preallocate that many memory locations.

Because the indexing of the QF affects the relationship between input indices and output indices, it is necessary to adopt some conventions. We must specify the support interval $[\alpha, \omega]$ of the filter sequence, and the values of the sequence $f = \{f(n)\}$ on that interval, namely $f(\alpha), f(\alpha+1), \ldots, f(\omega)$, which includes all of the nonzero values. In some cases we have a bit of freedom to shift the sequence: the free choice of M in Equation 5.19 is one example. We will adopt some standard method of storing filter and signal sequences, and refer to them collectively as *conventional indexing.*

The first assumption is that a filter sequence f supported in $[\alpha, \omega]$ will have its extreme indices satisfy

$$\alpha \leq 0 \leq \omega. \tag{5.89}$$

We can always arrange that both members of an orthogonal pair of QFs have this property, and we will incorporate this convention into the functions that generate filter specifications.

Conventional indexing will impose additional requirements depending upon the type of filter we are using. For a conjugate orthogonal pair h, g of Daubechies-type QFs, which will have equal and even support diameters $2R$, conventional indexing will also mean choosing $\alpha = 0$ and $\omega = 2R - 1$. This implies that $g(n) = (-1)^n h(2R - 1 - n)$.

For a symmetric or antisymmetric quadruplet h, g, h', g' of biorthogonal filter sequences, we only need to store half the coefficients for each filter sequence. Conventional indexing in this case will imply putting this determining half into $f(0), f(1), \ldots, f(\omega)$. However, we will also store the other half of the coefficients, the ones in $f(\alpha), f(\alpha+1), \ldots, f(-1)$, so that we can use the same convolution-decimation function for this as well as the nonsymmetric case. The symmetry implies the following relationships among the conventionally indexed arrays:

Symmetric about $n = 0$: $h(-n) = h(n)$ for $n = 1, 2, \ldots, \omega$: there must be an odd number of filter coefficients, $2\omega + 1$ in all, and $\alpha = -\omega$;

Symmetric about $n = -1/2$: $h(-n) = h(n-1)$ for $n = 1, 2, \ldots, \omega$: there must be an even number of filter coefficients, $2\omega + 2$ in all, and $\alpha = -\omega - 1$;

Antisymmetric about $n = -1/2$: $h(-n) = -h(n-1)$ for $n = 1, 2, \ldots, \omega$: there must be an even number of filter coefficients, $2\omega + 2$ in all, and $\alpha = -\omega - 1$.

Using symmetry to reduce the space and time requirements falls into the category of "speed tricks," discussed at the end of this chapter, and requires writing a special-purpose convolution-decimation function.

We will always assume that our QFs and input sequences are real-valued. This means that F and F^* can use the same filter sequence arrays, and that we can use ordinary Standard C floating-point arithmetic. It also implies that the output sequences will be real-valued. The filter sequence f can be packaged into a *prepared quadrature filter* or *PQF* data structure with these members:

- *PQF.F*, the original conventionally indexed filter sequence,
- *PQF.ALPHA*, the least index of the original filter sequence for which the filter coefficient is nonzero,
- *PQF.OMEGA*, the greatest index of the original filter sequence for which the filter coefficient is nonzero.

Additionally, it is useful to precompute and store the quantities used to correct for the phase shift of the filter:

- *PQF.CENTER*, the center of energy $c[f]$ of the filter sequence,
- *PQF.DEVIATION*, the maximum deviation $d[f]$ from linear phase.

Also, we will leave space for arrays of preperiodized coefficients, to be used in periodic convolution-decimation:

- *PQF.FP*, one preperiodized sequence for each even period $q > 0$ which is less than or equal to `PQF.OMEGA-PQF.ALPHA`.

Remark. We have a choice, whether to store the preperiodized sequence as a list of individual arrays or as one long concatenated array. In the latter method, which we shall adopt, it is necessary to write an indexing function which returns the starting offset of each preperiodized subarray.

To compute the center of energy and the deviation from linear phase of a filter sequence, we will use two utilities. These are more generally useful, and not significantly slower, if they do not assume that the sequence has unit norm.

Center of energy for an arbitrary sequence

```
coe( U, LEAST, FINAL ):
   Let ENERGY = 0
   Let CENTER = 0
   For I = LEAST to FINAL
      Let USQUARED = U[I]*U[I]
      CENTER += I*USQUARED
      ENERGY += USQUARED
   If ENERGY>0 then
      CENTER /= ENERGY
   Return CENTER
```

Deviation from linear phase for a filter sequence

```
lphdev( F, ALPHA, OMEGA ):
   Let ENERGY = 0
   For K = ALPHA to OMEGA
      ENERGY += F[K]*F[K]
   Let DEVIATION = 0
   If ENERGY>0 then
      Let SGN = -1
      For N = 1 to IFH(OMEGA-ALPHA)
         Let  FX = 0
         For K = N+ALPHA to OMEGA-N
            FX += K * F[K-N] * F[K+N]
         DEVIATION += SGN*FX
         Let SGN = -SGN
      Let DEVIATION = 2 * absval(DEVIATION) / ENERGY
  Return DEVIATION
```

Utilities to compute the preperiodized filter coefficients will be defined in the section on periodic convolution-decimation.

5.4.2 Aperiodic convolution-decimation

This algorithm can be applied to input sequences of arbitrary finite length. Suppose that f is a filter sequence, supported in $[\alpha, \omega]$ with the conventional indexing. Suppose too that u is a finitely supported sequence whose nonzero elements are $\{u(a), u(a+1), \ldots, u(0), \ldots, u(b)\}$ for $a \leq 0$ and $b \geq 0$. The convolution-decimation formula in the second half of Equation 5.1 then reduces to

$$Fu(i) = \sum_{j=\alpha}^{\omega} f(j)u(2i-j). \tag{5.90}$$

Notice that the summand will be zero if $2i - \alpha < a$ or $2i - \omega > b$. Hence, we need only compute the values of the convolution-decimation for indices $i \in \{a', \ldots, 1, 0, 1, \ldots, b'\}$, where

$$a' = \lceil (a+\alpha)/2 \rceil; \qquad b' = \lfloor (b+\omega)/2 \rfloor. \tag{5.91}$$

Now,

$$\lceil n/2 \rceil = \begin{cases} n/2, & \text{if } n \text{ is even,} \\ (n+1)/2, & \text{if } n \text{ is odd;} \end{cases} \qquad \lfloor m/2 \rfloor = \begin{cases} m/2, & \text{if } m \text{ is even,} \\ (m-1)/2, & \text{if } m \text{ is odd.} \end{cases}$$

Thus we can implement $\lceil n/2 \rceil$ and $\lfloor m/2 \rfloor$ by the following preprocessor macros:

```
#define ICH(n)             ((n)&1?((n)+1)/2:(n)/2)
#define IFH(m)             ((m)&1?((m)-1)/2:(m)/2)
```

Remark. In Standard C, integer division rounds the quotient toward zero. In other words, the following holds true for integer expressions n:

$$n/2 = \begin{cases} \lfloor n/2 \rfloor, & \text{if } n > 0, \\ \lceil n/2 \rceil, & \text{if } n < 0. \end{cases}$$

Now $b \geq 0 \Rightarrow b' \geq 0$, and also $a \leq 0 \Rightarrow a' \leq 0$ if we use conventionally indexed QFs. Hence, if we start with a sequence v whose support straddles zero, then we can use simple integer division to compute a', b' from a, b. But using `ICH()` and `IFH()` reduces the number of assumptions and is the safer method.

Equation 5.90 also shows that for each output index i, the only values of $j \in [\alpha, \omega]$ for which the summand will be nonzero are those which also satisfy $a \leq 2i - j \leq b$. Thus the range of the summation will be $j = \max\{\alpha, 2i - b\}$ to $j = \min\{\omega, 2i - a\}$. These endpoints can be computed with the aid of the two preprocessor macros defined in Chapter 2

From Equation 5.90, we obtain the following implementation of an aperiodic convolution-decimation function:

Aperiodic convolution-decimation: sequential output version

```
cdao( OUT, STEP, IN, A, B, F ):
   Let APRIME = ICH(A+F.ALPHA)
   Let BPRIME = IFH(B+F.OMEGA)
   For I = APRIME to BPRIME
      Let BEGIN = max( F.ALPHA, 2*I-B )
      Let END   = min( F.OMEGA, 2*I-A )
      For J = BEGIN to END
         OUT[I*STEP] += F.F[J]*IN[2*I-J]
```

This code finishes computing each successive value of OUT[] before affecting the next one. Notice that this code includes a parameter STEP, possibly different from 1, by which to increment the array OUT[].

If we wish to assign output values rather than to superpose them, we need only replace the increment operator in the OUT[] statement by an assignment operator. Let us call the resulting modification of cdao() by the name cdae(); it has the same parameter list.

Alternatively, we can interchange the order of the two "for" loops and imagine "spraying" values of the input array onto the output array. Then we must consider what output indices i are affected by each input value. From the first half of Equation 5.1, we see that

$$Fu(i) = \sum_{j=a}^{b} f(2i-j)u(j). \tag{5.92}$$

The summand will be zero unless $\alpha \leq 2i-j \leq \omega$, so that for fixed $j \in [a,b]$ we must compute the output for $\lceil (j+\alpha)/2 \rceil \leq i \leq \lfloor (j+\omega)/2 \rfloor$. Notice that these endpoints can be of either sign so that we are forced to use ICH() and IFH() rather than integer division. We get a second implementation of aperiodic convolution-decimation:

Aperiodic convolution-decimation: sequential input version

```
cdai( OUT, STEP, IN, A, B, F ):
   For J = A to B
      Let BEGIN = ICH(J+F.ALPHA)
      Let END   = IFH(J+F.OMEGA)
      For I = BEGIN to END
         OUT[I*STEP] += F.F[2*I-J]*IN[J]
```

This implementation finishes using each input value before reading its successor.

Which of the two implementations should be used in a particular case? If the output values cannot be buffered, or must be written successively to a stream, then we must use the *sequential output* version. If we are mixing the "superposition" and "assignment" versions of the function, then it is better to use the sequential output version since the order of summation will be the same in both versions. If we are reading from a stream with no possibility of backtracking, then we must use the sequential input version. The number of operations differs between versions, too, due to the overhead required to compute loop ranges and indices. These differences, however, are minor.

5.4.3 Adjoint aperiodic convolution-decimation

Using Equation 5.2 with a conventionally indexed real QF f and a sequence u supported in $[c, d]$ yields the following formula:

$$F^*u(j) = \sum_{i=c}^{d} f(2i - j)u(i). \tag{5.93}$$

The summand will be zero unless $\alpha \leq 2i - j \leq \omega$ and $c \leq i \leq d$. Hence, the output index j is confined to the interval $[c', d']$, where

$$c' = 2c - \omega; \qquad d' = 2d - \alpha. \tag{5.94}$$

For any such j, the summation will range from the index $i = \max\{\lceil (j+\alpha)/2 \rceil, c\}$ up to the index $i = \min\{\lfloor (j + \omega)/2 \rfloor, d\}$. Thus one version of the adjoint convolution-decimation algorithm can be implemented as follows:

Adjoint aperiodic convolution-decimation: sequential output

```
acdao( OUT, STEP, IN, C, D, F ):
   Let CPRIME = 2*C-F.OMEGA
   Let DPRIME = 2*D-F.ALPHA
   For J = CPRIME to DPRIME
      Let BEGIN = max( ICH(J+F.ALPHA ), C )
      Let END   = min( IFH(J+F.OMEGA ),  D )
      For I = BEGIN to END
         OUT[J*STEP] += F.F[2*I-J]*IN[I]
```

There is also a second version in this adjoint case. Namely, we can read the input values sequentially and build up the output values. But for each input index i in the range $[c, d]$, we will only have a (possibly) nonzero sum in Equation 5.93 if $\alpha \leq 2i - j \leq \omega$, or $2i - \omega \leq j \leq 2i - \alpha$ as in the code fragment below:

```
...
For I = C to D
   Let BEGIN = 2*I-F.OMEGA
   Let END   = 2*I-F.ALPHA
   For J = BEGIN to END
      OUT[J*STEP] += F.F[2*I-J]*IN[I]
...
```

But it is wasteful to compute the shifted endpoints for j only to unshift them in the innermost loop, so we replace j with $2i - j$ in the actual function implementation:

Adjoint aperiodic convolution-decimation: sequential input

```
acdai( OUT, STEP, IN, C, D, F ):
   For I = C to D
      For J = F.ALPHA to F.OMEGA
         OUT[(2*I-J)*STEP] += F.F[J]*IN[I]
```

If we wish to assign output values rather than to superpose them, we need only replace the increment operator in the `OUT[]` statement by an assignment operator. Let us call the resulting modification of `acdao()` by the name `acdae()`; it has the same parameter list as `acdao()` or `acdai()`.

As before, there are situations in which either of the two versions would be preferable. The criteria for the adjoint are the same as those for convolution-decimation.

5.4.4 Periodic convolution-decimation

We will restrict our attention to convolution followed by decimation by two. In the q-periodic case, we will assume that the period is divisible by two so that the two-decimated sequence has integer period $q/2$.

Preperiodized QFs

Starting with a single finitely supported and conventionally indexed filter sequence $f = \{f(\alpha), \ldots, f(\omega)\}$, we generate all the periodized sequences $f_2, f_4, \ldots$ with even periods q less than the support length $1+\omega-\alpha$ of f. This is defined by the following equation:

$$f_q(k) \stackrel{\text{def}}{=} \sum_{j=\lceil(\alpha-k)/q\rceil}^{\lfloor(\omega-k)/q\rfloor} f(k+jq), \qquad \text{for } k = 0, 1, \ldots, q-1. \tag{5.95}$$

First we must decide how to store the coefficients. For the moment, let us suppose that the PQF data structure's FP member is a single long array, and that the coefficients preperiodized to the even period `Q` begin at the offset `FQ = F.FP+PQFO(Q/2)`. We will define the offset calculation function `PQFO()` later.

The preperiodization calculation can be done with a simple utility program. The only computation we must do is to find which values of $k + jq$ fall in the support of the filter. We can take advantage of conventional indexing and note that the quantity $\alpha - k$ is never greater than zero for all $k = 0, \ldots, q-1$. Thus we can compute $\lceil (\alpha - k)/q \rceil = (\alpha - k)/q$ in Standard C by negative integer division. In a language not conforming to the rule that integer quotients round to the integer nearest to zero, we could insert a statement "`If K+J*Q<ALPHA then J+=1`" just before the "while" loop:

q-Periodization of a sequence

```
periodize( FQ, Q, F, ALPHA, OMEGA):
   For K = 0 to Q-1
      Let FQ[K] = 0
      Let J = (ALPHA-K)/Q
      While K+J*Q <= OMEGA
         FQ[K] += F[K+J*Q]
         J += 1
```

We have also assumed that the output array has been preallocated, although it need not be initialized with zeroes.

Periodic convolution-decimation with the "mod" operator

If u and f are q-periodic sequences, it is only necessary to store q values of each, namely $u(0), \ldots, u(q-1)$ and $f(0), \ldots, f(q-1)$. We use the second part of Equation 5.3 to get the periodized convolution-decimation formula, for $i = 0, 1, \ldots, q/2 - 1$:

$$Fu(i) = \begin{cases} \sum_{j=\alpha}^{\omega} f(j)u(2i - j \bmod q), & \text{if } q > \omega - \alpha; \\ \sum_{j=0}^{q-1} f_q(j)u(2i - j \bmod q), & \text{if } q \leq \omega - \alpha. \end{cases} \quad (5.96)$$

Notice that the summation uses different filter coefficients for different periods.

Let us assume that the filter sequence f has been packaged into the PQF data structure described earlier. To implement the periodic convolution-decimation, we

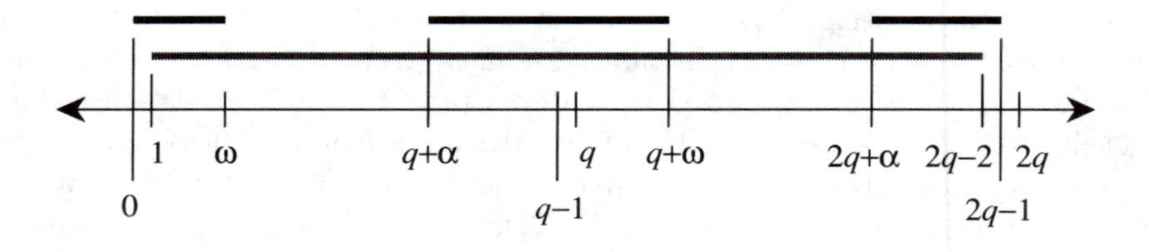

Figure 5.7: Index intervals $q + 2i - j \in [1, 2q - 2]$ and $2i - j \in [\alpha, \omega] \bmod q$.

can use the remainder or "mod" operator, which in Standard C is represented by "%". Notice how we add q to insure that the left argument of "mod" is nonnegative:

Modular convolution-decimation: sequential output

```
cdmo( OUT, STEP, IN, Q, F ):
   Let Q2 = Q/2
   If Q > F.OMEGA-F.ALPHA then
      Let FILTER = F.F
      For I = 0 to Q2-1
         For J = F.ALPHA to F.OMEGA
            OUT[I*STEP] += FILTER[J]*IN[(Q+2*I-J)%Q]
   Else
      Let FILTER = F.FP + PQFO(Q2)
      For I = 0 to Q2-1
         For J = 0 to Q-1
            OUT[I*STEP] += FILTER[J]*IN[(Q+2*I-J)%Q]
```

We can also interchange the order of summation and use the first part of Equation 5.3 to get a *sequential input* algorithm. There are two cases to consider:

- If $q \leq \omega - \alpha$, then $f_q(j)$ is defined for all $j = 0, 1, \ldots, q - 1$; we use the observation that $1 \leq q + 2i - j \leq 2q - 2$ for all $0 \leq j < q$ and all $0 \leq i < q/2$ to insure that the operands of mod are positive. This range of values is depicted by the gray line in Figure 5.7.

- If $q > \omega - \alpha$, then $f_q(n)$ will be nonzero for $0 \leq n < q$ only if $n \in [0, \omega] \cup [q + \alpha, q - 1]$. Now let $n = 2i - j \bmod q$; since $-q + 2 \leq 2i - j \leq q - 1$, we get three contiguous intervals on which $f_q(2i - j)$ is nonzero. These are depicted by the black lines in Figure 5.7.

In the first case, we can reuse the double loop from `cdmo()`, with the i and j summations interchanged so that we read input values sequentially. As before, we augment the input index $2i - j$ by q to present the "mod" operator with a positive argument.

The three i-intervals for the second of these two cases may be computed as follows:

$$\begin{aligned} -q+1 \le 2i-j \le -q+\omega \quad &\Rightarrow \quad \left\lceil \frac{j+1}{2} \right\rceil - \frac{q}{2} \le i \le \left\lfloor \frac{j+\omega}{2} \right\rfloor - \frac{q}{2}, \\ \alpha \le 2i-j \le \omega \quad &\Rightarrow \quad \left\lceil \frac{j+\alpha}{2} \right\rceil \le i \le \left\lfloor \frac{j+\omega}{2} \right\rfloor, \\ q+\alpha \le 2i-j \le q-2 \quad &\Rightarrow \quad \left\lceil \frac{j+\alpha}{2} \right\rceil + \frac{q}{2} \le i \le \left\lfloor \frac{j}{2} \right\rfloor + \frac{q}{2} - 1. \end{aligned} \tag{5.97}$$

Since $0 \le i \le \frac{q}{2} - 1$, we can narrow these i-intervals still further:

$$\begin{aligned} \max\left\{0, \left\lceil \frac{j+1}{2} \right\rceil - \frac{q}{2}\right\} \quad &\le \quad i \quad \le \quad \min\left\{\frac{q}{2} - 1, \left\lfloor \frac{j+\omega}{2} \right\rfloor - \frac{q}{2}\right\}, \\ \max\left\{0, \left\lceil \frac{j+\alpha}{2} \right\rceil\right\} \quad &\le \quad i \quad \le \quad \min\left\{\frac{q}{2} - 1, \left\lfloor \frac{j+\omega}{2} \right\rfloor\right\}, \\ \max\left\{0, \left\lceil \frac{j+\alpha}{2} \right\rceil + \frac{q}{2}\right\} \quad &\le \quad i \quad \le \quad \min\left\{\frac{q}{2} - 1, \left\lfloor \frac{j}{2} \right\rfloor + \frac{q}{2} - 1\right\}. \end{aligned} \tag{5.98}$$

Conventional indexing allows us to predict the values of some of the endpoints:

$$\begin{aligned} 0 \quad &\le \quad i \quad \le \quad \left\lfloor \frac{j+\omega}{2} \right\rfloor - \frac{q}{2}, \\ \max\left\{0, \left\lceil \frac{j+\alpha}{2} \right\rceil\right\} \quad &\le \quad i \quad \le \quad \min\left\{\frac{q}{2} - 1, \left\lfloor \frac{j+\omega}{2} \right\rfloor\right\}, \\ \left\lceil \frac{j+\alpha}{2} \right\rceil + \frac{q}{2} \quad &\le \quad i \quad \le \quad \frac{q}{2} - 1. \end{aligned} \tag{5.99}$$

Some of these intervals might be empty, and we must be sure that this will be correctly handled by the loop semantics, as it is in our pseudocode. We can precompute $q/2$, $\lceil (j+\alpha)/2 \rceil$, and $\lfloor (j+\omega)/2 \rfloor$, storing them in temporary variables, since they are reused throughout the calculation:

Modular convolution-decimation: sequential input

```
cdmi( OUT, STEP, IN, Q, F ):
   Let Q2  = Q/2
   If Q > F.OMEGA-F.ALPHA then
      Let FILTER = F.F
      For J = 0 to Q-1
         Let JA2 = ICH(J+F.ALPHA)
         Let JO2 = IFH(J+F.OMEGA)
         For I = 0 to JO2-Q2
            OUT[I*STEP] += FILTER[Q+2*I-J]*IN[J]
         For I = max(0,JA2) to min(Q2-1,JO2)
            OUT[I*STEP] += FILTER[2*I-J]*IN[J]
         For I = JA2+Q2 to Q2-1
            OUT[I*STEP] += FILTER[2*I-J-Q]*IN[J]
   Else
      Let FILTER = F.FP + PQFO(Q)
      For J = 0 to Q-1
         For I = 0 to Q2-1
            OUT[I*STEP] += FILTER[(Q+2*I-J)%Q]*IN[J]
```

Avoiding the "mod" operator

For $0 \le i \le \frac{q}{2} - 1$ and $0 \le j \le q - 1$, the quantity $2i - j$ lies between $-q + 1$ and $q-2$. Thus if we extend the periodized filter sequence $\{f_q(0), f_q(1), \ldots, f_q(q-1)\}$ by prepending $\{f_q(-q) = f_q(0), f_q(-q+1) = f_q(1), f_q(-q+2) = f_q(2), \ldots, f_q(-1) = f_q(q-1)\}$, we can avoid using the remainder operator.

The extended preperiodizations will be arrays of length $2+2, 4+4, \ldots, 2M+2M$, where $M = \lfloor(\omega - \alpha)/2\rfloor$. The total length of all arrays is $4(1 + 2 + \cdots + M) = 2M(M + 1)$. If we concatenate the arrays in increasing order of length, starting with the one of period two, then the "origin" of the subarray of length $q = 2m$ will be offset by $2m(m + 1) - 2m = 2m^2$ from the beginning of the array.

The following pair of preprocessor macros can be used to compute this offset and the total length:

Length and origin for concatenated, preperiodized filter arrays

```
#define PQFL(m)        (2*(m)*(m+1))
#define PQFO(m)        (2*(m)*(m))
```

The concatenated array `FP[]`, preallocated to length `PQFL(M)`, should contain $f_q(n)$ at offset `PQFO(Q/2)+N`, for $M = \lfloor(\omega - \alpha)/2\rfloor$ and $-q \le n < q$. The following utility performs the assignment:

Extension of a preperiodized QF

```
qfcirc(FP, F, ALPHA, OMEGA):
   Let M = IFH(OMEGA-ALPHA)
   For N = 1 to M
      Let Q = 2*N
      Let FQ = FP + PQFO(N)
      periodize( FQ, Q, F, ALPHA, OMEGA )
      For K = 0 to Q-1
         Let FQ[K-Q] = FQ[K]
```

After such preparation, we can eliminate the "mod" operator from the sequential input version of periodic convolution-decimation:

Periodic convolution-decimation: sequential input

```
cdpi( OUT, STEP, IN, Q, F ):
   Let Q2  = Q/2
   If Q > F.OMEGA-F.ALPHA then
      Let FILTER = F.F
      For J = 0 to Q-1
         Let JA2 = ICH(J+F.ALPHA)
         Let JO2 = IFH(J+F.OMEGA)
         For I = 0 to JO2-Q2
            OUT[I*STEP] += FILTER[Q+2*I-J]*IN[J]
         For I = max(0,JA2) to min(Q2-1,JO2)
            OUT[I*STEP] += FILTER[2*I-J]*IN[J]
         For I = JA2+Q2 to Q2-1
            OUT[I*STEP] += FILTER[2*I-J-Q]*IN[J]
   Else
      Let FILTER = F.FP+PQFO(Q2)
      For J = 0 to Q-1
         For I = 0 to Q2-1
            OUT[I*STEP] += FILTER[2*I-J]*IN[J]
```

In the sequential output version, we must again eliminate the "mod" operator from the $q > \omega - \alpha$ case. Again we use the first part of Equation 5.3 and assume conventional indexing, only this time we solve for the j-intervals. There might be as many as three of these:

$$
\begin{aligned}
-q+1 \leq 2i-j \leq -q+\omega &\Rightarrow 2i+q-\omega \leq j \leq 2i+q-1, \\
\alpha \leq 2i-j \leq \omega &\Rightarrow 2i-\omega \leq j \leq 2i-\alpha, \\
q+\alpha \leq 2i-j \leq q-2 &\Rightarrow 2i-q+2 \leq j \leq 2i-q-\alpha.
\end{aligned}
\tag{5.100}
$$

Since $0 \leq j \leq q-1$, we can narrow these j-intervals still further:

$$
\begin{array}{rcccl}
\max\{0, 2i+q-\omega\} & \leq & j & \leq & \min\{q-1, 2i+q-1\}, \\
\max\{0, 2i-\omega\} & \leq & j & \leq & \min\{q-1, 2i-\alpha\}, \\
\max\{0, 2i-q+2\} & \leq & j & \leq & \min\{q-1, 2i-q-\alpha\}.
\end{array}
\tag{5.101}
$$

We simplify this using the conventional indexing assumptions:

$$
\begin{array}{rcccl}
2i+q-\omega & \leq & j & \leq & q-1, \\
\max\{0, 2i-\omega\} & \leq & j & \leq & \min\{q-1, 2i-\alpha\}, \\
0 & \leq & j & \leq & 2i-q-\alpha.
\end{array}
\tag{5.102}
$$

We split up the j-summation accordingly, allowing for some of these intervals to be empty:

Periodic convolution-decimation: sequential output (1)

```
cdpo1( OUT, STEP, IN, Q, F ):
   Let Q2  = Q/2
   If Q > F.OMEGA-F.ALPHA then
      Let FILTER = F.F
      For I = 0 to Q2-1
         Let A2I = 2*I-F.ALPHA
         Let O2I = 2*I-F.OMEGA
         For J = 0 to A2I-Q
            OUT[I*STEP] += FILTER[2*I-J-Q]*IN[J]
         For J = max(0,O2I) to min(Q-1,A2I)
            OUT[I*STEP] += FILTER[2*I-J]*IN[J]
         For J = O2I+Q to Q-1
            OUT[I*STEP] += FILTER[Q+2*I-J]*IN[J]
   Else
      Let FILTER = F.FP+PQFO(Q2)
      For I = 0 to Q2-1
         For J = 0 to Q-1
            OUT[I*STEP] += FILTER[2*I-J]*IN[J]
```

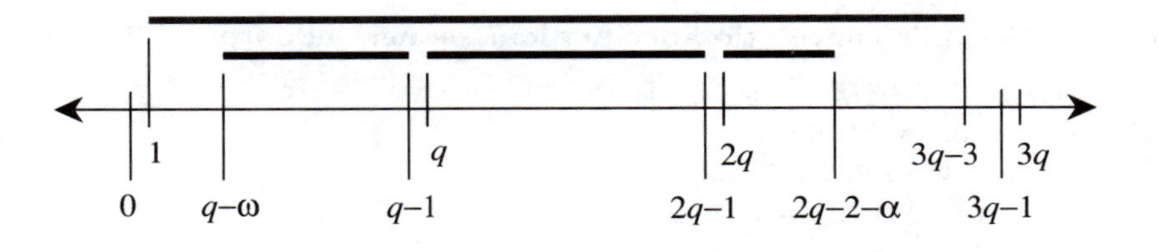

Figure 5.8: Index intervals $q + 2i - j \in [1, 2q - 2]$ and $j \in [\alpha, \omega]$ mod q.

We can also use the second part of Equation 5.3 to obtain a variant of this implementation. The ranges are now $0 \le i \le \frac{q}{2} - 1$ and $\alpha \le j \le \omega$, so that $1 \le q - \omega \le q + 2i - j \le 2q - 2 - \alpha \le 3q - 3$. Thus j-summation might occur over as many as three contiguous intervals, as depicted by the black lines in Figure 5.8. The formulas for the endpoints of these intervals are given below:

$$\begin{aligned} q - \omega \le q + 2i - j \le q - 1 &\quad\Rightarrow\quad 2i + 1 \le j \le 2i + \omega, \\ q \le q + 2i - j \le 2q - 1 &\quad\Rightarrow\quad 2i - q + 1 \le j \le 2i, \\ 2q \le q + 2i - j \le 2q - 2 - \alpha &\quad\Rightarrow\quad 2i + \alpha - q - 2 \le j \le 2i - q. \end{aligned} \tag{5.103}$$

We must also have $\alpha \le j \le \omega$, so the j-intervals are as follows:

$$\begin{aligned} \max\{\alpha, 2i + 1\} &\le j \le \min\{\omega, 2i + \omega\}, \\ \max\{\alpha, 2i - q + 1\} &\le j \le \min\{\omega, 2i\}, \\ \max\{\alpha, 2i + \alpha - q - 2\} &\le j \le \min\{\omega, 2i - q\}. \end{aligned} \tag{5.104}$$

Conventional indexing allows us to predict the values of some of the endpoints:

$$\begin{aligned} 2i + 1 &\le j \le \omega, \\ \max\{\alpha, 2i - q + 1\} &\le j \le \min\{\omega, 2i\}, \\ \alpha &\le j \le 2i - q. \end{aligned} \tag{5.105}$$

These j-intervals are arranged in reverse order; the last one is to the left of the middle one, which is to the left of the first one. Also, some of them might be empty.

We combine the summation over these intervals, again assuming that the loop syntax will handle the empty intervals correctly. We can save some effort by computing and retaining the values of endpoints between loop segments, since the segments are adjacent.

Periodic convolution-decimation: sequential output (2)

```
cdpo2( OUT, STEP, IN, Q, F ):
   Let Q2 = Q/2
   If Q > F.OMEGA-F.ALPHA then
      Let FILTER = F.F
      For I = 0 to Q2-1
         Let END   = 2*I-Q
         For J = F.ALPHA to END
            OUT[I*STEP] += FILTER[J]*IN[Q+2*I-J]
         Let BEGIN = max(F.ALPHA, END+1)
         Let END   = min(F.OMEGA, 2*I)
         For J = BEGIN to END
            OUT[I*STEP] += FILTER[J]*IN[2*I-J]
         Let BEGIN = 2*I+1
         For J = BEGIN to F.OMEGA
            OUT[I*STEP] += FILTER[J]*IN[2*I-J-Q]
   Else
      Let FILTER = F.FP+PQFO(Q2)
      For I = 0 to Q2-1
         For J = 0 to Q-1
            OUT[I*STEP] += FILTER[2*I-J]*IN[J]
```

Since the two implementations `cdpo1()` and `cdpo2()` perform the same calculation in a different order, they may be used to validate the arithmetic on a particular computer. Before doing that, they can help validate the competence of the programmer: the implementation is certainly incorrect if the two functions produce outputs that differ by more than a small multiple of the machine precision.

In an actual implementation, we simply choose a default version with a preprocessor macro:

Fix a default sequential-output convolution-decimation

```
#define cdpo cdpo2
```

Of course, `cdpo1()` would work just as well for the default.

When reusing arrays, on occasion it is more efficient to assign values into the output rather than to superpose them. For this we must use the sequential output version of convolution-decimation, but we replace the increment operator with an assignment operator in all statements with `OUT[]` in the left-hand side. Let us call the resulting function `cdpe()`; it has the same parameter list as `cdpi()` and `cdpo()`.

5.4.5 Adjoint periodic convolution-decimation

The adjoint of periodic convolution-decimation can also take advantage of the preperiodized and extended arrays. Here the filter f_q is used to produce an output array of length q from an input array of length $q/2$.

We will leave the "mod" implementation as an exercise; what follows are two implementations which avoid the "mod" operator. Consider first the sequential output case: for $j = 0, 1, \ldots, q-1$, we sum over i.

For $q \leq \omega - \alpha$, Equation 5.4 can be used directly in the implementation.

For $q > \omega - \alpha$, the summation splits into pieces just as in Figure 5.7, and the i-interval endpoints are given by Equations 5.99. Not surprisingly, the sequential output adjoint convolution-decimation looks much like sequential input convolution-decimation, only with i and j playing reversed roles:

Periodic adjoint convolution-decimation: sequential output

```
acdpo( OUT, STEP, IN, Q2, F ):
   Let Q = 2*Q2
   If Q > F.OMEGA-F.ALPHA then
      Let FILTER = F.F
      For J = 0 to Q-1
         Let JA2 = ICH(J+F.ALPHA)
         Let JO2 = IFH(J+F.OMEGA)
         For I = 0 to JO2-Q2
            OUT[I*STEP] += FILTER[Q+2*I-J]*IN[I]
         For I = max(0,JA2) to min(Q2-1,JO2)
            OUT[I*STEP] += FILTER[2*I-J]*IN[I]
         For I = JA2+Q2 to Q2-1
            OUT[I*STEP] += FILTER[2*I-J-Q]*IN[I]
   Else
      Let FILTER = F.FP+PQFO(Q2)
      For J = 0 to Q-1
         For I = 0 to Q2-1
            OUT[J*STEP] += FILTER[2*I-J]*IN[I]
```

Interchanging the i and j summations results in a sequential input version of the same algorithm. There are at least two ways to implement it. One is to use Equation 5.5 and formulas analogous to Equations 5.103 to 5.105. Developing the adjoints of those formulas is a good test of the understanding of the principles of circular convolution, and is left as an exercise.

Here too it is sometimes useful to assign output values rather than to superpose them. For this we can use the sequential output algorithm but with assignments rather than increments in all statements with `OUT[]` in their left-hand sides. Let us call the resulting function `acdpe()`; it has the same parameter list as `acdpo()`.

The other method uses Equation 5.4 and lets j run over the appropriate subsets of the indices $0, \ldots, q-1$. This adjoint is nearly identical with the sequential output version of convolution-decimation, except that the i and j variables exchange places as input and output array indices.

If $q \leq \omega - \alpha$, there is nothing to else to do. If $q > \omega - \alpha$, then we use the three j-intervals as given by Equation 5.102. The resulting function is given below.

Periodic adjoint convolution-decimation: sequential input

```
acdpi( OUT, STEP, IN, Q2, F ):
   Let Q = 2*Q2
   If Q > F.OMEGA-F.ALPHA then
      Let FILTER = F.F
      For I = 0 to Q2-1
         Let A2I = 2*I-F.ALPHA
         Let O2I = 2*I-F.OMEGA
         For J = 0 to A2I-Q
            OUT[J*STEP] += FILTER[2*I-J-Q]*IN[I]
         For J = max(0,O2I) to min(Q-1,A2I)
            OUT[J*STEP] += FILTER[2*I-J]*IN[I]
         For J = O2I+Q to Q-1
            OUT[J*STEP] += FILTER[Q+2*I-J]*IN[I]
   Else
      Let FILTER = F.FP+PQFO(Q2)
      For I = 0 to Q2-1
         For J = 0 to Q-1
            OUT[J*STEP] += FILTER[2*I-J]*IN[I]
```

5.4.6 Tricks

Validation

Since convolution-decimation and its adjoint are inverses if we use a dual pair of QFs, it is possible to validate the implementations by seeing if we get perfect reconstruction of test signals of varying lengths. A suite of signals should include ones that are short compared to the filter as well as ones that are long.

At the start, it is even useful to use the filter $\{\ldots,0,1,1,0,\ldots\}$ which, although it does not satisfy any of the biorthogonal QF conditions, nonetheless produces results that can be checked by hand. If the input signals consist of integers, then the output will also be integers. No convolution-decimation program should ever be used without performing such elementary tests.

After the convolution-decimation functions are validated, they can be used to validate QF sequences. Such sequences are themselves the product of calculations which might be in error, or else they are copied from sources like Appendix C and might contain typographical errors or misprints. It makes good sense to check whether alleged perfect reconstruction filters give perfect reconstructions.

Fast convolution via FFT

The complexity of convolution and decimation can be reduced by using the convolution theorem and the "fast" discrete Fourier transform. Because of the overhead of complex multiplication, this becomes cost-effective only for long QFs.

Embedded coefficients for embedded software

For fixed-use or embedded software programs, the choice of quadrature filter will sometimes be made at the time the convolution-decimation function is written. This foreknowledge can be used to speed up the innermost loops at a cost of lost generality. If the filter is not too long, then it is not difficult to hard-code a routine which embeds the filter coefficients as numerical constants within the code. In some complex instruction set computers (CISCs), the coefficient and the multiply instruction may be loaded simultaneously during a single clock cycle, eliminating the need for an extra memory access and index computation. The practical constraint on this trick is the size of the resulting executable program. Also, in some conforming Standard C compilers the preprocessor symbol table will overflow if filled with the preperiodized coefficients of a filter of length 30 or more. But it is wise to treat short filters and especially Haar–Walsh filters as a special case.

Unrolling the loops

It may also be advantageous to "unroll" the innermost loops in the convolution. This can reduce the overhead of initialization, testing the loop termination condition, and incrementing the loop index. It also provides complete information about the number of arithmetic expressions or "rvalues" ([98], p.36) at compilation time, which can aid the optimizing compiler to make best use of any available registers. To write one such file by hand is rather tedious, and if we need many such files it is almost essential to first write a short program-generating utility and then to produce them mechanically.

Using symmetry

If $f(n) = \pm f(-n-1)$ and the filter f is supported on $[-R, R-1]$, then we can rewrite the convolution-decimation formula as follows:

$$Fu(n) = \sum_{k=-R}^{R-1} f(k)u(2n+k) = \sum_{k=0}^{R-1} f(k)\left[u(2n+k) \pm u(2n-k-1)\right]. \quad (5.106)$$

This has the advantage of reducing $2R$ multiplications down to just R, while preserving the number ($2R$) of additions. It also reduces the storage requirements for the filter from $2R$ memory locations to just R. To take advantage of these reduced space and time requirements, we must write a special convolution-decimation function which uses Equation 5.106. What we gain in speed we lose in generality and code reusability, so it is commercial and real-time applications which will benefit most from such a tradeoff.

5.5 Exercises

1. Prove Equation 5.11, using the argument in Proposition 5.1.

2. Prove the estimate in Equation 5.51.

3. Prove that we can always conventionally normalize the high-pass filters in a biorthogonal set of CDFs, as claimed after Lemma 5.2.

4. Compute $H^*Hu(t)$, $G^*Gu(t)$, $H^*Gu(t)$, and $G^*Hu(t)$ for a pair H, G of orthogonal QFs acting on functions.

5. Prove Equation 5.23, the second part of Lemma 5.5.

6. Prove Equations 5.80 and 5.81.

7. Find $\psi_n(t)$ in the Shannon case. That is, compute the inverse Fourier transform of the function in Theorem 5.19. This is called the *Shannon wavelet*.

8. Write a pseudocode program that takes a low-pass QF sequence and produces a complete PQF data structure. Then write a second program that produces its conventionally indexed and normalized conjugate.

9. Implement adjoint periodic convolution-decimation using the "mod" operator. Compare its speed on your computer with the "non-mod" implementation.

10. Implement sequential output adjoint periodic convolution-decimation using Equation 5.5. Compare its results with the implementation of `acdpo()` given in the text.

11. Implement periodic convolution-decimation using FFT. For what length filters does this realize a speed gain?

12. Start with an aperiodic sequence supported in $[a, b]$ and apply the convolution-decimation F (supported in $[0, R-1]$) a total of L times. What is the support of the resulting filtered sequence? Now apply F^* L times to the filtered sequence. What is the final support?

Chapter 6

The Discrete Wavelet Transform

The field of Fourier analysis also includes the study of expansions of arbitrary functions in bases other than the exponential functions or the trigonometric functions. One of our goals will be to adapt appropriate basic waveforms to each particular class of functions, so that the coefficients of the expansion convey maximal information about the functions. To choose which bases are appropriate, it is necessary to establish some mathematical properties of those waveforms, such as their size, smoothness, orthogonality, and support. Some of these properties can best be determined with classical methods involving the Fourier transform.

Once we fix the properties of other orthogonal function systems, possibly using facts about sines, cosines, and exponentials, we can use the new transforms in contexts where traditional Fourier analysis does not work well, where they may work better. In this way, we can "bootstrap" a few deep results about a single basis into broad knowledge about a whole catalog of basic waveforms.

Such is the case with *wavelets.* These are functions which have prescribed smoothness, which are well localized in both time and frequency, and which form well-behaved bases for many of the important function spaces of mathematical analysis. What makes wavelet bases especially interesting is their *self-similarity*: every function in a wavelet basis is a dilated and translated version of one (or possibly a few) *mother functions.* Once we know about the mother functions, we know everything about the basis.

In the discrete case, we face the same complexity issue as for DFT since the wavelet transform matrix is not sparse in general. We solve it in the same way as FFT: we factor the discrete wavelet transform into a product of a few sparse matrices using the self-similarity property. The result is an algorithm that requires only $O(N)$ operations to transform an N-sample vector. This is the "fast" discrete wavelet transform of Mallat and Daubechies which we will now study in some detail.

6.1 Some wavelet basics

Unlike the Fourier transform, the discrete wavelet transform or *DWT* is not a single object. A block diagram of the generic "fast" DWT is depicted in Figure 6.1, but it really hides a whole family of transformations. Individual members of this family are determined by which QFs are chosen for H and G, and whether we treat the sequence x as periodic, finitely supported, or restricted to an interval.

6.1.1 Origins

Like most mathematical animals, the fast or factored DWT has many parents. It descends from a construction given by Strömberg [105], which itself generalizes work by Haar [54] and Franklin [49], and more generally from the Littlewood-Paley decomposition of operators and functions [88]. The *subband coding* construction of a basis by convolution and decimation has been used by electrical engineers for more than a decade, with the wavelet basis appearing as the *octave subband decomposition.* The two-dimensional discrete wavelet transform descends from the *Laplacian pyramid scheme* of Burt and Adelson [12]. Daubechies, Grossmann and Meyer [39] observed the connection between pyramid schemes and Littlewood-Paley expansions. Meyer noted the possibility of *smooth* and *orthogonal* wavelet bases. Mallat [71, 70] showed that the octave subband coding algorithm was a multiscale decomposition of arbitrary functions, related to the visual transformation proposed by Marr [76]. Daubechies [36] showed that with the proper choice of filters one could produce wavelets with *compact support*, any desired degree of *regularity*, and any number of *vanishing moments.* The algorithm depicted in Figure 6.1 is the one due to Mallat; the filters we will use in our implementations are the ones developed by Daubechies and her coworkers.

The input to every DWT consists of the sequence x in the topmost or *root* block, and the output consists of the sequences in the shaded *leaf* blocks. Traversing a line between two blocks requires convolution and decimation with a QF operator H or G. The natural data structure for this algorithm is a binary tree of arrays of coefficients. The input array corresponds to sequential samples from a time series and has its own natural ordering, but the ordering of the shaded output blocks and

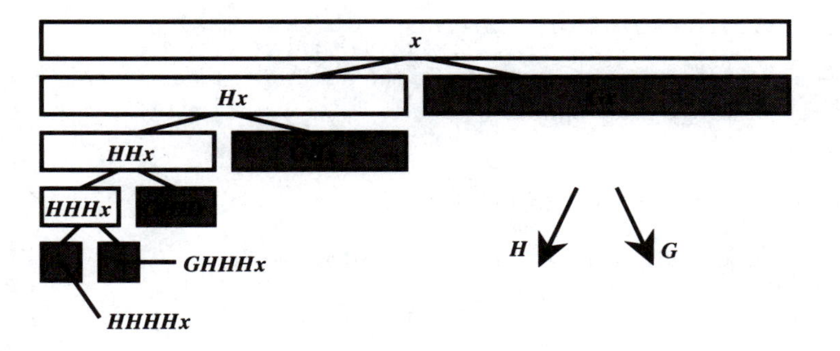

Figure 6.1: Discrete wavelet transform.

even the coefficients within them is a matter of convention. The numbers inside these output blocks have two indices:

- a *scale* or *level* which indicates how many applications of H or G were required to reach that block from the root;
- a *position* index within the block.

We will adopt an indexing convention which results in reasonably simple computer programs.

6.1.2 The DWT family

If we regard the input as a finitely supported infinite sequence, then the output will also be a finitely supported infinite sequence and we will have the *aperiodic DWT*. This transform has the simplest mathematical description, but its implementation requires more complicated indexing to keep track of the supports of the intermediate sequences. We can eliminate these indexing difficulties if we modify the operators H, G as in [20]; then we will have the *DWT on the interval*.

With an orthogonal pair of QFs H, G, we get the *orthogonal DWT*, while if H, G are biorthogonal with duals $H' \neq H, G' \neq G$ then we have the *biorthogonal DWT*.

If the input sequence is q-periodic, then the shaded blocks of the output will be periodic sequences and H, G can be considered periodized QFs. The subfamily of DWTs with this assumption will be called *periodic DWTs*. The input sequence for a q-periodic DWT consists of just q numbers.

6.1.3 Multiresolution analysis

We need some abstract definitions to fix the notation. A *multiresolution analysis* of $L^2(\mathbf{R})$, or MRA, is a chain of subspaces $\{V_j : j \in \mathbf{Z}\}$ satisfying the following conditions:

Containment: $V_j \subset V_{j-1} \subset L^2$ for all $j \in \mathbf{Z}$;

Decrease: $\lim_{j\to\infty} V_j = 0$, *i.e.*, $\bigcap_{j>N} V_j = \{0\}$ for all N;

Increase: $\lim_{j\to-\infty} V_j = L^2$, *i.e.*, $\bigcup_{j<N} V_j = L^2$ for all N;

Dilation: $v(2t) \in V_{j-1} \iff v(t) \in V_j$;

Generator: There is a function $\phi \in V_0$ whose translates $\{\phi(t-k) : k \in \mathbf{Z}\}$ form a Riesz basis for V_0.

Notice that $\{\phi(2^{-L}t - k) : k \in \mathbf{Z}\}$ is a Riesz basis for V_L. We may assume without loss that $\|\phi\| = 1$. An *orthogonal MRA* is one in which the Riesz basis for V_0 is in fact an orthonormal basis.

Remark. We have adopted the "Daubechies" indexing convention [37] for an MRA: $V_j \to L^2$ as $j \to -\infty$. In this convention, the sampling interval is the primary unit, the sampled signal belongs to V_0, and the scale of a wavelet, computed in sampling intervals, increases with j. This corresponds directly to the indices computed by our software implementations.

The alternative or "Mallat" indexing convention [71] reverses the sense of scale: $V_j \to L^2$ as $j \to +\infty$. There, the scale of the largest wavelet is the fundamental unit, and if the sampling interval is 2^{-L} then the signal belongs to V_L. This convention gives simple formulas if the signal has fixed size but must be considered at different resolutions.

The two conventions are equivalent and produce equal sets of coefficients; they differ only in how the coefficients are tagged for subsequent processing.

We conclude from the containment, dilation, and generator properties that a normalized ϕ satisfies a *two-scale equation*:

$$\phi(t) = \sqrt{2} \sum_{k \in \mathbf{Z}} h(k)\phi(2t-k) \stackrel{\text{def}}{=} H\phi(t). \tag{6.1}$$

Here $\{h(k)\}$ is a square-summable sequence of coefficients which defines a linear operator H. Note the resemblance of H to a quadrature filter.

If we define *complementary subspaces* $W_j = V_{j-1} - V_j$ so that $V_{j-1} = V_j + W_j$, then we can telescope the union in the "increase" property to write

$$L^2 = \sum_{j\in\mathbf{Z}} W_j. \tag{6.2}$$

The subspaces W_j are called *wavelet subspaces*, and Equation 6.2 is called a *wavelet decomposition* of L^2. W_0 has a Riesz basis generated by a function ψ satisfying the *wavelet equation:*

$$\psi(t) = \sqrt{2}\sum_{k\in\mathbf{Z}} g(k)\phi(2t-k) \stackrel{\text{def}}{=} G\phi(t); \qquad g(k) = (-1)^k\bar{h}(1-k). \tag{6.3}$$

The distinguished function ψ is called the *mother wavelet.* If ϕ or equivalently H is well-behaved, then $\{\psi(2^{-j}t-k) : k \in \mathbf{Z}\}$ gives a Riesz basis for W_j, and $\{2^{-j/2}\psi(2^{-j}t-k) : j,k \in \mathbf{Z}\}$ gives a Riesz basis for L^2.

In an orthogonal MRA the functions $\{2^{-j/2}\psi(2^{-j}t-k) : j,k \in \mathbf{Z}\}$ provide an orthonormal *wavelet basis* for L^2. In that case $W_j = V_{j-1} \cap V_j^{\perp}$, the decomposition $V_{j-1} = V_j \oplus W_j$ is an orthogonal direct sum, and

$$L^2 = \bigoplus_{j\in\mathbf{Z}} W_j \tag{6.4}$$

is a decomposition of L^2 into orthogonal subspaces. Orthogonal MRAs are obtained by using orthogonal pairs of QFs in the two-scale and wavelet equations. References [37] and [68] describe additional properties of orthogonal MRAs.

6.1.4 Sequences from functions

Let $x = x(t)$ be a function in $L^2(\mathbf{R})$, and let $\{\omega(p) : p \in \mathbf{Z}\}$ be the coefficients of its projection P_L onto V_L:

$$\begin{aligned}\omega(p) &\stackrel{\text{def}}{=} \langle \sigma_2^L \tau_p \phi, x\rangle = \int 2^{-L/2}\bar{\phi}(2^{-L}t-p)\,x(t)\,dt;\\ P_L x(t) &\stackrel{\text{def}}{=} \sum_p \omega(p)\sigma_2^L\tau_p\phi(t) = \sum_p \omega(p)\,2^{-L/2}\phi(2^{-L}t-p).\end{aligned}$$

Here ϕ is the nice bump function whose integer translates provide an orthonormal basis for the subspace V_L. The L^2 function given by this projection will be denoted $x_L = P_L x$. The error in this approximation, $\|x - x_L\|$, is controlled by the projection $I - P_L$ of x onto the orthogonal complement $V_L^{\perp}$ in L^2. This complementary projection is small if both x and ϕ are regular. We have the following results from Littlewood–Paley theory:

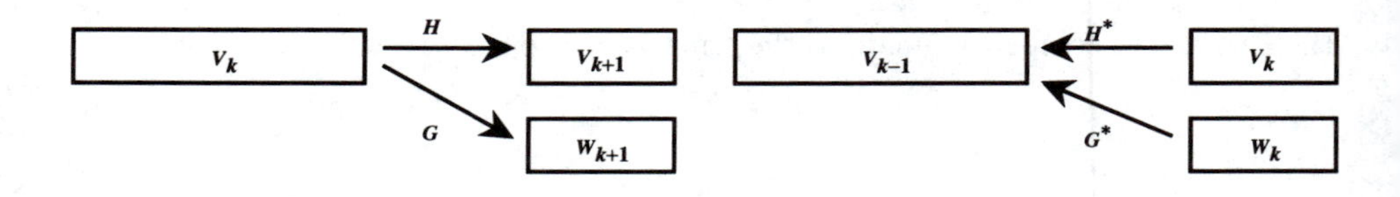

Figure 6.2: One step in a recursive DWT, one step in a recursive iDWT.

Proposition 6.1 *If both x and ϕ have d continuous derivatives, then there is a constant C such that $\|x - x_L\| < C2^{dL}\|x\|$.* □

Proposition 6.2 *If both x and ϕ have d continuous derivatives, and ϕ has vanishing moments:*

$$\int t^k \phi(t)\, dt = 0 \qquad \textit{for all } 0 < k < d,$$

then $\|x - 2^{L/2}\omega(n)\|_* < C\|x\|_* 2^{-dL}$, where $\|x\|_* = \sup\{|x(t)| : n < 2^L t < n+1\}$. □

Proofs of these facts may be found in the excellent two volume monograph by Meyer [79, 80]. Now, if we make a physical measurement of a continuously varying quantity $x = x(t)$, we are actually computing an average $\langle \varphi, x \rangle$ with an unknown function φ. In general, this *sampling function* or *instrument response* will differ from our wavelet scaling function ϕ. However, under the reasonable assumptions that both the quantity and the averaging function are smooth, we can apply Proposition 6.2 to conclude that our unknown average is close to a sample $x(t_n)$. A second application of Proposition 6.2 implies that we can approximate $\omega(n)$ with the sample $x(t_n)$. The quality of both approximations depends on the sampling interval 2^L, but if this is small enough then any physical measurements give good approximations to wavelet scaling coefficients.

6.2 Implementations

The DWT has a particularly simple recursive structure. There is a natural index for the depth of recursion, the number of filter applications, and the recursive steps are always the same. Suppose that we sample a function at uniform intervals and declare the sampling interval to be the unit. Then we can write V_0 for the space in which the signal is approximated, V_k for the approximation space k filter applications below the signal, and W_k for the wavelet coefficients at scale k below the signal. Figure 6.2 depicts one step in the DWT and inverse DWT, respectively.

The following is a skeleton implementation of the DWT:

Generic discrete wavelet transform

```
Allocate an array SUMS
Read the input into SUMS
Define MAXIMUM_LEVEL
Allocate MAXIMUM_LEVEL+1 arrays for DIFS
For L = 0 to MAXIMUM_LEVEL-1
   Filter SUMS at L into DIFS at L+1 with the high-pass QF
   Filter SUMS at L into SUMS at L+1 with the low-pass QF
Write SUMS into DIFS at 1+MAXIMUM_LEVEL
```

We can apply this to various concrete examples, in which we can be more specific about the output data structures and the convolution-decimation subroutine. We want a simple data indexing arrangement so we can write a simple nonrecursive implementation when it is advantageous to do so. For embedded software with fixed parameters, or for real-time signal processing applications, it may pay to eliminate the overhead associated to recursive function calls. Also, the nonrecursive versions can be implemented in a greater variety of programming languages.

The inverse (or adjoint) discrete wavelet transform has the following skeleton implementation:

Generic inverse discrete wavelet transform

```
Read the MAXIMUM_LEVEL
Allocate MAXIMUM_LEVEL+1 arrays for DIFS
Read the input into DIFS
Allocate an array SUMS
Let L = MAXIMUM_LEVEL
While L > 0,
   Adjoint-filter SUMS at L with the low-pass QF
   Assign the result into SUMS at L-1
   Adjoint-filter DIFS at L with the high-pass QF
   Superpose the result into SUMS at L-1
   Decrement L -= 1
Write SUMS into the output
```

The output data structures and the convolution-decimation subroutine used for DWT should match the ones used by iDWT. If H, H', G, G' are a set of biorthogonal filters, then we compute the inverse of the DWT computed with H, G using the filters H', G'.

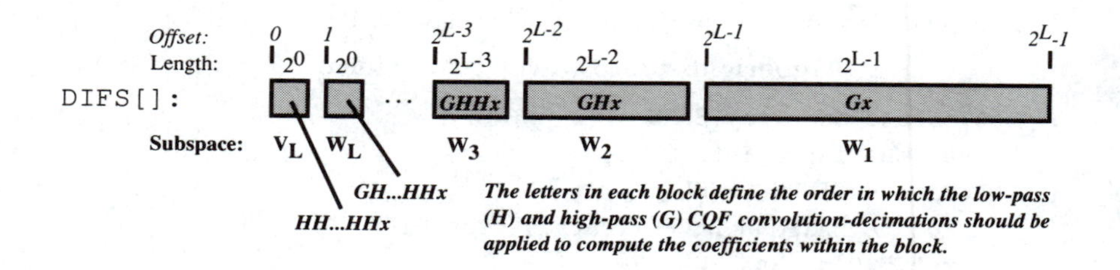

Figure 6.3: Arrangement of 2^L-periodic DWT of x into one array of length 2^L.

6.2.1 Periodic DWT and iDWT

This is the simplest case of the discrete wavelet transform. With 2^L inputs, we can decimate by two as many as L times. After s decimations, for $1 \leq s \leq L$, we will have 2^{L-s} coefficients to write into the output data structure. The array at level s corresponds to the wavelet subspace W_s; it is produced from the input array x by $s-1$ convolution-decimations with the low-pass filter H followed by one convolution-decimation with the high-pass filter G. In addition, there will be a single array V_L at level L which is the result of L convolution-decimations with the low-pass filter. The summed lengths of these $L+1$ arrays are

$$\overbrace{2^0}^{\text{scaling array}} + \overbrace{2^0 + 2^1 + \cdots + 2^{L-2} + 2^{L-1}}^{\text{wavelet arrays}} = 2^L,$$

and the output data structure can therefore be taken to be a single array of the same length as the input array.

Periodic DWT on $N = 2^L$ points

We adopt the convention of ordering the concatenated output arrays like V_L, W_L, W_{L-1}, W_{L-2}, ..., W_2, W_1. This is depicted in Figure 6.3. The convention results in the coefficients grouped by decreasing scale or equivalently by increasing bandwidth, where we take the bandwidth of a function to be its momentum uncertainty $\Delta\xi$ as in Equation 1.58. Since $\Delta\xi \propto \xi_0$ for wavelets, and ξ_0 is in some sense a "frequency" parameter, our convention produces coefficients in increasing order of frequency analogous to the Fourier transform.

We must compute coefficients of the form $H \cdots Hx$ as a side-effect of computing the DWT, but since they can be useful we will provide them as part of the output of our implementation. These numbers give the decomposition of x in the various

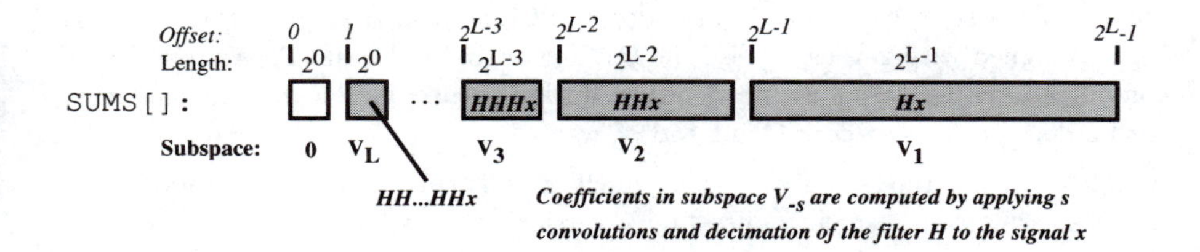

Figure 6.4: Averages computed during the 2^L-periodic DWT of x.

approximation spaces $V_1, V_2, \ldots, V_{L-1}, V_L$. The original signal x is taken to be coefficients from V_0. The space V_s in the 2^L-periodic case requires 2^{L-s} coefficients, so the concatenation of all these arrays will require an array of total length $2^{L-1} + \cdots + 2^1 + 2^0 = 2^L - 1$, as depicted in Figure 6.4. We have chosen to leave `SUMS[0]=0` so that the indexing formula is simple: the subarrays corresponding to subspaces V_s and W_s both begin at the offset 2^{L-s}.

We will use periodic convolution-decimation functions such as `cdpo()`, defined in Chapter 5. We can substitute the sequential input version for the sequential output version if that proves advantageous. Once we have implemented the convolution-decimation function, coding the DWT takes just a few lines. We have two versions: one which assumes that the input and output arrays are disjoint, and one which transforms "in place" by replacing the input array with the wavelet (or differences) coefficients. The following pseudocode implements the disjoint version:

Recursive disjoint L-level periodic DWT on $N = 2^L$ points

```
dwtpd0( DIFS, SUMS, IN, N, H, G ):
   If N > 1 then
      cdpo( DIFS+N/2, 1, IN, N, G )
      cdpo( SUMS+N/2, 1, IN, N, H )
      dwtpd0( DIFS, SUMS, SUMS+N/2, N/2, H, G )
   Else
      DIFS[0] += IN[0]
   Return
```

This function should be called after allocating three disjoint arrays `IN[]`, `SUMS[]` and `DIFS[]` with `N` $= 2^L$ members each, and reading the input into the array `IN[]`.

Note that it is necessary to preassign zeroes to all the elements of `SUMS[]`. While `SUMS[]` contains useful information at the end of the calculation, it is used as a work array and the result will not be meaningful if it contains nonzero elements at the outset. Likewise, the DWT will be *added* into `DIFS[]`, so that this array must also be preassigned with zeroes, or else the DWT will simply be superposed onto it. To compute a simple DWT we must either initialize the preallocated output arrays with zeroes or else use `cdpe()` instead of `cdpo()`.

Although it is the recursive version which we will evolve into more general DWTs, it is also easy to implement nonrecursive version of `dwtpd0()`:

Nonrecursive disjoint L-level periodic DWT on $N = 2^L$ points

```
dwtpd0n( DIFS, SUMS, IN, N, H, G ):
   While N > 1
      cdpo( DIFS+(N/2), 1, IN, N, G )
      cdpo( SUMS+(N/2), 1, IN, N, H )
      Let IN = SUMS+(N/2)
      N /= 2
   DIFS[0] += IN[0]
```

If the input and output arrays are to be the same, then we must use a scratch array to hold the low-pass filtered sequence. We can perform this "in place" transform with or without recursion; we use `cdpe()` instead of zeroing the scratch array:

Nonrecursive in place L-level periodic DWT on $N = 2^L$ points

```
dwtpi0n( DATA, WORK, N, H, G ):
   While N > 1
      cdpe( WORK+N/2, 1, DATA, N, G )
      cdpe( WORK, 1, DATA, N, H )
      N /= 2
      For I = 0 to N-1
         Let DATA[I] = WORK[I]
```

Recursive in place L-level periodic DWT on $N = 2^L$ points

```
dwtpi0( DATA, WORK, N, H, G ):
   If N > 1 then
      cdpe( WORK+N/2, 1, DATA, N, G )
      cdpe( WORK, 1, DATA, N, H )
      For I = 0 to N/2-1
         Let DATA[I] = WORK[I]
      dwtpi0( DATA, WORK, N/2, H, G )
   Return
```

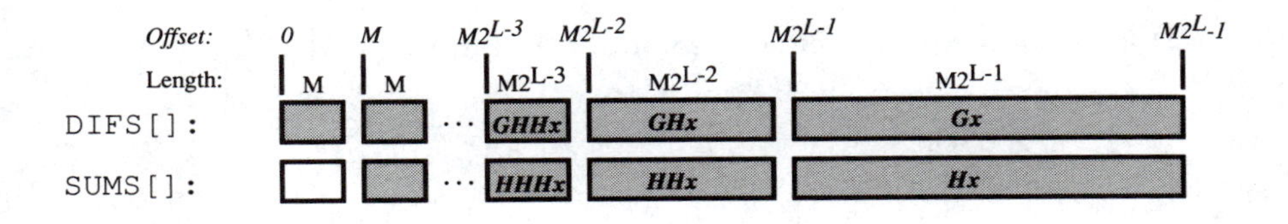

Figure 6.5: L-level DWT of an $M2^L$-periodic signal x.

Periodic L-level DWT on $M2^L$ inputs

A modification of the subroutine `dwtpd0()` permits us to compute the discrete wavelet transform down to level L, when the number of inputs is not a power of two but is nonetheless divisible by 2^L. The new function `dwtpd()` uses the level parameter `L` to control the depth of recursion. After `N` has been divided `L` times by two, it is passed to the last invocation of `dwtpd()` with the value M. At that invocation, the parameter `IN` is a pointer to element M of `SUMS[]`, so that the final loop copies `SUMS[`M`],...,SUMS[`$2M-1$`]` into `DIFS[0],...,DIFS[`$M-1$`]`. After the transform, the output will be the arrays `DIFS[]` and `SUMS[]`, laid out as in Figure 6.5.

Below is a pseudocode implementation of this transform:

Recursive disjoint L-level periodic DWT on $N = M * 2^L$ points

```
dwtpd( DIFS, SUMS, IN, N, L, H, G ):
   If L > 0 then
      cdpo( DIFS+N/2, 1, IN, N, G )
      cdpo( SUMS+N/2, 1, IN, N, H )
      dwtpd( DIFS, SUMS, SUMS+N/2, N/2, L-1, H, G )
   Else
      For K = 0 to N-1
         DIFS[K] += IN[K]
   Return
```

We have assumed that `IN[]`, `SUMS[]`, and `DIFS[]` are mutually disjoint arrays, and that `SUMS[]` starts out full of zeroes. The latter assumption can be eliminated by using `cdpe()` rather than `cdpe()`.

The nonrecursive version `dwtpdn()` is left as an exercise.

Likewise, modifying `dwtpi0()` allows us to perform the previous transformation in place. This is best done with convolution-decimations that assign rather than superpose, since the data and scratch arrays must both be overwritten:

Recursive in-place L-level periodic DWT on $N = M * 2^L$ points

```
dwtpi( DATA, WORK, N, L, H, G ):
   If L > 0 then
      cdpe( WORK+N/2, 1, DATA, N, G )
      cdpe( WORK, 1, DATA, N, H )
      For I = 0 to N/2-1
         Let DATA[I] = WORK[I]
      dwtpi( DATA, WORK, N/2, L-1, H, G )
   Return
```

This too has a nonrecursive implementation `dwtpin()`, also left as an exercise.

All of the functions `dwtpdn()`, `dwtpd()`, `dwtpin()`, and `dwtpi()` handle the trivial case `L` = 0 by simply copying the `N` input values into the array `DIFS[]`. They also handle the case `N` = 2^{L} correctly, thus generalizing `dwtpd0n()`, `dwtpd0()`, `dwtpi0n()`, and `dwtpi0()`. However, it is the programmer's responsibility to ensure that `N` is divisible by 2^{L}, for example by an `assert((N>>L)<<L == N)` statement.

Inverse of periodic DWT on $N = 2^L$ points

The inverse periodic DWT can be implemented nonrecursively as follows:

Nonrecursive disjoint L-level periodic iDWT on $N = 2^L$ points

```
idwtpd0n( OUT, SUMS, IN, N, H, G ):
   If N > 1 then
      SUMS += 1
      Let M = 1
      Let SUMS[0] = IN[0]
      IN += 1
      While M < N/2
         acdpo( SUMS+M, 1, SUMS, M, H )
         acdpo( SUMS+M, 1,  IN,  M, G )
         SUMS += M
         IN   += M
         M *= 2
      acdpo( OUT, 1, SUMS, N/2, H )
      acdpo( OUT, 1,  IN,  N/2, G )
   Else
      OUT[0] += IN[0]
```

We have used the adjoint periodic convolution-decimation function `acdpo()` defined in Chapter 5. We assume that the input to the inverse algorithm is arranged as in Figure 6.3, and that the input, output, and scratch arrays are all disjoint. The preallocated scratch array `SUMS[]` should be full of zeroes; it can be initialized by the calling program, or we can add zeroing loops to each of `idwtpd0()` and `idwtpd0n()`. The reconstructed values will be superposed onto the output array `OUT[]`; if we want just the values, then `OUT[]` too must be initialized with zeroes.

Alternatively, we could filter with the function `acdpe()`, which does the same computation as `acdpo()` but assigns the result to the output array rather than superposing it. This is how we will modify the functions `idwtpd0()` and `idwtpd0n()` so that they reconstruct the signal in place. We save memory because the signal overwrites its coefficients, but we will still use a scratch array as long as the signal:

Nonrecursive in place L-level periodic iDWT on $N = 2^L$ points

```
idwtpi0n( DATA, WORK, N, H, G ):
   Let M = 1
   While M < N
      acdpe( WORK, 1,  DATA,  M, H )
      acdpo( WORK, 1, DATA+M, M, G )
      For I = 0 to 2*M-1
         Let DATA[I] = WORK[I]
      M *= 2
```

Inverse L-level periodic DWT on $N = M2^L$ points

We assume first that the input and output arrays are disjoint:

Recursive disjoint L-level periodic iDWT on $N = M * 2^L$ points

```
idwtpd( OUT, SUMS, IN, N, L, H, G ):
   If L > 0 then
      N /= 2
      idwtpd( SUMS+N, SUMS, IN, N, L-1, H, G )
      acdpo( OUT, 1, SUMS+N, N, H )
      acdpo( OUT, 1,  IN +N, N, G )
   Else
      For K = 0 to N-1
         OUT[K] += IN[K]
   Return
```

We have assumed that the input to the inverse algorithm is arranged as in Figure 6.5. We have reused the adjoint periodic convolution-decimation function `acdpo()`. Thus it is again necessary that `SUMS[]` be initialized with all zeroes. The values of this periodic iDWT will be superposed onto `OUT[]`.

Alternatively, we could replace `acdpo()` with `acdpe()` and assign the results; then the contents of the output and scratch arrays will simply be overwritten. In the same spirit, we may also perform the iDWT in place to save some space. We will still use a scratch array of N elements:

Recursive in place L-level periodic iDWT on $N = M * 2^L$ points

```
idwtpi( DATA, WORK, N, L, H, G ):
   If L > 0 then
      Let M = N/2
      idwtpi( DATA, WORK, M, L-1, H, G )
      acdpe( WORK, 1,  DATA,  M, H )
      acdpo( WORK, 1, DATA+M, M, G )
      For I = 0 to N-1
         Let DATA[I] = WORK[I]
```

For the in place transforms, it is necessary to assume that `DATA[]` and `WORK[]` are disjoint arrays.

Both `idwtpd()` and `idwtpi()` have nonrecursive implementations `idwtpdn()` and `idwtpin()` which are left as exercises.

Once the DWT and iDWT are implemented, they can be validated by testing whether they are inverses. This should be done first with short QFs like the Haar–Walsh filters, using random signals of varying lengths. Also, one can cycle through all the versions of the convolution-decimations: sequential output and sequential input, modular and periodic, based on Equations 5.3 to 5.5.

6.2.2 Aperiodic DWT and iDWT

We assume that the input coefficient sequence is supported on the interval $c, \ldots, d$, so the signal has length $N = 1 + d - c$. In this case it does not matter whether N is a power of two or divisible by a high power of two. We must specify the number L of levels to decompose the signal, just as in the periodic $N = M2^L$ case. The chief new feature of this variant is that each time we apply H or G we obtain a new support interval, and the sum of the lengths of all the output arrays may in general be longer than the input array. The result is a collection of arrays of varying lengths, as seen in the "exploded" view in Figure 6.6.

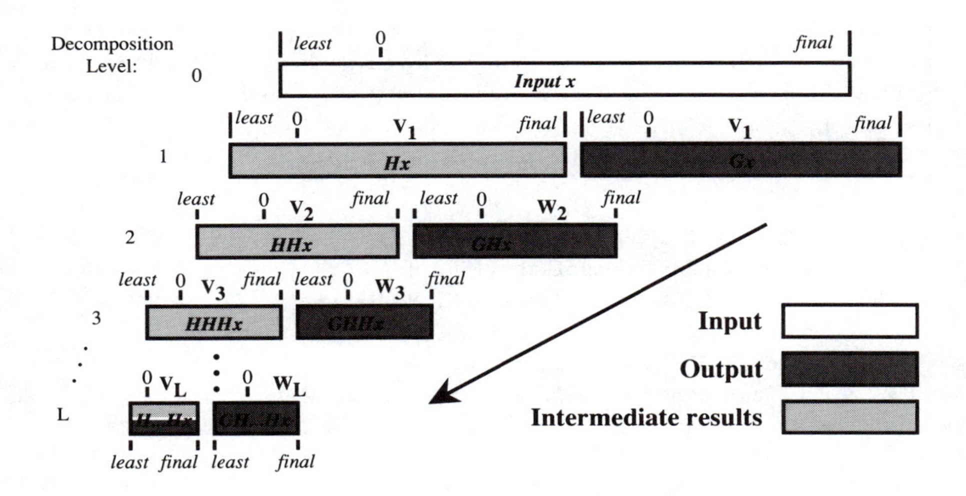

Figure 6.6: Intervals for the L-level aperiodic DWT.

We must therefore use a more complicated storage method and indexing function for the coefficients corresponding to the subspaces $V_1, \ldots, V_L$ and $W_1, \ldots, W_L$. Our chosen solution is to use the INTERVAL data structures defined in Chapter 2 to store the subspace descriptions while using long concatenated arrays `SUMS[]` and `DIFS[]` for the actual coefficients. The ORIGIN members in the output V_k and W_k INTERVALs will point into these long arrays, and their LEAST and FINAL members will be the endpoint offsets relative to those origins.

Counting coefficients, going down

In the aperiodic case, if the filter sequence f is supported in the interval $[\alpha, \omega]$ and the input signal sequence x is supported in the interval $[a, b]$, then the output sequence y is supported in the interval $[a', b']$, where $a' = \lceil (a + \alpha)/2 \rceil$ and $b' = \lfloor (b + \omega)/2 \rfloor$ are given by Equation 5.91. We can implement the maps $a \mapsto a'$ and $b \mapsto b'$ in terms of the preprocessor macros `ICH()` and `IFH()` of Chapter 5:

Interval endpoints after aperiodic convolution-decimation

```
#define cdaleast(I,F)      ICH(I.LEAST+F.ALPHA)
#define cdafinal(I,F)      IFH(I.FINAL+F.OMEGA)
```

For each level, we need to store the sequence produced by the G filter from the previous level's H sequence. We will first determine the interval endpoints of the V and W arrays, which are determined by the endpoints of the input signal interval. Suppose that V and W are preallocated arrays of L and $L+1$ INTERVALs, respectively:

V and W interval endpoints for aperiodic DWT

```
dwtaintervals( V, W, IN, L, H, G ):
   If L>0 then
      Let V.LEAST = cdaleast( IN, H )
      Let V.FINAL = cdafinal( IN, H )
      Let W.LEAST = cdaleast( IN, G )
      Let W.FINAL = cdafinal( IN, G )
      dwtaintervals( V+1, W+1, V, L-1, H, G )
   Else
      Let W.LEAST = IN.LEAST
      Let W.FINAL = IN.FINAL
  Return
```

Notice that we append a copy of V_L at the end of the W array, so that the single array DIFS[] will in fact hold the entire transform. This is why the W array must hold one extra INTERVAL.

We add up the lengths of the V intervals at levels 1 to L to get the total amount of space needed for the SUMS[] array, and the lengths of the W intervals plus V_L to get the total length of DIFS[]. Since the lengths of the two QFs H and G might be different, the aperiodic case has the added complication that the SUMS[] and DIFS[] arrays (and each of their subarrays) might be of different lengths. The respective lengths are computed and returned by the functions intervalstotal(V,L) and intervalstotal(W,L+1), with the function defined in Chapter 2. We allocate those arrays, then we assign pointers to the origins of the V and W subspaces so that the output of the DWTA will exactly fit. This can be done in many ways; we choose to arrange the coefficients in such a way that the coarsest scale or lowest frequency subspace is at the beginning of W, so that the coefficients are arranged in order of increasing frequency. This entails starting at the end of DIFS[].

Shifts to concatenate intervals from the end of the long array

```
#define shifttoorigin(I)         (-1-I.FINAL)
#define shifttonextinterval(I)   (I.LEAST-I.FINAL-1)
```

Or, by changing these shifts to their complements, we will end up with coefficients in decreasing order of frequency:

Shifts to concatenate intervals from the start of the long array

```
#define shifttoorigin(I)          (-I.LEAST)
#define shifttonextinterval(I)    (1+I.FINAL-I.LEAST)
```

For consistency, we arrange the subintervals in `SUMS[]` and `DIFS[]` in the same order, though we are under no obligation to do so:

Set ORIGIN pointers from V and W into long arrays

```
dwtaorigins( V, W, SUMS, DIFS, L ):
   If L>0 then
      Let V.ORIGIN = SUMS + shifttoorigin( V )
      Let W.ORIGIN = DIFS + shifttoorigin( W )
      SUMS += shifttonextinterval( V )
      DIFS += shifttonextinterval( W )
      dwtaorigins( V+1, W+1, SUMS, DIFS, L-1 )
   Else
      Let W.ORIGIN = DIFS + shifttoorigin( V )
   Return
```

Aperiodic DWT from an interval

Suppose that the input signal resides in an INTERVAL data structure. We compute the coefficients in the wavelet spaces $W_1, W_2, \ldots, W_L$ and the approximation spaces $V_1, V_2, \ldots, V_L$ and put the results into a long concatenated array while keeping track of the individual subspaces with two arrays of INTERVALs.

Most of the work is done by the aperiodic convolution-decimation function `cdao()` defined in Chapter 5. It takes an input array defined for a given interval of indices, convolves it with the given QF, and writes the computed values to an output array using a given increment. We can also use `cdai()`, the sequential input version of the same function. Notice that either function *adds* the computed values into the output array. This feature may be used to superpose more than one convolution-decimation operation into a single array, but it obliges us to make sure that intermediate arrays are preassigned with zeroes.

We will only consider a recursive implementation of the aperiodic algorithm. Naturally it is possible to eliminate the recursive function call with an appropriate indexing scheme.

For input, we need to specify the following:

- `IN[]`, the INTERVAL holding the original signal, which will not be changed,

- `V`, an array of L INTERVALs to hold the endpoints and origins of the scaling coefficient intervals;
- `W`, an array of $L+1$ INTERVALs to hold the endpoints and origins of the wavelet coefficient intervals plus the coarsest scaling interval;
- `L`, the number of levels of decomposition;
- `H,G`, the low-pass and high-pass QFs in PQF data structures.

Behind the arrays of pointers are two long arrays of coefficients:

- `SUMS[]`, an array of REALs long enough to hold all the scaling coefficients at all levels;
- `DIFS[]`, an array of REALs long enough to hold all the wavelet coefficients at all levels, plus the deepest level scaling coefficients.

We assume that these have been allocated and assigned using calls to the functions `dwtaintervals()`, `intervalstotal()`, and `dwtaorigins()`. The actual DWTA is then a matter of a few lines:

L-level aperiodic DWT from an INTERVAL

```
dwta( V, W, IN, L, H, G ):
   If L > 0
      cdao( V.ORIGIN, 1, IN.ORIGIN, IN.LEAST, IN.FINAL, H )
      cdao( W.ORIGIN, 1, IN.ORIGIN, IN.LEAST, IN.FINAL, G )
      dwta( V+1, W+1, V, L-1, H, G )
   Else
      For N = IN.LEAST to IN.FINAL
         Let W.ORIGIN[N] = IN.ORIGIN[N]
   Return
```

Notice that at the bottom level, we copy the input interval to the last of the wavelet intervals. In the typical case $L > 0$, this puts the coarsest scaling space V_L at the end of the wavelet coefficients array. In the trivial $L = 0$ case it simply copies the input sequence to the wavelet coefficients array. In all cases, it means that just the `DIFS[]` and `W[]` arrays are needed to reconstruct the signal.

With the Standard C array indexing convention, where arrays start at index zero, the subspace W_k will correspond to interval `W[K-1]`, V_k will correspond to `V[K-1]`, and the coarsest scaling subspace V_L will correspond to intervals `V[L-1]` and `W[L]`.

There is a need to justify including more than just the initial endpoints a and b in the output of the aperiodic DWT, since those two integers completely determine all the other offsets which can be recomputed whenever we desire. However, for the sake of generality we will not assume that wavelet coefficients come from a complete *aperiodic wavelet analysis* as done by `dwta()`. We will permit arbitrary superpositions of wavelets from arbitrary combinations of scale subspaces, in order to provide a more general inverse or *aperiodic wavelet synthesis* algorithm. We choose to make the output of the analysis conform to the more general description requirements of the synthesis.

Counting coefficients, coming up

The input to the inverse DWT consists of an array of INTERVALs, one for each level, corresponding to the largest scale approximation space V_L and the wavelet subspaces $W_L, \ldots, W_1$. The arrangement of these subspaces is depicted as unshaded boxes in Figure 6.7. We must also specify the number of INTERVALs in this list, *i.e.*, the number of levels L plus one. The total number of coefficients can be computed by adding up the lengths of the INTERVALs. The total length of the `DIFS[]` array filled by `idwta()` is the return value of `intervalstotal(W,L+1)`.

If x is the result of adjoint aperiodic convolution-decimation of the input sequence y with a filter sequence f supported in $[\alpha, \omega]$, and y is supported in the interval $[c, d]$, then x is supported in the interval $[c', d']$, where $c' = 2c - \omega$ and $d' = 2d - \alpha$ are given by Equation 5.94. These can be implemented with preprocessor macros:

Interval endpoints after adjoint aperiodic convolution-decimation

```
#define acdaleast(I,F)      (2*I.LEAST-F.OMEGA)
#define acdafinal(I,F)      (2*I.FINAL-F.ALPHA)
```

Since the aperiodic inverse DWT recursively reconstructs V_{k-1} from the adjoint convolution-decimations of both W_k and V_k, the temporary array V_{k-1} must be allocated with enough space for both. Namely, if the sequences of coefficients from W_k and V_k are supported in the intervals $[c_w, d_w]$ and $[c_v, d_v x]$, respectively, then the adjoint filter operation will produce a sequence for V_{k-1} supported on $[c', d']$, where

$$c' = \min\{2c_w - \omega_g, 2c_v - \omega_h\}; \qquad d' = \min\{2d_w - \alpha_g, 2d_v - \alpha_h\}. \tag{6.5}$$

Here h and g refer to the low-pass and high-pass QFs, respectively.

We can concatenate the array members of the INTERVALs into a single array, much as the function `dwta()` concatenates the subarrays of the output intervals into one contiguous array. The following function returns the total length of all the arrays representing $V_1, V_2, \ldots, V_L$ which will be used in the inverse aperiodic DWT. We have arranged to allow the reconstructed signal to wind up in a separate output array. We keep track of the interval endpoints and origins in the preallocated array `V` of L INTERVALs, and also compute and assign the endpoints of the output interval.

Compute interval endpoints and lengths for iDWTA

```
idwtaintervals( OUT, V, W, L, H, G ):
   If L>0 then
      idwtaintervals( V, V+1, W+1, L-1, H, G )
      Let OUT.LEAST = min( acdaleast(V,H), acdaleast(W,G) )
      Let OUT.FINAL = max( acdafinal(V,H), acdafinal(W,G) )
   Else
      Let OUT.LEAST = W.LEAST
      Let OUT.FINAL = W.FINAL
   Return
```

Once we have found the endpoints of the intermediate intervals, we can allocate an array `SUMS[]` long enough to hold the concatenated interval contents. The length is the return value of `intervalstotal(V,L)`. Then, we assign pointers into this long array to the ORIGIN members of the V intervals in such a way that all the scaling coefficients will exactly fit:

Concatenate V subspace arrays into SUMS for iDWTA

```
idwtaorigins( V, SUMS, L ):
   If L>0 then
      Let V.ORIGIN = SUMS + shifttoorigin( V )
      SUMS += shifttonextinterval( V )
      idwtaorigins( V, SUMS, V+1, L-1 )
   Return
```

We reuse the preprocessor macro that allows us to choose whether to arrange the data in increasing or decreasing order of frequency.

Inverse aperiodic DWT from an array of INTERVALs

For the inverse algorithm, we will use the adjoint aperiodic convolution-decimation function, or filter, defined in Chapter 5: `acdao()` takes a sequence supported in an interval, computes the aperiodic adjoint convolution-decimation with the desired

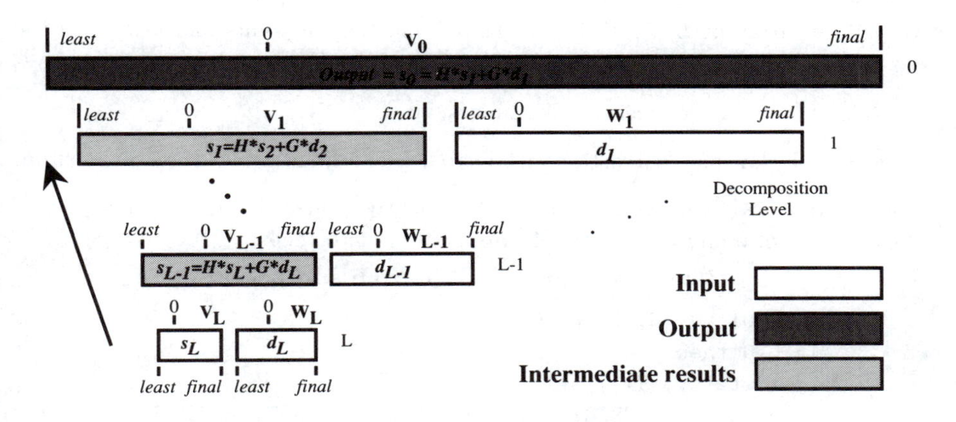

Figure 6.7: Input (blank) and output (shaded) for the L-level iDWTA.

filter, and adds the result into an output array using a given increment. It computes the range of indices which will be affected, but it assumes that the output array is already allocated to the proper length, which requires that the output range be calculated by the calling program as well.

To contain the output, we must supply two preallocated arrays:

- `SUMS[]`, for the scaling function coefficients in the intervals $V_1, \ldots, V_L$. The length is the return value of `idwtaintervals()`. This array is expected to be full of zeroes;

- `OUT.ORIGIN[]`, for the reconstructed signal. Indices `OUT.LEAST` to `OUT.FINAL` must be valid, so the length must be at least `1+OUT.FINAL-OUT.LEAST`. The reconstructed signal will be superposed onto this array.

A call to `idwtaintervals()`, `idwtaorigins()`, and `intervalstotal()` with the given W defines the lengths of these two arrays. Then we can allocate them in the usual way. The details are left as an exercise.

The other input parameters are the following:

- `OUT`, an interval with properly assigned LEAST and FINAL members, with its ORIGIN member pointing to a preallocated array for which `OUT.LEAST` to `OUT.FINAL` are valid indices;

- `V`, an array of L INTERVALs holding the endpoint offsets and origins of the scaling coefficient intervals reconstructed on the way up;

- `W`, an array of $L+1$ INTERVALs holding the endpoints and origins of the wavelet coefficient intervals plus the coarsest scaling interval;
- `L`, the number of levels of decomposition;
- `H`, `G`, the low-pass and high-pass QFs to use for adjoint convolution-decimation.

The arrays `DIFS[]` and `SUMS[]` are used implicitly, since the ORIGIN members of the elements of `W` and `W`, respectively, are pointers into them.

L-level aperiodic iDWT from subspaces

```
idwta( OUT, V, W, L, H, G ):
   If L>0 then
      idwta( V, V+1, W+1, L-1, H, G )
      acdao( OUT.ORIGIN, 1, V.ORIGIN, V.LEAST, V.FINAL, H )
      acdao( OUT.ORIGIN, 1, W.ORIGIN, W.LEAST, W.FINAL, G )
   Else
      For N = W.LEAST to W.FINAL
         Let OUT.ORIGIN[N] = W.ORIGIN[N]
   Return
```

It is left as an exercise to assemble these steps into a complete implementation of DWTA and iDWTA.

6.2.3 Remarks

We note that the only modification needed in order to use biorthogonal sets of QFs is that one pair should be used for analysis while the dual pair should be used for synthesis.

If one pair of orthogonal QFs is used for analysis while a different pair is used for synthesis, the result is a smooth orthogonal transformation of the input signal.

To implement the wavelet transform on intervals, it is necessary to use different convolution-decimation functions than the ones defined in Chapter 5. The main difference is that output values near the boundary of the interval are computed in a different manner than output values in the interior of the interval. A complete description of this modification may be found in [20].

6.3 Exercises

1. Suppose we use a QF with R coefficients to compute a periodic DWT of L levels on a signal of length $M2^L$. How many additions and multiplications

will be required? Test your result by timing one of the implementations.

2. Show that using the hat function defined in Equation 1.76 for ϕ generates an MRA. Find the operator H such that $\phi = H\phi$. Is it an OQF? Find its conjugates and duals.

3. Implement `dwtpdn()`, which is the nonrecursive version of `dwtpd()`. (Hint: look at `dwtpd0n()`.)

4. Implement `dwtpin()`, the nonrecursive version of `dwtpi()`. (Hint: look at `dwtpi0n()`.)

5. Implement `idwtpd0()`, the recursive version of `idwtpd0n()`.

6. Implement `idwtpi0()`, the recursive version of `idwtpi0n()`.

7. Implement `idwtpdn()`, the nonrecursive version of `idwtpd()`.

8. Implement `idwtpin()`, the nonrecursive version of `idwtpi()`.

9. Implement `dwtan()`, the nonrecursive version of `dwta()`.

10. Combine the functions defined so far to implement DWTA with a function that, given an array and its length, returns just a pointer to an array of INTERVALs describing the wavelet subspaces $W_1, \ldots, W_L$ and the scaling space V_L.

11. Combine the functions defined so far to implement iDWTA with a function that, given an array of intervals describing $W_1, \ldots, W_L$ and V_L, returns just an INTERVAL containing the reconstructed signal.

12. Implement `dwta()` so as to allocate intervals and coefficient arrays on the fly, rather than using long concatenated arrays.

Chapter 7

Wavelet Packets

Wavelet packets are particular linear combinations or superpositions of wavelets. They have also been called *arborescent wavelets* [87]. They form bases which retain many of the orthogonality, smoothness, and localization properties of their parent wavelets. The coefficients in the linear combinations are computed by a factored or recursive algorithm, with the result that expansions in wavelet packet bases have low computational complexity.

A *discrete wavelet packet analysis*, or *DWPA*, is a transformation into coordinates with respect to a collection of wavelet packets, while a *discrete wavelet packet transform* or *DWPT* is just the coordinates with respect to a basis subset. Since there can be many wavelet packet bases in that collection, the DWPA is not fully specified unless we describe the chosen basis. This *basis choice overhead* is reduced to a manageable amount of information precisely because wavelet packets are efficiently coded combinations of wavelets, which are themselves a fixed basis requiring no description.

Going the other way, a *discrete wavelet packet synthesis* or *DWPS* takes a list of wavelet packet coefficient sequences and superposes the corresponding wavelet packets into the output sequence. If the input is a basis set of components, then DWPS reconstructs the signal perfectly.

A discrete wavelet transform is one special case of a DWPA, and an inverse discrete wavelet transform is a special case of DWPS. Thus DWPS and DWPA implementations may be used instead of the specific DWT and iDWT implementations in Chapter 6. The more specific code, though, may have an advantage in speed or storage.

DWPA has its own rather complicated history. Wavelet packets are the basis functions behind *subband coding*, which has been used for more than a decade in both speech and image coding. Daubechies' construction of smooth orthonormal wavelets was generalized into the wavelet packet construction in [25] and [30].

7.1 Definitions and general properties

We shall use the notation and terminology of Chapters 5 and 6, and assume the results therein.

Let H, G be a conjugate pair of quadrature filters from an orthogonal or biorthogonal set. There are two other QFs H' and G', possibly equal to H and G, for which the maps H^*H' and G^*G' are projections on ℓ^2, and $H^*H' + G^*G' = I$.

7.1.1 Fixed-scale wavelet packets on R

Using the QFs H and G we recursively define the following sequence of functions:

$$\psi_0 \stackrel{\text{def}}{=} H\psi_0; \qquad \int_{\mathbf{R}} \psi_0(t)\, dt = 1, \tag{7.1}$$

$$\psi_{2n} \stackrel{\text{def}}{=} H\psi_n; \qquad \psi_{2n}(t) = \sqrt{2}\sum_{j\in\mathbf{Z}} h(j)\psi_n(2t-j), \tag{7.2}$$

$$\psi_{2n+1} \stackrel{\text{def}}{=} G\psi_n; \qquad \psi_{2n+1}(t) = \sqrt{2}\sum_{j\in\mathbf{Z}} g(j)\psi_n(2t-j). \tag{7.3}$$

The function ψ_0 is uniquely defined and can be identified with the function ϕ in [36] and Equation 5.68. The normalization condition implies that it has unit total mass. The function ψ_1 is the *mother wavelet* associated to the filters H and G. Its descendents ψ_n can be identified with the functions ψ_n of Equation 5.82. The collection of these functions, for $n = 0, 1, \ldots$, makes up what we will call the (unit-scale) *wavelet packets* associated to H and G. The recursive formulas in Equations 7.1 to 7.3 provide a natural arrangement in the form of a binary tree, as seen in Figure 7.1. We also define the *dual wavelet packets* $\{\psi_n' : n \geq 0\}$ using the dual QFs H' and G':

$$\int_{\mathbf{R}} \psi_0'(t)\, dt = 1; \qquad \psi_{2n}' \stackrel{\text{def}}{=} H'\psi_n'; \qquad \psi_{2n+1}' \stackrel{\text{def}}{=} G'\psi_n'. \tag{7.4}$$

If we view $a(j) \stackrel{\text{def}}{=} \sqrt{2}\,\psi_n(2t+j)$ as a sequence in j for (t,n) fixed, then we can rewrite Equations 7.2 and 7.3 as follows:

$$\psi_{2n}(t+i) \;=\; \sqrt{2}\sum_{j\in\mathbf{Z}} h(2i-j)\psi_n(2t+j) \;=\; Ha(i); \tag{7.5}$$

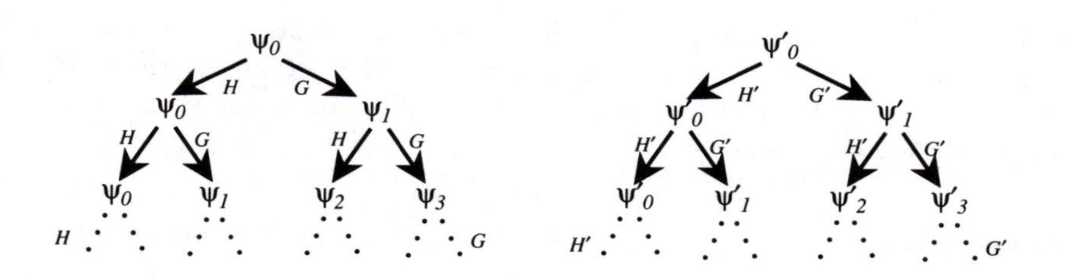

Figure 7.1: Fixed-scale wavelet packets.

$$\psi_{2n+1}(t+i) = \sqrt{2}\sum_{j\in\mathbf{Z}} g(2i-j)\psi_n(2t+j) = Ga(i). \tag{7.6}$$

Similar formulas hold for ψ'_n, using the sequences h' and g'. We can use them to prove that dual wavelet packets are the duals of wavelet packets:

Lemma 7.1 $\langle\psi_n(t-i), \psi'_m(t-j)\rangle_{L^2} = \delta(n-m)\delta(i-j)$.

Proof: First we establish the result for $n = m = 0$. Let $\mathbf{1}$ be the characteristic function of the interval $[0,1]$; then it is obvious that $\langle\mathbf{1}(t+i), \mathbf{1}(t+j)\rangle = \delta(i-j)$. If we define $\phi_n = H^n\mathbf{1}$ and $\phi'_n = (H')^n\mathbf{1}$, then

$$\begin{aligned}\langle\phi_{n+1}(t+i), \phi'_{n+1}(t+j)\rangle &= 2\int_{\mathbf{R}}\sum_{k,k'}\bar{h}(2i-k)h'(2j-k')\bar{\phi}_n(2t+k)\phi'_n(2t+k')\,dt \\ &= \sum_{k,k'}\bar{h}(2i-k)h'(2j-k')\delta(k-k') = \delta(i-j).\end{aligned}$$

Also, Equation 5.79 implies that $\int_{\mathbf{R}}\phi_n = 1$ for all $n > 0$. The same holds for ϕ'_n, so that each ϕ_n and ϕ'_n satisfies the unit mass condition. Thus by putting $\psi_0 = \lim_{n\to\infty}\phi_n$ and $\psi'_0 = \lim_{n\to\infty}\phi'_n$, we have

$$\langle\psi_0(t+i), \psi'_0(t+j)\rangle = \delta(i-j). \tag{7.7}$$

Now suppose that we have shown the result for all $i, j \in \mathbf{Z}$ but with $n, m < N$. Then by Equations 7.5 and 7.6 and a calculation virtually identical to the one resulting in Equation 7.7 for ψ_0, ψ'_0 we have

$$\langle\psi_{2n}(t+i), \psi'_{2m}(t+j)\rangle = \delta(n-m)\delta(i-j); \tag{7.8}$$

$$\langle\psi_{2n+1}(t+i), \psi'_{2m+1}(t+j)\rangle = \delta(n-m)\delta(i-j); \tag{7.9}$$

$$\langle\psi_{2n+1}(t+i), \psi'_{2m}(t+j)\rangle = 0; \tag{7.10}$$

$$\langle\psi_{2n}(t+i), \psi'_{2m+1}(t+j)\rangle = 0. \tag{7.11}$$

The first two follow from the duality of H, H' and G, G', while the third and fourth follow from the independence of H', G and G', H. This establishes the result for $0 \leq n, m < 2N$, and induction establishes it for all n. To complete the proof, we simply rename $i \leftarrow -i$ and $j \leftarrow -j$. □

Corollary 7.2 *The set $\{\psi_n(t-k) : n, k \in \mathbf{Z}, n \geq 0\}$ is linearly independent.*

Proof: If $x(t) = \sum_{n,k} a_{nk}\psi_n(t-k) \equiv 0$, then $a_{nk} = \langle \psi'_n(t-k), x(t)\rangle = 0$. □

Corollary 7.3 *If H, G are orthogonal QFs, then $\{\psi_n(t-k) : n, k \in \mathbf{Z}, n \geq 0\}$ is an orthonormal set.* □

Let us denote by Λ_n the closed linear span of the integer translates of the unit-scale wavelet packets ψ_n:

$$\Lambda_n \stackrel{\text{def}}{=} \left\{ x(t) = \sum_k c(k)\psi_n(t-k) \right\} \subset L^2(\mathbf{R}). \tag{7.12}$$

Since translations intertwine with QFs, Equations 7.1 to 7.3 imply that

$$\Lambda_{2n} = H\Lambda_n \quad \text{and} \quad \Lambda_{2n+1} = G\Lambda_n. \tag{7.13}$$

Notice that $\Lambda_0 = V_0$, the unit-scale approximation space defined by the low-pass filter H. We have a natural basis for each of these spaces:

Corollary 7.4 *The set $\{\psi_n(t-k) : k \in \mathbf{Z}\}$ is a basis for Λ_n.* □

Corollary 7.5 *If H, G are orthogonal QFs, then $\{\psi_n(t-k) : k \in \mathbf{Z}\}$ is an orthonormal basis for Λ_n.* □

We now show that putting the bases for $\Lambda_n, n \geq 0$ together yields a basis for $L^2(\mathbf{R})$. We do this by slightly adapting and translating to our notation the method of Coifman and Meyer [23], so as to include the biorthogonal case. Using the dual filters H' and G' and the exact reconstruction property we find:

$$\psi_n(t+j) = \frac{1}{\sqrt{2}} \sum_{i\in\mathbf{Z}} \bar{h}'(2i-j)\psi_{2n}\left(\frac{t}{2}+i\right) + \frac{1}{\sqrt{2}} \sum_{i\in\mathbf{Z}} \bar{g}'(2i-j)\psi_{2n+1}\left(\frac{t}{2}+i\right). \tag{7.14}$$

Thus if $x = x(t) = \sum_{k\in\mathbf{Z}} \bar{\lambda}_n(k)\psi_n(t+k)$, then we can use Equation 7.14 to write the following pair of series for x:

$$\begin{aligned} x(t) &= \frac{1}{\sqrt{2}}\sum_i \left(\sum_j \bar{h}'(2i-j)\bar{\lambda}_n(j)\right)\psi_{2n}\left(\frac{t}{2}+i\right) \\ &\quad + \frac{1}{\sqrt{2}}\sum_i \left(\sum_j \bar{g}'(2i-j)\bar{\lambda}_n(j)\right)\psi_{2n+1}\left(\frac{t}{2}+i\right) \qquad (7.15) \\ &= \sum_i \overline{H'\lambda_n(i)}\frac{1}{\sqrt{2}}\psi_{2n}\left(\frac{t}{2}+i\right) + \sum_i \overline{G'\lambda_n(i)}\frac{1}{\sqrt{2}}\psi_{2n+1}\left(\frac{t}{2}+i\right). \end{aligned}$$

The right-hand side is an expansion for x, using dilates of basis functions in Λ_{2n} and Λ_{2n+1} with coefficients $\overline{H'\lambda_n}$ and $\overline{G'\lambda_n}$, respectively. Equivalently, for any $x \in \Lambda_n$ we can write

$$x(t) = \frac{1}{\sqrt{2}}y\left(\frac{t}{2}\right) + \frac{1}{\sqrt{2}}z\left(\frac{t}{2}\right), \qquad \text{for } y \in \Lambda_{2n} \text{ and } z \in \Lambda_{2n+1}. \qquad (7.16)$$

If we define $\sigma x(t) = \frac{1}{\sqrt{2}}x\left(\frac{t}{2}\right)$, and $\sigma\Lambda_n = \{\sigma x : x \in \Lambda_n\}$, then Equation 7.16 shows that $\Lambda_n = \sigma H\Lambda_n + \sigma G\Lambda_n = \sigma\Lambda_{2n} + \sigma\Lambda_{2n+1}$, or

$$\Lambda_0 = \sigma\Lambda_0 + \sigma\Lambda_1 = \cdots = \sigma^k\Lambda_0 + \sigma^k\Lambda_1 + \cdots + \sigma^k\Lambda_{2^k-1}, \qquad (7.17)$$

and in general,

$$\Lambda_n = \sigma^k\Lambda_{2^k n} + \cdots + \sigma^k\Lambda_{2^k(n+1)-1}, \qquad \text{for all } n, k \geq 0. \qquad (7.18)$$

Likewise, for negative k we have

$$\sigma^k\Lambda_n = \Lambda_{2^{-k}n} + \cdots + \Lambda_{2^{-k}(n+1)-1}, \qquad \text{for all } n \geq 0,\ k \leq 0. \qquad (7.19)$$

The pieces of this decomposition expand to fill up all of L^2:

Theorem 7.6 *The functions $\{\psi_n(t-j) : j, n \in \mathbf{Z}, n \geq 0\}$ are a basis for $L^2(\mathbf{R})$.*

Proof: The functions are orthonormal and span $\Lambda_0 + \Lambda_1 + \cdots$, which contains $\sigma^{-k}\Lambda_0$ for every k. But $\sigma^{-k}\Lambda_0 = \sigma^{-k}V_0 = V_{-k}$, and $\int_{\mathbf{R}} \psi_0(t)\,dt = 1$ is enough to guarantee that $V_{-k} \to L^2(\mathbf{R})$ as $k \to \infty$. Thus $\{\psi_n(t-j) : n \geq 0, n, j \in \mathbf{Z}\}$ is complete. □

Corollary 7.7 *If H and G are orthogonal QFs, then $\{\psi_n(t-j) : j, n \in \mathbf{Z}, n \geq 0\}$ is an orthonormal basis for $L^2(\mathbf{R})$.* □

We remark that the collection $\{\psi_n : n \geq 0\}$ is a generalization of Walsh functions, and n can be taken as a "frequency" index. Thus $\{\Lambda_n : n \geq 0\}$ is a decomposition of L^2 into (possibly orthogonal) subspaces of unit-scale functions of different frequencies.

7.1.2 Multiscale wavelet packets on R

All of the functions ψ_n in the preceding section have a fixed scale, but we observe that multiple-scale decompositions of L^2 are also possible. We will refer to the translated, dilated, and normalized function $\psi_{sfp} \stackrel{\text{def}}{=} 2^{-s/2}\psi_f(2^{-s}t-p)$ as a *wavelet packet* of *scale index* s, *frequency index* f, and *position index* p. The wavelet packets $\{\psi_{sfp} : p \in \mathbf{Z}\}$ are a basis for $\sigma^s\Lambda_f$. If H, G are orthogonal QFs, then we will call them *orthonormal wavelet packets.*

The scale index gives a relative estimate of extent. The scaled wavelet packet's position uncertainty $\Delta x(\psi_{sfp})$, defined in Equation 1.57, is 2^s times the position uncertainty $\Delta x(\psi_f)$ of ψ_f. If H and G are conventionally indexed FIR filters supported in $[0, R]$, then ψ_f will be supported in the interval $[0, R]$ and ψ_{sfp} will be supported in the interval $[2^s p, 2^s(p+R)]$, which has 2^s times the diameter of the interval $[0, R]$.

In the orthogonal Haar–Walsh and Shannon examples, the frequency index is related by the Gray code permutation to sequency or mean frequency: see Equations 5.80 and 5.81. By analogy with Theorems 5.19 and 5.20 we will say that the *nominal frequency* of ψ_f is $\frac{1}{2}GC^{-1}(f)$. Dilation by 2^s scales this nominal frequency to $\frac{1}{2}2^s GC^{-1}(f)$.

The position index is related to the center of energy x_0 of the function ψ_0, adjusted for the phase shifts caused by the QFs. Consider first the unit-scale case: the coefficients $\{\lambda_0(n)\}$ of $x_0(t) = \psi_0(t+j)$ in Λ_0 consist of a single 1 at $n = j$, all others being 0. Thus $c[\lambda_0] = j$ in the notation of Equation 5.27. We then use the OQF phase shifts theorem (5.8) to compute the approximate center of energy of ψ_n from the sequence of filters determined by n. Then we multiply this by 2^{-s} to get the *nominal position* of the wavelet packet ψ_{sfp}.

The position, scale, and frequency are easy to define and compute for the example of *Haar–Walsh wavelet packets.* Let $H = \{\frac{1}{\sqrt{2}}, \frac{1}{\sqrt{2}}\}$ and $G = \{\frac{1}{\sqrt{2}}, \frac{-1}{\sqrt{2}}\}$ be the Haar–Walsh filters. Then

- $\psi_0 = \mathbf{1}_{[0,1]}$ will be the characteristic function of the interval $[0, 1]$;
- ψ_n will be supported in $[0, 1]$, where it takes just the values ± 1;
- ψ_n will have $GC^{-1}(n)$ zero-crossings, for a nominal frequency of $\frac{1}{2}GC^{-1}(n)$;

- $\psi_{knj}(t) = 2^{-k/2}\psi_n(2^{-k}t - j)$ will be supported in the interval $[2^k j, 2^k(j+1)]$, which has width 2^k.

We will denote the dual wavelet packet by ψ'_{sfp} since $\langle \psi_{sfp}, \psi'_{sfp}\rangle = 1$, but the duality is a bit more complicated here than in the fixed-scale case. We first define a *dyadic interval* $I_{kn} \subset \mathbf{R}$ by the formula

$$I_{kn} \stackrel{\text{def}}{=} \left[\frac{n}{2^k}, \frac{n+1}{2^k}\right[. \tag{7.20}$$

Omitting the right endpoint allows adjacent dyadic intervals to be disjoint. Notice that dyadic intervals are either disjoint or one of them contains the other.

There is a natural correspondence between dyadic subintervals and subspaces of L^2, namely

$$I_{sf} \longleftrightarrow \sigma^s \Lambda_f. \tag{7.21}$$

This correspondence is faithful and preserves independence; using dyadic *index intervals* to keep track of multiscale wavelet packets, we can generalize Lemma 7.1:

Lemma 7.8 *If the dyadic intervals $I_{s'f'}$ and I_{sf} are disjoint, or if $I_{s'f'} = I_{sf}$ but $p' \neq p$, then $\langle \psi_{sfp}, \psi'_{s'f'p'}\rangle = 0$.*

Proof: If $I_{s'f'} = I_{sf}$ then $s' = s$ and $f' = f$; rescaling by 2^s preserves the inner product which we compute as $\langle \psi_f(t-p), \psi'_f(t-p')\rangle = \delta(p-p')$ by Lemma 7.1.

If $s' = s$ but $f' \neq f$, we again rescale by s and compute the inner product $\langle \psi_f(t-p), \psi'_{f'}(t-p')\rangle = 0$ by the same lemma.

If $s' \neq s$, we may assume without loss that $s' < s$ and put $r = s - s' > 0$. Then ψ_{sfp} belongs to $\sigma^s \Lambda_f$, while by Equation 7.19,

$$\psi'_{s'f'p'} \in \sigma^{s'} \Lambda'_{f'} = \sigma^s \sigma^{-r} \Lambda'_{f'} = \sum_{n=2^r f'}^{2^r(f'+1)-1} \sigma^s \Lambda'_n.$$

If the dyadic intervals $I_{s'f'}$ and I_{sf} are disjoint, then every component Λ'_n, $2^r f' \leq n < 2^r(f'+1)$, is orthogonal to the single component Λ_f. □

We now refine the decomposition $L^2 = \sum_n \Lambda_n$ by allowing the scale to vary. Let $\mathcal{I}$ be a *disjoint dyadic cover* of $\mathbf{R}^+$, namely a collection of disjoint dyadic intervals I_{kn} whose union is the positive half line. To get a basis from a general dyadic decomposition, we need certain technical conditions on the scaling function ϕ and thus on its descendent wavelet packets. These conditions are described in [28]; they are independent of the dyadic cover. Let us say that the wavelet packets satisfying them are *well-behaved*.

Theorem 7.9 *If $\mathcal{I}$ is a disjoint dyadic cover of* $\mathbf{R}^+$, *then the well-behaved wavelet packets* $\{\psi_{sfp} : I_{sf} \in \mathcal{I}, p \in \mathbf{Z}\}$ *form a basis for* $L^2(\mathbf{R})$.

Proof: Since $\{\psi_{sfp} : p \in \mathbf{Z}\}$ is a basis for $\sigma^s \Lambda_f$ and since two such spaces have trivial intersection if their dyadic index intervals are disjoint, it suffices to show that $\{\sigma^s \Lambda_f : I_{sf} \in \mathcal{I}\}$ is dense in L^2.

First consider the special case in which $s \leq 0$ for all $I_{sf} \in \mathcal{I}$. Then by Equation 7.19, each I_{sf} corresponds to $\sigma^s \Lambda_f = \Lambda_{2^{-s}f} + \cdots + \Lambda_{2^{-s}(f+1)-1}$, and the subscripts for these spaces are exactly the integers in the index interval I_{sf}. Hence, if $\bigcup I_{sf} = \mathbf{R}^+$, we have $\sum \sigma^s \Lambda_f = \sum_{n=0}^{\infty} \Lambda_n$, whose closure is all of L^2.

Likewise, if $s \leq k$ for all intervals I_{sf} in the cover, then $\sum \sigma^s \Lambda_f = \sigma^k \sum_{n=0}^{\infty} \Lambda_n$. The closure of this sum is $\sigma^k L^2$, which is the same as L^2.

The general case requires that the wavelet packets be well-behaved and is proved in [28], Theorems 5 and 6. □

Theorem 7.9 may be called the *graph theorem* because disjoint dyadic covers might be viewed as graphs from partitions of $\mathbf{R}^+$ to bases for L^2. Plus, if we use orthogonal filters we get orthonormal *graph bases*:

Corollary 7.10 *If H and G are orthogonal QFs and $\mathcal{I}$ is a disjoint dyadic cover of* $\mathbf{R}$, *then* $\{\psi_{sfp} : p \in \mathbf{Z}, I_{sf} \in \mathcal{I}\}$ *is an orthonormal basis for* $L^2(\mathbf{R})$. □

Remark. All the cases encountered in practice satisfy the scale condition $s \leq L$ for some finite maximum L. Moreover, for real sampled signals we have a minimum scale: the sampling interval. Without loss of generality, we can make $s = 0$ the minimum or *sampling interval scale* and refer to L as the *depth of decomposition* of the wavelet packet basis. Then 2^{-L} will be the width of the smallest intervals in the dyadic cover.

An *(orthonormal) wavelet packet basis* of $L^2(\mathbf{R})$ is any (orthonormal) basis selected from among the functions ψ_{sfp}. Graph bases provide a large library of these, though they do not exhaust the possibilities in general. Some easily described examples of wavelet packet graph bases are the orthonormal *Walsh-type* basis $\Lambda_0 \oplus \Lambda_1 \oplus \cdots \oplus \Lambda_k \oplus \cdots$, the *subband basis* $\sigma^L \Lambda_0 \oplus \sigma^L \Lambda_1 \oplus \cdots \oplus \sigma^L \Lambda_n \oplus \cdots$ and the *wavelet basis* $\cdots \oplus \sigma^{-1} \Lambda_1 \oplus \Lambda_1 \oplus \sigma \Lambda_1 \oplus \cdots \oplus \sigma^k \Lambda_1 \oplus \cdots$.

If we restrict our attention to the unit dyadic subinterval $[0,1[$, we can characterize graph bases for the approximation space $V_0 \subset L^2(\mathbf{R})$ to which it corresponds. Since we can write $V_0 = \bigcup_{k=0}^{L} V_k$ for any $L \geq 0$, graph decompositions of this space to a finite depth L will consist of pieces of the coarser approximation spaces V_k, $0 \leq k \leq L$. These pieces are in one-to-one correspondence with the dyadic subintervals of the half-open interval $[0,1[$, and are defined for $s \geq 0$ and $0 \leq f < 2^s$.

If two dyadic intervals are distinct, their subspaces of V_0 are distinct, and if they are disjoint, the corresponding subspaces are independent. Thus if $\mathcal{I}$ is a disjoint dyadic cover of $[0,1[$, it corresponds to an (orthonormal) wavelet packet basis:

Corollary 7.11 *If $\mathcal{I}$ is a disjoint dyadic cover of $[0,1[$, then the wavelet packets $\{\psi_{sfp} : I_{sf} \in \mathcal{I}, p \in \mathbf{Z}\}$ form a basis for V_0.*

Proof: If $\bigcup_{I_{sf}\in\mathcal{I}} I_{sf} = [0,1[$, then $\sum_{I_{sf}\in\mathcal{I}} \sigma^s\Lambda_f = \sum_{n=0}^{2^s-1} \sigma^s\Lambda_n = \Lambda_0 = V_0$. □

Corollary 7.12 *If H and G are orthogonal QFs and $\mathcal{I}$ is a disjoint dyadic cover of $[0,1[$, then $\{\psi_{sfp} : p \in \mathbf{Z}, I_{sf} \in \mathcal{I}\}$ is an orthonormal basis for V_0.* □

7.1.3 Numerical calculation of wavelet packet coefficients

Now let $\{\lambda_{sf}(p) : p \in \mathbf{Z}\}$ be the sequence of inner products of a function $x = x(t)$ in $L^2(\mathbf{R})$ with the "backwards" basis functions in $\sigma^s\Lambda_f$:

$$\lambda_{sf}(p) \stackrel{\text{def}}{=} \langle x, \psi^<_{sfp}\rangle = \int_{\mathbf{R}} \bar{x}(t) 2^{-s/2}\psi_f(p - 2^{-s}t)\, dt. \tag{7.22}$$

Here $s, p \in \mathbf{Z}$, $f \geq 0$, and $\psi^<_{sfp}(t) \stackrel{\text{def}}{=} 2^{-s/2}\psi_f(p - 2^{-s}t)$. By duality, if $x = {\psi'}^<_{sfp'}$, we will have $\lambda_{sf}(p) = \delta(p - p')$. Similarly, the numbers $\bar{\lambda}_{sf}(p)$ are coefficients of the expansion of x in the functions of $\sigma^s\Lambda'_f$:

$$x(t) = \sum_p \bar{\lambda}_{sf}(p){\psi'}^<_{sfp}(t) \quad \Rightarrow \langle x, \psi^<_{sfp}\rangle = \lambda_{sf}(p). \tag{7.23}$$

We will also use the notation $\{\lambda\}$ for $\{\lambda_{00}\}$, the backwards inner products of $x(t)$ with the basis functions of $\sigma^0\Lambda_0 = V_0$. From these we may calculate the inner products of $x(t)$ with the wavelet packets in any space $\sigma^s\Lambda_f$, for $s > 0$ and $0 \leq f < 2^s$, by applying the operators H and G a total of s times. This is a consequence of the following:

Lemma 7.13 *The coefficient sequences $\{\lambda_{sf}\}$ satisfy the recursion relations*

$$\lambda_{s+1,2f}(p) = H\lambda_{sf}(p), \tag{7.24}$$
$$\lambda_{s+1,2f+1}(p) = G\lambda_{sf}(p). \tag{7.25}$$

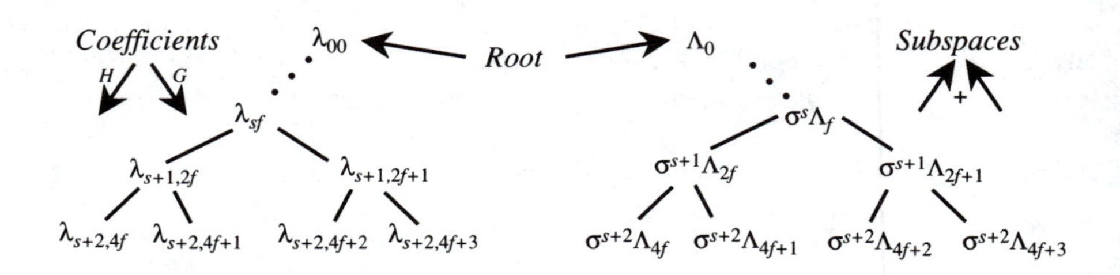

Figure 7.2: Wavelet packet analysis and synthesis.

Proof:

$$\begin{aligned}
\lambda_{s+1,2f}(p) &= \int_{\mathbf{R}} \bar{x}(t) 2^{\frac{-s-1}{2}} \psi_{2f}(p - 2^{-s-1}t)\, dt \\
&= \int_{\mathbf{R}} \bar{x}(t) 2^{\frac{-s-1}{2}} \left[2^{\frac{1}{2}} \sum_{j \in \mathbf{Z}} h(j) \psi_f \left(2[p - 2^{-s-1}t] - j\right) \right] dt \\
&= \sum_{j \in \mathbf{Z}} h(j) \int_{\mathbf{R}} \bar{x}(t) 2^{\frac{-s}{2}} \psi_f \left(2p - j - 2^{-s}t\right)\, dt \\
&= \sum_{j \in \mathbf{Z}} h(j) \lambda_{sf}(2p - j) \qquad = H\lambda_{sf}(p).
\end{aligned}$$

The proof for $\lambda_{s+1,2f+1}(p)$ is the same with h replaced by g. □

Remark. Using $\psi^{<}_{sfp}$ rather than ψ_{sfp} in the inner product is a technical trick that allows us to use the same filter convolution formulas for functions as for sequences.

The collection of wavelet packets $\{\psi_{sfp}\} = \{2^{s/2}\psi_f(2^s t - p)\}$ used in the inner products comprises a library of functions, with a natural organization deriving from Lemma 7.13. We can form them into a binary tree whose nodes are the spaces $\sigma^s \Lambda_f$. The root is $V_0 = \Lambda_0$, the leaves are $\sigma^L \Lambda_0, \ldots, \sigma^L \Lambda_{2^L-1}$, and the genealogy follows the right-hand diagram in Figure 7.2. Each node is the sum of its two immediate descendents, or *children*. If the QFs are orthogonal, the sum is an orthogonal direct sum. Likewise, if we start with a sequence of unit-scale wavelet packet coefficients $\lambda = \lambda_{00}$, then the multiscale wavelet packet coefficient sequences λ_{sf} form the nodes of the binary tree in the left-hand diagram of Figure 7.2.

To obtain the wavelet packet analysis of a function, we first find its coefficient sequence in the root subspace, then follow the branches of the wavelet packet coefficient tree to find the expansion in the descendent subspaces. Branches in the tree correspond to indices s, f and thus to sequences of the filters H and G. The correspondence may be computed using induction on Lemma 7.13. If we let $f = (f_{k-1} \cdots f_1 f_0)_2$ be the binary representation of the integer $f \in [0, 2^k - 1]$, we have the following:

Theorem 7.14 *For all $s \geq 0$ and $0 \leq f < 2^s$ we have*

$$\lambda_{sf} = F_0 \cdots F_{s-1} \lambda(p),$$

where $F_i = H$ if f_i is 0 and $F_i = G$ otherwise. □

The algorithm embodied in this theorem has low complexity in the following sense. Suppose we are given the unit-scale wavelet packet coefficients $\{\lambda(p) : p \in \mathbf{Z}\}$ of a function $x = x(t)$. Suppose that $|\lambda(p)|$ is negligibly small unless $|p| < N$. Then if we use finitely supported filters, each application of H or G to the coefficient sequence costs $O(N)$ multiply-adds. Then Theorem 7.14 gives a fast construction for all of the nonnegligible inner products $\lambda_{sf}(p)$: the complexity of producing all of them is $O(sN) \approx O(N \log N)$. This bounds the cost of a *wavelet packet analysis*, *i.e.*, discovering the coefficients $\{\lambda_{sf}(p)\}$ given $\{\lambda(p)\}$.

We can *synthesize* the root coefficients $\{\lambda\} = \{\lambda_{00}\}$ by applying the adjoints H'^* and G'^* of the dual QFs H' and G' to the sequence $\{\lambda_{sf}\}$. In the following results, we put $F_i'^* = H'^*$ and $F_i = H$ if the ith binary digit f_i of f is zero; otherwise $F_i'^* = G'^*$ and $F_i = G$:

Theorem 7.15 *For $s \geq 0$ and $0 \leq f < 2^s$ we can write $\psi'^{<}_{sfp} = \sum_{n \in \mathbf{Z}} \bar{\lambda}(n) \psi'^{<}_{00n}$, where the coefficient sequence $\{\lambda\}$ is given by*

$$\lambda = F'^*_{s-1} \cdots F'^*_1 F'^*_0 \mathbf{1}_p,$$

where $\mathbf{1}_p$ is the sequence of zeroes with a single "1" at index p. □

Note that by Lemma 7.1 a single wavelet packet ψ'_{sfp} is characterized by having a coefficient sequence with $\lambda_{sf}(p) = 1$ and $\lambda_{sf'}(p') = 0$ if $f \neq f'$ or $p \neq p'$.

Theorem 7.15 gives the following formula for ψ'_f in terms of ψ'_0 and the coefficient sequence λ derived from $\lambda_{sf} = \mathbf{1}_p$:

$$\psi'_f(t) = \sum_k \bar{\lambda}(k) 2^{s/2} \psi'_0 \left(2^s t + [k - p]\right). \tag{7.26}$$

A superposition of wavelet packets can be expanded using the coefficient sequence obtained from the superposition of elementary sequences. Let $\mathcal{B}$ be any subset of index triplets (s, f, p) which satisfy $s \geq 0$, $0 \leq f < 2^s$, and $p \in \mathbf{Z}$. Then a superposition of wavelet packets with these indices can be expanded as a sum in the approximation space V_0':

Corollary 7.16 *We can expand*

$$x = \sum_{(s,f,p)\in\mathcal{B}} \bar{\lambda}_{sf}(p)\psi'^{<}_{sfp} = \sum_{n\in\mathbf{Z}} \bar{\lambda}(n)\psi'^{<}_{00n}$$

using

$$\lambda = \sum_{(s,f,p)\in\mathcal{B}} F'^{*}_{s-1}\cdots F'^{*}_{1}F'^{*}_{0}\lambda_{sf}(p)\mathbf{1}_p,$$

where $\mathbf{1}_p$ is the sequence of zeroes with a single "1" at index p. □

By combining both the analysis and synthesis operations, we obtain projection operators:

Corollary 7.17 *Let $P : V_0' \to \sigma^s\Lambda_f'$ be defined by $Px(t) = \sum_p \bar{a}(p)\psi'^{<}_{00p}(t)$, where for each $x = \sum_p \bar{\lambda}(p)\psi'^{<}_{00p}$ we have*

$$a = F'^{*}_{s-1}\cdots F'^{*}_{1}F'^{*}_{0}F_0F_1\cdots F_{s-1}\lambda. \tag{7.27}$$

Then P is a projection. If $H = H'$ and $G = G'$ are orthogonal QFs, then P is an orthogonal projection. □

Each application of the finitely supported adjoint filters H'^* or G'^* also costs $O(N)$ multiply-adds. Thus *wavelet packet synthesis*, or reconstructing $\{\lambda_{00}\}$ given the coefficients $\{\lambda_{sf}(p)\}$, is also an algorithm of complexity $O(sN) \approx O(N \log N)$.

We will fix our attention onto the tree of coefficient sequences $\{\lambda_{sf}\}$, since that is what we compute in practice. It is useful to keep in mind a picture of this tree for a small example, such as the case of an eight-point sequence $\lambda(n) = x_n$, $n = 0, 1, \ldots, 7$, analyzed to three levels with Haar–Walsh filters. The complete tree may then be depicted as in Figure 7.3, as a rectangle of rows of blocks. The row number to the left of each row is the scale index s, starting with row 0 at the top or "root." The column number is a combination of the frequency index and the position index: the subscript within each block is the position index, while the number of the block counting from the left, which can be deduced from the sequence of letters within the block, provides the frequency index.

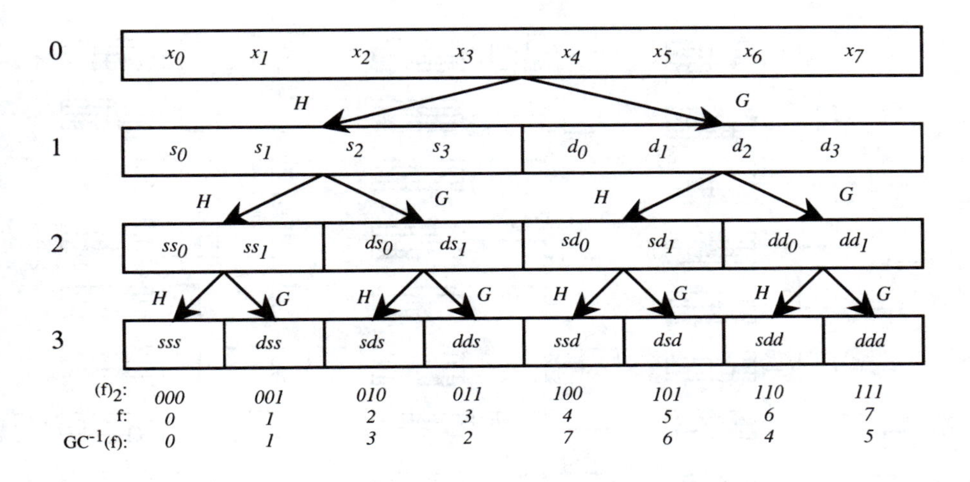

Figure 7.3: Naturally ordered Haar–Walsh wavelet packet coefficients, rank 8.

Observe that if we put the rectangle over the unit interval, then the blocks of coefficients lie over dyadic subintervals with the correspondence of Equation 7.21: the coefficient sequence λ_{sf} is in the block over index interval I_{sf}.

Each row of coefficients is computed from the row above it by one application of either H or G, which we think of as "summing" (s) or "differencing" (d) operations, respectively. Thus, for example, the subblock $\{ss_0, ss_1\}$ comes from the application of H to $\{s_0, s_1, s_2, s_3\}$, while $\{ds_0, ds_1\}$ comes similarly via G. Thus in the Haar–Walsh case, we have $ss_0 = \frac{1}{\sqrt{2}}(s_0 + s_1)$, $ss_1 = \frac{1}{\sqrt{2}}(s_2 + s_3)$, $ds_0 = \frac{1}{\sqrt{2}}(s_0 - s_1)$, and $ds_1 = \frac{1}{\sqrt{2}}(s_2 - s_3)$.

The two descendent H and G subblocks on row $n+1$ are determined by their mutual parent on row n, which conversely is determined by them through the adjoint anticonvolutions H'^* and G'^*. Let us draw the functions which correspond to the entries in our example rectangle. These are Haar–Walsh wavelet packets, displayed in Figure 7.4. Each waveform in the left half-column corresponds to the coefficient in the darkened block of the right half-column. Notice that there are 24 waveforms for this eight-dimensional space. There is more than a basis here, so we have the option of choosing a basis subset adapted to a particular signal or problem. In Chapter 8 we will discuss several selection methods.

Notice also that the number of oscillations in a waveform occasionally decreases as we move from left to right along a level of the tree. This is because the algorithm

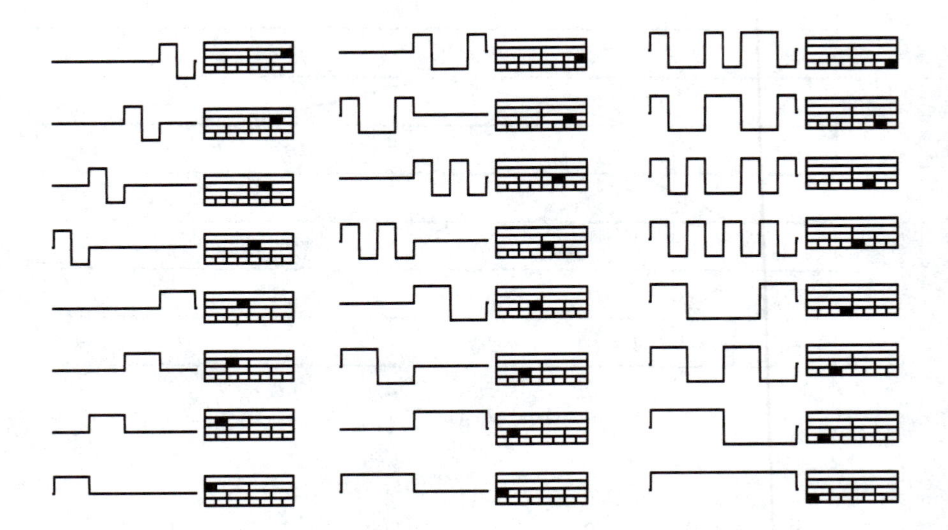

Figure 7.4: Haar–Walsh wavelet packets, rank 8.

produces a wavelet packet analysis in the *Paley* or *natural* order. It may be easily modified to produce a *sequency* ordered wavelet packet analysis, which contains the same coefficient sequences but rearranges them so that the nominal frequency of the associated wavelet packets increases as we read from left to right. Figure 7.5 depicts the permuted wavelet packet analysis on an eight point signal which if we use Haar–Walsh filters is equivalent to the Walsh transform in sequency order.

The rule for exchanging G and H to get the sequency ordered transform is to exchange H and G at each parent with odd sequency. It is deduced from Equations 5.80 and 5.81 as follows. Suppose that we wish to produce the wavelet packet analysis tree in such a way that the coefficients in the block I_{sf}, with level index s and frequency index f, are produced by the filter sequence of length s determined by $n = GC(f)$. This will ensure that the the fth node from the left at level s has nominal frequency $\frac{1}{2}GC^{-1}(n) = \frac{1}{2}f$, which is a nice monotonic function. This holds (trivially) for the top level 0, so we assume it is true for level s and proceed down by induction. For the next level $s+1$, we change the indices as follows:

- if $f = GC^{-1}(n)$ is even, then $2f = GC^{-1}(2n)$ and $2f+1 = GC^{-1}(2n+1)$, so we should use H to produce $\lambda_{s+1,2f}$ and G to produce $\lambda_{s+1,2f+1}$;
- if $f = GC^{-1}(n)$ is odd, then $2f = GC^{-1}(2n+1)$ and $2f+1 = GC^{-1}(2n)$, so

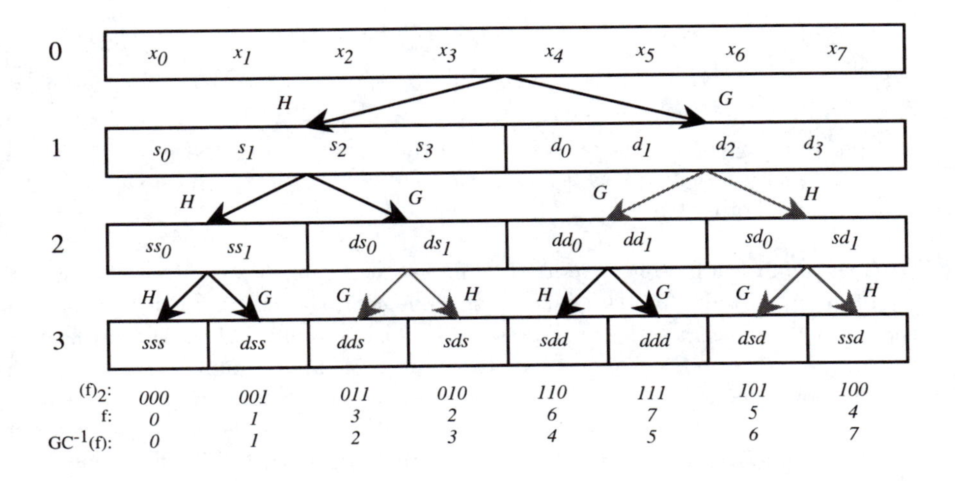

Figure 7.5: Sequency ordered wavelet packets, rank 8.

we should use G to produce $\lambda_{s+1,2f}$ and H to produce $\lambda_{s+1,2f+1}$.

Figure 7.5 indicates by gray arrows those filters which were switched to produce Gray coded frequency indices.

There are two drawbacks to performing the Gray code permutation in parallel with the wavelet packet analysis. The less serious is that it forces us to label each analysis with a marker indicating which of the two permutations we have used, thus adding some overhead. The more serious drawback is that it forces us to keep track of the current frequency index, which adds one parameter per dimension to the node data. The advantage of the sequency ordered analysis, of course, is that we can more easily interpret the frequency indices.

7.1.4 The discrete wavelet packet analysis family

Like the discrete wavelet transform or DWT, the discrete wavelet packet analysis is really a whole family of transforms. Each member of the DWPA family provides a decomposition of the input signal into the whole collection of its wavelet packet components, which are more than a basis set. For each analysis we must specify the following parameters:

- *L:* the number of levels of decomposition;
- *H and G:* the filters to use in descending a level;

- λ: the input sequence, which is taken to be wavelet scaling coefficients.

In addition, to obtain a discrete wavelet packet transform or DWPT, we must specify a basis subset. This can be done by giving a list of the scale, frequency, and position tags of the wavelet packets in the chosen basis, or else in the case of a graph basis the sequence of scales of the blocks in the graph, arranged in some predetermined order.

If the input is a finitely supported infinite sequence, then the output will also be a finitely supported infinite sequence; this is the *aperiodic DWPA*. This analysis has the disadvantage that it does not preserve rank: the number of output coefficients can in general be much larger than the number of input samples. Consider the simplest case, namely H and G are QFs of equal length, supported in $[\alpha, \omega]$ with $\omega - \alpha > 0$. Then a parent array supported in $[a, b]$ produces a pair of children supported in $[a', b']$ with $a' = \lceil (a+\alpha)/2 \rceil$ and $b' = \lfloor (b+\omega)/2 \rfloor$ as in Equation 5.91. The total length of the parent is $1 + b - a$; the total length of each child is

$$1 + b' - a' = 1 + \left\lfloor \frac{b+\omega}{2} \right\rfloor - \left\lceil \frac{a+\alpha}{2} \right\rceil \geq \frac{1+b-a}{2} + \frac{\omega - \alpha - 1}{2}. \tag{7.28}$$

Let $E \stackrel{\text{def}}{=} \omega - \alpha - 1$; this is 0 for Haar–Walsh filters, and positive for all others. Each child has at least $E/2$ more coefficients than half its parent. These in turn spawn their own extra descendents: the grandchildren contain at least $E/4 + E/2$ more coefficients than $1/4$ the parent. After L levels of decomposition, the total number of coefficients in a subspace is at least

$$\frac{1+b-a}{2^L} + \frac{E}{2^L} + \frac{E}{2^{L-1}} + \cdots + \frac{E}{2} = \frac{1+b-a}{2^L} + \frac{2^L - 1}{2^L} E. \tag{7.29}$$

This is a sharp lower bound. The total number of extra coefficients in the 2^L subspaces at level L will be at least $(2^L - 1)E$. The total number of coefficients needed in an L-level aperiodic DWPA is therefore at least

$$\sum_{s=0}^{L} [(1+b-a) + (2^s - 1)E] = (L+1)(1+b-a) + (2^{L+1} - L - 2)E. \tag{7.30}$$

If $E > 0$, this quantity grows very rapidly with the number of levels.

We get a sharp upper bound in Equation 7.28 by using $E + 2$ instead of E, so the total number of coefficients produced by an L-level aperiodic DWPA is at most

$$(L+1)(1+b-a) + (2^{L+1} - L - 2)(E+2). \tag{7.31}$$

If $N = 1 + b - a$ and we stop at $L \leq \log_2 N$, this tells us that the total number of coefficients in the complete aperiodic analysis will be $O(N[2E + \log_2 N])$. Of course, the exact number of extra coefficients can be different and must be computed in advance if we wish to preallocate all the space needed for the analysis. If the QFs H and G have different lengths, we have the additional complication that the support of the coefficient sequence in a subspace Λ_{sf} depends upon f and not just on s. The analytic formulas are complicated, but the supports themselves are quite simple to compute.

The number of coefficients can also grow during wavelet packet synthesis because adjoint aperiodic convolution-decimation with long filters more than doubles the length of the input. Let us once again assume for simplicity that H' and G' have equal support $[\alpha, \omega]$. If we start with a sequence λ_{sf} supported in $[c, d]$, one application of an adjoint QF will result in a sequence $\lambda_{s-1,\lfloor f/2 \rfloor}$, which has support in $[c', d']$ with $c' = 2c - \omega$ and $d' = 2d - \alpha$. These endpoints are calculated in Equation 5.94. Thus the length of the support grows from $1+d-c$ to $1+d'-c' = 2(1+d-c)+E$, where $E = \omega - \alpha - 1$ is defined above. After L levels, we will have a sequence in Λ_0 of support length

$$2^L(1 + d - c) + (2^L - 1)E. \tag{7.32}$$

A superposition of M wavelet packets from different subspaces $\sigma^s \Lambda'_f$ produces a sequence λ_{00} with as many as $M2^L(1 + E)$ coefficients, where $L = \max\{s\}$. There will be fewer coefficients if the wavelet packets overlap. Even with the greatest possible overlap, if the sequence λ_{Lf} contains M nonzero coefficients then the aperiodic DWPS will give a sequence λ_{00} which has length of order $2^L[M + E]$.

Remark. The projection operator of Corollary 7.17 can expand the support of a function by a rather large amount. In projecting onto a subset of the wavelet packet components, we may be discarding some which are needed for cancellation out beyond the nominal support of the function.

If the input sequence is q-periodic, then the output will consist of periodic sequences and H, G can be considered periodized QFs. The subfamily of DWPAs with this assumption will be called *periodic DWPAs*. The input sequence for a q-periodic DWPA consists of just q numbers.

The output of the DWPA will be a list of arrays, one for each sequence λ_{sf} for $0 \leq s < L$ and $0 \leq f < 2^s$. These can be arranged in a binary tree data structure with pointers, or else they may be concatenated into a single array in some predefined arrangement. The latter method works only if we know in advance the number and total length of all the output arrays.

With an orthogonal pair of QFs H, G, we get the *orthogonal DWPA*, while if H, G are biorthogonal with duals $H' \neq H, G' \neq G$ then we have the *biorthogonal DWPA*. Modification of the filters to give special treatment to the endpoints of the intervals gives the *DWPA on an interval.*

Likewise, we will distinguish the discrete wavelet packet transforms which consist of restrictions to particular basis subsets of these DWPAs.

In the other direction, we have the *periodic* and *aperiodic* discrete wavelet packet synthesis or DWPS, which can be *orthogonal* or *biorthogonal* or *on an interval.* If the input to these algorithms is a basis set of the appropriate wavelet packet components, then the output is a perfect reconstruction of the analyzed signal.

7.1.5 Orthonormal bases of wavelet packets

Because of the remarkable orthogonality properties which are a consequence of the orthogonal QF conditions, there are very many subsets of wavelet packets which constitute orthonormal bases for $V_0 = \sigma^0\Lambda_0$. Some of the most useful are the *orthonormal graph bases* given by Corollary 7.10. While these are not exhaustive, they form a very large library, are easy to construct and label, and are organized in such a way that they can be efficiently searched for extreme points of certain cost functions.

Suppose that we fix $L > 0$ and construct a tree of wavelet packet coefficients which has L additional levels below the root. From this tree, we may choose a subset of nodes which correspond to an orthonormal basis; the coefficient sequences in those nodes will be the coordinates in that basis.

The discrete wavelet transform provides one such basis. A three-level expansion corresponds to the decomposition $V_0 = \sigma^3\Lambda_0 \oplus \sigma^3\Lambda_1 \oplus \sigma^2\Lambda_1 \oplus \sigma^1\Lambda_1$. This is the subset of the wavelet packet tree labeled as shaded boxes in Figure 7.6. Since the frequency index of a wavelet coefficient is always one or zero, values which are fixed by GC^{-1}, it does not matter whether we compute the coefficients in Paley or sequency order.

Other choices give other orthonormal basis subsets. A single row of the rectangle corresponds to wavelet packets of equal scale, in loose analogy with windowed sines and cosines. In the Haar–Walsh case described above, the bottom level is exactly the Walsh basis, and intermediate levels are windowed Walsh bases at all dyadic window widths. Longer filters give smoother Walsh-like functions which are more like sines and cosines, though the analogy is never exact. The intermediate levels mix the resolvable frequencies together into subbands. For example, the complete level in Figure 7.7 corresponds to the decomposition $V_0 = \sigma^2\Lambda_0 \oplus \sigma^2\Lambda_1 \oplus \sigma^2\Lambda_2 \oplus \sigma^2\Lambda_3$.

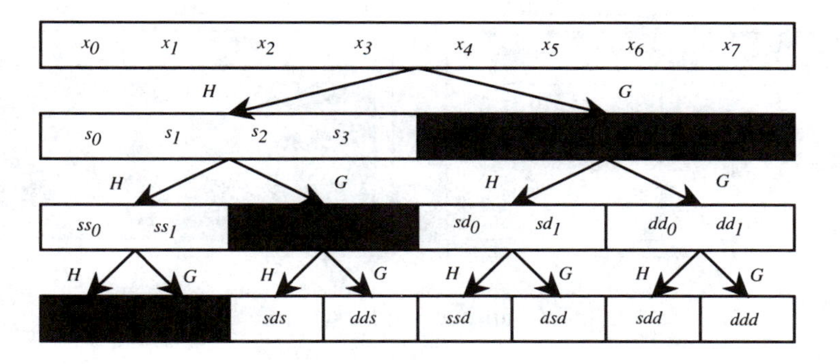

Figure 7.6: Discrete wavelet basis, rank eight.

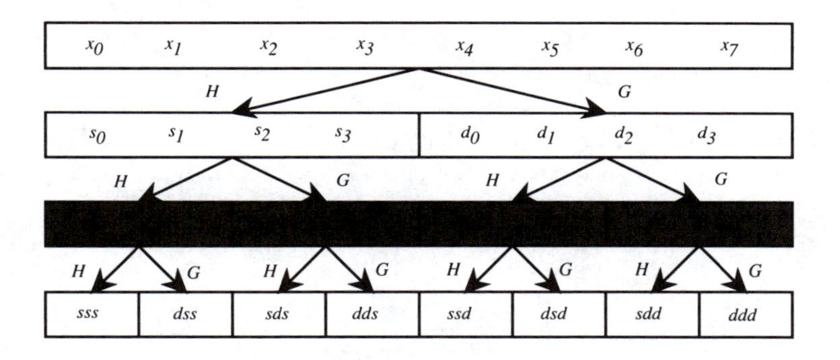

Figure 7.7: Fixed-level wavelet packet basis, Paley order, rank eight.

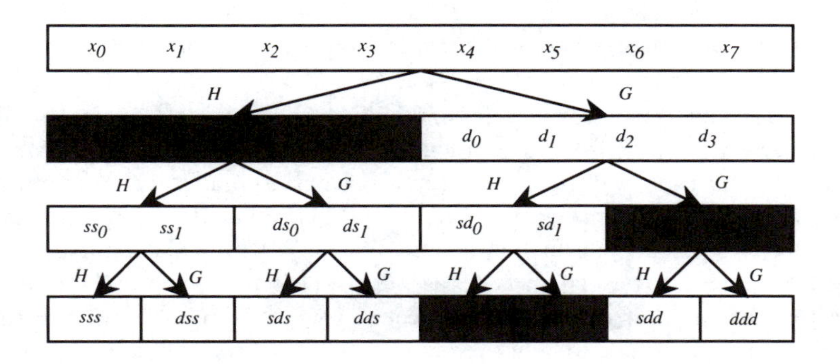

Figure 7.8: Another wavelet packet basis, Paley order, rank eight.

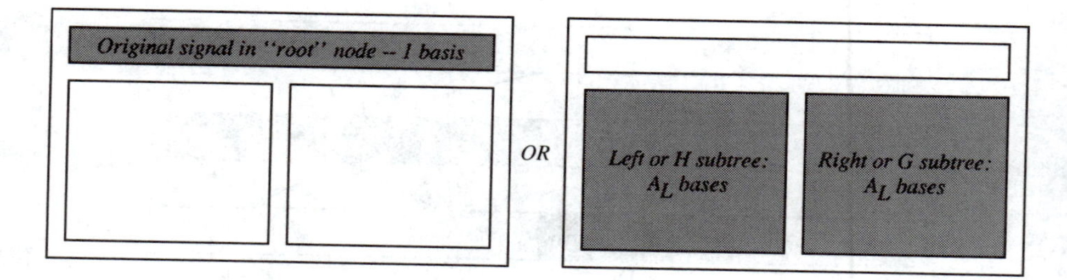

Figure 7.9: Counting graph bases recursively.

Any other subset corresponding to a disjoint dyadic decomposition will work as well, such as the basis $V_0 = \sigma^1\Lambda_0 \oplus \Lambda_4 \oplus \Lambda_5 \oplus \sigma^2\Lambda_3$ depicted in Figure 7.8.

The term "graph basis" originates in the observation that the bases described by Theorem 7.9 can be depicted as the graph of a function relating the scale index to the combined frequency and position indices.

By Corollary 7.11, every disjoint dyadic cover of $[0,1[$ corresponds to a basis for V_0. The number of such covers with subintervals no smaller than 2^{-L} may be counted by induction on L. Let A_L be the number of graph bases in the tree with $1 + L$ levels. Then $A_0 = 1$, and Figure 7.9 shows how A_{L+1} decomposes into two pieces. The left subtree is independent of the right subtree, so we have the following relation:

$$A_{L+1} = 1 + A_L^2. \tag{7.33}$$

A simple estimate then gives $A_{L+1} > 2^{2^L}$ for all $L > 1$.

If we start with a sequence of $N = 2^L$ nonzero coefficients, then we can decimate by two at least L times, which will provide more than 2^N bases. If the QFs are orthogonal, these will be orthonormal bases.

Graph bases are particularly useful and quite numerous, but they are not all the possible wavelet packet bases. In particular, if we use the Haar–Walsh filters, then there are orthonormal wavelet packet bases not corresponding to decompositions into whole blocks. This is proved in Chapter 10, Theorem 10.5. An example may be picked out by inspection in Figure 7.4. Figure 7.10 lists the complete library of seven Haar–Walsh wavelet packet bases for rank four, whereas there are only five graph bases at that rank. The two rectangles on the right in the top row do not correspond to disjoint dyadic covers of $[0,1[$. We will count the number of orthonormal wavelet packet bases in Chapter 10, after introducing the notion of information cells.

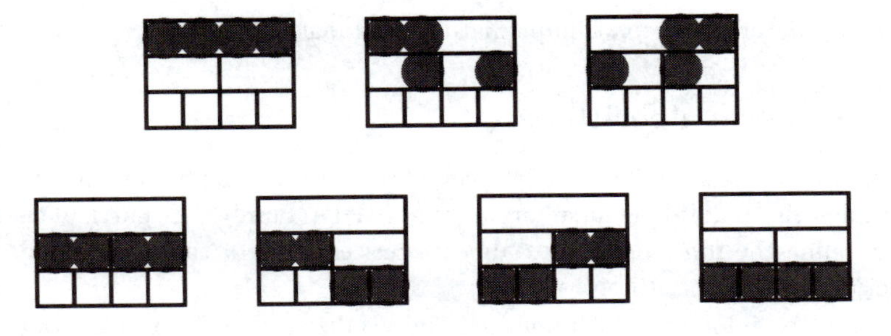

Figure 7.10: Haar–Walsh wavelet packet bases for $\mathbf{R}^4$.

7.2 Implementations

Wavelet packet analysis is perfectly recursive. Each newly computed wavelet packet coefficient sequence becomes the root of its own analysis tree, as depicted in the left half of Figure 7.11. The right half of the figure shows one step in the wavelet packet synthesis. This likewise is perfectly recursive, as we must first reconstruct both the H and G descendents of a node before reconstructing the node itself from its two descendents. The boxes in the figure are data structures which represent nodes in the tree.

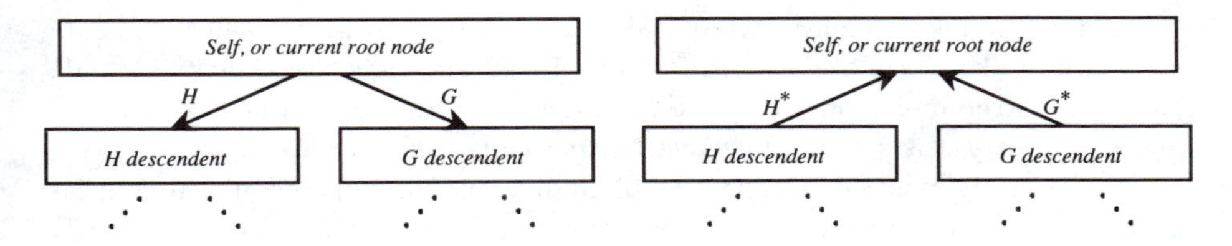

Figure 7.11: One step in a recursive DWPA and DWPS.

We can choose how much information will be stored in each node data structure. At the very least, we must store the array of wavelet packet coefficients. All other information can be kept in side tables, or can be computed from a few basic parameters like signal length and the depth of the analysis. This is what we shall do in the periodic case, where we have indexing conventions that give simple formulas. However, in the aperiodic case it is useful to keep handy the information about the length or support of each coefficient array, so we will use data structures for the coefficient spaces Λ_{sf} which contain that length information, the coefficient array itself, and pointers to the two immediate descendent subspaces.

7.2.1 Generic algorithms

We first sketch a skeleton DWTA implementation. It will be fleshed out later when we consider the individual members of the DWTA family. In particular, we will have to define the functions which allocate descendents of the proper size, as well as specify how to apply the filters.

To avoid pseudocode which contains uninitialized global variables, our generic implementation passes the filter structures HQF and GQF explicitly, as parameters. Of course they should ultimately be made into global static arrays which are initialized at some higher level.

Generic discrete wavelet packet analysis, natural order

```
dwpa0( SELF, LEVEL, HQF, GQF ):
   If LEVEL>0 then
      Filter SELF to its left descendent LEFT using H
      dwpa0( LEFT, LEVEL-1, HQF, GQF)
      Filter SELF to its right descendent RIGHT using G
      dwpa0( RIGHT, LEVEL-1, HQF, GQF)
```

The level index is used to terminate the recursion. This implementation can be used to produce a binary tree of linked data structures by side-effect. A data structure can be recognized as a leaf if both its left and right descendents are null. Each data structure is the root of its own aperiodic wavelet packet analysis tree, and the input consists of a data structure containing the sequence λ_{00}.

Alternatively, this code can be used to fill different parts of a preallocated linear array with periodic wavelet packet coefficients. Since the size of the tree is easily calculated from the signal length and the number of levels, there is no need to use null pointers to indicate leaves.

To illustrate the two implementations, we will use the binary tree structure in the aperiodic case, and the long array structure in the periodic case. Notice the close resemblance between the generic discrete wavelet packet analysis and the generic discrete wavelet transform.

We now jump forward a few steps to sketch a skeleton implementation of the discrete wavelet packet synthesis or DWPS algorithm. This code assumes that the input consists of a binary tree data structure in which all the coefficient sequence arrays are preallocated to the proper lengths needed to apply the adjoint QFs. The lengths are determined separately for each member of the DWPA family. The code also supposes that the data arrays are initially filled with zeroes except at the indices of the wavelet packets we wish to synthesize.

We also pass a level parameter to indicate how deeply we should go in our search for wavelet packets. In some cases it is not necessary to specify the number of levels of decomposition: if we are using data structures linked into a binary tree, and we want a complete superposition of all the wavelet packets, then recursion should terminate when both the left and right descendents are null pointers.

As in the generic DWPA, we pass the filters as parameters to avoid uninitialized global variables in the pseudocode. A complete implementation would pass the filter sequences more efficiently as global static variables; they would be set by a higher level calling routine or interface.

Generic discrete wavelet packet synthesis, natural order

```
dwps0( SELF, LEVEL, HQF, GQF ):
   If there is a left descendent LEFT then
      dwps0( LEFT, LEVEL-1, HQF, GQF )
      Adjoint filter LEFT to SELF using H
   If there is a right descendent RIGHT then
      dwps0( RIGHT, LEVEL-1, HQF, GQF )
      Adjoint filter RIGHT to SELF using G
```

Notice that, since wavelet packet analysis to depth L is redundant by a factor of $L+1$, a wavelet packet synthesis from the output of an L-level analysis reconstructs the original coefficient sequence multiplied by $1+L$. This is in contrast to the inverse discrete wavelet transform, which synthesizes the signal from a basis subset and thus reconstructs the signal perfectly.

When we use data structures for a wavelet packet analysis, we will suppose that they are the *binary tree node* or `BTN` data structures defined in Chapter 2. In addition we can put additional identifying and precomputed data into the tag. This might include a pointer to the parent node, the scale and frequency indices, flags describing whether we are using Gray code reordering, measurements of the coefficients such as the sum of their squares, identifiers for the filters, or other operators used to compute the coefficients, and as many other incidental quantities as might be useful for further processing.

The *wavelet packet coefficients* in the output arrays of the DWPA consist of amplitudes tagged with associated scale, frequency, and position indices. To hold this information we can use the `TFA1` data structures define in Chapter 2, which have level, block, and offset tags besides the amplitude member. For wavelet packets in the natural or Paley ordered analysis, the level tag is the nominal scale index, the block tag is the nominal frequency index in natural or Paley order, and the offset tag indicates at what index the amplitude is found in the data array of the node.

Scale and frequency are interrelated. A wavelet packet of scale s which is nominally 2^s samples in extent and which has ν oscillations per sampling interval has a nominal *centered frequency* of $\lfloor \frac{1}{2} + 2^s \nu \rfloor$. It therefore has $GC(\lfloor \frac{1}{2} + 2^{1+s}\nu \rfloor)$ as its (centered) nominal frequency index. This is a trivial calculation involving a shift and bitwise exclusive-or. Since the correspondence is so easy to compute, we will decree that henceforth all trees will be in natural order. We will do any necessary permuting *after* the coefficients are computed.

Likewise, given a nominal position p we need to determine the position index from the phase response of the QFs and the frequency index f. It is possible to shift the position indices in the data arrays of the nodes as we perform the analysis. But to follow our programming principles and keep different aspects of the calculation as independent as possible, we instead decree that henceforth all filter sequences and coefficient arrays will be conventionally indexed. We will do any necessary index shifting only after the coefficients are computed.

7.2.2 Periodic DWPA and DWPS

In the periodic case the length of a child's data array is half that of its parent node. The tree may therefore be stored as a single linear array, and we can use an indexing function to calculate the first index of the data array for any particular node. Figure 7.12 shows one possible arrangement of the these data arrays. With that arrangement, the coefficients array λ_{sf} for a signal of length N begins at an offset of $sN + f2^{-s}N$ in the linear array, which can be computed using the function `abtblock(N,S,F)`. Likewise, the length of the array λ_{sf} is $2^{-s}N$, which can be computed with the function `abtblength(N,S)`. The total length of the linear array for an L-level analysis of a signal of period N is $(L+1)N$. This can be allocated all at once prior to the analysis. Naturally we must assume that N is divisible by 2^L, that $0 \le s \le L$, and that $0 \le f < 2^s$. These conditions are tested by the range-checking function `tfa1sinabt()`.

Complete periodic DWPA

Now we modify the generic DWPA to use the data arrangement in Figure 7.12. Since the locations of the children and the lengths of their data arrays are completely determined by the three parameters N, s, f, it is not necessary to use data structures. Also, this code can just as well be implemented without recursion.

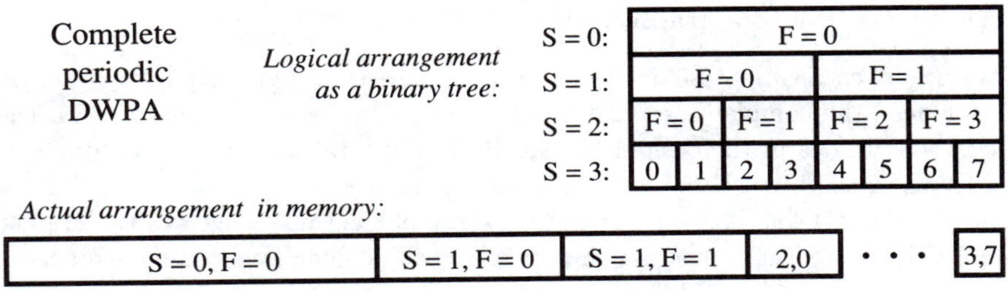

Complete periodic DWPA

Logical arrangement as a binary tree:

S = 0:	F = 0							
S = 1:	F = 0				F = 1			
S = 2:	F = 0		F = 1		F = 2		F = 3	
S = 3:	0	1	2	3	4	5	6	7

Actual arrangement in memory:

S = 0, F = 0	S = 1, F = 0	S = 1, F = 1	2,0	• • •	3,7

Figure 7.12: Arrangement of data structures in a periodic DWPA.

Complete periodic DWPA to a preallocated array binary tree

```
dwpap2abt0( DATA, N, MAXLEVEL, HQF, GQF ):
   For L = 0 to MAXLEVEL-1
      Let NPARENT = abtblength( N, L )
      For B = 0 to (1<<L)-1
         Let PARENT = DATA + abtblock( N, L, B )
         Let CHILD = DATA + abtblock( N, L+1, 2*B )
         cdpe( CHILD, 1, PARENT, NPARENT, HQF )
         Let CHILD = DATA + abtblock( N, L+1, 2*B+1 )
         cdpe( CHILD, 1, PARENT, NPARENT, GQF )
```

To complete the implementation, we must encase the basic transform in a short program which allocates an output array binary tree and copies the input array into its first row. The following function performs these tasks and returns a pointer to the output array binary tree:

Complete periodic DWPA

```
dwpap2abt( IN, N, MAXLEVEL, HQF, GQF ):
   Allocate an array of (MAXLEVEL+1)*N REALs at DATA
   For I = 0 to N-1
      Let DATA[I] = IN[I]
   dwpap2abt0( DATA, N, MAXLEVEL, HQF, GQF )
   Return DATA
```

Remark. The original signal survives this function unchanged. It should be clear how we could save space by omitting the first row of the output array binary tree.

Periodic DWPA to a graph basis

If we are given an input signal and the list of levels of a graph basis subset, we do not need the complete periodic DWPA; we only need to compute the wavelet packet amplitudes in the desired blocks. Since the total number of amplitudes in the graph equals the original number of signal samples, the transformation can be done "in place" using just one temporary array of the same length as the original signal. The computation and assignment can both be done by a recursive function:

Recursion for periodic "in place" DWPA to a hedge

```
dwpap2hedger( GRAPH, J, N, S, HQF, GQF, WORK ):
  If GRAPH.LEVELS[J]==S then
    J += 1
  Else
    Let PARENT = GRAPH.CONTENTS[J]
    cdpe(  WORK, 1, PARENT, N, HQF )
    cdpe( WORK+N/2, 1, PARENT, N, GQF )
    For I = 0 to N-1
       Let PARENT[I] = WORK[I]
    Let J = dwpap2hedger( GRAPH, J, N/2, S+1, HQF, GQF, WORK )
    Let GRAPH.CONTENTS[J] = PARENT + N/2
    Let J = dwpap2hedger( GRAPH, J, N/2, S+1, HQF, GQF, WORK )
  Return J
```

The counter J is incremented each time a pointer is added to the contents array; it keeps track of the end of the levels and contents arrays.

We initialize by allocating a contents array to hold pointers into the input and output array, placing it and the input levels array and its length into a HEDGE data structure, and allocating a scratch array as long as the input:

Periodic "in place" DWPA to a hedge

```
dwpap2hedge( IN, LENGTH, LEVELS, BLOCKS, HQF, GQF ):
  Let GRAPH = makehedge( BLOCKS, NULL, LEVELS, NULL )
  Let GRAPH.CONTENTS[0] = IN
  Allocate an array of LENGTH REALs at WORK
  dwpap2hedger( GRAPH, 0, LENGTH, 0, HQF, GQF, WORK )
  Deallocate WORK[]
  Return GRAPH
```

Remark. Using `dwpap2hedge()` with a hedge having an $L+1$ element levels list

$\{L, L, L-1, L-2, \ldots, 2, 1\}$ produces the periodic discrete wavelet transform. The in place transform is equivalent to the implementation in Chapter 6.

Periodic DWPS from arbitrary atoms

The DWPS can be performed using the array binary tree in Figure 7.12. We first assume that such an array has been preallocated with a zero at every location not containing a wavelet packet amplitude. The length and the maximum decomposition level of the signal to be synthesized must be known to determine the size of the tree. Then we adjoint filter from the bottom up, superposing the wavelet packets corresponding to the amplitudes sprinkled throughout the array binary tree, leaving the reconstructed signal in the first row:

Periodic DWPS from a prepared array binary tree

```
abt2dwpsp( DATA, N, MAXLEVEL, HQF, GQF ):
   Let L = MAXLEVEL
   While L>0
      Let NCHILD = abtblength( N, L )
      L -= 1
      For B = 0 to (1<<L)-1
         Let DPARENT = DATA + abtblock( N, L, B )
         Let DCHILD = DATA + abtblock( N, L+1, 2*B )
         acdpi( DPARENT, 1, DCHILD, NCHILD, HQF )
         Let DCHILD = DATA + abtblock( N, L+1, 2*B+1 )
         acdpi( DPARENT, 1, DCHILD, NCHILD, GQF )
```

We prepare the array binary tree by superposing amplitudes from a list of atoms, using **tfa1s2abt()**. Notice that amplitudes may be added to the same location more than once, and may also be added into the output locations in the first row:

Periodic DWPS from a list of atoms

```
tfa1s2dwpsp( ATOMS, NUM, N, HQF, GQF ):
   Let MAXLEVEL = ATOMS[0].LEVEL
   For I = 1 to NUM-1
      Let MAXLEVEL = max( MAXLEVEL, ATOMS[I].LEVEL )
   Allocate an array of N*(MAXLEVEL+1) 0s at DATA
   tfa1s2abt( DATA, N, ATOMS, NUM )
   abt2dwpsp( DATA, N, MAXLEVEL, HQF, GQF )
   Return DATA
```

We can enhance this code to perform range checking with **tfa1sinabt()**, to make sure the atoms can be accommodated in the tree. Also, we can save time by checking for totally zero children and not filtering them into the reconstruction. However, if the coefficients are uniformly sprinkled throughout the tree, then the complexity will remain $O(N \log N)$.

Using tfa1s2dwpsp() with a one element atoms list reconstructs a single periodic wavelet packet. This is one way to produce plots to visualize the functions underlying the wavelet packet transform, or to produce sample sequences which can be played as sounds through a digital to analog converter.

Periodic DWPS from a hedge

We can also superpose amplitudes from a hedge into an array binary tree using the function **hedge2abt()**. But it is better to take advantage of the equal lengths of a signal and its graph basis amplitudes array and reconstruct the signal "in place:"

Recursion for in place periodic DWPS from hedge

```
hedge2dwpspr( GRAPH, J, N, S, HQF, GQF, WORK ):
  If S < GRAPH.LEVELS[J] then
    Let J = hedge2dwpspr( GRAPH, J, N/2, S+1, HQF, GQF, WORK )
    Let LEFT = GRAPH.CONTENTS[J]
    Let J = hedge2dwpspr( GRAPH, J+1, N/2, S+1, HQF, GQF, WORK )
    Let RIGHT = GRAPH.CONTENTS[J]
    acdpe( WORK, 1, LEFT, N/2, HQF )
    acdpo( WORK, 1, RIGHT, N/2, GQF )
    For I = 0 to N-1
       Let LEFT[I] = WORK[I]
  Return J
```

The blocks counter J is incremented each time a hedge block is reconstructed from its descendents. We perform some initializations with a short calling program that returns a pointer to the start of the reconstructed signal:

Periodic in place DWPS from a hedge

```
hedge2dwpsp( GRAPH, LENGTH, HQF, GQF ):
  Allocate an array of LENGTH REALs at WORK
  hedge2dwpspr( GRAPH, 0, LENGTH, 0, HQF, GQF, WORK )
  Deallocate WORK[]
  Return GRAPH.CONTENTS[0]
```

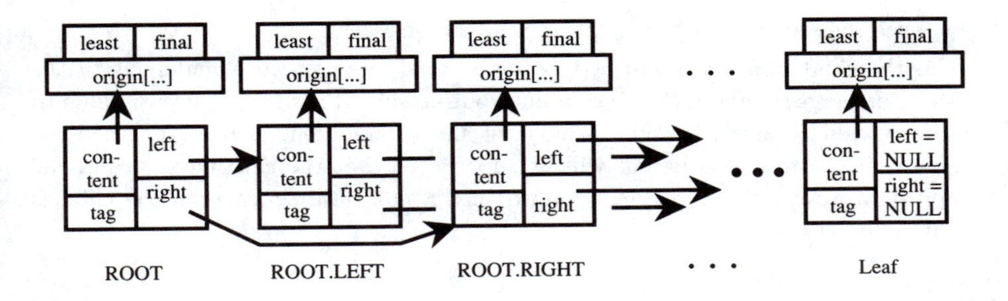

Figure 7.13: Arrangement of data structures in a DWPA.

Remark. Using `hedge2dwpsp()` with a hedge having an $L+1$ element levels list $\{L, L, L-1, L-2, \ldots, 2, 1\}$ produces the inverse periodic discrete wavelet transform. The result is equivalent to `idwtpd()`, defined in Chapter 6.

7.2.3 Aperiodic DWPA and DWPS

The aperiodic DWPA will allocate `BTN` data structures as needed and link them together into a binary tree. It will also allocate separate `INTERVAL` data structures for the nodes and fill them with wavelet packet coefficients. The result will be the arrangement of data structures depicted in Figure 7.13.

The data arrays in a node can be of any length and the least and final indices can be anything. We use Equation 5.91 to allocate and assign values to descendents. We use `cdai()` to compute the filter convolution and decimation. The following function performs the recursive step of convolving and decimating a parent interval into its immediate descendent or *child*:

Allocate and compute a child interval by aperiodic DWPA

```
cdachild( CHILD, PARENT, F ):
   If PARENT != NULL then
      If PARENT.ORIGIN != NULL
         Let CLEAST = cdaleast( PARENT, F )
         Let CFINAL = cdafinal( PARENT, F )
         Let CHILD = enlargeinterval( CHILD, CLEAST, CFINAL )
         cdai( CHILD.ORIGIN, 1, PARENT.ORIGIN,
                                      PARENT.LEAST, PARENT.FINAL, F )
   Return CHILD
```

We reuse the two macros `cdaleast()` and `cdafinal()` defined in Chapter 6 to compute the size needed to hold the parent's descendent.

If the child interval is null or too small, it will be created or enlarged by the call to `enlargeinterval()`. The calling routine should assign the return value to its own child pointer to take advantage of any reallocation. If it exists, then the values derived from the parent will be superposed over the existing values in the child data array. If the parent is null or has a null data array, then the child is returned unchanged.

Complete aperiodic wavelet packet analysis

Performing the recursive step to compute the contents of BTNs produces an entire subtree. Using the low-pass filter H to move left, and the high-pass filter G to move right results in the natural or Paley order:

Recursion to complete aperiodic DWPA subtree in natural order

```
dwpaa2btntr( NODE, LEVEL, HQF, GQF ):
   If LEVEL>0 then
      Let CHILD = cdachild( NULL, NODE.CONTENT, HQF )
      Let NODE.LEFT = makebtn( CHILD, NULL, NULL, NULL )
      dwpaa2btntr( NODE.LEFT, LEVEL-1, HQF, GQF )
      Let CHILD = cdachild( NULL, NODE.CONTENT, GQF )
      Let NODE.RIGHT = makebtn( CHILD, NULL, NULL, NULL )
      dwpaa2btntr( NODE.RIGHT, LEVEL-1, HQF, GQF )
   Return
```

To complete the implementation, we need to write a short interface function which copies an input array into a properly assigned BTN data structure, sets the maximum decomposition level, and selects the low-pass and high-pass QFs:

Complete aperiodic DWPA, natural order

```
dwpaa2btnt( IN, LEAST, FINAL, MAXLEVEL, HQF, GQF ):
   Let ROOT = makebtn( NULL, NULL, NULL, NULL )
   Let ROOT.CONTENT = makeinterval( IN, LEAST, FINAL )
   dwpaa2btntr( ROOT, MAXLEVEL, HQF, GQF )
   Return ROOT
```

The output is a binary tree of BTN data structures with variable-sized intervals as their contents. The leaves can be recognized from having null pointers instead of left or right descendents.

Aperiodic DWPA to a hedge

We can expand a signal in an aperiodic graph basis by first finding its complete DWPA and then extracting a hedge from the BTN tree with `btnt2hedge()`. Rather than fill the entire tree, we can save some time by developing the tree only to those nodes specified in the hedge. This is done with a recursive function that descends the tree only until it hits a target level:

Descent in an aperiodic DWPA to a hedge level

```
dwpaa2hedger( GRAPH, NODE, J, S, HQF, GQF ):
  If GRAPH.LEVELS[J]==S then
    Let GRAPH.CONTENTS[J] = NODE.CONTENT
    Let NODE.CONTENT = NULL
    J += 1
  Else
    Let CHILD = cdachild( NULL, NODE.CONTENT, HQF )
    Let NODE.LEFT = makebtn( CHILD, NULL, NULL, NULL )
    Let J = dwpaa2hedger( GRAPH, NODE.LEFT, J, S+1, HQF, GQF )
    Let CHILD = cdachild( NULL, NODE.CONTENT, GQF )
    Let NODE.RIGHT = makebtn( CHILD, NULL, NULL, NULL )
    Let J = dwpaa2hedger( GRAPH, NODE.RIGHT, J, S+1, HQF, GQF )
  Return J
```

We replace the content of each extracted node with a null pointer, so that we can later deallocate the tree without affecting the selected intervals. It is necessary to package the function with a few lines of interface:

Aperiodic DWPA to a hedge

```
dwpaa2hedge( LEVELS, BLOCKS, DATA, LEAST, FINAL, HQF, GQF ):
   Let ROOT = makebtn( NULL, NULL, NULL, NULL )
   Let ROOT.CONTENT = makeinterval( DATA, LEAST, FINAL )
   Let GRAPH = makehedge( BLOCKS, NULL, LEVELS, NULL )
   dwpaa2hedger( GRAPH, ROOT, 0, HQF, GQF )
   Let ROOT.CONTENT = NULL
   freebtnt( ROOT, freeinterval, free )
   Return GRAPH
```

Remark. Using `dwpaa2hedge()` with the levels list $\{L, L, L-1, L-2, \ldots, 2, 1\}$ produces the aperiodic discrete wavelet transform. This method should be compared with the aperiodic DWT implementation in Chapter 6, where we put all the amplitudes into a contiguous array and used INTERVALs just to keep track of the offsets. Here we allocate lots of small arrays in the form of INTERVALS and associate to each one its own index bounds.

Extracting wavelet packet amplitudes from BTN trees

When getting coefficients from the tree, the user should attempt to access only those nodes which are actually in the tree. Alternatively, the programmer can include some elementary range checks which are evaluated before a node is accessed. For example, in a DWPA down to level L, a valid node index pair s, f must satisfy the constraints $0 \leq s \leq L$ and $0 \leq f < 2^s$.

Some of this range checking can be avoided by taking advantage of the behavior of the utility functions defined in Chapter 2, which we use to extract BTNs. If `LEVEL` and `BLOCK` are valid for the complete wavelet packet analysis tree at `ROOT` produced by `dwpaa2btnr()`, then `btnt2btn(ROOT,LEVEL,BLOCK)` will return a pointer to the desired BTN structure. Otherwise, it will return a null pointer. We can handle most of the exceptions by treating null intervals as being completely zero.

If we get a non-null interval, then we can test whether the requested `OFFSET` is between the least and final indices by using the function `ininterval()`. If not, we can again suppose that the amplitude is zero. This is how the function `btnt2tfa1()` handles exceptional cases.

Altogether, our utilities behave as follows:

- If `ATOM` is a TFA1 data structure with valid level, block and offset indices for the tree at `ROOT`, then calling `btnt2tfa1(ATOM,ROOT)` gets the amplitude of the associated aperiodic wavelet packet;

- Otherwise, it puts zero into the amplitude member of `ATOM`.

This behavior may not be suitable for applications in which we must determine the wavelet packet amplitude which has not been previously computed or which is somehow missing from a partial DWPA tree. In those cases, it is necessary to compute additional amplitudes before extraction. Note that computing individual amplitudes is quite costly; wavelet packet analysis is efficient only because the large amount of arithmetic is spread over a large number of coefficients.

Aperiodic wavelet packet synthesis

To compute the wavelet packet coefficients of a parent interval from the expansions in one of its children, we must first make sure that the child exists and that the parent has enough room to store the output of the aperiodic adjoint convolution-decimation. If the parent interval is null, or if it is too small to accommodate the output of adjoint convolution and decimation, then it is reallocated and its new location is returned when the function exits. We reuse the two macros `acdaleast()` and `acdafinal()`, defined in Chapter 6, and the functions `enlargeinterval()` defined in Chapter 2, to compute the required size of the parent:

Aperiodic DWPS of parent interval from a descendent

```
acdaparent( PARENT, CHILD, F ):
   If CHILD != NULL then
      If CHILD.ORIGIN != NULL then
         Let LEAST = acdaleast( CHILD, F )
         Let LEAST = min( PARENT.LEAST, LEAST )
         Let FINAL = acdafinal( CHILD, F )
         Let FINAL = max( PARENT.FINAL, FINAL )
         Let PARENT = enlargeinterval( PARENT, LEAST, FINAL )
         acdai( PARENT.ORIGIN, 1, CHILD.ORIGIN,
                                       CHILD.LEAST, CHILD.FINAL, F )
   Return PARENT
```

For the aperiodic DWPS, we start with a BTN tree containing some nonempty nodes representing nonzero aperiodic wavelet packet amplitudes. To build the synthesis tree, we start by allocating a root BTN data structure to hold the reconstructed signal. This can be done using `makebtn()`, defined in Chapter 2.

Next we link together whole branches from the root leading to nonempty nodes, using the utility function `btn2branch()`. This function returns a pointer to a target node and allocates any intermediate nodes along the branch, including the target node itself if that is necessary. It assigns null pointers to any unused children in the allocated nodes.

Finally, we write the amplitudes into the target nodes' INTERVAL structures. This may require first enlarging some arrays to accommodate the new offsets.

We would use `tfa1s2btnt()` to perform these three tasks for an array of atoms, which would produce a binary tree of BTN data structures with the following properties:

- Every leaf node has null pointers in its left and right members,
- Empty BTNs, *i.e.*, those with no nonzero amplitudes, have null pointers instead of content,
- Every nonempty BTN has a valid INTERVAL as its content.

To synthesize, we recursively reconstruct the content intervals in the nodes between the leaves and the root. Intervals in the intermediate nodes have irregular lengths in general and might need to be enlarged to accommodate partial reconstructions of the signal. Ultimately, we reconstruct the coefficient sequence at the root of the tree:

Aperiodic DWPS from a BTN tree, natural order

```
btnt2dwpsa( ROOT, HQF, GQF ):
   If ROOT != NULL then
      btnt2dwpsa( ROOT.LEFT, HQF, GQF )
      If ROOT.LEFT != NULL then
         Let ROOT.CONTENT = acdaparent( ROOT.CONTENT,
                                        ROOT.LEFT.CONTENT, HQF )
      btnt2dwpsa( ROOT.RIGHT, HQF, GQF )
      If ROOT.RIGHT != NULL then
         Let ROOT.CONTENT = acdaparent( ROOT.CONTENT,
                                        ROOT.RIGHT.CONTENT, GQF )
   Return
```

The reconstructed signal occupies only the INTERVAL content of the root BTN. We can call `freebtnt()` to deallocate the unused remainder of the tree.

Remark. The expected number of memory locations used in a wavelet packet synthesis is $O(N \log N)$ even if we start with at most a basis subset of $O(N)$ nonzero amplitudes. This is because we must compute the intermediate coefficients between the root sequence and the leaf sequences. If wavelet packet coefficients are approximately equally distributed throughout the tree, then about half the leaves will be at levels deeper than half the maximum level and we will have to allocate at least half of the entire tree. But we can save some of this memory by deallocating nodes as soon as we are done with them; this modification of `btnt2dwpsa()` is left as an easy exercise.

Aperiodic DWPS from a list of atoms

Starting from an arbitrary array of TFA1s, we can produce a partial BTN tree with the utility function `tfa1s2btnt()` defined in Chapter 2. Then we reconstruct the root node, whose interval content is the reconstructed signal, extract that interval, and deallocate the remainder of the tree:

Aperiodic DWPS from atoms

```
tfals2dwpsa( ATOMS, NUM, HQF, GQF ):
   Let ROOT = makebtn( NULL,NULL, NULL, NULL )
   tfals2btnt( ROOT, ATOMS, NUM )
   btnt2dwpsa( ROOT, HQF, GQF )
   Let OUT = ROOT.CONTENT
   Let ROOT.CONTENT = NULL
   freebtnt( ROOT, freeinterval, free )
   Return OUT
```

Aperiodic DWPS from a hedge

Starting from a hedge or graph basis set, we can produce a partial BTN tree with the utility function `hedge2btnt()` defined in Chapter 2. We apply `btnt2dwpsa()` to reconstruct the root node, extract the signal from the root interval, then tidy up:

Aperiodic DWPS from a hedge

```
hedge2dwpsa( GRAPH, HQF, GQF ):
   Let ROOT = makebtn( NULL,NULL, NULL, NULL )
   hedge2btnt( ROOT, GRAPH )
   btnt2dwpsa( ROOT, HQF, GQF )
   Let OUT = ROOT.CONTENT
   Let ROOT.CONTENT = NULL
   freebtnt( ROOT, freeinterval, free )
   Return OUT
```

Remark. Using `hedge2dwpsa()` from a hedge with levels list $\{L, L, L-1, L-2, \ldots, 2, 1\}$ produces the aperiodic inverse discrete wavelet transform. Unlike the implementation of aperiodic DWT in Chapter 6, where we kept all partial results in a pair of contiguous arrays, here we use lots of small intermediate arrays in the BTN branches and their INTERVALS.

7.2.4 Biorthogonal DWPA and DWPS

No change is necessary to make the DWPA and DWPS algorithms work in the biorthogonal case. If we use filters H and G to analyze, we simply use the dual filters H' and G' to synthesize.

The underlying functions are not orthogonal in L^2, nor are their sample sequences orthogonal in ℓ^2, unless the QFs are self-dual, *i.e.*, orthogonal. However, since the functions have compact support, sufficiently separated biorthogonal wavelet packets really are orthogonal, and even pairs of nearby functions have small inner product.

7.3 Exercises

1. Prove the second part of Lemma 7.13, namely that $\lambda_{s+1,2f+1}(p) = G\lambda_{sf}(p)$.
2. Complete the induction and prove Theorem 7.14.
3. Write a program that uses the function `hedge2abt()` to superpose amplitudes from a hedge into an array binary tree and then performs the periodic discrete wavelet packet synthesis.
4. Modify the function `btnt2dwpsa()` to deallocate BTN structures as soon as their parents are reconstructed.
5. Explain how the functions `hedge2dwpsp()` and `dwpap2hedge()` know when to stop looking for new blocks.

Chapter 8

The Best Basis Algorithm

When there is a choice of bases for the representation of a signal, then it is possible to seek the best one by some criterion. If the choice algorithm is sufficiently cheap, then it is possible to assign each signal its very own adapted basis, or *basis of adapted waveforms*. The chosen basis carries substantial information about the signal; the chosen waveforms are a good match for the signal. If the basis description is efficient, then that information has been compressed.

Let $\mathcal{B}$ be a collection of (countable) bases for a (separable) Hilbert space X. Some desirable properties for $\mathcal{B}$ to have are:

- Speedy computation of inner products with the basis functions in $\mathcal{B}$, to keep the expansion complexity low;
- Speedy superposition of the basis functions, to keep the reconstruction complexity low;
- Good spatial localization, so we can identify the portion of a signal which contributes a large component;
- Regularity, or good frequency localization, so we can identify oscillations in the signal;
- Independence, so that not too many basis elements match the same portion of the signal.

The first property makes sense for finite-rank approximations; it holds for factored, recursive transformations like "fast" DFT or the wavelet transform. The second property holds for fast orthogonal transformations, whose inverses have identical complexity. The spatial localization property requires compactly supported or at least rapidly decreasing functions; in the sampled finite-rank case, good spatial localization means that each basis element is supported on just a few clustered samples. Good frequency localization means good localization for the Fourier transform of the signal, or of the DFT in the finite-rank case. For example, test functions are simultaneously well-localized in frequency and space.

Wavelets have attracted intense study because a compactly supported smooth wavelet basis has all five desirable properties. Wavelet packet bases and localized trigonometric bases possess them too, and they constitute huge collections of basis from which we may pick and choose.

Given a vector $x \in X$, we can expand it in each of the bases in $\mathcal{B}$, looking for a representation with two desirable properties: only a relatively tiny number of coefficients in the expansion should be nonnegligible, and adding up the magnitudes of the individually negligible coefficients should give a negligible collective. A rapidly decreasing coefficient sequence has these properties, as does a sequence with a rapidly decreasing rearrangement. Thus, we will seek a basis in which the coefficients, when rearranged into decreasing order, decrease as rapidly as possible. We will use tools from classical harmonic analysis to measure rates of decrease. These tools include *entropy* as well as other *information cost functions*.

If X is finite dimensional, its complete set of orthonormal bases is compact, hence there is a global minimum for every continuous information cost M. Unfortunately, this minimum will not be a rapidly computable basis in general, nor will the search for a minimum be of low complexity. Therefore, we will restrict our attention to discrete sets of bases sprinkled around the orthogonal group which have the aforementioned desirable properties. Namely, in the rank-N case we want each basis in $\mathcal{B}$ to have an associated transform of complexity $O(N \log N)$ or better, for which the inverse transform also has complexity at most $O(N \log N)$. We also insist that $\mathcal{B}$ be organized so that the search for a global minimum of M converges in $O(N)$ operations.

The next section is devoted to means of measuring these desirable properties; the rest of the chapter describes an algorithm for attaining them.

8.1 Definitions

8.1.1 Information cost and the best basis

Before we can define an optimum representation we need to have a notion of *information cost*, or the expense of storing the chosen representation. So, define an *information cost functional* on sequences of real (or complex) numbers to be any real-valued functional M satisfying the additivity condition below:

$$M(u) = \sum_{k \in \mathbf{Z}} \mu(|u(k)|); \qquad \mu(0) = 0. \tag{8.1}$$

Here μ is a real-valued function defined on $[0, \infty)$. We suppose that $\sum_k \mu(|u(k)|)$ converges absolutely; then M will be invariant under rearrangements of the sequence u. Also, M is not changed if we replace $u(k)$ by $-u(k)$ for some k, or, in the case of complex-valued sequences u, if we multiply the elements of the sequence by complex constants of modulus 1. We take M to be real-valued so that we can compare two sequences u and v by comparing $M(u)$ and $M(v)$.

For each $x \in X$ we can take $u(k) = B^*x(k) = \langle b_k, x \rangle$, where $b_k \in B$ is the kth vector in the basis $B \in \mathcal{B}$. In the finite-rank case, we can think of b_k as the kth column of the matrix B, which is taken with respect to a standard basis of X. The information cost of representing x in the basis B is then $M(B^*x)$. This defines a functional $\mathcal{M}_x$ on the set of bases $\mathcal{B}$ for X:

$$\mathcal{M}_x : \mathcal{B} \to \mathbf{R}; \qquad B \mapsto M(B^*x). \tag{8.2}$$

This will be called the *M-information cost of x in the basis B.*

We define the *best basis* for $x \in X$, relative to a collection $\mathcal{B}$ of bases for X and an information cost functional M, to be that $B \in \mathcal{B}$ for which $M(B^*x)$ is minimal. If we take $\mathcal{B}$ to be the complete set of orthonormal bases for X, then $\mathcal{M}_x$ defines a functional on the group $\mathbf{O}(X)$ of orthogonal (or unitary) linear transformations of X. We can use the group structure to construct *information cost metrics* and interpret our algorithms geometrically.

We can define all sorts of real-valued functionals M, but the most useful are those that measure concentration. By this we mean that M should be large when elements of the sequence are roughly the same size and small when all but a few elements are negligible. This property should hold on the unit sphere in ℓ^2 if we are comparing orthonormal bases, or on a spherical shell in ℓ^2 if we are comparing Riesz bases or frames.

Some examples of information cost functionals are:

- *Number above a threshold*

  We can set an arbitrary threshold ϵ and count the elements in the sequence x whose absolute value exceeds ϵ. *I.e.*, set

$$\mu(w) = \begin{cases} |w|, & \text{if } |w| \geq \epsilon, \\ 0, & \text{if } |w| < \epsilon. \end{cases}$$

This information cost functional counts the number of sequence elements needed to transmit the signal to a receiver with precision threshold ϵ.

- *Concentration in ℓ^p*

  Choose an arbitrary $0 < p < 2$ and set $\mu(w) = |w|^p$ so that $M(u) = \|\{u\}\|_p^p$. Note that if we have two sequences of equal energy $\|u\| = \|v\|$ but $M(u) < M(v)$, then u has more of its energy concentrated into fewer elements.

- *Entropy*

  Define the *entropy* of a vector $u = \{u(k)\}$ by

$$\mathcal{H}(u) = \sum_k p(k) \log \frac{1}{p(k)}, \tag{8.3}$$

  where $p(k) = |u(k)|^2/\|u\|^2$ is the normalized energy of the kth element of the sequence, and we set $p \log \frac{1}{p} = 0$ if $p = 0$. This is the entropy of the probability distribution function (or *pdf*) given by p. It is not an information cost functional, but the functional $l(u) = \sum_k |u(k)|^2 \log(1/|u(k)|^2)$ is. By the relation

$$\mathcal{H}(u) = \|u\|^{-2} l(u) + \log \|u\|^2, \tag{8.4}$$

  minimizing l over a set of equal length vectors u minimizes $\mathcal{H}$ on that set.

- *Logarithm of energy*

  Let $M(u) = \sum_{k=1}^{N} \log |u(k)|^2$. This may be interpreted as the entropy of a Gauss-Markov process $k \mapsto u(k)$ which produces N-vectors whose coordinates have variances $\sigma_1^2 = |u(1)|^2, \ldots, \sigma_N^2 = |u(N)|^2$. We must assume that there are no unchanging components in the process, *i.e.*, that $\sigma_k^2 \neq 0$ for all $k = 1, \ldots, N$. Minimizing $M(u)$ over $B \in \mathbf{O}(X)$ finds the *Karhunen–Loève basis* for the process; minimizing over a "fast" library $\mathcal{B}$ finds the best "fast" approximation to the Karhunen–Loève basis.

8.1.2 Entropy, information, and theoretical dimension

Suppose that $\{x(n)\}_{n=1}^{\infty}$ belongs to both L^2 and $L^2 \log L$. If $x(n) = 0$ for all sufficiently large n, then in fact the signal is finite-dimensional. Generalizing this notion, we can compare sequences by their rate of decay, *i.e.*, the rate at which their elements become negligible if they are rearranged in decreasing order.

We define the theoretical dimension of a sequence $\{x(n) : n \in \mathbf{Z}\}$ to be

$$d = \exp\left(\sum_n p(n) \log \frac{1}{p(n)}\right) \tag{8.5}$$

where $p(n) = |x(n)|^2/\|x\|^2$. Note that $d = \exp \mathcal{H}(x)$ where $\mathcal{H}(x)$, defined in Equation 8.3 above, is the entropy of the sequence x.

Elementary theorems on the entropy functional

Suppose that we have a sequence $\{p(n) : n = 1, 2, \ldots\}$ with $p(n) \geq 0$ for all n and $\sum_n p(n) = 1$. This sequence may be interpreted as a probability distribution function (pdf) for the sample space $\mathbf{N}$. Define

$$H(p) = \sum_{n=1}^{\infty} p(n) \log \frac{1}{p(n)} \tag{8.6}$$

to be the *entropy* of this pdf. As usual, we take $0 \log(1/0) \stackrel{\text{def}}{=} 0$ to make the function $t \to t \log \frac{1}{t}$ continuous at zero.

We now prove some elementary properties of entropy and theoretical dimension. The first is that entropy is a convex functional on pdfs:

Proposition 8.1 *If p and q are pdfs and $0 \leq \theta \leq 1$, then $\theta p + (1-\theta)q$ is a pdf, and $H(\theta p + (1-\theta)q) \geq \theta H(p) + (1-\theta)H(q)$.*

Proof: We first check that $\theta p + (1-\theta)q$ is a pdf. But $0 \leq \theta p(n) + (1-\theta)q(n) \leq 1$ for every $n = 1, 2, \ldots$, and

$$\sum_{n=1}^{\infty} \Big(\theta p(n) + (1-\theta)q(n) \Big) = \theta \sum_{n=1}^{\infty} p(n) + (1-\theta) \sum_{n=1}^{\infty} q(n) = \theta + (1-\theta) = 1.$$

Next, observe that $H(p) = \sum_n f(p(n))$, where $f(t) = t \log \frac{1}{t}$. Note that $f(t)$ is concave downward on the interval $[0,1]$, since $f''(t) = -\frac{1}{t}$. Thus $f(\theta t + (1-\theta)s) \geq \theta f(t) + (1-\theta) f(s)$. We let $t = p(n)$ and $s = q(n)$ and sum both sides of the inequality over $n = 1, 2, \ldots$ to get the result. □

Corollary 8.2 *If p_i is a pdf for $i = 1, 2, \ldots, M$ and $\sum_{i=1}^{M} \alpha_i = 1$ with $\alpha_i \in [0,1]$ for each i, then $q = \sum_{i=1}^{M} \alpha_i p_i$ is a pdf, and $H(q) \geq \sum_{i=1}^{M} \alpha_i H(p_i)$.* □

The entropy of a tensor product of two pdfs is the sum of their entropies:

Proposition 8.3 *If p and q are pdfs over $\mathbf{N}$, then $p \otimes q \stackrel{\text{def}}{=} \{p(i)q(j) : i, j \in \mathbf{N}\}$ is a pdf over $\mathbf{N}^2$ and $H(p \otimes q) = H(p) + H(q)$.*

Proof: We first check that $p \otimes q$ is a pdf. But clearly $0 \leq p(i)q(j) \leq 1$, and

$$\sum_{i \in \mathbf{N}} \sum_{j \in \mathbf{N}} p(i)q(j) = \left(\sum_{i \in \mathbf{N}} p(i) \right) \left(\sum_{j \in \mathbf{N}} q(j) \right) = 1.$$

Next we directly calculate $H(p \otimes q)$:

$$\begin{aligned}
H(p \otimes q) &= \sum_{i \in \mathbf{N}} \sum_{j \in \mathbf{N}} p(i)q(j) \log \frac{1}{p(i)q(j)} \\
&= \sum_{i \in \mathbf{N}} \sum_{j \in \mathbf{N}} p(i)q(j) \left[\log \frac{1}{p(i)} + \log \frac{1}{q(j)} \right] \\
&= \sum_{i \in \mathbf{N}} p(i) \left(\sum_{j \in \mathbf{Z}} q(j) \right) \log \frac{1}{p(i)} + \sum_{j \in \mathbf{Z}} \left(\sum_{i \in \mathbf{Z}} p(i) \right) q(j) \log \frac{1}{q(j)} \\
&= \sum_{i \in \mathbf{Z}} p(i) \log \frac{1}{p(i)} + \sum_{j \in \mathbf{Z}} q(j) \log \frac{1}{q(j)} \quad = \quad H(p) + H(q).
\end{aligned}$$

□

Lemma 8.4 *If p and q are pdfs, then $\sum_k p(k) \log \frac{1}{p(k)} \leq \sum_k p(k) \log \frac{1}{q(k)}$, with equality if and only if $p(k) = q(k)$ for all k.*

Proof: Since $\log t$ is a convex function, its graph lies below the tangent line at $t = 1$. We thus have $\log t \leq t - 1$ with equality if and only if $t = 1$. But then $\log(q(k)/p(k)) \leq (q(k)/p(k)) - 1$ with equality if and only if $p(k) = q(k)$, so that by multiplying by $p(k)$ and summing over k we get

$$\sum_k p(k) \log \frac{q(k)}{p(k)} \leq \sum_k (q(k) - p(k)) = 0, \Rightarrow \sum_k p(k) \log \frac{1}{p(k)} \leq \sum_k p(k) \log \frac{1}{q(k)},$$

with equality if and only if $p(k) = q(k)$ for all k. □

Proposition 8.5 *Suppose that p is a pdf over an N-point sample space. Then $0 \leq H(p) \leq \log N$, where $H(p) = 0$ if and only if $p(i) = 1$ for a single i with $p(n) = 0$ for all $n \neq i$, and $H(p) = \log N$ if and only if $p(1) = p(2) = \cdots = p(N) = \frac{1}{N}$.*

Proof: It is clear that $H(p) \geq 0$ with equality if and only if $p(i) = 0$ for all but one value of i, since $t \log \frac{1}{t} \geq 0$ for all $t \in [0, 1]$ with equality iff $t = 0$ or $t = 1$.

By Lemma 8.4, we have $H(p) = \sum_k p(i) \log \frac{1}{p(i)} \leq \sum_k p(i) \log N = \log N$, with equality iff $p(i) = 1/N$ for all $i = 1, 2, \dots, N$. □

Define a *stochastic matrix* to be a matrix $A = (a(i,j) : i, j = 1, 2, \dots, N)$ satisfying

- $a(i,j) \geq 0$ for all $i, j = 1, 2, \dots, N$;
- $\sum_{i=1}^N a(i,j) = 1$ for all $j = 1, 2, \dots, N$.

If p is a pdf on N points, and we write p as a column vector

$$p = \begin{pmatrix} p(1) \\ p(2) \\ \vdots \\ p(N) \end{pmatrix},$$

then Ap is also a pdf.

We will say that A is *doubly stochastic* if in addition to the above two conditions it also satisfies:

- $\sum_{j=1}^N a(i,j) = 1$ for all $i = 1, 2, \dots, N$.

Applying a doubly stochastic matrix to a pdf results in another pdf whose probabilities are averages of the first. The new pdf has higher entropy:

Proposition 8.6 *If $A = (a(i,j) : i, j = 1, 2, \dots, N)$ is a doubly stochastic matrix and p is a pdf, then Ap is a pdf, and $H(Ap) \geq H(p)$.*

Proof: This is a direct calculation. Putting $f(t) = t \log \frac{1}{t}$, we have:

$$\begin{aligned} H(Ap) &= \sum_{i=1}^N f([Ap](i)) = \sum_{i=1}^N f\left(\sum_{j=1}^N a(i,j)p(j)\right) \\ &\geq \sum_{i=1}^N \sum_{j=1}^N a(i,j) f(p(j)) \qquad \text{since } f \text{ is concave downwards,} \\ &= \sum_{j=1}^N \left(\sum_{i=1}^N a(i,j)\right) f(p(j)) = \sum_{j=1}^N f(p(j)) = H(p). \end{aligned}$$

□

Concentration and decreasing rearrangements

More rapidly decreasing sequences have lower entropy, in a sense made precise below. Given a sequence $\{p(i) : i = 1, 2, \ldots\}$, we define its *partial sum sequence* $S[p]$ by the formula $S[p](k) \stackrel{\text{def}}{=} \sum_{i=1}^{k} p(i)$. We will put $S[p](0) = 0$ to simplify some of the formulas. If p is a pdf, it is elementary to show that $0 \leq S[p](k) \leq 1$ for all $k = 0, 1, \ldots$, that $S[p]$ is a nondecreasing sequence, and that $S[p](k) \to 1$ as $k \to \infty$.

Given a sequence $q = \{q(i) : i = 1, 2, \ldots\}$, we define its *decreasing rearrangement* q^* to be the result of renumbering the coefficients so that $q^*(i) \geq q^*(i+1)$ for all $i = 1, 2, \ldots$. If q is a pdf then so is q^*. The decreasing rearrangement is uniquely defined, though the permutation used to obtain it from q may not be unique.

Decreasing rearrangements are extremal for certain functionals. For example,

Lemma 8.7 *$S[q^*](k) \geq S[q](k)$ for all $k = 0, 1, \ldots$* □

Since q^* decreases, its running averages must also decrease, so we have

Lemma 8.8 *If $j \leq k$ then $jS[q^*](k) \leq kS[q^*](j)$.* □

We can apply Lemma 8.8 three times and conclude that the partial sums of a decreasing rearrangement are *convex*:

Lemma 8.9 *If $i \leq j \leq k$, then $(k-i)S[q^*](j) \geq (k-j)S[q^*](i) + (j-i)S[q^*](k)$.* □

We will say that a pdf p is *more concentrated* than a pdf q if $S[p^*] \geq S[q^*]$. More concentrated pdfs have lower entropy:

Proposition 8.10 *If p and q are pdfs and $S[p^*] \geq S[q^*]$, then $H(p) \leq H(q)$.*

Proof: To simplify the notation, we will assume without loss of generality that $q = q^*$ and $p = p^*$. If $S[p] \geq S[q]$, then $1 - S[p] \leq 1 - S[q]$. Since q decreases, we know that $\log[q(j)/q(j+1)] \geq 0$ for each $j = 1, 2, \ldots$, and we can define $q(0) = 1$ so that the formula holds for $j = 0$ as well. Then for each $j = 0, 1, \ldots$, we have

$$\Big(1 - S[p](j)\Big) \log \frac{q(j)}{q(j+1)} \leq \Big(1 - S[q](j)\Big) \log \frac{q(j)}{q(j+1)}. \tag{8.7}$$

Now $1 - S[p](j) = \sum_{i=j+1}^{\infty} p(i)$, so that Inequality 8.7 implies

$$\sum_{j=0}^{\infty} \sum_{i=j+1}^{\infty} p(i) \log \frac{q(j)}{q(j+1)} \leq \sum_{j=0}^{\infty} \sum_{i=j+1}^{\infty} q(i) \log \frac{q(j)}{q(j+1)}.$$

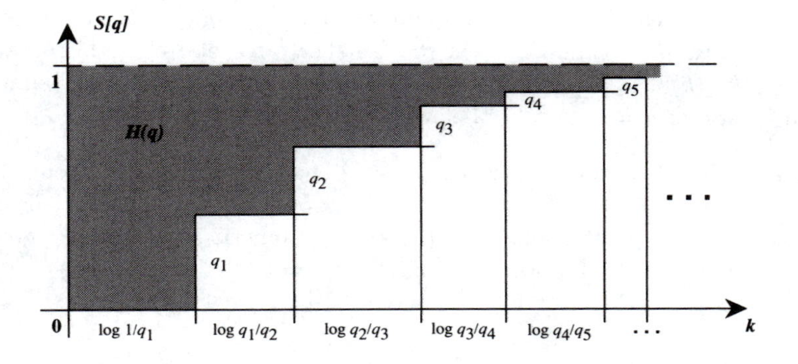

Figure 8.1: Entropy is the area above the partial sums.

Interchanging the order of summation yields

$$\sum_{i=1}^{\infty} p(i) \sum_{j=0}^{i-1} \log \frac{q(j)}{q(j+1)} \leq \sum_{i=1}^{\infty} q(i) \sum_{j=0}^{i-1} \log \frac{q(j)}{q(j+1)}.$$

The inner sum telescopes:

$$\sum_{i=1}^{\infty} p(i) \log \frac{1}{q(i)} \leq \sum_{i=1}^{\infty} q(i) \log \frac{1}{q(i)}.$$

The quantity on the right is $H(q)$. By Lemma 8.4, the quantity on the left is bigger than $H(p)$, which completes the proof. □

Remark. The relationship between entropy and the rate of increase of the partial sum of a decreasing pdf is shown in Figure 8.1. If $q = q^*$ is a decreasing sequence, then the widths of all the rectangles are nonnegative and the shaded area above $S[q]$ is equal to $H(q)$.

A similar result holds for the sum of the rth powers of a pdf with $r < 1$:

Lemma 8.11 *If p and q are pdfs with $S[p*](k) \geq S[q*](k)$ for all $k = 1, 2, \ldots$, then $\sum_{k=1}^{\infty} p(k)^r \leq \sum_{k=1}^{\infty} q(k)^r$ for all $0 < r < 1$.*

Proof: Since the sum of all the rth powers is invariant under rearrangements, we may assume without loss that $q = q^*$ and $p = p^*$. But then consider Figure 8.1, only with intervals of length $q(1)^{r-1}$, $q(2)^{r-1} - q(1)^{r-1}$, ... along the horizontal axis. Then the area above the histogram will be $\sum_{k=1}^{\infty} q(k)^r$. If we use $p(k)$ for the height of the kth rectangle, then the (smaller) area above the histogram will be $\sum_{k=1}^{\infty} p(k)q(k)^{r-1} \geq \sum_{k=1}^{\infty} p(k)^r$ by an argument similar to that in Lemma 8.4, since the graph of $y = x^r$ also lies below its tangents. □

Remark. The choice of r for which $\sum_k p(k)^r$ best shows the more concentrated pdf is of course $r = 0$, though that choice is impossible to implement. However, David Donoho has pointed out that the entropy functional essentially gives the linear approximation to $\sum_k p(k)^0$ starting from $r = 1$. Namely, let $f(r) = \sum p(k)^r$ for a pdf p. Then $f(1) = 1$ and we can write

$$f(0) \approx f(1) - f'(1) \cdot 1 = 1 - \frac{d}{dr} \sum_k p(k)^r \Big|_{r=1} = 1 + \sum_k p(k) \log 1/p(k). \tag{8.8}$$

8.2 Searching for the best basis

A *library* of bases in a linear space X is a collection whose elements are constructed by choosing individual basis subsets from a large set of vectors. The totality of these vectors must span X, but we expect to have far more than a linearly independent set. That way we can exploit the combinatorics of choosing independent subsets to obtain plenty of individual bases.

Suppose that X is a (separable) Hilbert space and we pick a collection $\{b_\alpha : \alpha \in T\} \subset X$ of unit vectors in X which together span X. Here α runs over an index set T which we will specify below for some examples of such collections. From this collection of vectors we can pick various subsets $\{b_\alpha : \alpha \in T_\beta\} \subset X$, where β runs over some other index set, and then form the subspace $V_\beta \subset X$ which is their (closed) linear span:

$$V_\beta \stackrel{\text{def}}{=} \text{span}\{b_\alpha : \alpha \in T_\beta\}. \tag{8.9}$$

For example, take $X = R^N$ with $N = 2^L$, and take the vectors to be the complete set of Haar–Walsh wavelet packets on N points. Then the index set T contains all the triplets $\alpha = (s, f, p)$ with $0 \leq s \leq L$, $0 \leq f < 2^s$, and $0 \leq p < 2^{L-s}$. Thus $b_\alpha = b_{(s,f,p)}$ is the wavelet packet with scale index s, frequency index f and position index p. The subspaces corresponding to blocks of wavelet packets in the complete wavelet packet analysis down to level L may be identified with the pairs $\beta = (s, f)$ with $0 \leq s \leq L$, $0 \leq f < 2^s$. Then $T_\beta = T_{(s,f)} = \{\alpha = (s, f, p) : 0 \leq p < 2^{L-s}\}$. The library is the set of Haar–Walsh wavelet packet graph bases.

8.2.1 Library trees

Any library of bases for a space X can be made into a *tree* by partially ordering the collection of subspaces V_β using inclusion. Namely, we will say that $V_{\beta'} \leq V_\beta \iff V_\beta \subset V_{\beta'}$. Subspaces are "smaller" in this partial order if they are closer to the whole space X, which forms the *root* or unique minimal element of the resulting tree. The indices β inherit the partial ordering; we may say that $\beta' \leq \beta \iff V_{\beta'} \leq V_\beta$. This structure becomes interesting if we can construct indices β for which this partial order is natural and easy to compute.

In the L-level Haar–Walsh case, we can use dyadic subintervals of $[0,1[$ as "indices" for the wavelet packet subspaces: $\beta = (s,f)$ corresponds to the interval

$$I_{sf} \stackrel{\text{def}}{=} \left[\frac{f}{2^s}, \frac{f+1}{2^s}\right[. \tag{8.10}$$

Then $\beta' \leq \beta \iff \beta \subset \beta'$, and the inclusion relation for indices mirrors the inclusion relation for subspaces. The root subspace X is indexed by the whole interval $[0,1[$. Furthermore, subspaces corresponding to disjoint index intervals are orthogonal:

$$\beta \cap \beta' = \emptyset \Rightarrow V_\beta \perp V_{\beta'}. \tag{8.11}$$

Notice that $I_{sf} = I_{s+1,2f} \cup' I_{s+1,2f+1}$, where $\cup'$ means disjoint union, and the corresponding subspaces give an orthogonal decomposition:

$$V_{sf} = V_{s+1,2f} \oplus V_{s+1,2f+1}. \tag{8.12}$$

Thus the Haar–Walsh library is organized like a homogeneous binary tree and it is easy to characterize the orthogonal basis subsets. This tree structure is so useful that we will formalize it and seek other libraries that have it, using the Haar–Walsh example as the model.

8.2.2 Fast searches for minimum information cost

If the library is a tree with finite depth L, then we can find the best basis by computing the information cost of a vector v in each node of the tree and comparing children to parents starting from the bottom. This is a low complexity search since the additive property of additive information cost functions means that each node is examined only twice: once as a child and once as a parent. To search an L-level decomposition of an $N = 2^L$-sample periodic signal therefore requires just $O(N)$ comparisons.

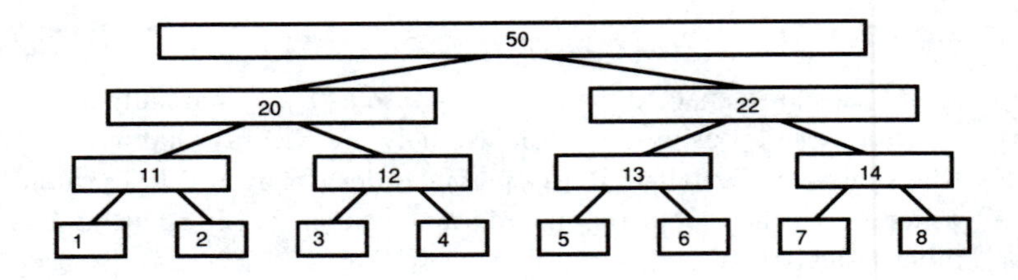

Figure 8.2: Sample information costs in a library tree.

The algorithm to search the tree is perfectly recursive and finds a best basis subset as follows. Denote by B_{kn} the standard basis for the node of the decomposition tree indexed by the dyadic interval I_{kn}. Now fix k and suppose that for all $0 \leq n < 2^k$ we have chosen from the library a basis A_{kn} for v restricted to the span of B_{kn}. Then we will choose $A_{k-1,n}$ for $0 \leq n < 2^{k-1}$ to minimize the information cost function. Let $M_B \stackrel{\text{def}}{=} M(B^*_{k-1,n}v)$ and $M_A \stackrel{\text{def}}{=} M(A^*_{k,2n}v) + M(A^*_{k,2n+1}v)$ be the M-information costs of v when expanded in the vectors of $B_{k-1,n}$ and $A_{k,2n} \oplus A_{k,2n+1}$, respectively. To define $A_{k-1,n}$, we choose the cheaper expansion:

$$A_{k-1,n} = \begin{cases} A_{k,2n} \oplus A_{k,2n+1}, & \text{if } M_A < M_B, \\ B_{k-1,n}, & \text{otherwise.} \end{cases} \tag{8.13}$$

Then we have the following:

Proposition 8.12 *The algorithm in Equation 8.13 yields the best basis for each fixed vector $v \in V_{00}$, relative to M and the library tree $\mathcal{B}$.*

Proof: This can be shown by induction on L. For $L = 0$, the library contains only the standard basis for V_{00} which wins the best basis contest unopposed.

Now suppose that the algorithm yields the best basis for any tree of depth L, and let $A = A_{00}$ be the basis for V_{00} which it choses for a tree of depth $L+1$. If A' is any basis for V_{00}, then either $A' = B_{00}$ or $A' = A'_0 \oplus A'_1$ is a direct sum of bases for V_{10} and V_{11}. Let A_{10} and A_{11} denote the best bases in these subspaces, which are found by the algorithm in the L-level left and right subtrees of the root. By the inductive hypothesis, $M(A_i^* v) \leq M(A_i'^* v)$ for $i = 0, 1$. By Equation 8.13, $M(A^*v) \leq \min\{M(B^*_{00}v), M(A^*_{01}v) + M(A^*_{11}v)\} \leq M(A'^*v)$. □

To illustrate the algorithm, consider the following example expansion into a three-level wavelet packet library tree. We have placed numbers representing information costs inside the nodes of the tree in Figure 8.2. We start by marking all the bottom nodes, as indicated by the asterisks in Figure 8.2. Their total information cost is an initial value which we will try to reduce. The finite number of

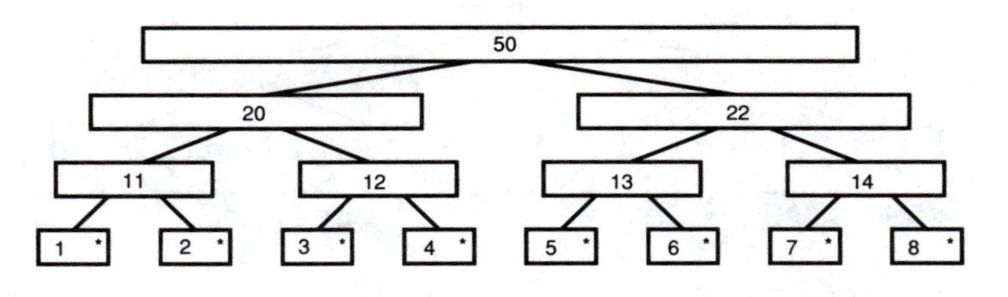

Figure 8.3: Step 1 in the best basis search: mark all bottom nodes.

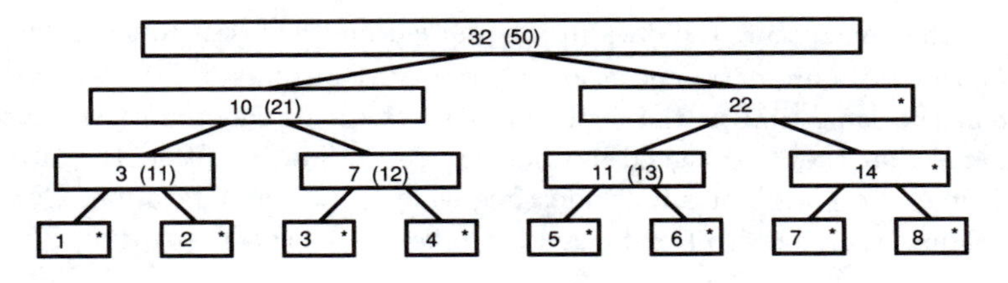

Figure 8.4: Step 2 in the best basis search: mark all nodes of lower cost.

levels in the tree guarantees that the algorithm will terminate. Whenever a parent node is of lower information cost than the children, we mark the parent with an asterisk. If the children have lower information cost, we do not mark the parent, but we assign the lower total information cost of the children to the parent. This inductive step prevents our having to examine any node more than twice: once as a child, and once as a parent. In Figure 8.4, the former information costs of adjusted nodes are displayed in parentheses. This marking may be done at the same time as the coefficient values in the nodes are calculated, if the construction of the tree proceeds in depth-first order. In that case, the information cost of the best (graph) basis below each node is known before the recursive descent into that node returns. The depth-first search algorithm terminates because the tree has finite depth. By induction, the topmost node (or the "root" of the tree) will return the minimum information cost of any basis subset below itself, namely, the best basis.

Finally, after all the nodes have been examined, we take the topmost marked nodes, which constitutes a basis by Theorem 7.9. These best basis nodes are displayed as shaded blocks in Figure 8.5. When the topmost marked node is encountered, the remaining nodes in its subtree are pruned away.

The best basis search is most easily implemented as a recursive function. It splits naturally into two pieces: searching for the nodes which comprise the best

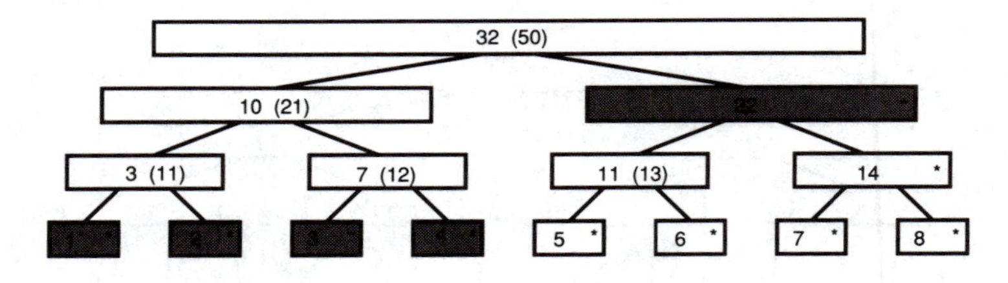

Figure 8.5: Step 3 in the best basis search: retain topmost marked nodes.

basis, and then extracting the data in the marked nodes closest to the root.

The coefficients are extracted into an output list in some order. This may also be done at the same time as the depth first search for topmost marked nodes. We may assume that they are tagged by their node of origin, *i.e.*, that they come out as quadruples (a, f, s, p) of amplitude, frequency, scale, and position. From the original signal, therefore, to this tagged list takes $O(N \log N)$ operations.

8.2.3 Adapted waveform analysis meta-algorithm

Yves Meyer [83] introduced the notion of a *meta-algorithm* for analysis. Given a *complicated function or operator* (like a sampled speech signal) and an *analysis objective* (such as separating the individual words), we proceed by determining the quantities to measure in order to achieve the objective (such as the intervals which contain particular frequencies). Next we choose a library of waveforms whose members have properties well matched to the function and the properties to be measured (similar smoothness, similar frequency spreading). We can choose from a rich bounty of recently discovered libraries, but this selection can be based on just a few simple criteria. We cannot in general be terribly selective because at the start we expect to have little knowledge about the function. Finally, we adapt a basis from the chosen library to represent the function, minimizing some measure of information cost which defines our notion of efficiency (such as total number of nonnegligible elements in a Gabor expansion of the signal).

We call the meta-algorithm an *adapted waveform analysis*. The result it produces is a choice of *efficient representatives*, the waveforms which match the complicated function well, and a *description sequence* of coordinates with respect to those waveforms, from which we can reconstruct a good approximation to the function. The choice of representatives might contain information useful for feature detection, though it can be ill-conditioned in the sense that nearby functions might produce very different choices. The description sequence can be used in further computations with the function; if the underlying basis is orthonormal or nearly so and we hold the choice of representatives fixed, then these numbers will be well-conditioned.

8.3 Implementation

The main ingredient in a best basis search algorithm is the library tree of coefficients produced by computing the inner products of the input signal with the various analyzing functions. We will suppose in this exposition that such expansions have been carried out and that we have at our disposal various arrays of coefficients. We must also keep track of how the coefficients assemble into basis subsets.

8.3.1 Information cost functionals

The properties of additive information cost functionals allow us to evaluate the information cost of an array by looking at each element individually. Thus, all the implementations will take an array and its length as input parameters and will contain a loop over the elements of the array.

Additive cost functions

We can measure the cost of a sequence of coefficients by the *number exceeding a threshold*. We choose a threshold ϵ and simply count how many elements of the input array have absolute values above this threshold. It uses the predefined function `absval()` which returns the absolute value of its argument:

Threshold information cost functional

```
thresh( U, LEAST, FINAL, EPSILON ):
   Let COST = 0
   For K = LEAST to FINAL
      If absval(U[K]) > EPSILON then
         COST += 1
   Return COST
```

We can compare expansions for *concentration in* ℓ^p by summing the pth powers of their coefficients, for $0 < p < 2$. Below is an example implementation for $p = 1$; it also uses the predefined function `absval()`:

ℓ^1 information cost functional

```
l1norm( U, LEAST, FINAL ):
   Let COST = 0
   For K = LEAST to FINAL
      COST += absval( U[K] )
   Return COST
```

For $p \neq 1$ we must test for the singularity $|u(k)| = 0$. We implement this using the predefined `exp()` and `log()` functions as we would in Standard C; in FORTRAN we could use the exponentiation operator `**` instead.

ℓ^p information cost functional

```
lpnormp( U, LEAST, FINAL, P ):
   Let COST = 0
   For K = LEAST to FINAL
      Let ABSU = absval( U[K] )
      If ABSU > 0 then
         COST += exp( P*log( ABSU ) )
   Return COST
```

Entropy itself is not an information cost functional, since it requires that the sequence be normalized and is thus not additive. Instead, we use the $-\ell^2 \log \ell^2$ "norm" which does satisfy the additivity condition in Equation 8.1 required of an information cost functional and is monotonic with the entropy of a sequence. Below is an example implementation:

$-\ell^2 \log \ell^2$ or "entropy" information cost functional

```
ml2logl2( U, LEAST, FINAL ):
   Let COST = 0
   For K = LEAST to FINAL
      Let USQ =  U[K] * U[K]
      If USQ > 0 then
         COST -= USQ * log( USQ )
   Return COST
```

Other cost functions

We will have occasion to use cost functions which are not additive, *i.e.*, which do not satisfy Equation 8.1.

The entropy of a Gauss–Markov process is the sum of the *logarithms of the variances* of its components. If the components have zero mean, then these variances may be approximated by the energies of the components. An example implementation is below.

$\log \ell^2$ or Gauss–Markov information cost functional

```
logl2( U, LEAST, FINAL ):
   Let COST = 0
   For K = LEAST to FINAL
      COST += log( U[K] * U[K] )
   Return COST
```

A simple Standard C function for computing *theoretical dimension* can be made by building upon the function `ml2logl2()` defined above. We need both the energy and the $\ell^2 \log \ell^2$ "norm" of the sequence, and we compute them both at once to avoid having to square each array element twice.

Theoretical dimension of a sequence

```
tdim( U, LEAST, FINAL ):
   Let MU2LOGU2 = 0
   Let ENERGY = 0
   For K = LEAST to FINAL
      Let USQ =  U[K] * U[K]
      If USQ > 0 then
         MU2LOGU2 -= USQ * log( USQ )
         ENERGY += USQ
   If ENERGY > 0 then
      Let THEODIM =  ENERGY * exp( MU2LOGU2 / ENERGY )
   Else
      Let THEODIM = 0
   Return THEODIM
```

8.3.2 Extracting a basis subset

We need two pieces of information to reconstruct a signal: the *basis description* for its expansion, and the *coefficients* of the signal with respect to that basis. There are many forms in which these may be encoded, and the most appropriate form depends on the application. For example, the form best suited to transmission of compressed data will squeeze out all redundancy at the expense of easy reconstruction. On the other hand, the form best suited for manipulation and computation will generally contain some excess information as tags for the coefficients, to reduce the complexity of computation by easily identifying the components.

Thus we will devise ways to produce two kinds of output. One kind will imitate the `HEDGE` data structure introduced in Chapter 4, in which the basis description is a separate and very compact piece of extra information from which with some work we can reconstruct the locations of the coefficients in the original library tree. We envision storing or transmitting the entire basis description, which must be exact, plus the entire coefficient list to some degree of approximation.

The other kind will produce `TFA1` data structures, also introduced in Chapter 4. In these we assign a tag to each coefficient which locates it in the library tree. In this way, any number of the coefficients to be kept without degrading our ability to reconstruct that portion of the signal.

In either case, the first step is to identify which nodes in the library tree belong to the basis subset.

Getting the best basis

We perform a recursive search for the best basis in library tree using an additive information cost function. This information cost function must be chosen in advance; it can be specified with a preprocessor definition:

Choose "entropy" as the information cost

```
#define infocost  ml2logl2
```

To use one of the information cost functions which has a different parameter list, the definition must include the value of the missing parameter:

Choose "$\ell^{1/4}$" as the information cost

```
#define infocost(IN,LEAST,FINAL)  lpnormp(IN,LEAST,FINAL,0.25)
```

The recursive step is to compare the information cost of the parent with the sum of the information costs of the best bases of its children. If the parent BTN data structure has lower information cost than any of the descendents, then the level of that node is appended to the levels description of the best basis, the content of that node is appended to the array of contents pointers, and the previously built description of the best basis below the parent is flushed. This is done by resetting the blocks counter back to the number of blocks encountered up to the parent, thus ignoring the descendents encountered afterwards. The blocks counter is incremented by one to account for the parent node. Then we return the parent's information cost for comparison with its ancestors'.

If the parent BTN loses, we preserve the recursively developed descendent best basis in the hedge and return the sum of the information costs of the best bases for the children. We proceed this way from the bottom nodes or *leaves* of the tree, which are best because they have no descendents, until we rise to the *root* node.

The information costs will be stored in the tag members of the BTN data structures. To make the search as generic as possible, we will assume that the information cost tags of the input BTN tree have been precomputed, using some other function tailored to the particular content arrays of amplitudes. The following pseudocode function performs the search from such a tree; note that it assumes that `GRAPH.BLOCKS` is zero at the start of the search.

Best basis hedge from a BTN tree with cost tags

```
costs2bbasis( GRAPH, ROOT, LEVEL ):
   If ROOT.LEFT==NULL && ROOT.RIGHT==NULL then
      Let GRAPH.LEVELS[GRAPH.BLOCKS] = LEVEL
      Let GRAPH.CONTENTS[GRAPH.BLOCKS] = ROOT.CONTENT
      GRAPH.BLOCKS += 1
      Let BESTCOST = ROOT.TAG
   Else
      Let BLOCKS = GRAPH.BLOCKS
      Let COST = 0
      If ROOT.LEFT != NULL then
         COST += costs2bbasis( GRAPH, ROOT.LEFT, LEVEL+1 )
      If ROOT.RIGHT != NULL then
         COST += costs2bbasis( GRAPH, ROOT.RIGHT, LEVEL+1 )
      If ROOT.TAG>COST then
         Let BESTCOST = COST
      Else
         Let BESTCOST = ROOT.TAG
         Let GRAPH.BLOCKS = BLOCKS
         Let GRAPH.LEVELS[GRAPH.BLOCKS] = LEVEL
         Let GRAPH.CONTENTS[GRAPH.BLOCKS] = ROOT.CONTENT
         GRAPH.BLOCKS += 1
   Return BESTCOST
```

Remark. The recursion termination condition is that both the left and right children are null pointers. This allows us to search partial BTN trees as well as complete trees to a given level.

To find the best basis in a BTN tree, we first fill its tag members with the information costs using the following function:

Add costs tags to a BTN tree of INTERVALs

```
btnt2costs( ROOT ):
   If ROOT != NULL then
      Let I = NODE.CONTENT
      Let ROOT.TAG = infocost(I.ORIGIN, I.LEAST, I.FINAL)
      btnt2costs( ROOT.LEFT )
      btnt2costs( ROOT.RIGHT )
   Return
```

We need a few lines of initialization, then we can use the generic function `costs2bbasis()` to search the tagged tree for the the best basis hedge:

Return a best basis hedge for a BTN tree of INTERVALs

```
btnt2bbasis( ROOT, MAXLEVEL ):
   btnt2costs( ROOT )
   Let GRAPH = makehedge( 1<<MAXLEVEL, NULL, NULL, NULL )
   Let GRAPH.BLOCKS = 0
   costs2bbasis( GRAPH, ROOT, 0 )
   Return GRAPH
```

We next turn our attention to array binary trees. We reuse the BTN search function after putting the information costs into a separate binary tree of BTN data structures:

Build a costs BTN tree from an array binary tree

```
abt2costs( DATA, LENGTH, MAXLEVEL ):
  Let COSTS = makebtnt( MAXLEVEL )
  For LEVEL = 0 to MAXLEVEL
    For BLOCK = 0 to (1<<LEVEL)-1
      Let ORIGIN = DATA + abtblock( LENGTH, LEVEL, BLOCK )
      Let BLENGTH = abtblength( LENGTH, LEVEL )
      Let CNODE = btnt2btn( COSTS, LEVEL, BLOCK )
      Let CNODE.TAG = infocost( ORIGIN, 0, BLENGTH-1 )
      Let CNODE.CONTENT = ORIGIN
  Return COSTS
```

After finding the best basis hedge with `costs2bbasis()`, we get rid of the auxiliary BTN costs tree. Notice that we deallocate the nodes, but preserve the contents:

Return the best basis hedge from an array binary tree

```
abt2bbasis( DATA, LENGTH, MAXLEVEL ):
   Let COSTS = abt2costs( DATA, LENGTH, MAXLEVEL )
   Let GRAPH = makehedge( 1<<MAXLEVEL, NULL, NULL, NULL )
   Let GRAPH.BLOCKS = 0
   costs2bbasis( GRAPH, COSTS, 0 )
   freebtnt( COSTS, NULL, free )
   Return GRAPH
```

Getting the best level basis

We first define a function which returns the information cost of an entire level:

Information cost of one level in a BTN tree with cost tags

```
levelcost( ROOT, LEVEL ):
  Let COST = 0
  For BLOCK = 0 to (1<<LEVEL)-1
    Let NODE = btnt2btn( ROOT, LEVEL, BLOCK )
    COST += NODE.TAG
  Return COST
```

We get the *best level* basis from a BTN tree with cost tags, using a single step of the bubble sorting algorithm:

Find the best level in a BTN tree with cost tags

```
costs2blevel( GRAPH, ROOT, MINLEVEL, MAXLEVEL ):
  Let BESTLEVEL = MINLEVEL
  Let BESTCOST = levelcost( ROOT, MINLEVEL )
  For LEVEL = MINLEVEL+1 to MAXLEVEL
    Let COST = levelcost( ROOT, LEVEL )
    If COST<BESTCOST then
      Let BESTCOST = COST
      Let BESTLEVEL = LEVEL
  Let GRAPH.BLOCKS = 1<<BESTLEVEL
  For BLOCK = 0 to GRAPH.BLOCKS-1
    Let GRAPH.LEVELS[BLOCK] = BESTLEVEL
    Let GRAPH.CONTENTS[BLOCK] = btnt2btn(ROOT,BESTLEVEL,BLOCK)
  Return BESTCOST
```

We loop over the levels, computing the information cost of each, and return the hedge of nodes from the least costly level.

The preceding function also allows us to specify the deepest and shallowest levels to be included in the search. It assumes that the shallowest specified level is no greater than the deepest; this should be tested with an `assert()` statement. We also assume that all the levels between the shallowest and deepest are complete; this can be tested by asserting that `NODE` is non-null in the `levelcost()` function.

We break ties by preferring the shallower level, on the grounds that deeper decomposition levels cost more to compute. Notice that we initialize the candidate best information cost with the cost of the shallowest level. This is because additive information cost functions can take any real value, so there is no *a priori* maximum information cost.

The combinations needed to implement `btnt2blevel()` and `abt2blevel()` are left as exercises.

Extracting atoms

To extract an array of `TFA1` data structures corresponding to the best basis, we use the utility functions defined in Chapter 2. We use `intervalhedge2tfa1s()` to produce a list of atoms from a hedge containing INTERVALs, and we use `abthedge2tfa1s()` to produce atoms from a hedge containing pointers into an array binary tree.

8.3.3 Extracting a branch

We can also select a branch from the root to a leaf in a BTN tree so as to minimize the information cost of the contents of the nodes along the branch. This can be done by building a BTN tree of costs tags, then recursively adding the cost of a parent to each of its two children as we descend the tree from its root. When we are done, the tag member of each leaf node in the costs tree will contain the total cost of the branch from the root to that leaf. The implementation of this *best branch* algorithm is left as an exercise.

Wavelet registration

We illustrate the best branch method with a specific application: an algorithm for finding a *translation invariant* periodic discrete wavelet transform. This procedure finds which periodic shift of a signal produces the lowest information cost. It was partially described by Beylkin in [8]; he earlier observed that computing the DWTPs of all N circulant shifts of an N-point periodic signal requires computing only

$N \log_2 N$ coefficients. If we build a complete BTN tree with information cost tags computed from from appropriate subsets of the shifted coefficients, then the best complete branch will give a representation of the circulant shift which yields the lowest cost DWTP. After solving the technical problem of ties, the computed shift can be used as a *registration point* for the signal.

The first step is to build a BTN tree of the information costs of the wavelet subspaces computed with all circulant shifts. We accumulate the cost of a branch into its leaf at the same time that we compute the coefficients.

The following function assumes that the Q-periodic input signal is $Q+1$ elements long and has its first element IN[0] duplicated in its last location IN[Q]. It also assumes that the output array is at least $Q+\log_2 Q$ elements long, to accommodate all the intermediate outputs which have lengths $Q/2^k + 1$, $k = 1, \ldots, \log_2 Q$. Note that we also allocate an empty initial BTN data structure to serve as the root of the costs tree:

Costs of circulant shifts for DWTP, accumulated by branch

```
shiftscosts( COSTS, OUT, Q, IN, HQF, GQF ):
  If Q>1 then
    Let Q2 = Q/2
    Let COSTS.LEFT = makebtn( NULL, NULL, NULL, COSTS.TAG )
    cdpe( OUT, 1,  IN,  Q, GQF )
    COSTS.LEFT.TAG += infocost( OUT, 0, Q2-1 )
    Let COSTS.RIGHT = makebtn( NULL, NULL, NULL, COSTS.TAG )
    cdpe( OUT, 1, IN+1, Q, GQF )
    COSTS.RIGHT.TAG += infocost( OUT, 0, Q2-1 )
    cdpe( OUT, 1,  IN,  Q, HQF )
    Let OUT[Q2] = OUT[0]
    shiftscosts( COSTS.LEFT, OUT+Q2+1, Q2, OUT, HQF, GQF )
    cdpe( OUT, 1, IN+1, Q, HQF )
    Let OUT[Q2] = OUT[0]
    shiftscosts( COSTS.RIGHT, OUT+Q2+1, Q2, OUT, HQF, GQF )
  Return
```

The function shiftscosts() accumulates the costs of a branch as we descend. Then the information cost of a 2^L point DWPT shifted by T will be found in the node at level L whose block index is the bit-reverse of T. We extract these values using one of the utilities defined in Chapter 2, use a bubble sort to find the least one while searching in bit-reversed order, and return its index. This finds the least circulant shift which yields the minimal information cost:

Least circulant shift with minimal DWTP cost

```
registrationpoint( COSTS, MAXLEVEL ):
   Let LEAF = btnt2btn( COSTS, 0, MAXLEVEL )
   Let BESTCOST = LEAF.TAG
   Let BESTSHIFT = 0
   For SHIFT = 1 to (1<<MAXLEVEL)-1
      Let BLOCK = br( SHIFT, MAXLEVEL )
      Let LEAF = btnt2btn( COSTS, BLOCK, MAXLEVEL )
      Let SHIFTCOST = LEAF.TAG
      If SHIFTCOST<BESTCOST then
         Let BESTCOST = SHIFTCOST
         Let BESTSHIFT = SHIFT
   Return BESTSHIFT
```

To *register* a periodic signal, we compute the registration point and then circularly shift the signal so that the registration point becomes index zero:

Shift in place by least amount to get minimal DWTP cost

```
waveletregistration( IN, L, HQF, GQF ):
   Let Q = 1<<L
   Allocate an array of Q+1 REALs at DATA
   For I = 0 to Q-1
      Let DATA[I] = IN[I]
   Let DATA[Q] = IN[0]
   Allocate an array of Q+log2(Q) REALs at WORK
   Let COSTS = makebtn( NULL, NULL, NULL, NULL )
   shiftscosts( COSTS, WORK, Q, DATA, HQF, GQF )
   Let REGPOINT = registrationpoint( COSTS, L )
   If REGPOINT>0 then
     For I = 0 to REGPOINT-1
        Let DATA[Q+I-REGPOINT] = IN[I]
     For I = REGPOINT to Q-1
        Let DATA[I-REGPOINT] = IN[I]
     For I = 0 to Q-1
        Let IN[I] = DATA[I]
   freebtnt( COSTS )
   Deallocate WORK[] and DATA[]
   Return REGPOINT
```

It is also possible to avoid the use of a BTN tree by directly writing the costs of circulant-shifted DWTPs to an array, but this is left as an exercise.

Wavelet registration works because the information cost of the wavelet subspace W_k of a 2^L-periodic signal is a 2^k-periodic function for each $0 \leq k \leq L$. Thus the tag in the node at level k, block n is the information cost of W_k with a circulant shift by $n' \pmod{2^k}$, where n' is the length k bit-reversal of n. A branch to a leaf node at block index n contains the wavelet subspaces $W_1, \ldots, W_L$ of the DWTP with shift n'. The scaling subspace V_L in the periodic case always contains the unweighted average of the coefficients, which is invariant under shifts.

We can define a *shift cost function* for a 2^L-periodic signal to be the map $f(n) = c_{n'L}$, the information cost in the tag of the costs BTN at level L and block index n', the bit-reverse of n.

Two 2^L-signals whose principal difference is a circulant shift can be compared by cross-correlating their shift cost functions. This is an alternative to traditional cross-correlation of the signals themselves, or multiscale cross-correlation of their wavelet and scaling subspaces as done in [58].

8.4 Exercises

1. Prove Lemma 8.7.

2. Prove Lemma 8.8.

3. Prove the following corollary to Lemma 8.8: If $k \leq k'$ and $j \leq j'$, then

$$\frac{S[p^*](k) - S[p^*](j)}{k - j} \geq \frac{S[p^*](k') - S[p^*](j')}{k' - j'}.$$

4. Using the result of Exercise 3, prove Lemma 8.9.

5. Write a pseudocode implementation of the function `btnt2bbhedge()`, which computes the information costs of the actual BTN tree of INTERVALs and builds a complete hedge containing the best basis for that tree. (Hint: Model your function after `costs2bbasis()`.)

6. Implement `btnt2blevel()`, a function that returns a complete hedge describing the best level basis found between two specified levels in a tree of BTNs containing INTERVALs.

7. Write a pseudocode implementation of the function `btnt2blhedge()`, which builds a complete hedge containing the best level basis for the actual BTN

tree of INTERVALs. (Hint: Model your function after `costs2blevel()` and write a new version of `levelcost()` which uses the original BTN tree.)

8. Implement `abt2blevel()`, a function that returns a complete hedge describing the best level basis found between two specified levels in an array binary tree.

9. It is simpler to find the best level in an array binary tree directly from the tree, since the amplitudes are stored contiguously by level. Write a pseudocode implementation of `abt2blhedge()` which fills a preallocated hedge with a complete description, including levels and contents, of the best level graph basis in an array binary tree.

10. Implement the wavelet registration algorithm without using a BTN tree, by directly superposing the costs of wavelet subspaces under shifts into an array.

11. Implement the best branch algorithm: take a costs BTN tree and recursively accumulate branch costs into the leaf nodes, then search the leaves for the minimal cost tag.

Chapter 9

Multidimensional Library Trees

Both the local trigonometric and the conjugate quadrature filter algorithms extend to multidimensional signals. We consider three methods of extension. The first two consist of *tensor products* of one-dimensional basis elements, combined so that the d-dimensional basis function is really the product of d one-dimensional basis functions: $b(x) = b(x_1, \ldots, x_d) = b_1(x_1) \cdots b_d(x_d)$. Such tensor product basis elements are called *separable* because we can factor them across sums and integrals, to obtain a sequence of d one-dimensional problems by treating each variable separately.

Consider the two-dimensional case for simplicity. Let $E = \{e_k : k \in I\}$ and $F = \{f_k : k \in J\}$ be bases for $L^2(\mathbf{R})$, where I and J are index sets for the basis elements. Then we can produce two types of separable bases for $L^2(\mathbf{R}^2)$:

- $E \otimes F \stackrel{\text{def}}{=} \{e_n \otimes f_m : (n, m) \in I \times J\}$, the separable tensor product of the bases E and F;

- $\{e_n \otimes f_m : (n, m) \in B\}$, a basis of separable elementary tensors from a *basis subset* B which is not necessarily all of $I \times J$.

We will introduce a geometric organization for the index set in the second case, to simplify the characterization of *basis subsets* in the multidimensional case. We will devote most of our time to separable bases, since they typically provide low complexity transforms which in addition can be run on several processors in parallel.

The diagram in Figure 9.1 depicts a decomposition of the unit square using two iterations of a generic two-dimensional, separable splitting algorithm. Notice that it produces a *quadtree*, or homogeneous tree with four children at each node,

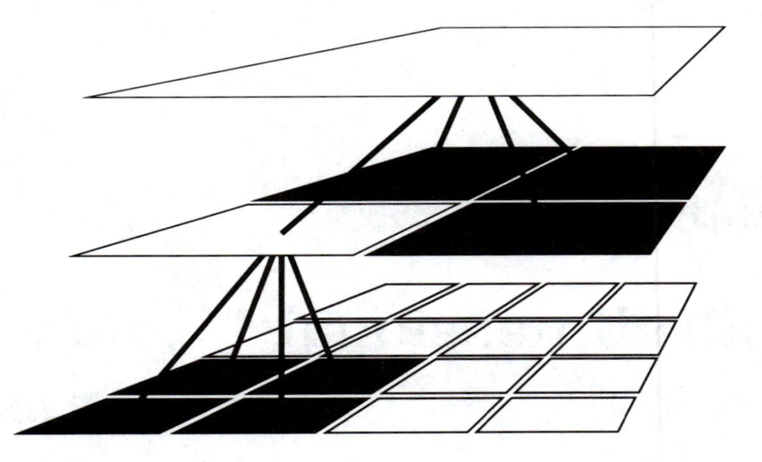

Figure 9.1: Two-dimensional quadtree decomposition: levels zero, one, and two.

if we consider the squares to be nodes. In the wavelet packet case, applying the CQFs recursively yields a homogeneous tree-structured decomposition organized by frequency. In the block trigonometric transform case, successively dividing the domain yields a segmentation into dyadic subblocks.

Alternatively, we can construct a basis for $L^2(\mathbf{R}^d)$ by treating it as an abstract space, ignoring the extra structure of "d-dimensionality" and performing a splitting using nonseparable d-dimensional convolution-decimations or other orthogonal projections. Such nonseparable algorithms are much more general and hence harder to characterize and implement, so we will mention them only in passing.

9.1 Multivariable splitting operators

Speaking abstractly, we have a splitting algorithm whenever we have a family $F_1, \ldots, F_m$ of operators on a Hilbert space X, with dual operators $F_1', \ldots, F_m'$, such that

$$F_j' F_i^* = F_i F_j'^* = \delta(i-j)I \quad \text{and} \quad F_1'^* F_1 + \cdots + F_m'^* F_m = I. \tag{9.1}$$

If $F_i = F_i'$ for all i, then the second equation is an orthogonal decomposition. In previous chapters we have fixed $m = 2$ and $X = L^2(\mathbf{R})$ or one of its approximation space V_0, but the formulas remain valid in more general contexts. This is largely a matter of language, as explained in [116].

If $x = (x_1, \ldots, x_d)$ is a coordinate in $\mathbf{R}^d$, then one way to realize the operator F is to use *quadratures* with a *kernel* $f = f(x)$:

$$Fu(x) = \sum_y f(x, y)u(y) \stackrel{\text{def}}{=} \sum_{y_1} \cdots \sum_{y_d} f(x_1, \ldots, x_d, y_1, \ldots, y_d)u(y_1, \ldots, y_d).$$

Such an operator generalizes a quadrature filter, for which $d = 1$ and $f(x, y) = h(2x - y)$ in the low-pass case. The complexity of computing a general Fu of this form is proportional to the volume of the support of f. When f is more or less isotropic and supported on a d-cube of side N, then we will pay $O(N^d)$ for each value $Fu(x)$, or $O(N^{2d})$ for all the values if u has support comparable to f.

We thus specialize to the separable case in which the quadrature kernel is a product of d functions of one variable. Then the summation can be performed in each variable separately, so each output value costs us at most $O(dN)$ operations, and the total work is $O(dN^{d+1})$. If the one-dimensional kernels have some special properties and factor into sparse matrices, then the complexity will be still lower. If the one-dimensional transforms are expansions in library trees, then we automatically obtain a library tree for the multidimensional product.

The two kinds of separable tree constructions we will consider are wavelet packets and local trigonometric functions. The associated splitting operators are tensor product QFs, tensor products of Fourier sine and cosine transforms, and tensor products of the unitary folding and unfolding operators.

9.1.1 Tensor products of CQFs

Let H and G be a conjugate pair of quadrature filters determined respectively by the one-dimensional sequences h and g. We define the *tensor product* $H \otimes G$ of H and G to be the following operator on bivariate sequences $u = u(x, y)$:

$$(H \otimes G)\, u(x, y) = \sum_{i,j} h(i)g(j)u(2x-i, 2y-j) = \sum_{i,j} h(2x-i)g(2y-j)u(i, j). \quad (9.2)$$

We define the adjoint of the tensor product by $(H \otimes G)^* \stackrel{\text{def}}{=} H^* \otimes G^*$, which is given by the following formula:

$$(H^* \otimes G^*)\, u(x, y) = \sum_i \sum_j \bar{h}(2x-i)\bar{g}(2y-j)u(i, j). \quad (9.3)$$

Notice that the first "factor" of the tensor product acts on the first variable, and so on. We can write $H \otimes G = H_x G_y$, which is a product of one-dimensional QFs each acting in just that variable noted in the subscript.

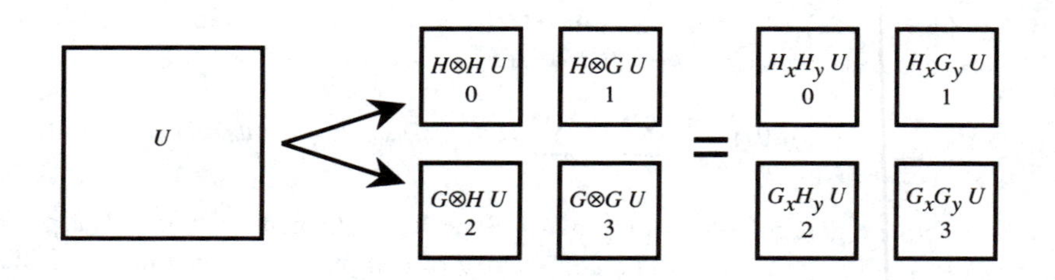

Figure 9.2: One level of wavelet packet decomposition with tensor-product CQFs.

These definitions naturally generalize to an arbitrary number d of dimensions. For example, if $\mathbf{x} = (x_1, \ldots, x_d)$ and $F^i \in \{H, G\}$ for $i = 1, \ldots, d$, then we can define a d-dimensional tensor product filter as follows:

$$\begin{aligned}(F^1 \otimes \cdots \otimes F^d)\, u(\mathbf{x}) &\stackrel{\text{def}}{=} \sum_{k_1} \cdots \sum_{k_d} f_1(k_1) \cdots f_d(k_d) u(2x_1 - k_1, \ldots, 2x_d - k_d) \\ &= F^1_{x_1} F^2_{x_2} \cdots F^d_{x_d} u(\mathbf{x}). \qquad (9.4)\end{aligned}$$

We will concentrate on the two-dimensional case, since it displays all the ingredients of the more general case without being unnecessarily complex.

If H and G are a pair of orthogonal CQFs, then $H \otimes H = H_x H_y$, $H \otimes G = H_x G_y$, $G \otimes H = G_x H_y$, and $G \otimes G = G_x G_y$ are a family of four orthogonal filters for $\mathbf{R}^2$. As such, they split the input signal into four descendents. We can number these 0 for $H_x H_y$, 1 for $H_x G_y$, 2 for $G_x H_y$, and 3 for $G_x G_y$. Each two-dimensional filter decimates by two in both in the x and y directions, so it reduces the number of coefficients by four. The coefficients in the various boxes of this picture are calculated by the application of the filters as labeled in Figure 9.2.

Note that the numbering corresponds to $\epsilon_1 \epsilon_2$ in binary, where

$$\epsilon_1 = \begin{cases} 0, & \text{if the } x\text{-filter is } H, \\ 1, & \text{if the } x\text{-filter is } G; \end{cases} \qquad \epsilon_2 = \begin{cases} 0, & \text{if the } y\text{-filter is } H, \\ 1, & \text{if the } y\text{-filter is } G. \end{cases}$$

For convenience and simplicity of description we will regard two-dimensional signals or *pictures* as periodic in both indices, and we shall assume that the x-period (the "height" $N_x = 2^{n_x}$) and the y-period (the "width" $N_y = 2^{n_y}$) are both positive integer powers of two, so that we can always decimate by two and get an integer period. The space of such pictures may be decomposed into a partially ordered set $\mathbf{W}$ of subspaces $W(m, n)$ called *subbands*, where $m \geq 0$, and $0 \leq n < 4^m$. These are the images of orthogonal projections composed of products of convolution-decimations. Denote the space of $N_x \times N_y$ pictures by $W(0, 0)$ (it is $N_x \times N_y$

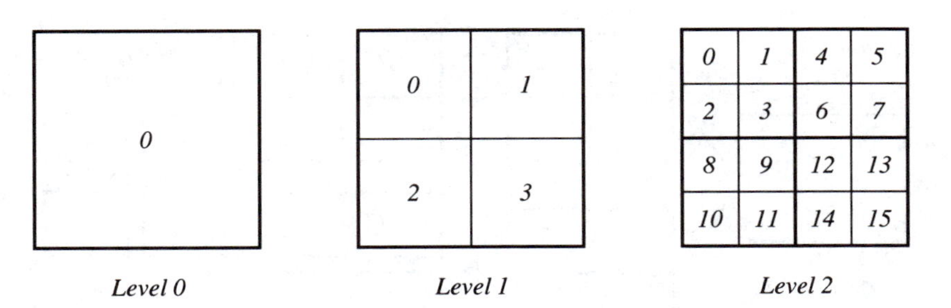

Figure 9.3: Two levels of subband decomposition.

dimensional), and define recursively

$$W(m+1, 4n+i) = F_i^* F_i W(m,n) \subset W(0,0), \qquad \text{for } i = 0,1,2,3. \tag{9.5}$$

The orthogonality condition on the CQFs implies that the projections from $W(0,0)$ onto $W(m,n)$ are orthogonal, *i.e.*, they conserve energy. The subspace $W(m,n)$ is $(N_x 2^{-m}) \times (N_y 2^{-m})$-dimensional. These subspaces may be partially ordered by inclusion. Subspaces of a space are called its *descendents*, while the first generation of descendents will naturally be called *children*. By specializing the orthogonality condition to the four tensor product CQFs, we obtain the relation

$$W = F_0^* F_0 W \oplus F_1^* F_1 W \oplus F_2^* F_2 W \oplus F_3^* F_3 W. \tag{9.6}$$

The right hand side contains all the children of W.

The subspaces $W(m,n)$ are called *subbands*, and this transform is the first step in *subband coding* image compression, which we describe in greater detail in Chapter 11, Section 11.1. If $S \in W(0,0)$ is a picture, then its orthogonal projection onto $W(m,n)$ can be computed in the standard coordinates of $W(m,n)$ by the formula $F_{(1)} \cdots F_{(m)} W(0,0)$, where each filter $F_{(i)}$ is one of $F_0, \ldots, F_3$, and the sequence of these filters $F_{(1)} \cdots F_{(m)}$ is determined uniquely by n. Therefore we can compute the orthogonal projections of $W(0,0)$ onto the complete tree of subspaces simply by recursively convolving and decimating.

Figure 9.3 shows the root picture and two generations of descendent subbands, which are labeled by index n in $W(m,n)$. If we had started with a picture of $Z \times Z$ pixels, then we could have repeated this decomposition process $\log_2(Z)$ times.

Notice that the frequency numbering within a particular level is generated by the following simple rule, applied recursively:

> Number the block at the upper left, proceed to the upper right neighbor, then the lower left neighbor, and finally the lower right neighbor.

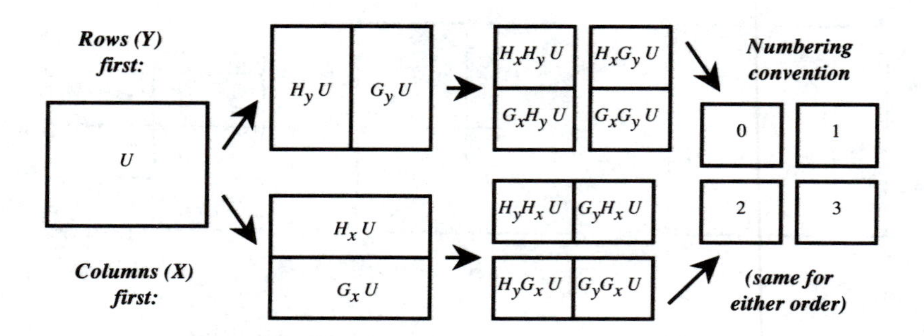

Figure 9.4: Separable two-dimensional conjugate quadrature filtering.

Here a "block" is a collection of 4^s cubes, with $s = 0, 1, 2, \ldots, L$. We start at the top with the original signal and its four children, descending to level L while keeping to the upper left.

The two summations of Equations 9.2 and 9.3 may be performed individually as depicted in Figure 9.4, where we first apply the filters in the y-direction and then in the x-direction. The x and y filterings commute in the separable case, so we may perform the summations in either order. But since arrays are usually presented as lists of rows it is most convenient to start with the y-direction. This influences our convention for numbering the descendents.

In general, for d-dimensional signals, we will use a 2^d member family of filters $F_{\epsilon_1} \otimes \cdots \otimes F_{\epsilon_d}$, where $\epsilon_i \in \{0, 1\}$ for $i = 1, \ldots, d$, $F_0 = H$, and $F_1 = G$. We will label the output of these separable filters with the binary number $\epsilon_1 \epsilon_2 \ldots \epsilon_d$. The wavelet packets produced by such filterings are products of one-dimensional wavelet packets $W_n(\mathbf{x}) = \prod_{i=1}^d W_{n_i}(x_i)$, together with their isotropic dilations and translations to arbitrary lattice points.

Inner products with multidimensional wavelet packets are computed from averages at the smallest scale, just as in the one-dimensional case. The coefficients may be organized into a stack of d-dimensional cubes, and there is a result analogous to the graph theorem (7.9) which characterizes basis subsets.

9.1.2 Tensor products of DTTs and LTTs

A d-dimensional Fourier basis element is a tensor product of d one-dimensional elements: $e^{2\pi i x \cdot \xi} = e^{2\pi i x_1 \xi_1} \times \cdots \times e^{2\pi i x_d \xi_d}$. Thus the Fourier transform, its discrete version, and all of the derived trigonometric transforms like DCT and DST can be computed separately in each dimension. In the bivariate q-periodic sequence case,

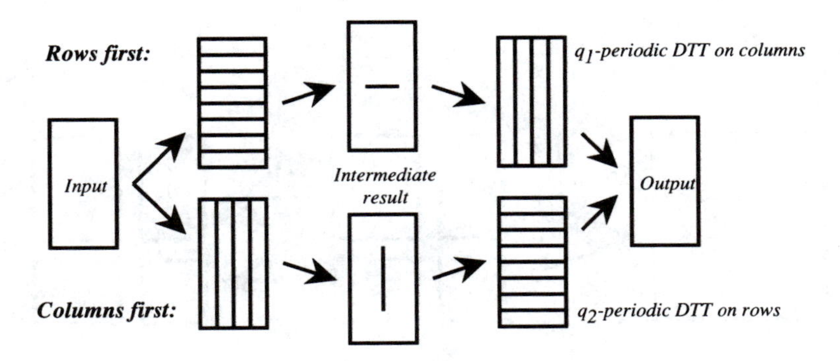

Figure 9.5: Two-dimensional discrete trigonometric transform.

this is equivalent to replacing the double sum with an iterated summation:

$$\sum_{x_1,x_2=0}^{q-1} e^{2\pi i(x\cdot k)/q} u(x_1,x_2) = \sum_{x_1=0}^{q-1} e^{2\pi i x_1 k_1/q} \left(\sum_{x_2=0}^{q-1} e^{2\pi i x_2 k_2/q} u(x_1,x_2) \right). \quad (9.7)$$

If the bivariate sequence is q_1-periodic in the first index and q_2-periodic in the second index, then the Fourier basis element will be $e^{2\pi i x_1 k_1/q_1} e^{2\pi i x_2 k_2/q_2}$ and the sums in Equation 9.7 will contain q_1 and q_2 terms, respectively. If we view one period of such a bivariate sequence as a (height) q_1 by (width) q_2 rectangle of coefficients, then the DFT can be applied in two steps as depicted in Figure 9.5. The same procedure may be used for any of the discrete trigonometric transforms (DTTs) such as DCT-I or DST-IV.

We can perform a two-dimensional DTT on four smaller subblocks of the signal by first cutting the original signal into four pieces and then performing the DTT on each piece. This is depicted in Figure 9.6. That procedure may be applied recursively to obtain a decomposition into smaller and smaller blocks, if we make a copy of each subblock after it is cut out but before we apply the DTT.

To obtain the two-dimensional local trigonometric transform (LTT) on subblocks, we must first "fold" the signal in both the x and y directions and restrict it into subblocks as depicted in Figure 9.7. The shaded regions indicate the coefficients influenced by the folding operator.

Suppose that the signal itself is the result of previous folding and restriction. Then the four subblocks satisfy the proper boundary conditions in both the x and y directions, so that applying the two-dimensional DTT yields local trigonometric

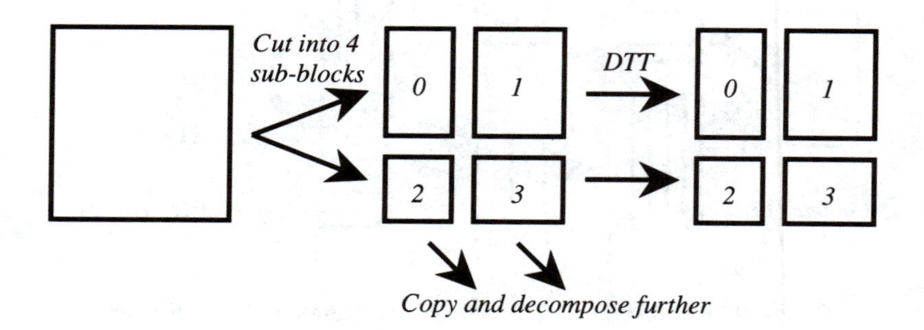

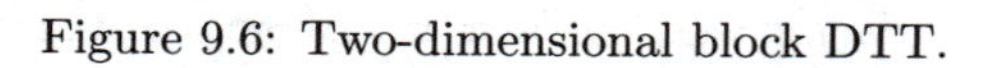
Figure 9.6: Two-dimensional block DTT.

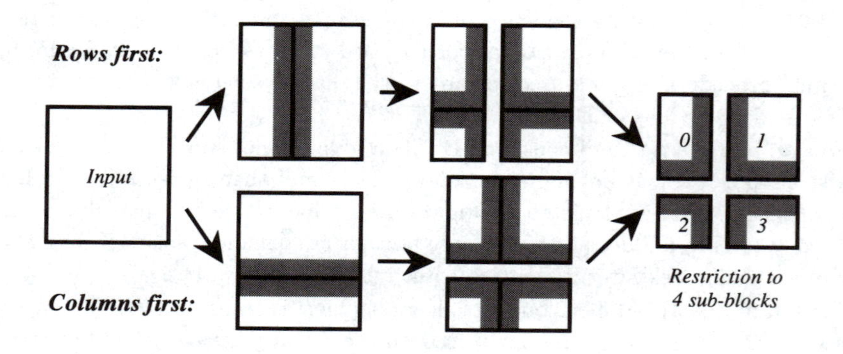

Figure 9.7: Two-dimensional folding in two steps.

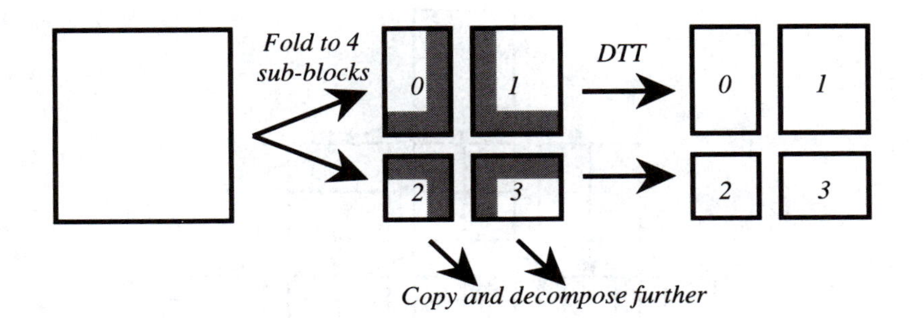

Figure 9.8: Two-dimensional discrete trigonometric transform.

transform coefficients. This combination of folding, restriction, and trigonometric transform is depicted in Figure 9.8.

As in the block DTT case, if we wish to perform a recursive decomposition of the signal into smaller and smaller pieces we must copy the subblocks after folding and restriction but before the DTT.

9.1.3 Complexity of the d-dimensional best basis algorithm

Suppose for simplicity that we start with a d-dimensional periodic signal of size $N = 2^L \times \ldots \times 2^L = 2^{Ld}$. This can be developed down to level L, at which point each subspace will have a single element. Each level requires $O(N)$ operations, where the constant is proportional to the product of d and the length of the conjugate quadrature filters. The total complexity for calculating the wavelet packet coefficients—*i.e.*, correlating with the entire collection of modulated bumps—is $O(N \log N)$.

Let A_L be the number of covers in the tableau from levels 0 to L in the d-dimensional case. Then $A_0 = 1$ and $A_{L+1} = 1 + A_L^{2^d}$, so we have an estimate that $A_{L+1} \geq 2^{2^{Ld}}$. In the periodic case with a signal sampled at $N = 2^{Ld}$ points, this implies that there are more than 2^N bases. The tree contains $1 + 2^d + \ldots + (2^d)^L = (2^{(L+1)d} - 1)/(2^d - 1) = O(N)$ subblocks, each of which is examined at most three times in the search for a best basis. This shows that the extraction of a best basis has complexity $O(N)$ with small constant.

Reconstruction from an arbitrary basis is of the same complexity, since the algorithm is an orthogonal transformation.

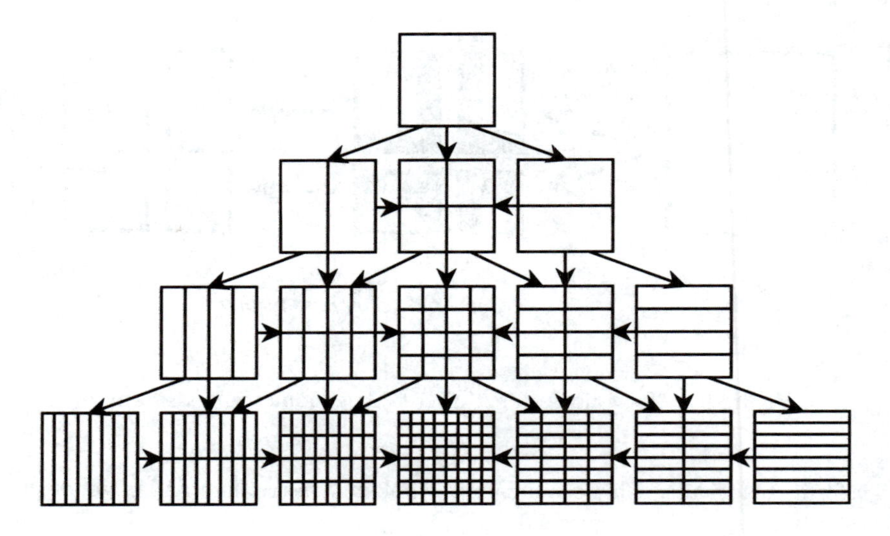

Figure 9.9: All tensor products of wavelet packets.

9.1.4 Anisotropic dilations in multidimensions

We have so far considered only those multidimensional wavelets which have the same scale in each direction. If we consider bases of wavelet packets whose scales are different in different directions, then the number of coefficients and the number of bases increase substantially. Figure 9.9 shows one way to organize the calculation of two-dimensional wavelet packet coefficients. The arrows represent filter convolution-decimations in one of the dimensions, X or Y. The diagram shows that there will be $O(N[\log N]^2)$ coefficients in the two-dimensional case. More generally, it can be proved that there will be $O(N[\log N]^d)$ coefficients in the d-dimensional case.

With anisotropic dilations, wavelet packet coefficients no longer form a tree. Each subspace has d parents, one for each filter, as illustrated in the two-dimensional case by the genealogy in Figure 9.10. The resulting structure can be made into an inhomogeneous tree by including duplicate copies of the parent, one for each grandparent, recursively for each generation. The result is a very large tree even for the 4×4 case, as the diagram in Figure 9.11 illustrates.

Each node within this dependency tree is indexed by the string of convolution-decimation operators that produced it, *i.e.*, $G_m \ldots G_2G_1$, where $m \leq 2L$ and where G_i is one of $F_0(X)$, $F_0(Y)$, $F_1(X)$, or $F_1(Y)$. Order is important, and there can be no more than L convolution-decimations in each direction X or Y. Counting these

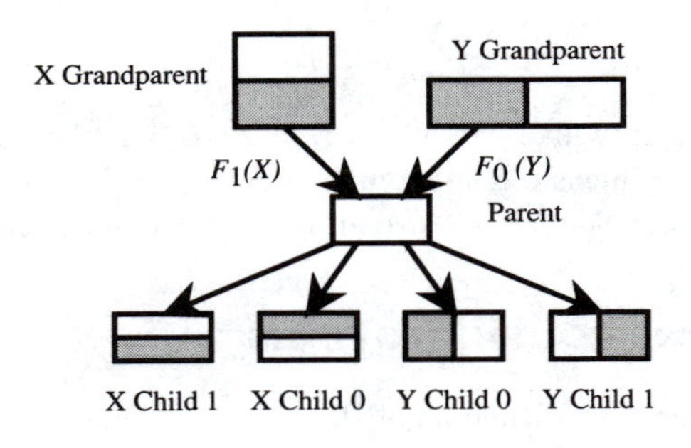

Figure 9.10: Genealogy of tensor products of wavelet packets.

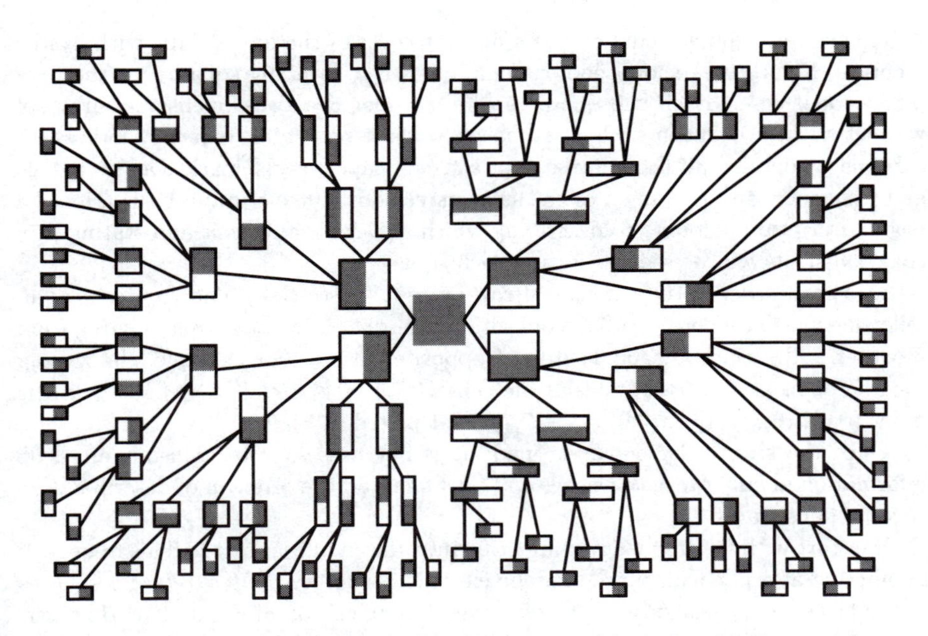

Figure 9.11: The anisotropic tree for a 4×4 signal includes duplicate nodes.

gives

$$\sum_{x=0}^{L}\sum_{y=0}^{L}\frac{(x+y)!}{x!\,y!}2^x 2^y. \tag{9.8}$$

This can be shown to grow like N^2, where $N = 2^L \times 2^L$ is the number of samples in the picture. In higher dimensions the combinatorial growth is even worse, suggesting that tensor product wavelet packets are an impractical generalization.

9.2 Practical considerations

Certain practical issues arise when algorithms are implemented which have little to do with the mathematics of the transformation. We shall consider the spreading of compactly supported wavelets in the nonperiodic case, and the memory requirements of the algorithm.

9.2.1 Labeling the bases

A basis in the d-dimensional case is a disjoint cover of the unit d-cube with dyadic d-cubes. Figure 9.13 shows one example of such a cover by squares, in the two-dimensional case. Each subsquare can be considered a two-dimensional array of wavelet packet coefficients which all have the same scale and frequency indices. A subsquare contains all the translates of a fixed-shape wavelet packet at that scale and frequency, and the union of the subsquares constitutes a graph basis. Hence a basis description amounts to describing which squares are present and stating how the coefficients will be scanned out of each square.

One scan order is the traditional row-by-row, left-to-right, and top-to-bottom, following the Occidental reading convention. Another is *zig-zag* order starting from a corner and tacking diagonally to the opposite corner. For example, the zig-zag scan through an 8×8 array visits the cells $\{(i,j) : 0 \leq i < X, 0 \leq j < Y\}$ in the order (0,0), (0,1), (1,0), (2,0), ..., (7,7), as depicted in Figure 9.12.

Once the scan order choice is made it is not necessary to transmit any side information about translations. We will thus focus on the problem of describing the cover of subsquares.

We start with the frequency numbering introduced in Figure 9.4. This frequency numbering at a particular level may be extended to an *encounter order* for squares in the basis cover, as shown in the two-level decomposition of Figure 9.3. However, if the basis contains subsquares from different depths of decomposition, then the encounter order number of a square is no longer related to its frequency index. Each square will additionally carry a level or scale index. Figure 9.13 shows these three

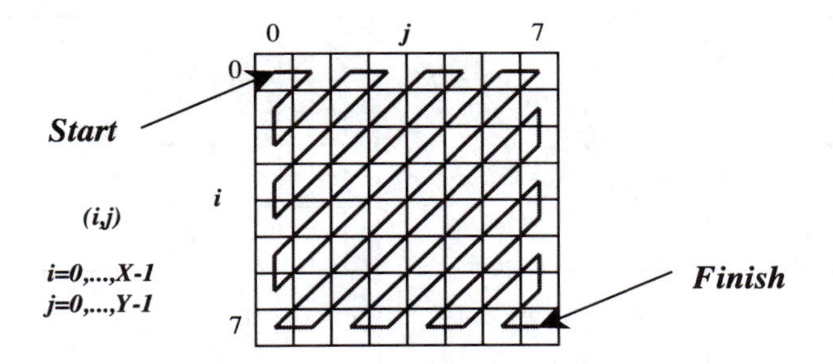

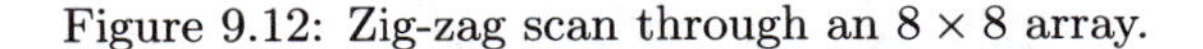
Figure 9.12: Zig-zag scan through an 8×8 array.

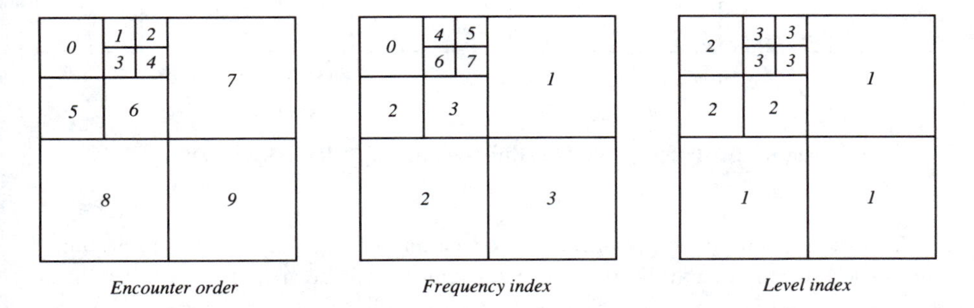

Figure 9.13: Numbering an example two-dimensional wavelet packet basis.

quantities—the encounter order, the frequency index, and the level index—for an example two-dimensional wavelet packet basis. We can choose some combination of these indices to label the basis.

One way to describe a basis is to give the list of ordered pairs (s, f) corresponding to the scale-frequency combinations in the subsquares. Each level s of the wavelet packet decomposition has its own frequency numbering, and the ordered pair (s, f) uniquely defines both the size and the position of an individual square. In this convention, the basis of Figure 9.13 would be described by the list $\{(2,0),(3,4),(3,5),(3,6),(3,7),(2,2),(2,3),(1,1),(1,2),(1,3)\}$, given in any order. We take no advantage of the information contained in the ordering of the list of subsquares, and thus this method does not give the most efficient basis description. Indeed, after an L-level decomposition we may have to transmit as many as 4^L pairs of numbers, with each pair requiring $2L + \log_2 L$ bits, in order to furnish a complete basis description.

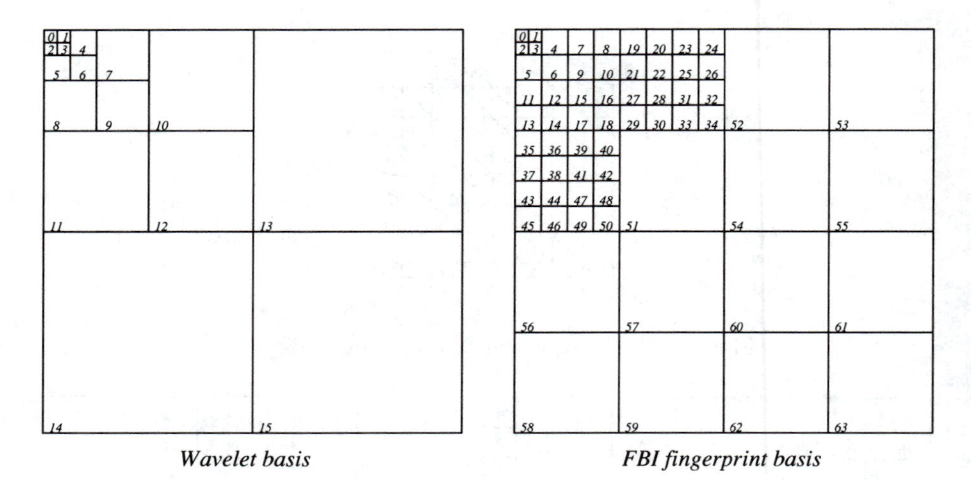

Figure 9.14: Example two-dimensional wavelet packet bases.

To take advantage of the encounter order numbering scheme, we simply agree that descriptive information about the subsquares will be transmitted in the order that the subsquares are encountered. We may choose an encounter order that corresponds to the way the basis is developed, to simplify using the resulting data. The encounter order convention we have adopted corresponds to recursive or depth-first development of the wavelet packet basis with the children traversed in the order upper-left, upper-right, lower-left, lower-right. This choice mimics the scan order within a block and is hallowed by tradition, though any other convention would do as well. To describe a basis, we must simply give the levels of the subsquares in their encounter order; this fixes the level where the depth-first recursion terminates and uniquely defines the squares. To describe the example basis in Figure 9.13, we simply give the ordered list of levels $(2, 3, 3, 3, 3, 2, 2, 1, 1, 1)$. Such a scheme uses at most 4^L numbers of $\log_2 L$ bits each to describe a basis, which is an order of magnitude less that the previous method. We therefore declare it to be the standard method, and call it the *levels list* basis description method.

Figure 9.14 shows the encounter order number for a pair of common and useful wavelet packet bases. On the left is a five-level wavelet basis; it is described by the levels list $(5, 5, 5, 5, 4, 4, 4, 3, 3, 3, 2, 2, 2, 1, 1, 1)$. Like all wavelet bases, it starts with a quadruple of Ls, then contains triples of each successively shallower level up to level one. On the right is the basis chosen by the FBI for its "WSQ" gray-scale fingerprint image compression algorithm [60]. This is a hand-tuned custom decomposition developed after experimentation with the best bases chosen for a training set of fingerprints. Its basis description is a list of four fives followed by 47 fours followed by 13 twos.

Naturally, if the transmitter and receiver agree on a fixed basis, it is not necessary to include that information in the transmission. However, if we will be using many custom bases, we can write a general wavelet packet analysis and synthesis function which takes data plus a basis description as input and produces either the custom decomposition or the reconstruction from such a decomposition. The alternative is to write a new function for each basis.

9.2.2 Saving memory

Memory requirements for d-dimensional signals grow rapidly with d. If the signal can be decimated by two down to level L, then there must be at least 2^{Ld} samples all together. If d is large so that $N = 2^{Ld}$ is enormous even for moderate L, it may be impractical to store all LN wavelet packet coefficients. This problem can be overcome by trading off space for time, namely discarding the computed coefficients as soon as their information cost has been recorded. This requires that the coefficients be calculated recursively in depth-first preorder, and that the memory used by a parent subspace be freed as soon as the information cost of that parent has been compared to the cost of the best basis among its descendents. Figure 9.15 depicts an intermediate stage in such an algorithm in the one-dimensional case.

We can prove that it takes no more than $\frac{2^d}{2^d-1}N$ memory locations (plus a constant overhead depending on d) to find a best basis for an N-sample signal of dimension d. The tradeoff is that each wavelet packet coefficient may have to be computed twice, at worst doubling the computation time.

The reconstruction algorithm is naturally organized to use no more than $2N$ memory locations for a signal of N samples.

9.3 Implementations

To escape the combinatorial explosion which threatens most multidimensional transforms, we rely on one-dimensional algorithms as much as possible. In particular, we will use rectangular arrays and transform along one dimension at a time. We are aided in this by the indexing conventions of typical general purpose computers, which store data contiguously in memory along the direction of the "most rapidly changing index." In Standard C, the last index is always the most rapidly changing.

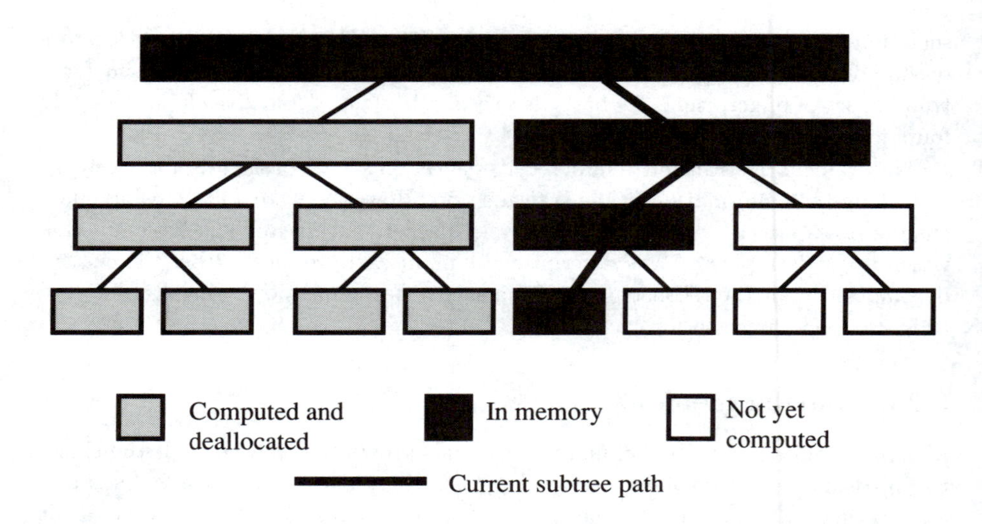

Figure 9.15: Space-saving best basis search.

Some special-purpose computers have extra instructions for quickly performing certain multidimensional transformations such as transposition, or can perform arithmetic on many operands simultaneously. We will not dwell on the special techniques needed to squeeze the best performance out of such machines.

9.3.1 Transposition

We have invested some effort in writing our convolution-decimation functions and our folding functions so that they may be used to perform a transposition merely by writing output sequentially with an increment. We exploit this feature when we have enough memory to store disjoint input and output arrays. The simple problem of implement just transposition while copying to a disjoint array is left as an exercise.

However, if we have only enough memory to store a single copy of the array, then all transforms including transposition must be performed in place. To begin, consider the two-dimensional case. Let the parameters X and Y be the numbers of "rows" and "columns" of an array, respectively. For the easy case $X = Y$, transposition just exchanges pairs of array elements. If $X \neq Y$, we must first determine if each element of the array has already been moved to its proper place:

In place two-dimensional transposition

```
xpi2( DATA, X, Y ):
    For N = 1 to X*Y-2
       If DATA[N] has not been permuted then
          Let TEMP = DATA[N]
          Let TARGET = N
          Let SOURCE = (N*Y) % (X*Y-1)
          While SOURCE > N
             Let DATA[TARGET] = DATA[SOURCE]
             Let TARGET = SOURCE
             Let SOURCE = (SOURCE*Y) % (X*Y-1)
          Let DATA[TARGET] = TEMP
```

`DATA[]` is a one-dimensional array masquerading as a two-dimensional array by way of the indexing formula: we treat `DATA[I*Y+J]` as if it were `DATA[I][J]`.

We will assume that a d-dimensional array is stored in our computer's memory as a one-dimensional sequence of consecutive memory locations, with an indexing formula that converts d-tuples into offsets into the array. In Standard C, the convention is that the last (or d^{th}) index changes fastest as we move along the sequence. For example, if a three-dimensional array A has dimensions $x \times y \times z$ and we index it with the triple $[i][j][k]$, then the offset formula is

$$A[i][j][k] = A[i\,yz + j\,z + k]. \tag{9.9}$$

If we move to the location $A + i_0\,yz + j_0\,z$, then the next z locations will be the elements $A[i_0][j_0][0], \ldots, A[i_0][j_0][z-1]$. This arrangement is suitable for applying one-dimensional transformations along the k axis.

To perform a separable three-dimensional transform using a one-dimensional function, we apply the function along the k-axis of A, then transpose A so that the j-index changes fastest and apply the function again, then transpose A so that the i-index changes fastest and apply the function a third time, then transpose A a final time so that its k-index changes fastest as at the outset. We can use the same transposition function for each step, one which cyclically permutes the three indices $[i][j][k] \to [k][i][j] \to [j][k][i] \to [i][j][k]$, which we shall call a *cyclic transposition.* Two applications of the same cyclic transposition moves us from left to right in Figure 9.16.

To perform the transposition we must replace the element indexed by $[i][j][k]$ with the element indexed by $[k][i][j]$. The two offsets are $n = i\,yz + j\,z + k$ and

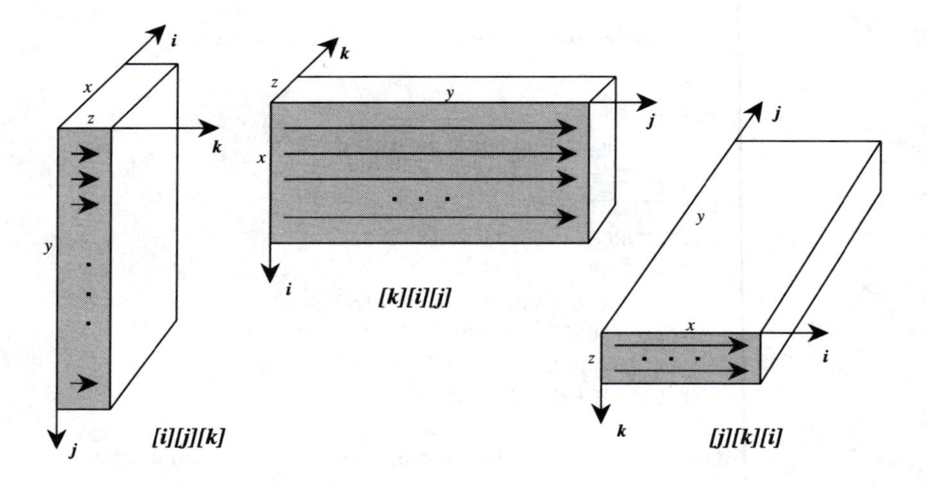

Figure 9.16: Cyclic transpositions.

$n' = k\,xy + i\,y + j$, respectively, and we note that

$$n' = n\,xy \pmod{xyz - 1}; \qquad n = n'\,x \pmod{xyz - 1}. \tag{9.10}$$

We further note that xyz is the *volume* of the three-dimensional array A, and that xy is the *surface area* of the shaded face in the $[k][i][j]$-indexed middle orientation of Figure 9.16. With these notions, this transposition formula generalizes easily to the d-dimensional case. Let $x_1, x_2, \ldots, x_d$ be the dimensions of the array, and let $i_1, i_2, \ldots, i_d$ be the index variables. Suppose that the array A is presented with indexing $A[i_1][i_2]\ldots[i_d]$. We compute the volume $V = x_1 \cdots x_d$ and the area $S = x_1 \cdots x_{d-1} = V/x_d$, and then replace the element at offset n with the one at n', where

$$n' = n\,x_d \pmod{V - 1}; \qquad n = n'\,S \pmod{V - 1}. \tag{9.11}$$

To do this, we can loop over n ranging from 0 to $V-1$, and only perform the cycle of exchanges if n has never before been moved. Afterwards we must cyclically permute the list of lengths so that the indexing algorithm is correct for the new arrangement $A[i_d][i_1]\ldots[i_{d-1}]$. If these lengths were originally presented as the ordered d-tuple $(x_1, x_2, \ldots, x_d)$, the output should be the d-tuple $(x_d, x_1, x_2, \ldots, x_{d-1})$. Such a function requires $O(V)$ operations, and d applications returns us to the original arrangement of A.

Below is an implementation of the d-dimensional cyclic transposition in place:

In place cyclic d-dimensional transposition

```
xpid( DATA, LEN, D ):
   Let VOLUME = LEN[0]
   For K = 1 to D-1
      VOLUME *= LEN[K]
   For N = 1 to VOLUME-2
      If DATA[N] has not been permuted then
         Let TEMP = DATA[N]
         Let TARGET = N
         Let SOURCE = ( N*LEN[D-1] ) % (VOLUME-1)
         While SOURCE > N
             Let DATA[TARGET] = DATA[SOURCE]
             Let TARGET = SOURCE
             Let SOURCE = ( SOURCE*LEN[D-1] ) % (VOLUME-1)
         Let DATA[TARGET] = TEMP
   Let LTEMP = LEN[D-1]
   For K = 1 to D-1
      Let LEN[D-K] = LEN[D-K-1]
   Let LEN[0] = LTEMP
```

9.3.2 Separable convolution-decimation

Our predefined convolution-decimation functions expect the input array to be contiguous, but they can write the output array with any regular increment. We choose this increment so as to transpose the output array, in the two-dimensional case.

We start by applying the low-pass filter H_y along the rows of `IN[]`, writing the output to the scratch array `WORK[]` in such a way that it ends up transposed. We then successively apply the low-pass filter H_x and high-pass filter G_x to `WORK[]` as shown in Figure 9.17. This yields `OUT0[]` and `OUT2[]`. Then we redo the procedure as shown in Figure 9.18 to get `OUT1[]` and `OUT3[]`. The input array is read twice, once when it is low-pass filtered along rows and then again when it is high-pass filtered along the rows. Each time we get a half-width intermediate result, stored in the scratch array `WORK[]`. Hence the scratch array can be half the size of the largest input array. But since we write to the output arrays in between these two readings of `IN[]`, we must assume that the input and output arrays are disjoint.

The result is a separable two-dimensional periodic convolution-decimation of the input array. The input array is a list of rows, *i.e.*, `IN[0]`,`IN[1]`,...,`IN[IY-1]`

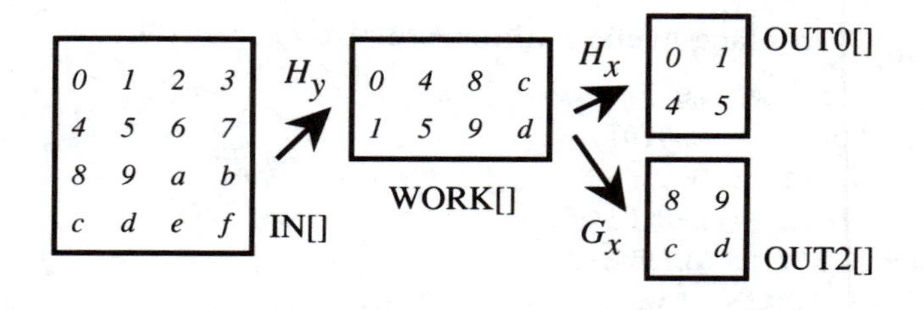

Figure 9.17: First step in two-dimensional separable filtering.

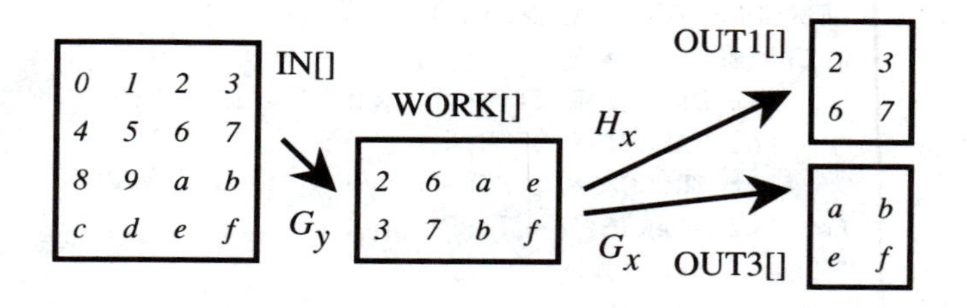

Figure 9.18: Second step in two-dimensional separable filtering.

are the contiguous elements of the first row of `IN[]`. The scratch array `WORK[]` of `IX*OY` elements stores intermediate results, and the output arrays `OUT0[]`, `OUT1[]`, `OUT2[]`, and `OUT3[]` receive the four children of the input in the conventional order 0, 1, 2, 3, described in Figure 9.4.

Disjoint separable two-dimensional periodic convolution-decimation

```
scdpd2( OUT0, OUT1, OUT2, OUT3, IN, IX, IY, WORK, HQF, GQF ):
   Let OY = IY/2
   For I=0 to IX-1
      cdpe( WORK+I, IX, IN+I*IY, IY, HQF )
   For I=0 to OY-1
      cdpo( OUT0+I, OY, WORK+I*IX, IX, HQF )
      cdpo( OUT2+I, OY, WORK+I*IX, IX, GQF )
   For I=0 to IX-1
      cdpe( WORK+I, IX, IN+I*IY, IY, GQF )
   For I=0 to OY-1
      cdpo( OUT1+I, OY, WORK+I*IX, IX, HQF )
      cdpo( OUT3+I, OY, WORK+I*IX, IX, GQF )
```

We pass a pair of conjugate QFs as two data structures HQF and GQF, which are the low-pass and high-pass filters, respectively. The output arrays are written as lists of rows. Note that the number of output rows and columns is determined by the number of input rows IX and columns IY. We assume that these dimensions are both even, though they need not be equal. Note also that the output is superposed rather than assigned to OUTO[], etc., because the convolution-decimation function cdpo() superposes rather than assigns.

By replacing every cdpo() with cdpo(), we obtain a a variant scdpe2() of the disjoint algorithm which assigns values to the output arrays rather than superposing them. Alternatively, we can implement the same transform in place. We then have to allocate twice the space (IX*IY elements) for the WORK[] array in order to process all of the input in Figures 9.17 and 9.18 before writing to the output. This allows us to overwrite the memory locations occupied by DATA[] with the four output arrays:

In place separable two-dimensional periodic convolution-decimation

```
scdpi2( DATA, IX, IY, WORK, HQF, GQF ):
   Let OY = IY/2
   Let N  = OY*IX/2
   Let WORK1 = WORK
   Let WORK2 = WORK + OY*IX
   For I=0 to IX-1
      cdpe( WORK1+I, IX, DATA+I*IY, IY, HQF )
      cdpe( WORK2+I, IX, DATA+I*IY, IY, GQF )
   For I=0 to OY-1
      cdpe( DATA+ I,     OY, WORK1+I*IX, IX, HQF )
      cdpe( DATA+2*N+I, OY, WORK1+I*IX, IX, GQF )
      cdpe( DATA+ N +I, OY, WORK2+I*IX, IX, HQF )
      cdpe( DATA+3*N+I, OY, WORK2+I*IX, IX, GQF )
```

9.3.3 Separable adjoint convolution-decimation

We restrict our attention to the two-dimensional case. The adjoint operation recovers the parent from its four children. We first reconstruct the low-pass filtered columns by applying H_y^* to IN0[] and G_y^* to IN1[], writing the output to the scratch array WORK[] in such a way that it ends up transposed. Then we can apply H_x^* to WORK[] to obtain a partial output. This is shown in Figure 9.19. We then reconstruct the high-pass filtered columns in transposed form by applying H_y^* to IN2[] and G_y^* to IN3[] and using the right increment to write the intermediate

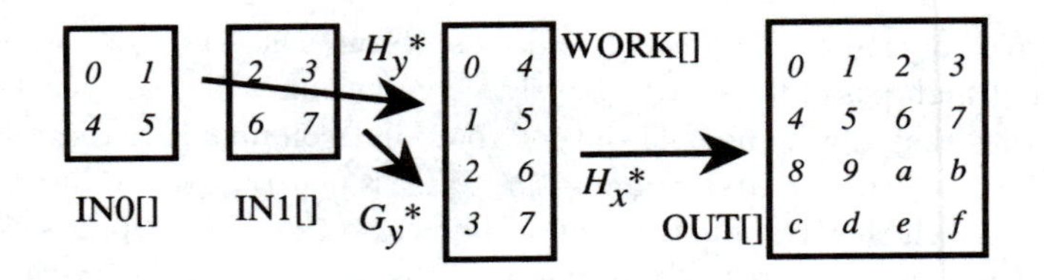

Figure 9.19: First step in two-dimensional separable adjoint filtering.

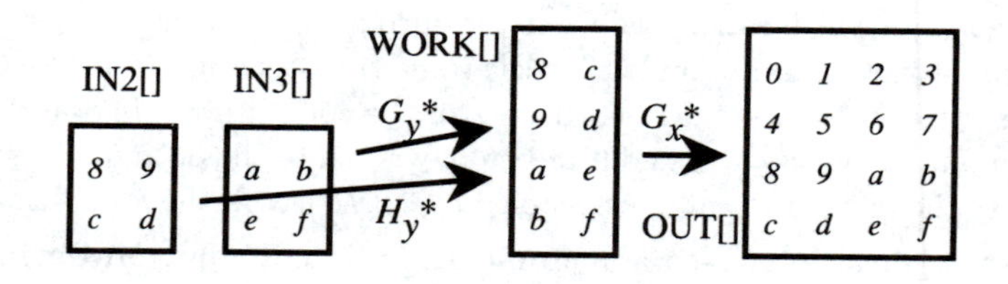

Figure 9.20: Second step in two-dimensional separable adjoint filtering.

result into `WORK[]`. Finally, we apply G_x^* to `WORK[]` to superpose the rest of the output onto `OUT[]`, as shown in Figure 9.20.

The following implements the transformation:

Disjoint separable two-dimensional periodic adjoint convolution-decimation

```
sacdpd2( OUT, IN0, IN1, IN2, IN3, IX, IY, WORK, HQF, GQF ):
   Let OY = 2*IY
   For I=0 to IX-1
      acdpe( WORK+I, IX, IN0+I*IY, IY, HQF )
      acdpo( WORK+I, IX, IN1+I*IY, IY, GQF )
   For I=0 to OY-1
      acdpo( OUT+I, OY, WORK+I*IX, IX, HQF )
   For I=0 to IX-1
      acdpe( WORK+I, IX, IN2+I*IY, IY, HQF )
      acdpo( WORK+I, IX, IN3+I*IY, IY, GQF )
   For I=0 to OY-1
      acdpo( OUT+I, OY, WORK+I*IX, IX, GQF )
```

We assume that the input consists of four arrays of equal size, each having `IX` rows and IY columns, and that they are in the conventional order of Figure 9.4. We further assume that the input arrays are presented as lists of rows, and that they are disjoint with the first 4*IX*IY elements of the output array. Finally, we need a predefined scratch array WORK[] which has enough space to hold two of the input arrays, or 2*IY*IX memory locations. These arrangements are the responsibility of the calling program.

By using acdpe() instead of the first acdpo() to the output array, we obtain a variant sacdpe2() which assigns output values rather than superposing them. But as in the convolution-decimation case, we also can perform the operation "in place" if we allocate 4*IX*IY memory locations for the WORK[] array:

In place separable two-dimensional periodic adjoint convolution-decimation

```
sacdpi2( DATA, IX, IY, WORK, HQF, GQF ):
   Let OY = 2*IY
   Let N = IY*IX
   Let WORK1 = WORK
   Let WORK2 = WORK + OY*IX
   For I=0 to IX-1
      acdpe( WORK1+I, IX, DATA   +  I*IY, IY, HQF )
      acdpo( WORK1+I, IX, DATA+ N +I*IY, IY, GQF )
      acdpe( WORK2+I, IX, DATA+2*N+I*IY, IY, HQF )
      acdpo( WORK2+I, IX, DATA+3*N+I*IY, IY, GQF )
   For I=0 to OY-1
      acdpe( DATA+I, OY, WORK1+I*IX, IX, HQF )
      acdpo( DATA+I, OY, WORK2+I*IX, IX, GQF )
```

Remark. We can test for null input arrays prior to calling acdpo() or acdpe() in the adjoint algorithm, to handle the case of a totally zero child. Thus modified, the algorithm can be used to reconstruct a parent from a single child. Alternatively, we can extract pieces of the adjoint algorithm to superpose the children one by one.

Likewise, we can omit calling cdpo() or cdpe() in the convolution-decimation algorithm if the output array is null; this test can be used to produce a single child from an input array. Or, we can divide the convolution-decimation function into individual pieces to produce the children one by one. But the individual methods take twice as long per output coefficient as treating all four children at once, with no savings in temporary storage.

9.3.4 Separable wavelet packet bases

We now turn to the problem of computing the expansion of a signal into a basis of separable wavelet packets. The tensor product QFs implemented in the preceding section must be called in the proper order to recursively develop the desired coefficients.

If the basis is known in advance, then we will follow a definite path and develop the expansion in place. If we want an expansion in the tensor product of two one-dimensional wavelet packet bases, then we can do one-dimensional transforms in place on all the rows, transpose in place, then do one-dimensional wavelet packet transforms on all the columns and transpose back to the original configuration. The outcome will be the same array containing the various subspaces in segments of rows and columns.

If we seek a two-dimensional graph basis, then the outcome will be a portion of the complete quadtree of wavelet packet subspaces. The coefficients will be conventionally arranged into concatenated, contiguous segments of the output array, one for each node. In the periodic case, the total length of the output will be the same as the length of the input so this transform can be performed in place. In the aperiodic case, we may have more output coefficients than input coefficients, but it is still possible to use the conventional output arrangement. In either case, we can use a HEDGE data structure for the output. We can assign a side array which contains the coefficients and simply put pointers to this array into the contents members of the hedge.

If we seek the best graph basis, then it will at various times be necessary to store more intermediate results than there are input or output values. Thus we will need some scratch storage in addition to the space for the final choice of coefficients. We will implement the two-dimensional version using the memory-saving algorithm of Figure 9.15, as a compromise between speed of computation and economy of memory. This is because the best basis algorithm has practical applications in image compression.

The best basis of wavelet packets

The main ingredient is a recursive function which filters a signal into its immediate descendents, then compares the entropy of the best basis expansions of those descendents against its own. The signal must be retained until the best bases of its four children are computed and written into the output array. Then, if the signal's information cost is lower than that of its children, the signal is written to the output array over the previously written children. We can obtain such a data flow with the disjoint separable convolution-decimation functions `scdpd2()` or `scdpe2()`. We also keep track of the position within the output array and the levels encounter list, so that we can overwrite them if the children turn out to be losers:

Best basis of separable two-dimensional periodic wavelet packets

```
bbwp2( GRAPH, KD, IN, X,Y, S, L, WK, H, G ):
  Let XY = X*Y
  Let COST = infocost( IN, 0, XY-1 )
  If  S<L then
    Let BLOCK = GRAPH.BLOCKS
    For J = 0 to 3
      Let K[J] = KD + J*XY/4
    Let KD = K[3]  + XY/4
    scdpe2( K[0], K[1], K[2], K[3], IN, X,Y, WK, H, G)
    Let KCOST = 0
    For J = 0 to 3
      KCOST += bbwp2(GRAPH,KD, K[J], X/2,Y/2, S+1, L, WK,H,G)
    If KCOST < COST then
      Let COST = KCOST
    Else
      Let GRAPH.LEVELS[BLOCK] = S
      Let OUT = GRAPH.CONTENTS[BLOCK]
      For I = 0 to XY-1
        Let OUT[I] = IN[I]
      Let GRAPH.BLOCKS = BLOCK+1
      Let GRAPH.CONTENTS[GRAPH.BLOCKS] = OUT + XY
  Else
    Let GRAPH.LEVELS[GRAPH.BLOCKS] = L
    Let OUT = GRAPH.CONTENTS[GRAPH.BLOCKS]
    For I = 0 to Y*X-1
      Let OUT[I] = IN[I]
    GRAPH.BLOCKS += 1
    Let GRAPH.CONTENTS[GRAPH.BLOCKS] = OUT + XY
  Return COST
```

The input consists of a preallocated HEDGE data structure to hold the graph, with its first contents member pointing to a preallocated output array of length `X*Y`; a separate output array `KD[]` of length at least $\frac{4}{3}$`X*Y`; the input array `IN[]`; the number of columns `Y` and rows `X`; the current level `S`; the maximum level `L`, a working array WK[] for transposition; and the low and high-pass PQFs.

The contents array of the hedge should be allocated with $2^L + 1$ memory locations, one more than the maximum number of blocks in any L-level graph basis, since it will get one extra pointer to the first location after the last block of coefficients. GRAPH.CONTENTS[0] should be an array with at least X*Y locations, since it will be filled with the best basis coefficients.

We can enhance this code in several ways. The children can be permuted before convolution-decimation so that the frequency indices are in sequency rather than Paley order. This requires passing one additional parameter, the "descendent number" of the current node, which is 0, 1, 2, or 3. This is defined by the convention in Figure 9.4. To get sequency ordering, we replace array K[I] with K[J] in the call to scdpd2(), where I = J^DNO is the bitwise exclusive-or of the child's descendent number and the parent's.

The arrays WK[] and KD[] can share space, since the amount of room needed by scdpd2() for transposition shrinks as we descend the tree and use up space for development. The total number of memory locations needed for WK[]+KD[] is 3*X*Y/2.

The function bbwp2() can be called from a short "main" program that allocates the hedge, the scratch arrays and assigns global variables, so the parameter lists in the recursive calls are not so long. It can also clean up after the transform by deallocating the scratch arrays and shortening the hedge arrays to their proper lengths.

Custom wavelet packet bases

If the basis to be developed is known, then it is possible to compute the expansion of a signal "in place" using scdpi2():

Separable two-dimensional custom basis of periodic wavelet packets

```
cbwp2( GRAPH, LEVEL, IX, IY, WORK, H, G ):
  If LEVEL < GRAPH.LEVELS[GRAPH.BLOCKS]  then
    scdpi2( GRAPH.CONTENTS[GRAPH.BLOCKS], IX,IY, WORK, H, G )
    For K = 0 to 3
       cbwp2( GRAPH, LEVEL+1, IX/2, IY/2, WORK, H, G )
  Else
    Let GRAPH.CONTENTS[GRAPH.BLOCKS+1] =
                       GRAPH.CONTENTS[GRAPH.BLOCKS] + IX*IY
    GRAPH.BLOCKS += 1
  Return
```

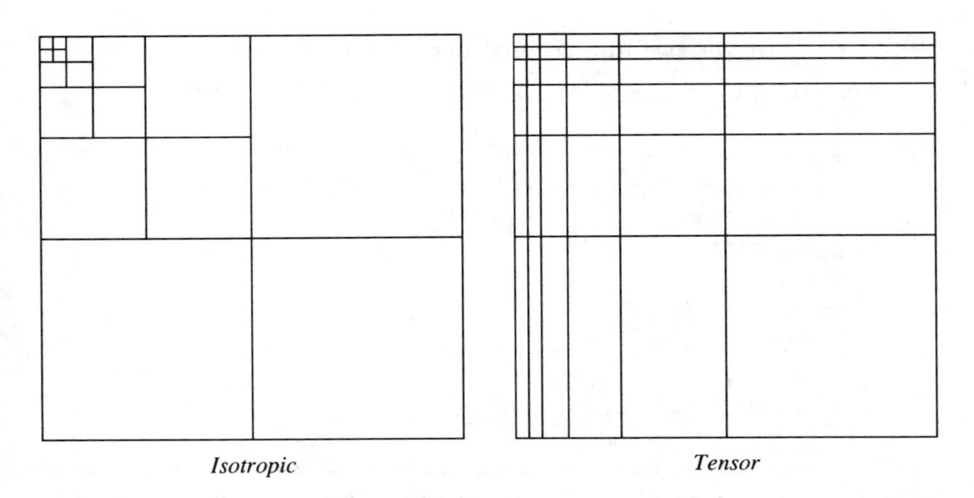

Figure 9.21: Two kinds of separable two-dimensional wavelet bases.

The array targeted by the hedge contents serves as both the input and the output for this transform; it must be preallocated with `IX*IY` elements. In computing these memory requirements, we depend heavily on the conservation of the number of coefficients in the periodic wavelet packet case.

The wavelet basis is a special case. We can use `cbwp2()` with a levels list of four Ls followed by three $L-1$s, three $L-2$s, and so on up to three ones, where L is the maximum level of the decomposition. This produces the isotropic two-dimensional wavelet basis expansion of the input signal, as in the left part of Figure 9.21.

Tensor products of wavelet bases

Alternatively, we can compute the wavelet basis which is a tensor product of two one-dimensional wavelet bases, whose decomposition is represented by the right part of Figure 9.21. Notice that in that case it is not necessary to use the same number of levels in the X and Y bases.

The tensor wavelet basis does not correspond to a quadtree, so we will compute and store it in a different manner. Namely, we will perform a one-dimensional periodized discrete wavelet transform to level `YLEVEL` on each row, transpose in place, then perform a one-dimensional periodized discrete wavelet transform to level `XLEVEL` on each row which is now really a column of the original array. The pseudocode below implements this transformation:

Tensor two-dimensional periodic wavelet basis

```
dwtpt2(DATA, IX,IY, XLEVEL,YLEVEL, HX,GX, HY,GY, WORK):
   Let I = 0
   While I < IY*IX
      dwtpi( DATA+I, WORK, IY, YLEVEL, HY, GY )
      I += IY
   xpi2( DATA, IX, IY )
   Let I = 0
   While I <  IY*IX
      dwtpi( DATA+I, WORK, IX, XLEVEL, HX, GX )
      I += IX
   xpi2( DATA, IY, IX )
```

The resulting transformation leaves the data in the same one-dimensional array, but with a different interpretation for the coefficients. A simple generalization works for the d-dimensional case. We assume that the data is a d-dimensional rectangle of sides LENGTHS[0] $\times \cdots \times$ LENGTHS[D $-$ 1]. So as not to obscure the main idea, we use the same filters HQF, GQF in each dimension:

Tensor d-dimensional periodic wavelet basis

```
dwtptd( DATA, D, LENGTHS, LEVELS, HQF,GQF, WORK ):
  Let VOLUME = LENGTHS[0]
  For K = 1 to D-1
    VOLUME *= LENGTHS[K]
  For K = 1 to D
    Let I = 0
    While I < VOLUME
      dwtpi( DATA+I, WORK, LENGTHS[D-1], LEVELS[D-K], HQF,GQF )
      I += LENGTHS[D-1]
    xpid( DATA, LENGTHS, D )
```

9.3.5 Separable folding and unfolding

We implement separable folding and unfolding very much like separable convolution-decimation. The main difference is that we intend to perform an additional step, the discrete cosine transform, after which the data will be unsuitable for further decomposition.

We will examine only two-dimensional folding; separable two-dimensional unfolding and higher dimensional versions are left as exercises. The parameters x_0, x_1, y_0, and y_1 give the number of rows and columns in the four subblocks, while r_x and r_y are the rising cutoff functions to use in the X and Y directions:

Disjoint separable two-dimensional cosine folding

```
fdc2(OUT0,OUT1,OUT2,OUT3, IN, X0,X1,Y0,Y1, WORK, RX,RY)
  Let IX = X0 + X1
  Let IY = Y0 + Y1
  Let IPTR = IN + Y0
  Let WPTR = WORK + Y0*IX
  For I = 0 to IX-1
    fdcn( WPTR+I, IX, IPTR, IPTR, Y0, RY )
    IPTR += IY
  Let WPTR = WORK + X0
  OUT0 += Y0*X0
  For I = 0 to Y0-1
    fdcn( OUT0+I, Y0, WPTR, WPTR, X0, RX )
    fdcp( OUT2+I, Y0, WPTR, WPTR, X1, RX )
    WPTR += IX
  Let IPTR = IN + Y0
  Let WPTR = WORK
  For I = 0 to IX-1
    fdcp( WPTR+I, IX, IPTR, IPTR, Y1, RY )
    IPTR += IY
  Let WPTR = WORK + X0
  OUT1 += X0*Y1
  For I=0 to Y1-1
    fdcn( OUT1+I, Y1, WPTR, WPTR, X0, RX )
    fdcp( OUT3+I, Y1, WPTR, WPTR, X1, RX )
    WPTR += IX
```

9.4 Exercises

1. Implement the d-dimensional tensor product wavelet transform so as to allow rectangular data sets of varying side-length, different filters, and different levels of decomposition in each dimension. Call it `dwtpvd()`.

2. Implement `dwtpi2()`, the two-dimensional isotropic DWTP, by transposing

in place.

3. Describe how to modify `dwtpt2()`, `dwtptd()`, `dwtpvd()`, and `dwtpi2()` to obtain their inverses `idwtpt2()`, `idwtptd()`, `idwtpvd()`, and `idwtpi2()`.

4. Implement `xpd2()`, two-dimensional transposition and copying from an input array to a disjoint output array.

5. Modify `bbwp2()` to produce a d-dimensional best basis of wavelet packets.

6. Modify `cbwp2()` to produce a d-dimensional custom wavelet packet basis.

7. Implement `udc2()`, two-dimensional separable cosine-polarity unfolding.

8. Describe how to modify `fdc2()` and `udc2()` to obtain the sine-polarity versions `fds2()` and `uds2()`. How could these be modified to use different polarities along different dimensions?

9. Implement d-dimensional separable folding.

10. Implement d-dimensional separable unfolding.

Chapter 10

Time-Frequency Analysis

We now turn to the problem of decomposing one-dimensional signals so as to illuminate two important properties: localization in time of transient phenomena, and presence of specific frequencies. Our starting point will be Heisenberg's inequality, which limits how precisely we can compute these properties.

The main tool will be expansion in orthonormal bases whose elements have good time-frequency localization. Features in this context are just the basis elements which contribute large components to the expansion; they are detectable from their size. Alternatively, we can look for combinations of large components, or of not-so-large components that share similar time or frequency location. The localization of the basis elements does most of our work for us; when we find a large component, we can mark the time-frequency location of its basis element to build a time-frequency picture of the analyzed signal.

10.1 The time-frequency plane

The *time-frequency plane* is a two-dimensional space useful for idealizing two measurable quantities associated to transient signals. A signal may be represented in this plane in a number of ways.

10.1.1 Waveforms and time-frequency atoms

The game is to choose the best representation for a given signal, and our strategy will be to decomposing a signal into pieces called *time-frequency atoms*, then drawing idealized representations of these atoms in the plane.

Suppose that ψ is a modulated waveform of finite total energy, and suppose that both the position and momentum uncertainties of ψ are finite:

$$\triangle x(\psi) < \infty; \qquad \triangle\xi(\psi) < \infty. \tag{10.1}$$

These quantities are defined in Equations 1.57 and 1.58. Finite $\triangle x$ requires that on average $\psi(x)$ decays faster than $|x|^{-3/2}$ as $|x| \to \infty$. Finite $\triangle\xi$ requires that ψ is smooth, in the sense that ψ' must also have finite energy. Note that every function ψ belonging to the Schwartz class S satisfies Equation 10.1.

If ψ gives the instantaneous value of a time-varying signal, then it is reasonable to speak of *time* and *frequency* rather than position and momentum, especially since both pairs of quantities are related by the Fourier transform. We will say then that ψ is *well localized in both time and frequency* if the product of its time and frequency uncertainties is small. A musical note is an example of a time-frequency atom. It may be assigned two parameters, duration and pitch, which correspond to time uncertainty and frequency. A third parameter, location in time, can be computed from the location of the note in the score, since traditional music is laid out on a grid of discrete times and frequencies. We may name these three parameters *scale*, *frequency*, and *position*, to abstract them somewhat from the musical analogy.

Heisenberg's inequality (Equations 1.56 and 1.59) imposes a lower bound on the *Heisenberg product*: $\triangle x \, \triangle\xi \geq \frac{1}{4\pi} \approx 0.08$. We need not be too precise about what we mean by "small" in this context; it is enough to have a Heisenberg product of about one. We will call such functions *time-frequency atoms.* Not every Schwartz function is a time-frequency atom, but each one may be written as a linear combination of "unit" time-frequency atoms using rapidly decreasing coefficients:

Theorem 10.1 *For each $\psi \in \mathcal{S}$, there is a sequence $\{\phi_n : n = 1, 2, \ldots\} \subset \mathcal{S}$ of time-frequency atoms and a sequence of numbers $\{c_n : n = 1, 2, \ldots\}$ such that:*

1. *$\psi(t) = \sum_{n=1}^{\infty} c_n \phi_n(t)$, with uniform convergence;*
2. *$\|\phi_n\| = 1$ for all $n \geq 1$;*
3. *$\triangle x(\phi_n) \, \triangle\xi(\phi_n) < 1$ for all $n \geq 1$;*
4. *For each $d > 0$ there is a constant M_d such that $|c_n| n^d \leq M_d < \infty$ for all $n \geq 1$.* □

This theorem is proved by constructing the Littlewood-Paley decomposition of the given function ψ. Let us call a function ψ a *time-frequency molecule* if it satisfies the four conditions of Theorem 10.1. Notice that the last condition implies $\{c_n\}$ is absolutely summable.

The theorem states that all functions in S are time-frequency molecules. It also states that time-frequency atoms are dense in the Schwartz class. Since the Schwartz class in turn is dense in many other function spaces, we see that less regular functions can be decomposed into time-frequency atoms, though in general the coefficients $\{c_n\}$ will not decay rapidly. We can now place into context the surprising discovery by Yves Meyer [78] that a single sequence of orthonormal time-frequency atoms works for all Schwartz functions, and thus for many useful function spaces:

Theorem 10.2 *There is a sequence $\{\phi_n : n = 1, 2, \ldots\} \subset \mathcal{S}$ of time-frequency atoms with the following properties:*

1. $\|\phi_n\| = 1$ *for all* $n \geq 1$;
2. *If* $m \neq n$, *then* $\langle \phi_m, \phi_n \rangle = 0$;
3. $\triangle x(\phi_n)\ \triangle \xi(\phi_n) < 1$ *for all* $n \geq 1$.
4. *The set* $\{\phi_n : n = 1, 2, \ldots\}$ *is dense in S.*

Also, for each $\psi \in \mathcal{S}$ there is a sequence of numbers $\{c_n : n = 1, 2, \ldots\}$ such that $\psi(t) = \sum_{n=1}^{\infty} c_n \phi_n(t)$ with uniform convergence, and for each $d > 0$ there is a constant M_d such that $|c_n| n^d \leq M_d < \infty$ for all $n \geq 1$. □

Meyer's theorem permits characterizing function spaces solely in terms of the rate of decay of positive sequences [50, 79, 80], and vastly simplifies calculating properties of operators such as continuity.

In an orthogonal *adapted waveform analysis*, the user is provided with a collection of standard libraries of waveforms—called *wavelets*, *wavelet packets*, and *windowed trigonometric waveforms*—which can be combined to fit specific classes of signals. All these functions are time-frequency atoms. In addition, it is sometimes useful to consider orthogonal libraries of functions which have large or unbounded Heisenberg product, such as *Haar–Walsh functions*, *block sines* and *block cosines*.

Examples of such waveforms are displayed in Figure 10.1.

Nonorthogonal examples of time-frequency atoms are easy to construct by modifying smooth bump functions. Suppose ϕ has finite Heisenberg product, e.g., take ϕ to be $O(t^{-2})$ as $|t| \to \infty$ and suppose ϕ' is continuous and $O(t^{-1})$ as $|t| \to \infty$. Then ϕ might not be in the Schwartz class S, but it will be good enough for many practical applications. We define the *dilation*, *modulation*, and *translation* operators on ϕ by $\sigma^s\phi \stackrel{\text{def}}{=} \sigma_2^s\phi(t) = 2^{-s/2}\phi(2^{-s}t)$, $\mu_f\phi(t) = e^{2\pi i f t}\phi(t)$, and $\tau_p\phi(t) = \phi(t-p)$, respectively. If ϕ has small Heisenberg product, then the collection of dilated, modulated and translated ϕ's are also time-frequency atoms since the transformations

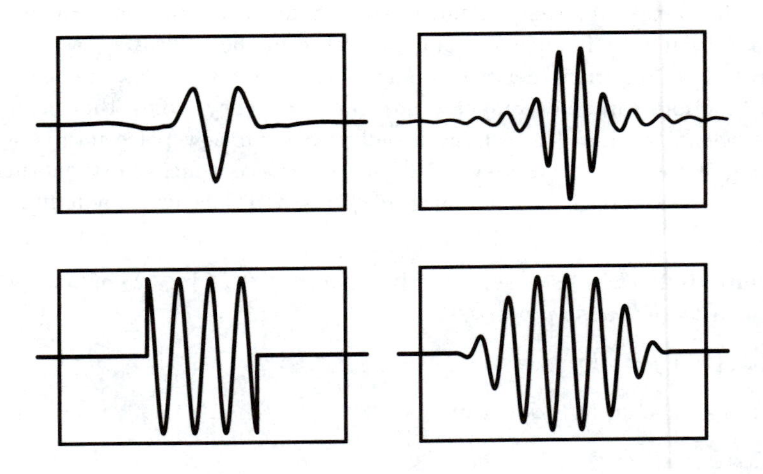

Figure 10.1: Example waveforms: wavelet, wavelet packet, block cosine, and local cosine functions.

σ, μ, τ conserve the Heisenberg product. They also conserve the energy of ϕ, so we may assume that the waveforms are all unit vectors in L^2, *i.e.*, they all have unit energy. If $\triangle x(\phi) = 1$, $\xi_0(\phi) = 0$ and $x_0(\phi) = 0$, then applying $\sigma_2^s, \mu_f, \tau_p$ moves these parameters to 2^s, f, and p, respectively.

In our analyses, we will say that the component of a function u at s, f, p is the inner product of u with the modulated waveform whose parameters are s, f, p. If the component is large, we may conclude that u has considerable energy at scale s near frequency f and position p.

We need to modify our notion of these parameters when we use real-valued time-frequency atoms, because the uncertainty results will be misleading. Real-valued function have Hermitean-symmetric Fourier transforms: if $u(x) = \bar{u}(x)$, then $\hat{u}(-\xi) = \overline{\hat{u}(\xi)}$. Such a function must have $\xi_0(u) = 0$ no matter how much it oscillates, just because $\xi\,|\hat{u}(\xi)|^2$ is an odd function. Another notion of frequency is needed in that case. For example, since $\int_0^\infty |\hat{u}|^2 = \int_{-\infty}^0 |u|^2 = \|u\|^2/2$, we could restrict our attention to the "positive" frequencies and use

$$\xi_0^+ \stackrel{\text{def}}{=} \frac{\left(2\int_0^\infty \xi|\hat{u}(\xi)|^2\,d\xi\right)^{1/2}}{\left(\int_0^\infty |\hat{u}(\xi)|^2\,d\xi\right)^{1/2}}. \tag{10.2}$$

This is equivalent to projecting the function u onto the Hardy space H^2 prior to calculating its power spectrum's center. The orthogonal projection $P: L^2 \mapsto H^2$ is defined by $\widehat{Pu}(\xi) = \mathbf{1}_{\mathbf{R}^+}(\xi)\, u(\xi)$, and the function Pu is called the *analytic signal* associated to the signal u.

If u is real-valued, then the frequency uncertainty $\Delta\xi(Pu)$, computed with $\xi_0^+ = \xi_0(Pu)$, is never larger than $\Delta\xi(u)$ computed with $\xi_0 = \xi_0(u)$. Unfortunately, P destroys decay, so that even a compactly supported u might have $\Delta x(Pu) = \infty$. The hypothesis that $\Delta x(Pu) \leq \Delta x(u) < \infty$ implies, by the Cauchy–Schwarz inequality, that both $u \in L^1$ and $Pu \in L^1$. Now if $Pu \in L^1$, then $\widehat{Pu}(\xi)$ must be continuous at $\xi = 0$, by the Riemann–Lebesgue lemma. This requires that $\hat{u}(0) = 0$.

10.1.2 The idealized time-frequency plane

We now consider an abstract two-dimensional signal representation in which time and frequency are indicated along the the horizontal and vertical axes, respectively. A waveform is represented by a rectangle in this plane with its sides parallel to the time and frequency axes, as seen in Figure 10.2. Let us call such a rectangle an *information cell.* The time and frequency of a cell can be read, for example, from the coordinates of its lower left corner. The uncertainty in time and the uncertainty in frequency are given by the width and height of the rectangle, respectively. Since the time and frequency positions are uncertain by the respective dimensions of the cell, it does not matter whether the nominal frequency and time position is taken from the center or from a corner of the rectangle. The product of the uncertainties is the area of the cell; it cannot be made smaller than the lower bound $1/4\pi$ given by Heisenberg's inequality.

Three waveforms with information cells of approximately minimal area are drawn schematically in the signal plot at the bottom of Figure 10.2. The two at the left have small time uncertainty but big frequency uncertainty, with low and high modulation, respectively. Since they are evidently orthogonal, we have chosen to draw their information cells as disjoint rectangles. The wider waveform at the right has smaller frequency uncertainty, so its information cell is not so tall as the ones for the narrower waveforms. It also contains more energy, so its cell is darker than the preceding two. Notice that each information cell sits above its (circled) portion of the signal in this idealization.

The amplitude of a waveform can be encoded by darkening the rectangle in proportion to its waveform's energy. The idealized time-frequency plane closely resembles a musical score, and the information cells play the role of notes. However, musical notation does not indicate the pitch uncertainty by the shape of a note; for a particular instrument, this is determined by the duration of the note and the timbre of the instrument. Likewise, musical notation uses other means besides darkening

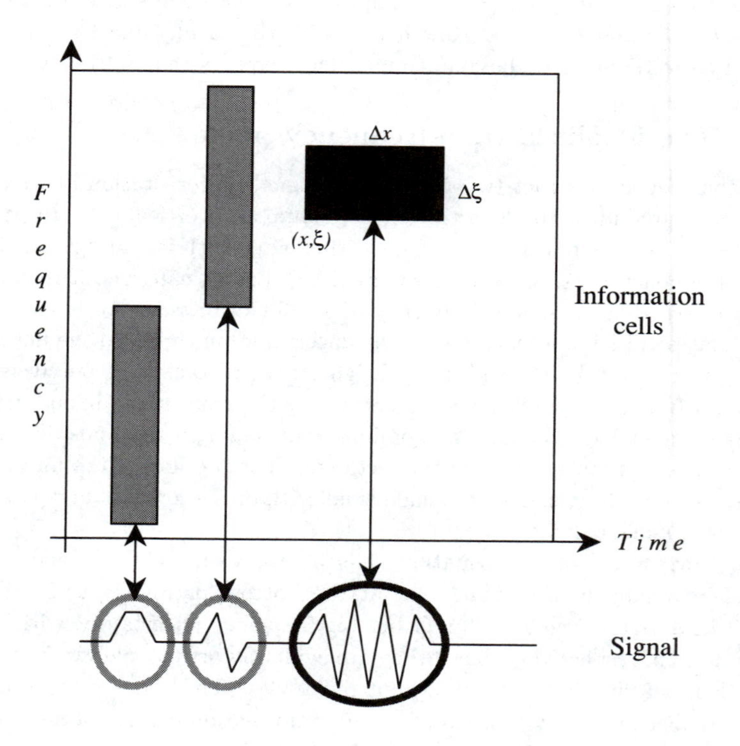

Figure 10.2: Information cells in the time-frequency plane.

the notes to indicate amplitude.

Heisenberg's uncertainty principle for continuous waveforms implies that the area of an information cell can never be less than $1/4\pi$. Only the *Gaussian function* $g(t) = e^{-\pi t^2/2}$, suitably dilated, modulated, and translated, has the minimal information cell area. The other atoms are not too far off, though, and we will avoid the many restrictions of the Gaussian by relaxing the minimality condition. The only price we will have to pay is that a single atom might in practice require a few of the approximate atoms.

We now discuss in detail how to perform the approximate time-frequency analysis. If we have a signal of only finitely many points, then we can construct a finite version of the time-frequency plane. We will treat the signal sequence $\{a_N(k) : k = 0, 1, \ldots, N-1\}$ to be the coefficients of the function with respect to a particular N-member synthesis family Φ_N of time-frequency atoms:

$$f_N(t) = \sum_{k=0}^{N-1} a_N(k)\phi_{N,k}(t). \tag{10.3}$$

For any such finite signal approximation, the information cells will be confined to a finite region of the time-frequency plane, and their corners will lie on a discrete set of points determined by the sampling interval. If the signal is uniformly sampled at N points and we take the unit of length to be one sampling interval, then the width of the visible and relevant portion of the time-frequency plane is N. If $f \in L^2([0,1])$ and we use an N-dimensional approximation spanned by N time-frequency atoms which are translates of a single time-frequency atom ϕ supported in $[0,1]$, then Equation 10.3 specializes to the following:

$$f_N(t) = \sum_{k=0}^{N-1} a_N(k)\phi(Nt-k). \tag{10.4}$$

The signal may then be represented by adjacent information cells lined up at the grid points $\{\frac{k}{N} : 0 \leq k < N\}$, with equal areas and with the kth cell shaded to indicate its amplitude $a(k)$. The cells will be disjoint if the function ϕ is orthogonal to its translates by integers, *i.e.*, if $k \neq j \Rightarrow \int \phi(Nt-k)\phi(Nt-j)\,dt = 0$.

The Fourier exponential functions $1, e^{2\pi i x/N}, \ldots, e^{2\pi i(N-1)x/N}$ form an orthogonal basis for all such N-sampled functions. If our basic oscillating function is written as $e^{2\pi i \frac{fx}{N}}$, this means that the frequency index f ranges over the values $0, 1, \ldots, N-1$, so there are N discrete values for the frequency index and we may introduce N equally spaced points on the frequency axis to account for these. Thus the smallest region that contains all possible cells for a signal of length N must be

N time units wide by N frequency units tall, for a total area of N^2 time-frequency units.

If N is even, we may use the equivalent numbering $-\frac{N}{2}, \ldots, -1, 0, 1, \ldots, \frac{N}{2} - 1$ for the frequency indices. Notice that, since we cannot distinguish the exponential of frequency f from the one at $N - f$ just by counting oscillations, there are really only $N/2$ distinguishable frequencies in the list. This, suitably rigorized, is called the *Nyquist theorem*; the maximum distinguishable or *Nyquist frequency* for this sampling rate is $N/2$ oscillations in N units, or $1/2$.

10.1.3 Bases and tilings

A family of time-frequency atoms with uniformly bounded Heisenberg product may be represented by information cells of approximately equal area. A basis of such atoms corresponds to a cover of the plane by rectangles; an orthonormal basis may be depicted as a cover by disjoint rectangles. Certain bases have characterizations in terms of the shapes of the information cells present in their cover of the time-frequency plane. For example, the *standard basis* or *Dirac basis* consists of the cover by the tallest, thinnest patches allowed by the sampling interval and the underlying synthesis functions. The Dirac basis has optimal time localization and no frequency localization, while the Fourier basis has optimal frequency localization, but no time localization. These two bases are depicted in Figure 10.3.

The Fourier transform may be regarded as a rotation by 90° of the standard basis, and as a result the information cells are transposed by interchanging time and frequency. We may note that it is also possible to apply an element of the *Hermite group* [48] (also called the *angular Fourier transform*) to obtain information cells which make arbitrary angles with the time and frequency axes. This transform is a pseudodifferential operator with origins in quantum mechanics; it is formally represented by $A_t \stackrel{\text{def}}{=} \exp(-itH)$ where t is the angle from the horizontal which we wish to make with our rotated atoms, and H is the selfadjoint Hamiltonian operator obtained by quantization of the harmonic oscillator equation:

$$H = \frac{1}{2}\left(\frac{d^2}{dx^2} + x^2\right). \tag{10.5}$$

Then A_t is the evolution operator which produces the wave function $u(x,t) = A_t u(x,0)$ from an initial state $u(x,0)$, assuming that the function u evolves to satisfy the Schrödinger equation

$$\frac{du}{dt} + Hu = 0. \tag{10.6}$$

Figure 10.3: Dirac and Fourier bases tile the time-frequency plane.

We bring all this up mainly to apprise the reader that many ideas used in the time-frequency analysis of signals have their roots in quantum mechanics and have been studied by physicists and mathematicians for several generations.

Windowed Fourier or trigonometric transforms with a fixed window size correspond to covers with congruent information cells whose width $\triangle x$ is proportional to the window width. The ratio of frequency uncertainty to time uncertainty is the aspect ratio of the information cells, as seen in Figure 10.4

The wavelet basis is an octave-band decomposition of the time-frequency plane, depicted by the covering on the left in Figure 10.5. A wavelet packet basis gives a more general covering; the one on the right in Figure 10.5 is appropriate for a signal containing two almost pure tones near 1/3 and 3/4 of the Nyquist frequencies, respectively. Tilings which come from graph basis in library trees built through convolution and decimation must always contain complete rows of cells, since they first partition the vertical (frequency) axis and then fill in all the horizontal (time) positions.

An adapted local trigonometric transform tiles the plane like the left part of Figure 10.6. Such bases are transposes of the wavelet packet bases, since wavelet packets are related to local trigonometric functions by the Fourier transform. The tilings corresponding to graphs in a local trigonometric library tree must contain complete columns, since these first segment the horizontal (time) axis and then represent all the vertical positions (or frequencies) within each segment.

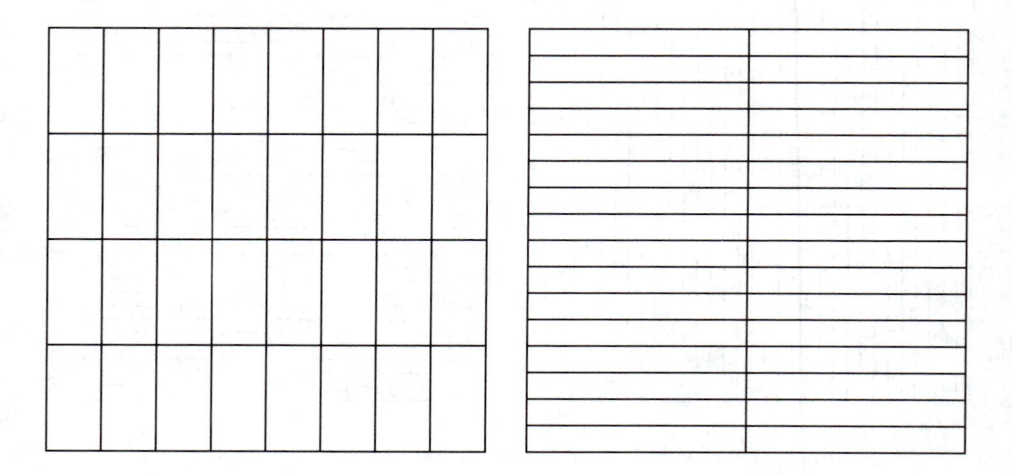

Figure 10.4: Windowed Fourier bases tile the time-frequency plane.

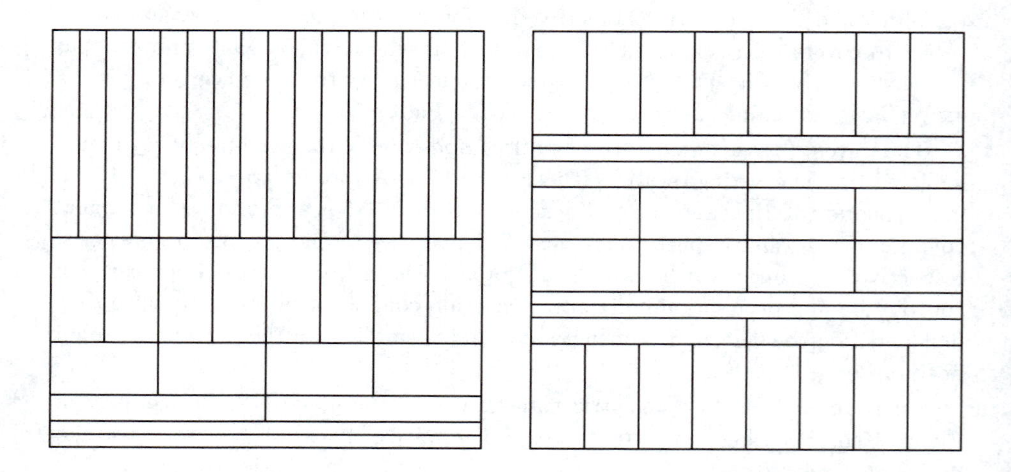

Figure 10.5: Wavelet and wavelet packet bases tile the time-frequency plane.

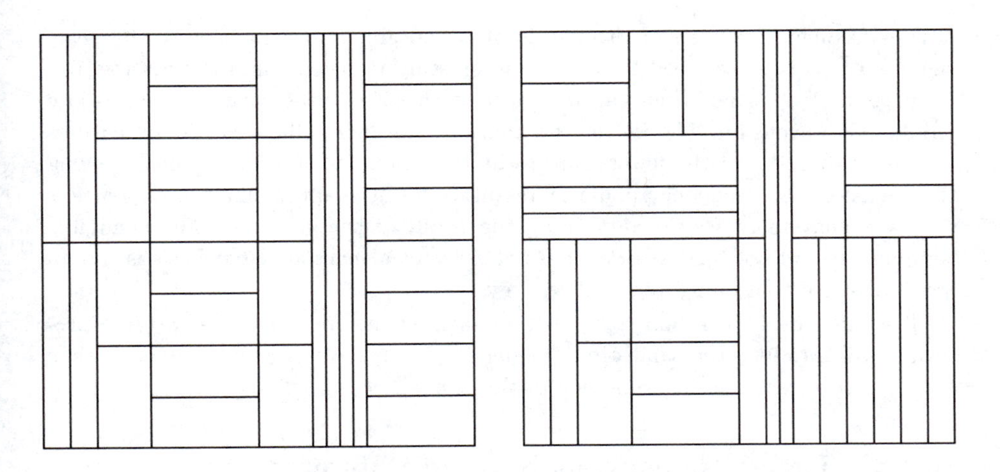

Figure 10.6: Adapted local trigonometric tiling and a general dyadic tiling.

The right part of Figure 10.6 shows a more general tiling of the time-frequency plane, one which does not correspond to a graph in either a wavelet packet or local trigonometric library tree. Such a decomposition is achievable using the Haar quadrature filters and picking additional basis subsets besides graphs, or else by combining the local trigonometric and wavelet packet bases on a segmented signal. The number of such tilings is an order of magnitude greater than the number of graph wavelet packet bases.

10.1.4 Analysis and compression

We can interpret choosing an entropy-minimizing basis from a library of time-frequency atom bases in two ways. First, this fits a cover to the signal so as to minimize the number of dark information cells. The *compression* of the sampled signal achieved by such a fitted decomposition and a quantization threshold below which information cells are invisibly light is the ratio of the total area of the time-frequency plane ($N \times N$ for a signal sampled at N points) divided by the total area of the dark visible information cells (each of area N). Different constraints on the library result in different covers: the best basis tiling allows rectangles of all aspect ratios, while the best level or adapted window basis fits a cover of equal aspect ratio rectangles to the signal.

In the second interpretation, we imagine fitting together the smallest number of time-frequency atoms so as to obtain an acceptable approximation of the given signal. To do this, we need to find the interesting transients and the notable frequencies of the signal, then capture each of those features with an appropriate modulated waveform. The better the analysis, the more efficiently these features will be described and the fewer will be the total number of atoms. Then drawing the idealized time-frequency plane representation of the approximating sequence is simply a convenient way of describing the result of the analysis. The negligible components will not be drawn, as it is not relevant which particular basis is chosen for a subspace containing negligible energy.

If we use an information cost function such as entropy, then the two interpretations will produce the same time-frequency picture. Analysis of a signal is thus equivalent to lossy data compression of that signal.

10.1.5 Time-frequency analysis with library trees

The scale, frequency, and position indices of an element in a library tree of wavelet packets or local trigonometric functions can be used to draw an information cell in the time-frequency plane. We now derive the formulas for the nominal values of x_0, ξ_0, $\triangle x$, and $\triangle\xi$ for such functions.

Wavelet packets

Let ψ_{sfp} be the wavelet packet with scale index s, frequency index f, and position index p. We use conventional indexing for the filters and Paley or natural ordering for the frequency. Let us further suppose that the signal consists of $N = 2^L$ equally spaced samples, and that the library tree contains a complete wavelet packet analysis down to level L. Then we have $0 \leq s \leq L$, $0 \leq f < 2^s$, and $0 \leq p < 2^{L-s}$.

The scale parameter s gives the number of levels of decomposition below the original signal. Each application of convolution and decimation doubles the nominal width, so we set $\triangle x = 2^s$. With the usual assumption that $\triangle x \cdot \triangle\xi \approx N$, we can thus assign $\triangle\xi = 2^{L-s}$.

The frequency parameter must first be corrected by using the inverse Gray code permutation, so we compute $f' = GC^{-1}(f)$. This produces an index which is again in the range 0 to $2^s - 1$. The lower left-hand corner of the information cell should then be placed at vertical position $\triangle\xi \cdot f' = 2^{L-s} f'$.

The position parameter p needs to be shifted to correct for the (frequency-dependent) phase response of quadrature filters. We can use the corollary to the phase shifts theorem (Corollary 5.9) to compute the amount by which to shift the horizontal location of the information cell:

$$p' = 2^s p + (2^s - 1)c[h] + (c[g] - c[h])f''. \tag{10.7}$$

Here $c[h]$ and $c[g]$ are the centers of energy of the low-pass and high-pass QFs h and g, respectively, and f'' is the bit-reverse of f considered as an s-bit binary integer. The result will be inaccurate by at most the deviation ϵ_h of the filter h from linear phase, as defined in Theorem 5.8. If the wavelet packet coefficients were computed using periodized convolution-decimation, then we should replace the equal sign in Equation 10.7 with congruence modulo 2^L and take p' in the range 0 to $2^L - 1$.

Since the horizontal position p' of the information cell is uncertain by $\triangle x = 2^s$, we may as well slide its lower left-hand corner horizontally left to the nearest multiple of 2^s below p'. To emphasize this uncertainty, we will use a position index p'' in the range 0 to $2^{L-s} - 1$, with each integer value representing an interval starting at an integer multiple of $\triangle x = 2^s$:

$$p'' = \lfloor p'/2^s \rfloor = \left\lfloor p + (1 - 2^{-s})c[h] + 2^{-s}(c[g] - c[h])f'' \right\rfloor . \tag{10.8}$$

These conventions, in the periodic case, produce a disjoint tiling which exactly fills the $N \times N = 2^L \times 2^L$ square time-frequency plane:

Theorem 10.3 *If $B = \{(s, f, p) \in B\}$ is the index subset of a wavelet packet graph basis for a 2^L-point signal, then the collection of rectangles*

$$\{[2^s p'', 2^s(p'' + 1)[\times[2^{L-s} f', 2^{L-s}(f' + 1)[: (s, f, p) \in B\} \tag{10.9}$$

is a disjoint cover of the square $[0, 2^L[\times[0, 2^L[$. □

By our discussions in Chapters 5 and 7, these information cells will be located approximately where they should be to describe the time-frequency content of their wavelet packets.

Adapted local trigonometric functions

For this discussion, it makes no difference whether we use local cosines or local sines. The one is obtained from the other merely by reversing the direction of time, which has no effect on the geometry of information cells. We again suppose that the signal consists of $N = 2^L$ equally spaced samples, and that the library tree contains all the local trigonometric analyses to level L, with windows of size $2^L, 2^{L-1}, \ldots, 1$. The basis functions will be indexed by the triplet (s, f, p), and we will have $0 \leq s \leq L$, $0 \leq f < 2^{L-s}$, and $0 \leq p < 2^s$.

The scale parameter s again gives the number of decompositions of the original signal window into subwindows. Each subdivision halves the nominal window width, so we set $\triangle x = 2^{L-s}$. With the usual assumption that $\triangle x \cdot \triangle \xi \approx N$, we can thus assign $\triangle \xi = 2^s$.

The position index p numbers the adjacent windows starting with zero at the left edge of the signal. Thus the information cell should be drawn over the horizontal (time) interval $I_{sp} \stackrel{\text{def}}{=} [2^{L-s}p, 2^{L-s}(p+1)[$.

One local cosine basis for the subspace over the subinterval I_{sp} consists of the (DCT-II) functions $\cos \pi \left(f + \frac{1}{2}\right) n/2^{L-s} = \cos \pi 2^s \left(f + \frac{1}{2}\right) n/N$, where $n \in I_{sp}$ and $0 \leq f < 2^{L-s}$, multiplied by the window function subordinate to I_{sp}. These orthonormal basis functions are precisely defined in Chapter 4. These have nominal frequencies $2^s \left(f + \frac{1}{2}\right)$, so we will draw the associated information cell alongside the interval $[2^s f, 2^s(f+1)[$ on the vertical (or frequency) axis. Local cosine bases developed from DCT-III or DCT-IV as well as local sine bases can be depicted in exactly the same way.

A graph basis set of triplets produces a disjoint cover of such information cells:

Theorem 10.4 *If $B = \{(s, f, p) \in B\}$ is the index subset of an adapted local trigonometric graph basis for a 2^L-point signal, then the collection of rectangles*

$$\{[2^{L-s}p, 2^{L-s}(p+1)[\times[2^s f, 2^s(f+1)[: (s, f, p) \in B\} \tag{10.10}$$

is a disjoint cover of the square $[0, 2^L[\times[0, 2^L[$. □

These information cells will be located in the time-frequency plane at the appropriate locations for their position and frequency content. Notice how local trigonometric functions have scale, frequency, and position indices that are easier to interpret than wavelet packets; no Gray coding, bit-reversal or phase shifting is needed.

Arbitrary tilings with Haar–Walsh functions

The Haar-Walsh wavelet packet library has a special property not shared by libraries of smoother wavelet packets. Namely, we can put all its waveforms into one-to-one correspondence with dyadic information cells which are located where they should be, but in such a way that every disjoint cover corresponds to an orthonormal basis. The total number of disjoint covers of $2^L \times 2^L$ by dyadic rectangles—*i.e.*, those whose coordinates are of the form $n2^j$ for integers n and j—is much greater than the number of graph bases for a 2^L-point signal analyzed to level L.

We define the correspondence as follows: Let ψ_{sfp} be the Haar–Walsh wavelet packet on $N = 2^L$ points with scale index $0 \leq s \leq L$, Paley order frequency index

$0 \leq f < 2^s$, and unshifted position index $0 \leq p < 2^{L-s}$. We associate to it the rectangle

$$\psi_{sfp} \leftrightarrow R_{sfp} \stackrel{\text{def}}{=} [2^s p, 2^s(p+1)[\times[2^{L-s}\tilde{f}, 2^{L-s}(\tilde{f}+1)[\subset [0,N[\times[0,N[. \quad (10.11)$$

Here $\tilde{f} = GC^{-1}(f)$ is the inverse Gray code permutation of f, which adjusts the vertical location of R_{sfp} so it is proportional to the number of oscillations of the wavelet packet ψ_{sfp}. Since $c[h] = 1/2$ and $c[g] - c[h] = 0$ for the Haar–Walsh filters, we need not shift or bit-reverse the position index p to get the actual horizontal location of the information cell. This puts R_{sfp} where it should be to describe the location and oscillation of ψ_{sfp}. We remark that two rectangles will be disjoint in the sequency ordering of their frequency indices if and only if they are disjoint in the Paley ordering.

Theorem 10.5 *The Haar–Walsh wavelet packets* $\{\psi_{sfp} : (s,f,p) \in B\}$ *form an orthonormal basis of* $\mathbf{R}^N$ *if and only if the dyadic rectangles* $\{R_{sfp} : (s,f,p) \in B\}$ *form a disjoint cover of* $[0,N[\times[0,N[$.

Proof: Since the area of R_{sfp} is N while the area of $[0,N[\times[0,N[$ is N^2, it suffices to show that two rectangles R_{sfp} and $R_{s'f'p'}$ are disjoint if and only if ψ_{sfp} and $\psi_{s'f'p'}$ are orthogonal. Since exactly N rectangles fit into the square, and the space is N-dimensional, we must have a basis set of wavelet packets.

Let $I_{sp} = [2^s p, 2^s(p+1)[$ and $I_{s'p'} = [2^{s'}p', 2^{s'}(p'+1)[$ be the supports of ψ_{sfp} and $\psi_{s'f'p'}$, respectively, and consider the two cases:

$s = s'$: then ψ_{sfp} and $\psi_{sf'p'}$ will be orthogonal if and only if $f \neq f'$ or $f = f'$ but $p \neq p'$, one of which will be true if and only if the congruent rectangles R_{sfp} and $R_{sf'p'}$ are disjoint. In this case, both ψ_{sfp} and $\psi_{sf'p'}$ are members of the same single-level or subband basis, which is an orthonormal graph basis.

$s \neq s'$: then ψ_{sfp} and $\psi_{s'f'p'}$ will be orthogonal if and only if

$$\int_{I_{sp}\cap I_{s'p'}} \psi_{sfp}(t)\psi_{s'f'p'}(t)\,dt = 0. \quad (10.12)$$

But two dyadic intervals either are disjoint or one contains the other. If $I_{sp} \cap I_{s'p'} = \emptyset$, then $R_{sfp} \cap R_{s'f'p'} = \emptyset$ since the rectangles share no horizontal coordinates. If the overlap is nonempty, then we may suppose without loss that $s < s'$ and thus $I_{sp} \subset I_{s'p'}$. We can define the *nominal frequency intervals* $J_{sf} \stackrel{\text{def}}{=} [2^{L-s}\tilde{f}, 2^{L-s}(\tilde{f}+1)[$ and $J_{s'f'} = [2^{L-s'}\tilde{f}', 2^{L-s'}(\tilde{f}'+1)[$ of ψ_{sfp} and $\psi_{s'f'p'}$, respectively. Since these too are dyadic intervals, and $L - s > L - s'$, we have that either $J_{sf} \cap J_{s'f'} = \emptyset$, in which case R_{sfp} and $R_{s'f'p'}$ are disjoint, or $J_{s'f'} \subset J_{sf}$, in which case R_{sfp} and $R_{s'f'p'}$ overlap.

If J_{sf} and $J_{s'f'}$ are disjoint, then ψ_{sfp} and $\psi_{s'f'p'}$ come from different branches of the wavelet packet tree and can be embedded in a graph basis, hence must be orthogonal. Thus in all cases where R_{sfp} and $R_{s'f'p'}$ are disjoint, the functions ψsfp and $\psi_{s'f'p'}$ are orthogonal.

We now check the last remaining subcase, in which R_{sfp} and $R_{s'f'p'}$ overlap. But then, the function ψ_{sfp} is supported in one of the $2^{s'-s}$ adjacent subintervals of length 2^s contained in $I_{s'p'}$, and $\psi_{s'f'p'}$ is a direct descendent of ψ_{sfp} in the wavelet packet tree. From Equations 7.2 and 7.3 with the filters $h = \{\frac{1}{\sqrt{2}}, \frac{1}{\sqrt{2}}\}$ and $g = \{\frac{1}{\sqrt{2}}, -\frac{1}{\sqrt{2}}\}$, we observe that $\psi_{s'f'p'} = \pm 2^{(s-s')/2}\psi_{sfp}$ on each of those adjacent intervals. Thus the inner product in Equation 10.12 will be $2^{(s-s')/4} \neq 0$.

The normalization part of the theorem is free, since all Haar–Walsh wavelet packets have unit norm. □

Coifman and Meyer have remarked that the proof works with smooth time-frequency atoms as well, as long as we use Haar–Walsh filters to perform the frequency decompositions:

Corollary 10.6 *If ϕ is any time-frequency atom which is orthogonal to its integer translates, and X is the vector space spanned by $\{\phi(t-n) : 0 \leq n < 2^L\}$, then every disjoint tiling of the square $[0, 2^L[\times[0, 2^L[$ by dyadic rectangles corresponds to an orthonormal basis for X made of time-frequency atoms.* □

This corollary may be used to build plenty of smooth orthonormal bases in a smooth approximation space. We iterate longer orthogonal filters to produce smooth sampling functions ϕ of fixed scale, then use the Haar–Walsh filters to do a fixed finite number of frequency decompositions. We thus avoid using Haar–Walsh wavelet packets, which are not even time-frequency atoms since discontinuous functions have infinite frequency uncertainty. However, the Heisenberg product of these hybrid wavelet packets will blow up as the number of levels L of decomposition increases.

The idea of decoupling the underlying sampling functions $\phi_{N,k}$ from the filters used to decompose the time-frequency plane has also appears in the PhD thesis of Hess-Nielsen [55]. He found that for each smooth sampling function ϕ, number of decomposition levels L and desired filter length R, there is an optimal set of OQFs of length R which maximizes the average frequency localization of the wavelet packets. The localization criterion is energy within the desired band J_{sf} for the wavelet packet ψ_{sfp}.

10.2 Time-frequency analysis of basic signals

We now take certain canonical signals and analyze them in various wavelet packet libraries with a program written for a desk top computer. The user selects a conjugate quadrature filter from a list of 17 at the right, and the "mother wavelet" determined by that filter is displayed in the small square window at the lower right. The signal is plotted in the rectangular window at bottom, and the time-frequency plane representation is drawn in the large main square window.

The first signal is a pair of well-separated modulated Gaussians. Figure 10.7 shows the output of an approximate analysis with "D 20" wavelet packets. It should be noticed how several dark information cells lie above each modulated bump in the signal plot. In addition, there is quite a bit of variation in the amplitude of adjacent information cells, plus the centers of the cells do not exactly line up with the perceived centers of the analyzed bumps. These artifacts devolve from the rigidity of the grid pinning down the information cells and the imperfect time-frequency localization of the approximate atoms. They can never be eliminated, though the deviations can be made as small as the time and frequency uncertainty of the analysis and synthesis functions.

10.2.1 Benefits of adaption

Whether we decide to decompose in frequency first or in time first, it is still necessary to choose a window size appropriate to the analysis. We can visualize our information cost function as the amount of area occupied by dark information cells, or the number of nonnegligible waveforms, in the time-frequency analysis of a signal. Lots of white space means a low information cost; most of the components have negligible energy, so that that the signal energy is concentrated into just a few waveforms.

Very short (time) windows are most efficient for sharp impulses, while long windows correspond to information cells all of which contain some of the energy in the impulse, as seen in Figure 10.8. Conversely, long windows are more efficient than short ones for nearly continuous tones, as depicted in Figure 10.9, because they correlate well with nearly periodic behavior of long duration. This comparison illustrates how it can be useful to examine the signal in many window sizes at once and then to adapt a representation to maximize efficiency.

In these comparisons, we only examined *fixed-level* basis subsets, taken from one level of a complete wavelet packet tree. We used "V 24" OQFs for these examples.

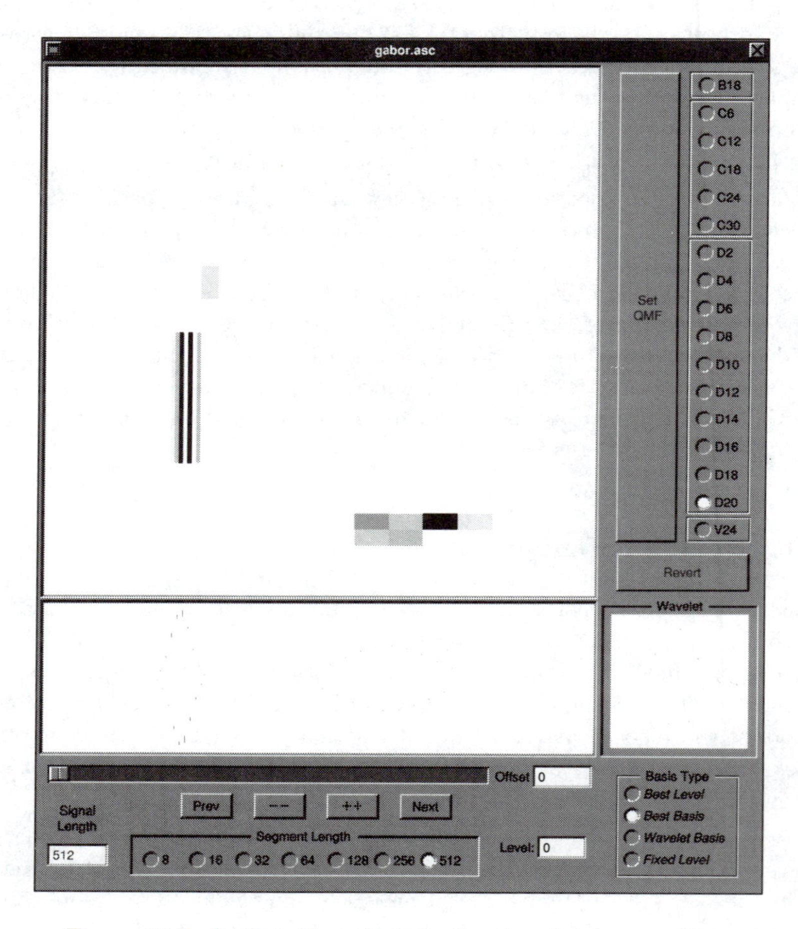

Figure 10.7: An actual analysis in the time-frequency plane.

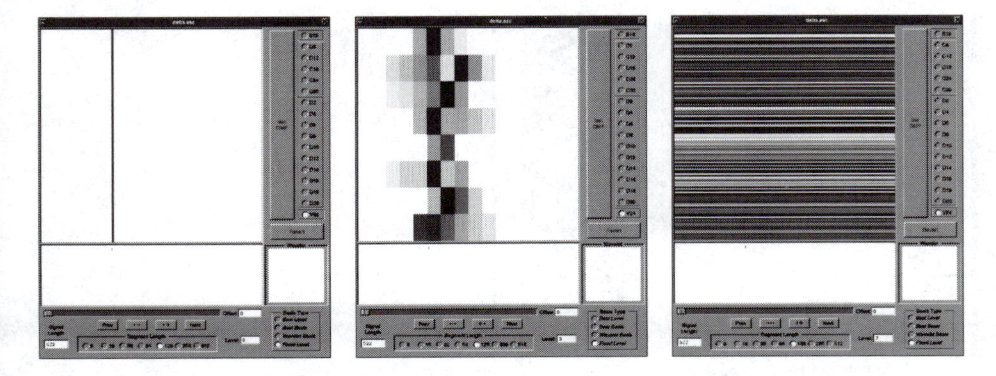

Figure 10.8: Time-frequency analyses of an impulse at increasing window sizes.

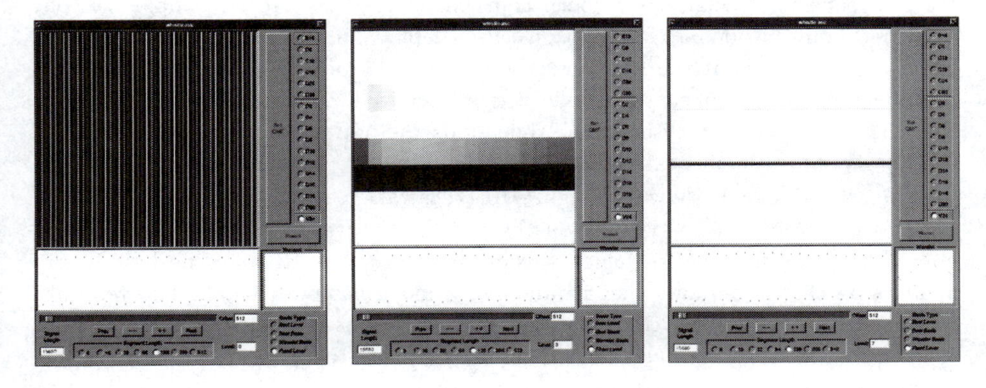

Figure 10.9: Time-frequency analyses of a pure tone at increasing window sizes.

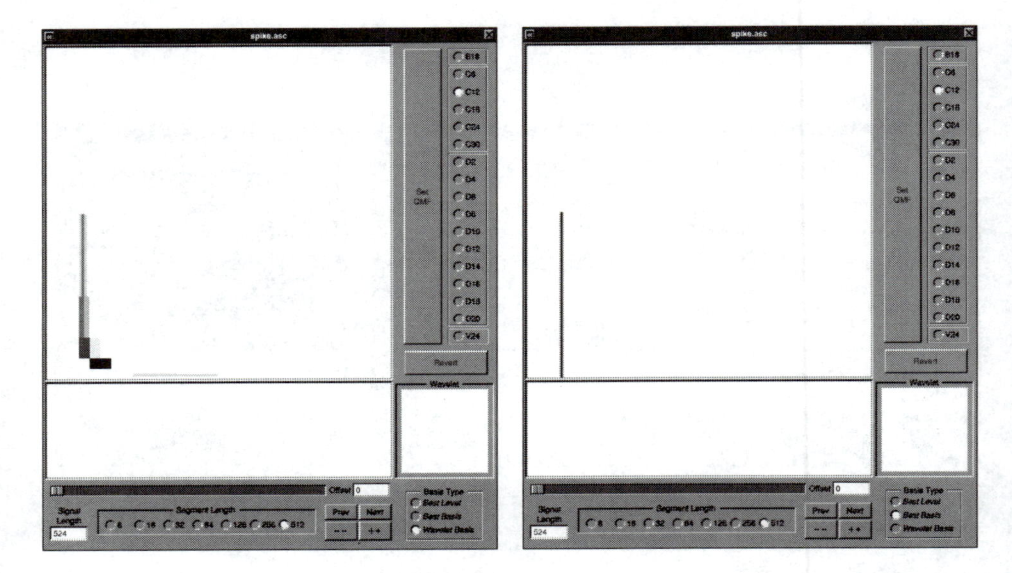

Figure 10.10: Fast transient in the wavelet and best wavelet packet bases.

10.2.2 Wavelets versus wavelet packets

We now consider a relatively smooth transient, spread over seven samples out of 512. Its decomposition into information cells is depicted in Figure 10.10. Notice that the wavelet analysis at the left correctly localizes the peak in the high frequency components, but is forced to include poorly localized low frequency elements as well. The best basis analysis on the right finds the optimal representation within the library, which in this case is almost a single wavelet packet.

The second signal, shown in Figure 10.11, is taken from a recording (at 8012 samples per second) of a person whistling. Here the wavelet basis on the left is only able to localize the frequency within an octave, even though the best basis analysis on the right shows that it falls into a much narrower band. The vertical stripes among the wavelet information cells may be used to further localize the frequencies, but the best basis decomposition performs this analysis automatically.

Let us now combine the transient and periodic parts in different ways. In the example in Figure 10.12, we take a damped oscillator which receives an impulse, and decompose the resulting solution in the Dirac and wavelet bases. The Dirac basis analysis indicates the envelope of the "ping" since the eye has a tendency to average the amount of dark gray in the time-frequency plane. The wavelet decomposition

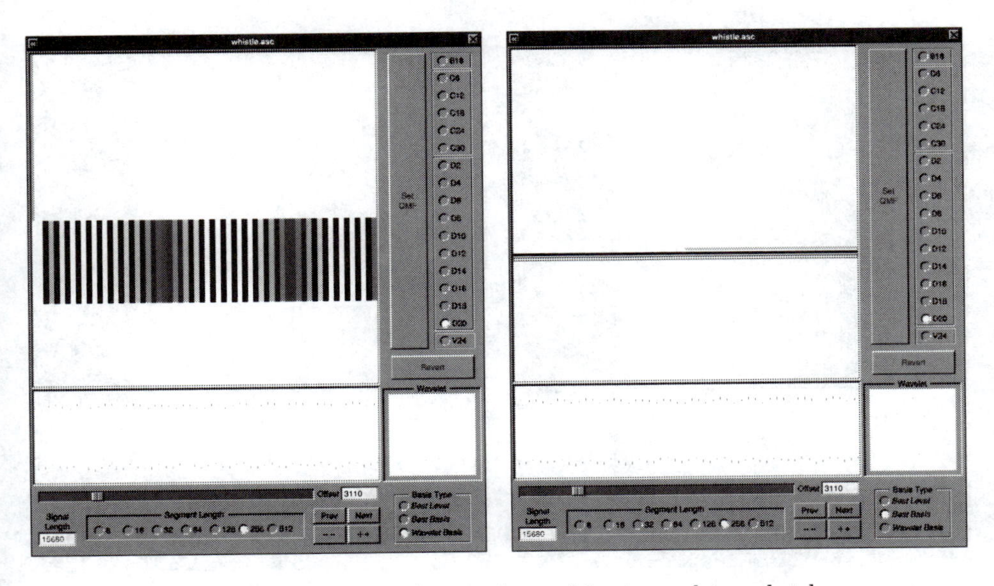

Figure 10.11: Whistle in the wavelet and best wavelet packet bases.

locates the discontinuity at the impulse with large coefficients at all scales. The exponential decay of the amplitude is visible in both analyses.

Figure 10.13 shows the damped oscillator in the best level and best basis wavelet packet representations. Both are more efficient than the wavelet or Dirac representations. Both find the resonant frequency of the oscillator, though the best basis analysis is more precise. The number of dark information cells is quite small. Here we use the "D 20" filters.

10.2.3 Chirps

We will call an oscillatory signal with increasing modulation by the name *chirp*. We note that some other authors prefer chirps to have decreasing modulation. Figure 10.14 shows two examples: the functions $\sin(250\pi t^2)$ and $\sin(190\pi t^3)$ on the interval $0 < x < 1$, sampled 512 times. The modulation increases linearly and quadratically, respectively. The information cells form a line and a parabolic arc, respectively. In the wavelet packet best basis analysis of the linear chirp, most of the information cells have the same aspect ratio, which is appropriate for a line. In the best basis analysis of the quadratic chirp, the information cells near the zero-slope portion have smaller aspect ratio than those near the large-slope portion.

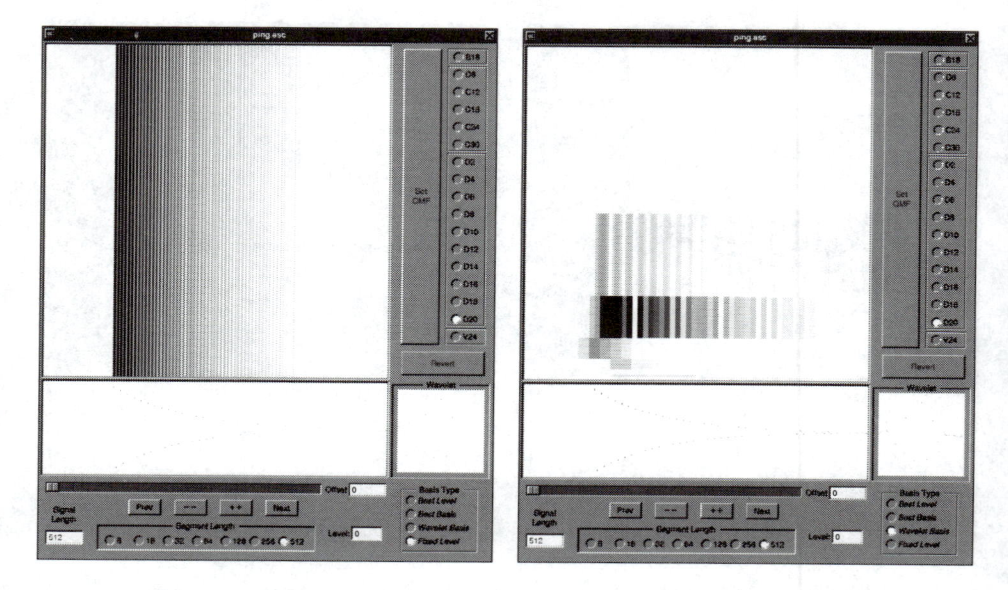

Figure 10.12: Damped oscillator in the Dirac and wavelet bases.

Figure 10.13: Damped oscillator in the best level and best wavelet packet bases.

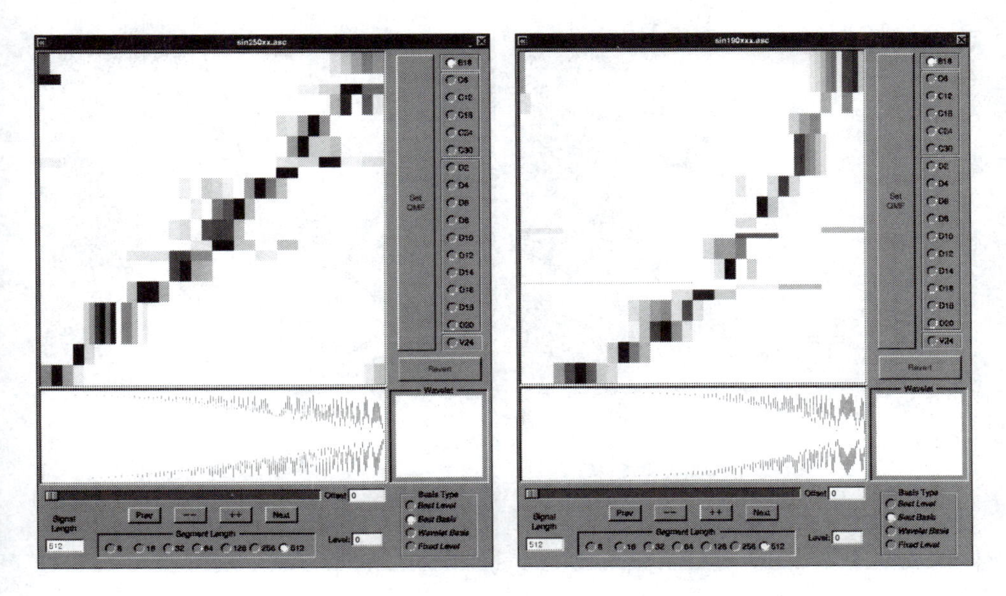

Figure 10.14: Linear and quadratic chirps in their best wavelet packet bases.

Such a time-frequency analysis can separate superposed chirps. Figure 10.15 shows a pair of linear chirps, differing either by modulation law or phase. Both are functions on the interval $0 < t < 1$, sampled 512 times. On the left is the function $\sin(250\pi t^2) + \sin(80\pi t^2)$ analyzed in the best wavelet packet basis. Note that the milder slope chirp is represented by Heisenberg boxes of lower aspect ratio. On the right is $\sin(250\pi t^2) + \sin(250\pi(t - \frac{1}{2})^2)$, analyzed by best level wavelet packets. The downward-sloping line comes from the aliasing of negative frequencies.

Aliasing appears in both the time-frequency plane and the signal plot, since the plotting algorithm connects adjacent points with a line and must assume that the phase advances by less than half a turn between them. The signals in Figure 10.16 show how linear chirps exceeding the Nyquist frequency result in "reflected" lines.

10.2.4 Speech signals

Figure 10.17 shows portions of speech signals sampled 8012 times per second and decomposed using "D 10" OQFs, as one might do in practice. The patterns are quite complicated, though we might notice how much compression is achieved in the sense that relatively little of the time-frequency plane is covered by dark information cells.

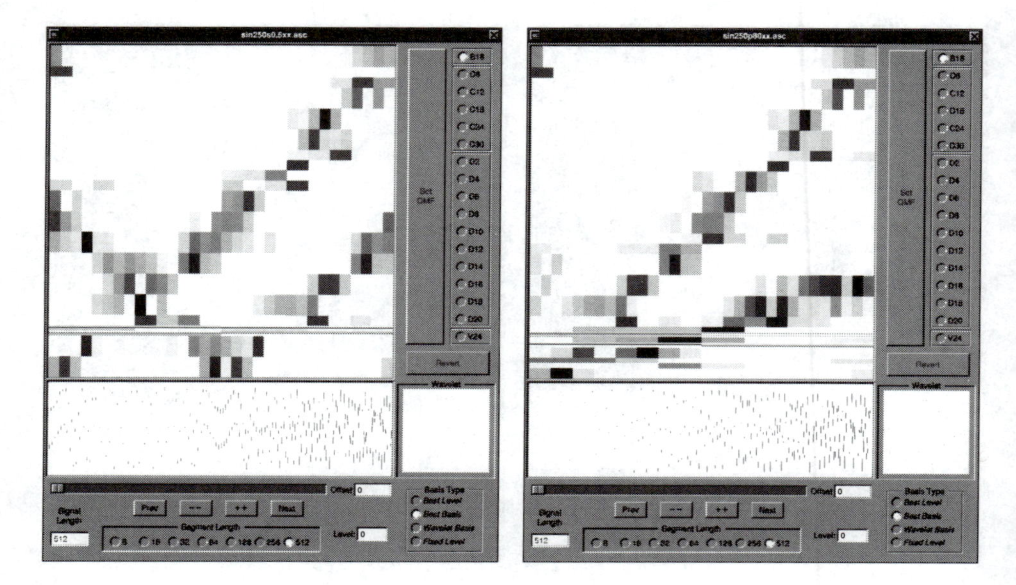

Figure 10.15: Superposed chirps.

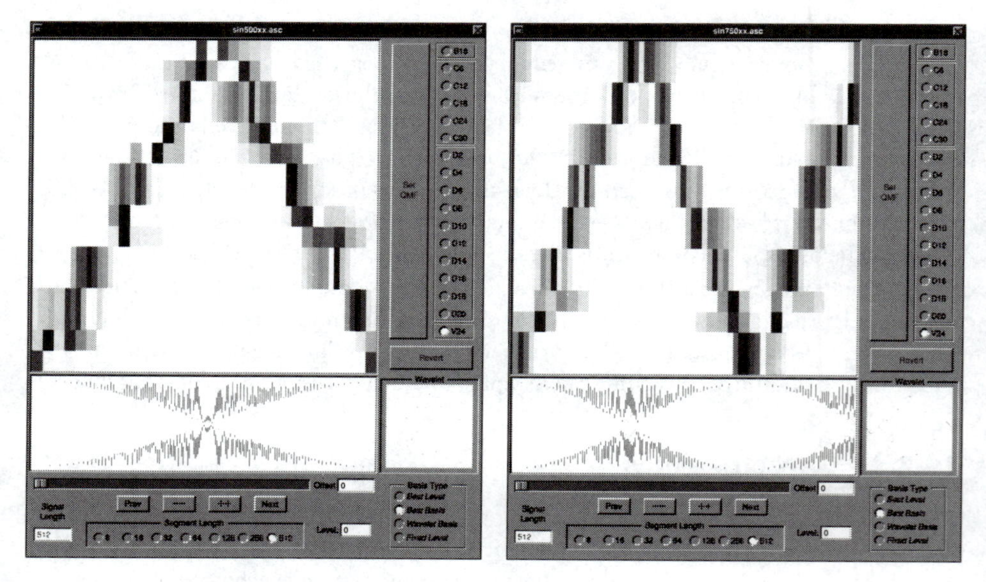

Figure 10.16: Aliasing in chirps to two and three times the Nyquist frequency.

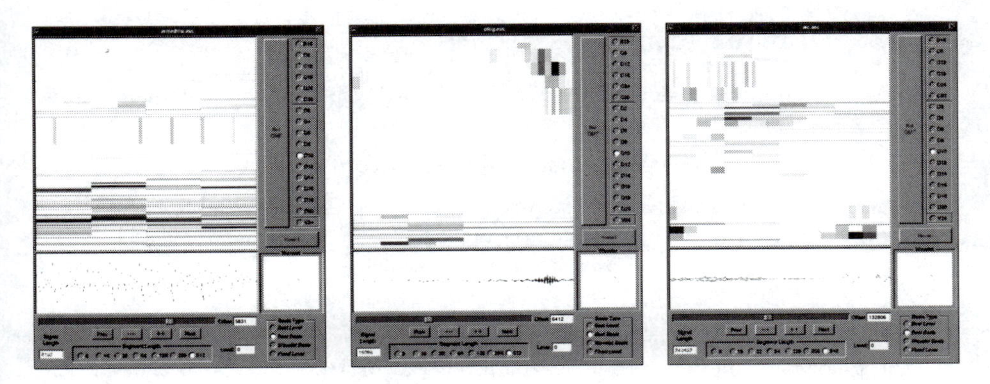

Figure 10.17: Wavelet packet best basis analysis of speech.

10.3 Implementation

To display the idealized time-frequency plane and the information cells associated to an image, it is necessary to perform three operations:

- Generate the desired signal samples, or read them from a file;
- Compute the expansion of the signal in a particular basis of time-frequency atoms;
- Draw shaded rectangles on some output device.

We will assume that the first step has been performed and that the signal samples are in an array of REAL data elements, at the indices 0 through $M2^L - 1$ for some positive integers M, L. We will then perform a decomposition down to level L. The assumption that the number of samples is divisible by a reasonably large power of two is not necessary, but it simplifies the implementation considerably.

We will confine our attention to orthonormal libraries of time-frequency atoms so that we can represent the signal with disjoint information cells. We will further restrict ourselves to wavelet packets and local trigonometric functions for the time-frequency analysis. The user must choose which one of these to apply to a given signal, as well as any associated parameters such as quadrature filter or cutoff function. One point of the analysis is to choose an appropriate time-frequency decomposition based on visual feedback, but we must keep the number of choices reasonable so that we will have an understandable interface. Thus we will only have a few QFs and a few choices of rising cutoff functions, with the option of adding additional QFs and cutoffs and even additional libraries in the future.

10.3.1 Drawing primitives

To draw the rectangles, we will create a file of PostScript drawing commands in a file, using the following primitives:

Basic PostScript operators

```
XVAL YVAL moveto     XO YO translate          XVAL XSCALE mul
XNEW YNEW lineto     closepath stroke         AMPL setgray fill
showpage             /MACRO { ... commands ... } def
```

These commands are defined and fully explained in [59].

The following function sets up normalized drawing coordinates so that an information cell and its gray level can always be in the range $[0.0, 1.0]$. It works by writing a prologue which states the total dimensions of the time-frequency plane and defining two PostScript macros 'xunits' and 'yunits', which scale the normalized values to the actual display dimensions:

Write prologue, scale and bounding box dimensions

```
epsprologue(PSFILE, BBXMIN, BBYMIN, BBXMAX, BBYMAX)
  Write "%!" to PSFILE
  Write "%%Pages: 0 1" to PSFILE
  Write "%%BoundingBox: BBXMIN BBYMIN BBXMAX BBYMAX" to PSFILE
  Write "%%EndComments" to PSFILE
  Write "%%BeginSetup" to PSFILE
  Let XSCALE = BBXMAX-BBXMIN-4
  Write "/xunits { XSCALE mul } def" to PSFILE
  Let YSCALE = BBYMAX-BBYMIN-4
  Write "/yunits { YSCALE mul } def" to PSFILE
  Write "% Outline the rectangle" to PSFILE
  Let X = BBXMIN+1
  Let Y = BBYMIN+1 and write "X Y moveto" to PSFILE
  Let Y = BBYMAX-1 and write "X Y lineto" to PSFILE
  Let X = BBXMAX-1
  Let Y = BBYMAX-1 and write "X Y lineto" to PSFILE
  Let Y = BBYMIN+1 and write "X Y lineto" to PSFILE
  Write "closepath stroke" to PSFILE
  Let X = BBXMIN+2
  Let Y = BBYMIN+2 and write "X Y translate" to PSFILE
  Write "%%EndSetup" to PSFILE
```

The prologue fixes where on the page the time-frequency plane will be drawn, and provides some comments. It also draws a rectangle one pixel inside the bounding box and one pixel outside the plot area. It can be used to outline an area in which the signal will be plotted and another area in which the information cells will be plotted. Then, after all the plotting commands are written, we can write an epilogue to cause the page to be printed out:

Write epilogue to draw the page and end the file

```
epsepilogue(PSFILE):
  Write "% end of EPS commands" to PSFILE
  Write "showpage" to PSFILE
  Write "% all done" to PSFILE
```

10.3.2 Plotting the signal samples

To plot the signal as a polygon — *i.e.*, to plot its samples and "connect the dots" — we use a series of short programs, each of which writes instructions for a single plot segment to an output file:

Write basic PostScript statements to a file

```
epslineto(PSFILE, XVAL, YVAL):
   Write "XVAL xunits YVAL yunits lineto" to PSFILE

epsmoveto(PSFILE, XVAL, YVAL):
   Write "XVAL xunits YVAL yunits moveto" to PSFILE

epsstroke(PSFILE):
   Write "stroke" to PSFILE

epstranslate(PSFILE, XPTVAL, YPTVAL):
   Write "XPTVAL YPTVAL translate" to PSFILE
```

The page coordinates of the signal portion of the time-frequency display are a matter of taste. We choose to put the signal plot below the time-frequency representation, as shown in Figure 10.2. Since this is the same location for all like-sized pages, we can use preprocessor macros to set the bounding box dimensions. The numbers given here are coordinates in PostScript points of the lower left (LLXS,LLYS) and upper right (URXS,URYS) corners of the bounding box which contains the plotted signal:

Dimensions for a signal plot on $8.5'' \times 11''$ paper

```
#define LLXS 72
#define LLYS 72
#define URXS 528
#define URYS 254
```

We need a prologue, one line per sample, and an epilogue to ink the drawing:

Normalize and plot a sampled signal as a polygon

```
plotsig(PSFILE, SIGNAL, LENGTH):
  epsprologue( PSFILE,  LLXS, LLYS, URXS, URYS )
  epstranslate( PSFILE, 0.0, (URYS - LLYS)/2.0 )
  Let NORM = 0.0
  For N = 0 to LENGTH-1
     Let NORM = max( absval(SIGNAL[N]), NORM )
  If NORM == 0.0
      epsmoveto( PSFILE, 0.0, 0.0 )
      epslineto( PSFILE, 1.0, 0.0 )
  Else
      Let NORM = 0.45 / NORM
      Let YVAL = SIGNAL[0]*NORM
      epsmoveto( PSFILE, 0.0, YVAL )
      If LENGTH == 1
         epslineto( PSFILE, 1.0, YVAL )
      Else
         Let XVAL = 0.0
         Let INCR = 1.0 / (LENGTH - 1.0)
         For N = 1 to LENGTH-1
            XVAL += INCR
            Let YVAL = SIGNAL[N]*NORM
            epslineto( PSFILE, XVAL, YVAL )
      epsstroke( PSFILE )
  epsepilogue( PSFILE )
```

10.3.3 Plotting the time-frequency plane

The following function writes commands to a file which will cause the rectangle [XMIN, XMAX] $\times$ [YMIN, YMAX] to be drawn and filled to level GRAY. Notice that in

PostScript, the gray level indicates how much light hits the page, so that to indicate a fully dark cell the gray level should be 0.0. We call this function once for each nonnegligible information cell.

Plot a single shaded information cell

```
epsfrect(PSFILE, XMIN, YMIN, XMAX, YMAX, GRAY):
  Write "% begin new rectangle:" to PSFILE
  Write "XMIN xunits YMIN yunits moveto" to PSFILE
  Write "XMIN xunits YMAX yunits lineto" to PSFILE
  Write "XMAX xunits YMAX yunits lineto" to PSFILE
  Write "XMAX xunits YMIN yunits lineto" to PSFILE
  Write "closepath" to PSFILE
  Write "GRAY setgray fill" to PSFILE
```

We must begin with a prologue that centers the density plot in the top part of an $8.5'' \times 11''$ page. The following definitions put the time-frequency plot directly above the signal plot, at the same horizontal scale:

Dimensions for a time-frequency plot on $8.5'' \times 11''$ paper

```
#define LLXA  72
#define LLYA 264
#define URXA 528
#define URYA 720
```

We must also define the machine-dependent threshold of visibility for the plotted information cells:

Minimum discernible gray-scale, with white = 0

```
#define MINGRAY 0.01
```

Remark. Information cells whose amplitudes result in a shading lighter than MINGRAY will not be visible, so there is no reason to draw them. The threshold is a fixed fraction of the maximum amplitude or some function thereof, depending upon the chosen relationship between amplitude and gray-scale. Such thresholding amounts to compression, and the more we boost the largest amplitude relative to the rest, the more compression we will obtain.

To plot time-frequency atoms in a list, we use a loop. The following function produces PostScript commands to draw a nominal time-frequency density plot from an array of TFA1 data structures. It writes the results to the specified file.

Normalize and plot an idealized time-frequency plane

```
tfa1s2ps( PSFILE, SAMPLES, ATOMS, NUM ):
  epsprologue( PSFILE, LLXA, LLYA, URXA, URYA )
  Let ANORM = 0.0
  For N = 0 to NUM-1
     Let ANORM = max( absval(ATOMS[N].AMPLITUDE), ANORM )
  If ANORM > 0.0
     Let ANORM  = 1.0 / ANORM
     For N = 0 to NUM-1
        Let AMPL  = ATOMS[N].AMPLITUDE
        Let GRAY  = ANORM * AMPL
        If GRAY > MINGRAY then
           Let WIDTH = ( 1<<ATOMS[N].LEVEL ) / SAMPLES
           Let XMIN = WIDTH * ATOMS[N].OFFSET
           Let XMAX = XMIN + WIDTH
           Let HEIGHT = 1.0 / WIDTH
           Let YMIN = HEIGHT * ATOMS[N].BLOCK
           Let YMAX = YMIN + HEIGHT
           epsfrect( PSFILE, XMIN, YMIN, XMAX, YMAX, 1.0-GRAY )
  epsepilogue( PSFILE )
```

We are assuming that the level, block, and offset indices are the nominal scale, frequency, and position, respectively, as in the wavelet packet case. To obtain these numbers, it is necessary to correct the phase shift and frequency permutation induced by filtering (in the wavelet packet case), and interchange the role of block index and offset in the local cosine case. These computations are left as exercises.

10.3.4 Computing the atoms

Computing a time-frequency analysis proceeds in four steps:

- Extract a smooth interval from the signal;
- Write the segment as a sum of time-frequency atoms;
- Calculate the positions, frequencies, and time-frequency uncertainties of the atoms;
- Plot the information cells at their respective positions.

In addition, it is necessary to choose the library of time-frequency atoms to use in the analysis. The two cases we have considered, wavelet packets and local trigonometric functions, require different interpretations for their scale, position, and frequency indices and must be done separately. The first step, cutting out a smooth segment of the signal, is the same for both libraries. We use the method described in Chapter 4, namely smooth local periodization, since we will be using periodized time-frequency atoms. It does not matter whether we use local sines (`lpds()`) or local cosines:

Local periodization to a disjoint interval, cosine polarity

```
lpdc( OUT, IN, N, RISE ):
   fdcp(  OUT,  1,  IN,   IN,  N/2, RISE )
   fdcn( OUT+N, 1, IN+N, IN+N, N/2, RISE )
   uipc( OUT+N, OUT, RISE )
```

Before calling this function, we must ensure that the array `IN[]` is defined at all indices from `RISE.LEAST` to `N+RISE.FINAL`. In particular, we must make sure that we have at least `RISE.FINAL` previous samples before our first offset into the input array. This can be done by padding the input signal with zeroes, or else by including a conditional statement which tests if we are at the beginning and then acts appropriately. Likewise, we must also test for the end of the signal or else pad the end with extra zeroes.

Notice that the function `lpdc()` copies the transformed signal samples and does not affect the signal itself. This must be taken into account when we reassemble the signal.

10.4 Exercises

1. Use the DFT on a suitably long periodic interval to approximate $\triangle x \cdot \triangle \xi$ for "D 4" and "D 20" wavelets. Do the same for the "C 6" and "C 30" wavelets.

2. Write a program that reads a sequence of N integers in the range 1 to 88, which represent keys on a piano, and then writes an array of $1024N$ eight-bit samples that can be played through a digital-to-analog converter at the standard CODEC rate of 8012 samples per second. Centered over each adjacent interval of length 1024 should be a wavelet packet (or local cosine) whose frequency corresponds to that of the piano note. Use your program to play J. S. Bach's "Prelude No. 1 in C" from *The Well-Tempered Clavier.* Try different QFs (or different rising cutoffs) to compare timbres. (Hint: adjacent piano

notes have frequencies in the ratio $\sqrt[12]{2} : 1$, and $\sqrt[12]{2} = 1.059463\ldots \approx 18/17$.)

3. Consider the signal $u_p = \{u(k) = 1 + p\delta(k) : 0 \leq k < N = 2^L\}$, which is constantly one except at sample zero where it is $1 + p$.

   (a) Find the best basis time-frequency representation when $p = 0$.

   (b) Find the best basis time-frequency representation when $p = 1,000,000$.

   (c) Show that $\|u_p - u_q\| = |p - q|$.

   (d) Show that for every $\epsilon > 0$ there are p and q with $\|u_p - u_q\| < \epsilon$ but with the best wavelet packet bases for u_p and for u_q sharing no common element.

4. Write a pseudocode function that reads an input array of TFA1s and the length of their periodic signal, then produces an output array of TFA1s with corrected level, block and offset, members assuming it is a local cosine.

5. Write a pseudocode function like the previous one, only assuming that the input TFA1s are wavelet packets. Note that the QFs must be specified as well.

6. Compare the time-frequency plane analyses of the 512-point signal

$$\left\{ \sin\left(190\pi \left[\frac{j}{512}\right]^3\right) : 0 \leq j < 512 \right\}$$

in the local cosine case and in the wavelet packet case with C 6 and C 24 filters.

Chapter 11

Some Applications

We will briefly discuss the numerous applications of adapted wavelet analysis, some of which are just being explored. These include image compression, fast numerical methods for principal factor analysis and matrix application, acoustic signal processing and compression, and de-noising.

Many of these applications were first described in survey articles [27, 25, 29, 26, 32, 31, 115, 81, 112], while others depend on novel numerical algorithms which were individually analyzed [114, 111, 118, 9, 8, 11]. Still others were introduced as methods used for numerical experiments [34, 46, 51, 57, 110, 113] with specific signal processing problems. In a few cases we will go down a somewhat different path than the one described in the literature. For example, we will examine a fast approximate version of matching pursuit [72], and we will describe best basis transform coding in addition to various quantization methods to be applied after a fixed transform [77, 60].

There are plenty of other applications which are beyond the scope of this book. For example, we will not discuss applications such as [63, 73, 53], which depend on the continuous wavelet transform, or ones which depend on spline rather than QF or local trigonometric constructions of adapted wavelets [14, 41].

11.1 Picture compression

We shall first consider the problem of storing, transmitting, and manipulating digital electronic images. Because of the file sizes involved, transmitting images will always consume large amounts of bandwidth, and storing images will always require hefty resources. Because of the large number N of pixels in a high resolution image, manipulation of digital images is unfeasible without low complexity algorithms, *i.e.*, $O(N)$ or $O(N \log N)$. Our methods, based on wavelets and the Fourier transform, are firmly grounded in harmonic analysis and the mathematical theory of function spaces. They combine effective image compression with low complexity image processing.

11.1.1 Digital pictures

An image consists of a large-dimensional vector whose components we will call "pixels." These represent measurements of total light intensity or *gray-scale*, or perhaps the intensity of light in a primary band of color such as red, green, or blue. In practice these pixels take only discrete values, and only in a finite range. Typical values for this range are $0, 1, \ldots, 255$ (eight bits per pixel) or $-1024, -1023, \ldots, 1022, 1023$ (11 bits per pixel). The initial determination of this vector of values introduces at least three types of errors:

- Measurement error: to create a digitized image requires making physical measurements, which are always uncertain to some degree depending upon the measuring device;

- Irregular sampling error: each pixel has a definite location in space, and each measurement is made at a definite location in space, but these two locations might not be the same.

- Sampling quantization error: no values can be represented other than the discrete values allowed to the pixels. This is the error which is introduced by rounding the measured sample value to one of these allowed pixel values.

After determining the pixel values, it is not generally possible to recover the measurement error or the irregular sampling error. While these can be determined, they play no role in the subsequent processing. Errors with physical origins do not affect the precision of the pixel values, only their accuracy. On the other hand, sampling quantization error plays a larger role because it limits the precision of the coding scheme.

Sampling band-limited functions

A digitally sampled image can only represent a band-limited function. Nyquist's theorem implies that there is no way of resolving spatial frequencies higher than half the pixel pitch. Band-limited functions are smooth: the Paley–Wiener theorem implies that they are entire analytic, which means that at each point they can be differentiated arbitrarily often and the resulting Taylor series converges arbitrarily far away. In practice we also require that the first few derivatives be small relative to the size of the function.

Since digitally sampled images faithfully reproduce the originals as far as our eyes can tell, we may confidently assume that our images are in fact smooth and well-approximated by band-limited functions. It is a deep observation that "natural" pictures have a characteristic power spectrum, namely the energy at spatial frequency ω drops off like a negative power of ω as $\omega \to \infty$. This can be exploited by so called *fractal* image synthesis algorithms [6] which perturb an image with texture so as to create a particular power spectrum envelope. Such textures appear remarkably realistic. The functions so produced can be made band-limited simply by cutting off the power spectrum at a fixed point, and the visible result is an imperceptible smoothing of the very finest textures.

Smoothness is correlation

Another notion of "smoothness" in this context is the correlation of adjacent pixels when they are treated as random variables. Here we need some definitions from probability theory. Let $X = \{X_n : n = 1, \ldots, N\} \subset \mathbf{R}^d$ be an ensemble of vectors. Write

$$E(X) \stackrel{\text{def}}{=} \frac{1}{N} \sum_{n=1}^{N} X_n \tag{11.1}$$

for the average vector in the ensemble, i.e., the *expectation* of x over the set X.

Let $\sigma(X) \subset \mathbf{R}^d$ be the vector of the *standard deviations* of the coefficients of X. Namely,

$$\sigma(X)(k) = \left(\frac{1}{N} \sum_{n=1}^{N} [X_n(k) - E(X)(k)]^2 \right)^{1/2}. \tag{11.2}$$

We define the *variance ellipsoid* of an ensemble X to be the ellipsoid centered at $E(X) \in \mathbf{R}^d$, with semiaxes $\sigma(X)(1), \sigma(X)(2), \ldots, \sigma(X)(d)$ aligned with the d coordinate axes. Its volume is $\omega_d \times [\sigma(X)(1)] \times [\sigma(X)(2)] \times \cdots \times [\sigma(X)(d)]$, where ω_d is the surface area of the unit sphere in $\mathbf{R}^d$. As seen in Figure 11.1, the volume of the variance ellipsoid depends a great deal upon the choice of coordinate axes. The so-called *Karhunen–Loève* coordinates, defined below, minimize this volume.

If the ensemble X is fixed forever, then we can assume without loss of generality that $\frac{1}{N}\sum_{n=1}^{N} X_n(k) = 0$ for all $k = 1, 2, \ldots, d$, namely that $E(X) = 0$, because this can always be arranged by subtracting the average vector $E(X)$ from each of

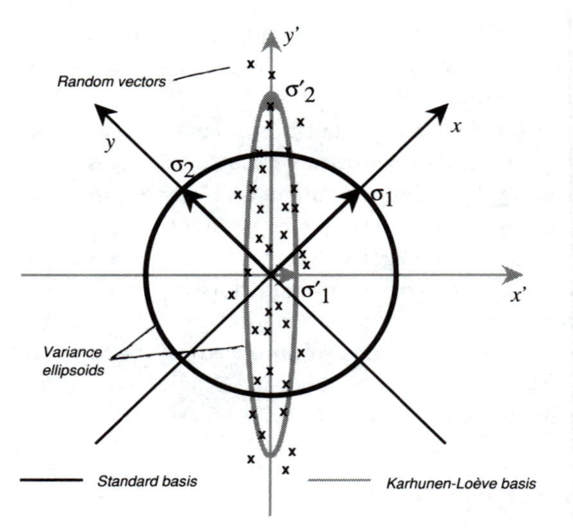

Figure 11.1: The variance ellipsoids for the standard and Karhunen–Loève bases.

$X_1, X_2, \ldots, X_N$. It results in a simpler formula for $\sigma(X)$:

$$E(X) = 0 \Rightarrow \sigma(X)(k) = \left(\frac{1}{N} \sum_{n=1}^{N} X_n(k)^2 \right)^{\frac{1}{2}}. \tag{11.3}$$

Then the variance ellipsoid is centered at zero.

We write $\mathrm{Var}(X)$ for the total variance of the ensemble X. This is the sum of the squares of the coordinates in the variance vector $\sigma(X) \in \mathbf{R}^d$. In other words, $\mathrm{Var}(X) = \|\sigma(X)\|^2 \stackrel{\text{def}}{=} \sum_{k=1}^{d} \sigma(X)(k)^2$, or

$$\mathrm{Var}(X) = \sum_{k=1}^{d} \left[\frac{1}{N} \sum_{n=1}^{N} X_n(k)^2 - \left(\frac{1}{N} \sum_{n=1}^{N} X_n(k) \right)^2 \right]. \tag{11.4}$$

The *correlation coefficient* $C(X, Y)$ of two random variables is defined by the following equation:

$$C(X, Y) = \frac{E(XY) - E(X)E(Y)}{\sqrt{(E(X^2) - E(X)^2)\,(E(Y^2) - E(Y)^2)}}. \tag{11.5}$$

Here $E(XY) = \frac{1}{N} \sum_{n=1}^{N} X_n Y_n$ denotes the expectation of the variable xy, etc. We have $0 \le C(X, Y) \le 1$, with $C(X, Y) = 0$ if X and Y are independent random variables, *i.e.*, if knowledge of X confers no knowledge of Y. Likewise, $C(X,Y) = 1$ if and only if $X = Y$ almost surely.

We will show that the correlation coefficient for adjacent pixels is controlled by smoothness. For simplicity, consider the one-dimensional case with a periodic N-pixel picture. Suppose that $x(n)$ is the pixel at location n, where $n \in \{1, 2, \ldots, N\}$. We treat $\{1, 2, \ldots, N\}$ as the sample space, and define two random variables $X_n = x(n)$ and $Y_n = x(n+1)$, namely adjacent pixel values, where $x(N+1) \stackrel{\text{def}}{=} x(1)$. Without loss of generality, we may assume that $\frac{1}{N}\sum_{n=1}^{N} x(n) = 0$. Then $E(X) = E(Y) = 0$ and $E(X^2) = E(Y^2)$, which simplifies the denominator:

$$\sqrt{[E(X^2) - E(X)^2]\,[E(Y^2) - E(Y)^2]} \;=\; E(X^2) \;\geq\; 0.$$

Now suppose that $x = \{x(n)\}$ is *numerically smooth*, which means that the following assumption is valid:

$$0 \leq \delta \stackrel{\text{def}}{=} \max_{1 \leq n \leq N} \frac{|x(n+1) - x(n)|}{|x(n)| + |x(n+1)|} \ll 1. \tag{11.6}$$

Roughly speaking, this hypothesis guarantees that $\Delta x(n) \stackrel{\text{def}}{=} x(n+1) - x(n)$ remains small relative to $|x(n)|$ for all $n = 1, 2, \ldots, N$. Then we have the following lower estimate for the numerator of $C(X,Y)$:

$$\begin{aligned}
x(n+1) &= x(n) + \Delta x(n) \\
&\Rightarrow x(n+1) \geq x(n) - \delta\left(|x(n)| + |x(n+1)|\right) \\
&\Rightarrow x(n+1)x(n) \geq x(n)^2 - \delta\left(|x(n)|^2 + |x(n+1)||x(n)|\right); \\
x(n) &= x(n+1) - \Delta x(n) \\
&\Rightarrow x(n) \geq x(n+1) - \delta\left(|x(n)| + |x(n+1)|\right) \\
&\Rightarrow x(n+1)x(n) \geq x(n+1)^2 - \delta\left(|x(n)||x(n+1)| + |x(n+1)|^2\right).
\end{aligned}$$

We now average these two inequalities:

$$\begin{aligned}
x(n+1)x(n) &\geq \frac{1}{2}\left(x(n)^2 + x(n+1)^2\right) - \frac{\delta}{2}\left(|x(n)| + |x(n+1)|\right)^2 \\
&\geq \frac{1}{2}\left(x(n)^2 + x(n+1)^2\right) - \delta\left(|x(n)|^2 + |x(n+1)|^2\right) \\
&\qquad \Rightarrow E(XY) \geq E(X^2) - 2\delta E(X^2).
\end{aligned}$$

We can therefore estimate the correlation coefficient as follows:

$$1 \geq C(X,Y) \geq \frac{E(X^2) - 2\delta E(X^2)}{E(X^2)} = 1 - 2\delta. \tag{11.7}$$

Similarly, if we interpret $x(N+n)$ as $x(n)$ for $n = 1, 2, \ldots, k$ and $1 \le k \le N$, then the correlations between more distant pixels will also be close to one:

Proposition 11.1 *If* $\frac{1}{N}\sum_{n=1}^{N} x(n) = 0$ *and* x *is numerically smooth, then for* $X_n = x(n)$ *and* $Y_n = x(n+k)$ *we have* $1 \ge C(X,Y) \ge 1 - 2k\delta$. □

Since $C(X,Y) = C(Y,X)$, we can allow negative values of k just by substituting $|k|$. Notice too that $1 - 2|k|\delta \approx (1-\delta)^{2|k|} \stackrel{\text{def}}{=} r^{|k|}$, where $r = (1-\delta)^2 \approx 1$.

If the pixels are highly correlated, then there is a lower rank description of the image which captures virtually all of the independent features. In transform coding, we seek a basis of these features, in which the coordinates are less highly correlated or even uncorrelated. In this new basis most of the variation takes place in many fewer coordinates. When these few are approximated to sufficient precision to meet the distortion requirements, we get data compression.

11.1.2 Transform coding image compression

A particular application of the adapted waveform methods discussed so far is transform coding picture compression. The generic transform coding compression scheme is depicted in Figure 11.2. It consists of three pieces:

- *Transform:* An invertible or "lossless" transform which decorrelates the mutually dependent parts of the image;
- *Quantize:* A quantizer which replaces transform coefficients with (small) integer approximations. In theory, all of the distortion in "lossy" compression is introduced at this stage;
- *Remove redundancy:* An invertible redundancy remover, or *entropy coder*, which rewrites the stream of transform coefficients into a more efficient alphabet to asymptotically approach the information-theoretic minimum bitrate.

These three pieces are depicted in Figure 11.2.

To recover an image from the coded, stored data, we invert the steps in Figure 11.2 as shown in Figure 11.3. The first and third blocks of the compression algorithm are exactly invertible in exact arithmetic, but the "Unquantize" block does not in general produce the same amplitudes that were given to the "Quantize" block during compression. The errors thus introduced can be controlled both by the fineness of the quantization (which limits the maximum size of the error) and by favoritism (which tries to reduce the errors for certain amplitudes at the expense of greater errors for others).

We will explore each of these components in turn.

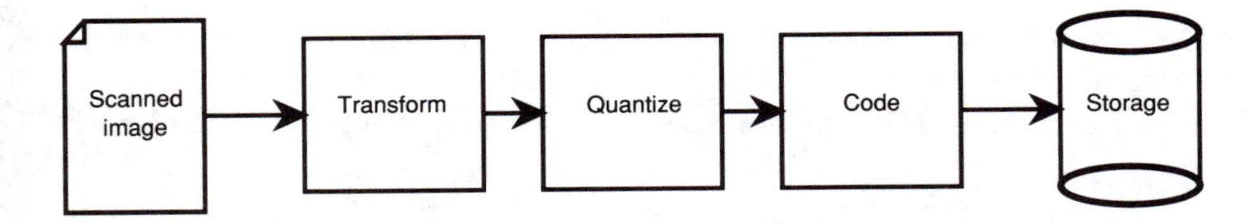

Figure 11.2: Generic transform coding image compression device.

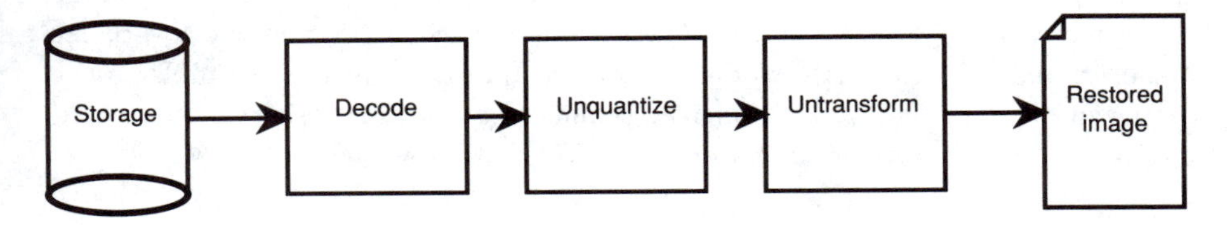

Figure 11.3: Inverse of generic transform coder: decoder.

Decorrelation by transformation

The first block of a transform coder applies an invertible change of coordinates to the image. We think of this transformation as implemented in real arithmetic, with enough precision to keep the truncation error below the quantization error introduced by the original sampling. The output of this block will be treated as a vector of real numbers. We will also treat the output of the decoding algorithm, which produces an image from coded data, as a vector of real numbers. If we consider the original pixels to be a vector of real numbers which happen to take only discrete values, then we can measure the *error of reconstruction* or the *lossiness* of the compression by comparing those two vectors.

We will consider an image coding algorithm to be *lossless* if no pixel of the restored image vector differs from the corresponding pixel of the original image by more than the sampling quantization error. Thus we have the following proposition:

Proposition 11.2 *For any fixed number of bits per pixel and fixed number of pixels, every transform coding scheme utilizing a continuous transform is lossless if the quantization is sufficiently fine.* □

If the transformation is linear, then having a fixed finite picture size guarantees that the transformation is continuous. If the transformation is orthogonal, then there is a very simple relationship between the quantization error and the reconstruction error which we shall compute below.

We will consider six pixel transformations which have proven useful in decorrelating smooth pictures.

Karhunen–Loève. The *autocovariance matrix* for an ensemble X is defined by

$$A \stackrel{\text{def}}{=} E(\tilde{X} \otimes \tilde{X}); \qquad A(i,j) = \frac{1}{N}\sum_{n=1}^{N} \tilde{X}_n(i)\tilde{X}_n(j). \tag{11.8}$$

Here we have taken $\tilde{X}_n \stackrel{\text{def}}{=} X_n - E(X)$ to be the original vector with the average vector subtracted. Thus $E(\tilde{X}) = 0$. The matrix coefficient $A(i,j)$ is the covariance of the ith and j^{th} coordinate of the random vector $\tilde{X}$, using the ensemble as the sample space. The matrix A is evidently symmetric. It is also positive (some would say positive semidefinite) since for every vector $Y \in \mathbf{R}^d$ we have

$$\begin{aligned} \langle Y, AY\rangle &= \sum_{i=1}^{d}\sum_{j=1}^{d} Y(i)A(i,j)Y(j) \\ &= \frac{1}{N}\sum_{n=1}^{N}\sum_{i=1}^{d}\sum_{j=1}^{d} Y(i)\tilde{X}_n(i)\tilde{X}_n(j)Y(j) \\ &= \frac{1}{N}\sum_{n=1}^{N}\langle Y, \tilde{X}_n\rangle^2 \ \geq\ 0. \end{aligned}$$

Let us now fix an image size—say height H and width W, so that $d = H \times W$ pixels—and treat the individual pixels as random variables. Our probability space will consist of some finite collection of pictures $\mathcal{S} = \{S_1, S_2, \ldots, S_N\}$, where N is a big number. The intensity of the ith pixel $S(i)$, $1 \leq i \leq d$, is a random variable that takes a real value for each individual picture $S \in \mathcal{S}$.

As in the analysis of the smooth one-dimensional function in Proposition 11.1, we can compute the correlations of adjacent pixels over the probability space. We may assume without loss of generality that $E(S) = 0$, *i.e.*, $\frac{1}{N}\sum_{n=1}^{N} S_n(i) = 0$ for all i. If $X_n = S_n(i)$ is the ith pixel value of the nth picture and $Y_n = S_n(j)$ is the jth pixel value of the nth picture, then

$$C(X,Y) = \frac{\frac{1}{N}\sum_{n=1}^{N} S_n(i)S_n(j)}{\left(\frac{1}{N}\sum_{n=1}^{N} S_n(i)^2\right)^{1/2}\left(\frac{1}{N}\sum_{n=1}^{N} S_n(j)^2\right)^{1/2}}. \tag{11.9}$$

Of course $i = (i_x, i_y)$ and $j = (j_x, j_y)$ are multi-indices. The numerator of this expression is the autocovariance matrix of the collection of pictures $\mathcal{S}$. If the collection is *stationary*, *i.e.*, if it contains all translates of each picture, then the variance of each pixel will be the same. Thus the denominator will be constant in i and j, and the autocovariance matrix and matrix of correlation coefficients will be indistinguishable.

This notion of adjacent pixel correlation is different from the one defined in Equation 11.5. In that case, we computed the correlation of two pixels a fixed distance apart, averaging over all the shifts of a single function, whereas here we are computing the correlation of pixels at two fixed locations, averaging over a bunch of different functions. The single function case is naturally stationary, or shift-invariant: the correlation coefficient depends only on the distance $|i-j|$ between the pixels. We can always make the second case stationary by enlarging an arbitrary $\mathcal{S}$ to include all (periodized) shifts of its pictures. This adds no new information, but it lets us estimate the autocovariance matrix with our model of the correlation coefficients. It also guarantees that A will be a convolution matrix: there will be some function f such that $A(i,j) = f(|i-j|)$. We can use absolute values because $A(i,j) = A(j,i)$ is a symmetric real matrix.

Because of adjacent pixel correlation in smooth pictures, the autocovariance matrix A will have off-diagonal terms. However, A can be diagonalized because it is real and symmetric (see [2], Theorem 5.4, p.120, for a proof of this very general fact). Also, we have shown that A is positive semidefinite so all of its (real) eigenvalues are nonnegative. We can write K for the orthogonal matrix that diagonalizes A; then K^*AK is diagonal, and K is called the *Karhunen–Loève transform*, or alternatively the *principal orthogonal decomposition*. The columns of K are the vectors of the Karhunen–Loève basis for the collection S, or equivalently for the matrix A. The number of positive eigenvalues on the diagonal of K^*AK is the actual number of uncorrelated parameters, or degrees of freedom, in the collection of pictures. Each eigenvalue is the variance of its degree of freedom. K^*S_n is S_n written in these uncorrelated parameters: we can achieve compression by just transmitting this lower number of coordinates.

Unfortunately, the above method is not practical because of the amount of computation required. For typical pictures, d is in the range 10^4 to 10^6. To diagonalize A and find K requires $O(d^3)$ operations in the general case. Furthermore, to apply K^* to each picture requires $O(d^2)$ operations in general. We will discuss a lower-complexity approximate version of the algorithm in Section 11.2. However, we can make some simplifying assumptions to get a rapidly computable substitute for the full Karhunen–Loève transform. We will assume that the autocovariance matrix for a collection $\mathcal{S}$ of smooth pictures is of the form

$$A(i,j) = r^{|i-j|}, \tag{11.10}$$

where the adjacent pixel correlation coefficient r satisfies $0 < 1 - r \ll 1$. The

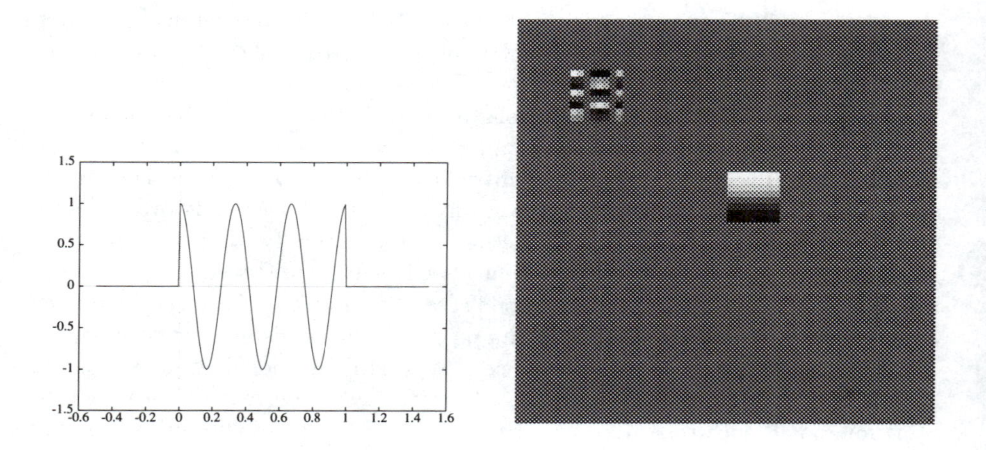

Figure 11.4: Example DCT basis functions in one and two dimensions.

expression $|i - j|$ should be interpreted as $|i_r - j_r| + |i_c - j_c|$, where i_r and i_c are respectively the row and column indices of pixel i, and similarly for j. Experience shows that this is quite close to the truth for small sections of large collections of finely sampled smooth pictures.

DCT methods and JPEG. In the limit $d \to \infty$, it is possible to compute the Karhunen–Loève basis exactly for the autocovariance matrix in Equation 11.10. In that case A is the matrix of a two-dimensional convolution with an even function, so it is diagonalized by the two-dimensional discrete cosine transform (DCT). The basis functions for this transform in one and two dimensions are displayed in Figure 11.4. This limit transform can be used instead of the exact Karhunen–Loève basis; it has the added advantage of being rapidly computable via the fast DCT derived from the fast Fourier transform. The Joint Photographic Experts Group (JPEG) algorithm [1, 109] uses this transform and one other simplification. d is limited to 64 by taking 8×8 subblocks of the picture.

Lapped orthogonal or local trigonometric transforms. Chopping a picture into 8×8 blocks creates artifacts along the block boundaries which must be ameliorated downstream. Or, we can fix this problem within the transform portion of the coder. Rather than use disjoint blocks as in JPEG, it is possible to use "localized" or "lapped" (but still orthogonal) discrete cosine functions which are supported on overlapping patches of the picture. These *local cosine transforms* (LCT, as in [24]) or *lapped orthogonal transforms* (LOT, as in [74]) are modifications of DCT and are

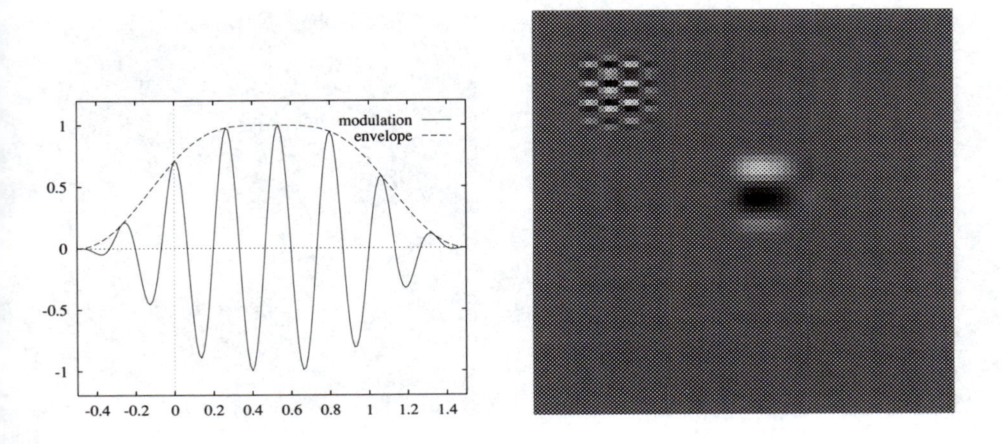

Figure 11.5: Example LCT basis functions in one and two dimensions.

described in Chapter 4. The basis functions chiefly differ from those of the DCT-IV transform, which is the discrete cosine transform using half-integer grid points and half-integer frequencies, in that they are multiplied by a smooth window. Figure 11.5 shows one and two dimensional versions of such functions, with d chosen large enough so that the smoothness is evident. Overlapping functions of this form are orthogonal because of the half-integer frequency indices. In Chapter 4 we described how to use arbitrary periodic functions, albeit of a different form. The formulas for the smooth overlapping basis functions in two dimensions are tensor products of the formulas in one dimension; they are described in Chapter 9.

Rather than calculate inner products with the sequences ψ_k, we can preprocess data so that standard fast DCT-IV algorithms may be used. This may be visualized as "folding" the overlapping parts of the bells back into the interval; the formulas are described in Chapter 4, and they require just $2d$ operations to implement for an d-pixel picture. LOT therefore has the same order of complexity as DCT.

Adapted block cosines and local cosines. We can also build a library tree of block LCT bases (or block DCT bases) and search it for the minimum of some cost function. The chosen *best LCT basis* will be a patchwork of different sized blocks, adapted to different sized embedded textures in the picture. An example of two such functions is shown in Figure 11.6.

In addition to the transform coefficients, we will also have to transmit a description of which blocks were chosen. This is discussed in Chapter 9, Section 9.2.1.

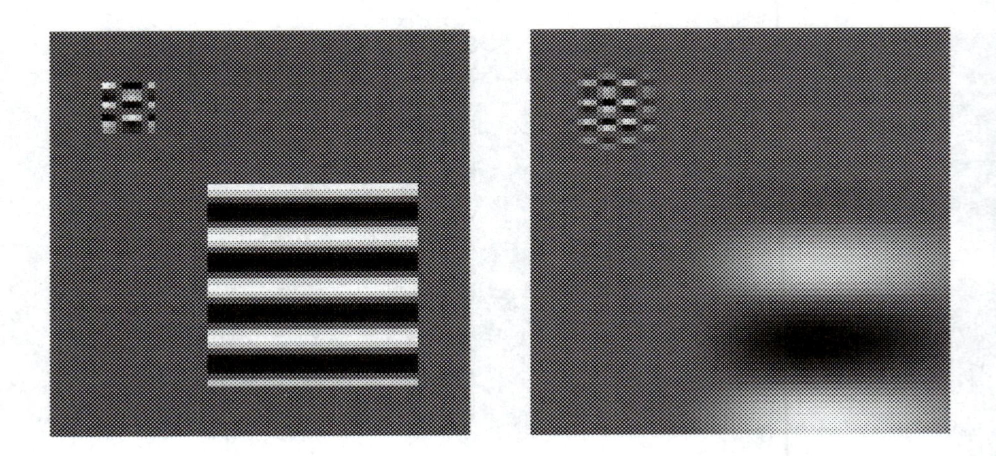

Figure 11.6: Example adapted DCT and LCT basis functions in two dimensions.

Custom subband coding. We can also choose the underlying basis functions to be products of one-dimensional wavelets and wavelet packets. When these are combined into functions of two variables, they produce the two-dimensional wavelets and wavelet packets shown in Figures 11.7 and 11.8. Superpositions of such basis functions produce textures and other image features, as shown in Figure 11.9.

The underlying functions are distinguishable by their spatial frequencies which are extracted by filtering. A picture S may be thus divided into orthogonal spatial frequency *subbands* by repeated application of a pair of digital filters, one high-pass and one low-pass, with mutual orthogonality properties. If H and G are respectively low-pass and high-pass conjugate quadrature filters (CQFs) defined on one-dimensional signal sequences, then we can define four two-dimensional convolution-decimation operators in terms of H and G, namely the tensor products of the pair of conjugate quadrature filters: $F_0 \stackrel{\text{def}}{=} H \otimes H$, $F_1 \stackrel{\text{def}}{=} H \otimes G$, $F_2 \stackrel{\text{def}}{=} G \otimes H$, $F_3 \stackrel{\text{def}}{=} G \otimes G$. These convolution-decimations have adjoints F_i^*, $i = 0, 1, 2, 3$. Their orthogonality relations are those of Equation 9.1, specialized to four operators.

Since the conjugate quadrature filters H and G form a partition of unity in the Fourier transform or wavenumber space, the same is true for the separable filters F_i. They can be described as nominally dividing the support set of the Fourier transform $\hat{S}$ of the picture into dyadic squares. If the filters were perfectly sharp, then this would be literally true, and the children of S would correspond to the four dyadic subsquares one scale smaller. We illustrate this in Figure 11.10.

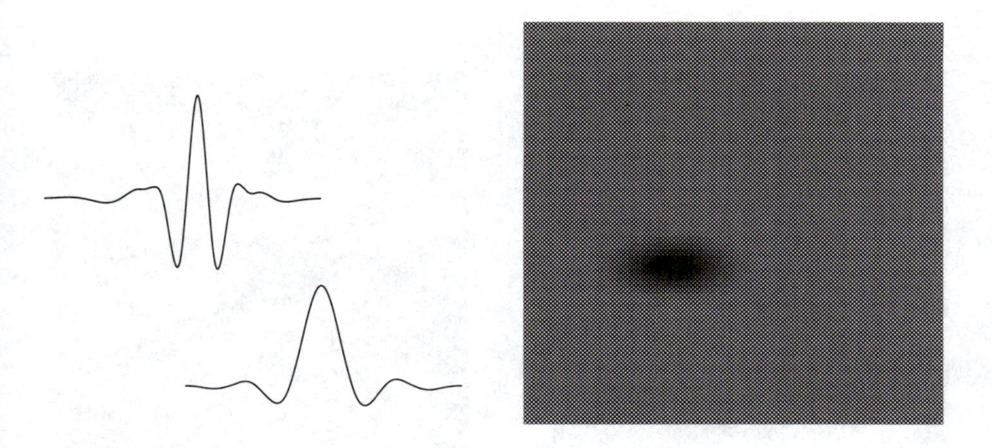

Figure 11.7: Wavelets in one and two dimensions.

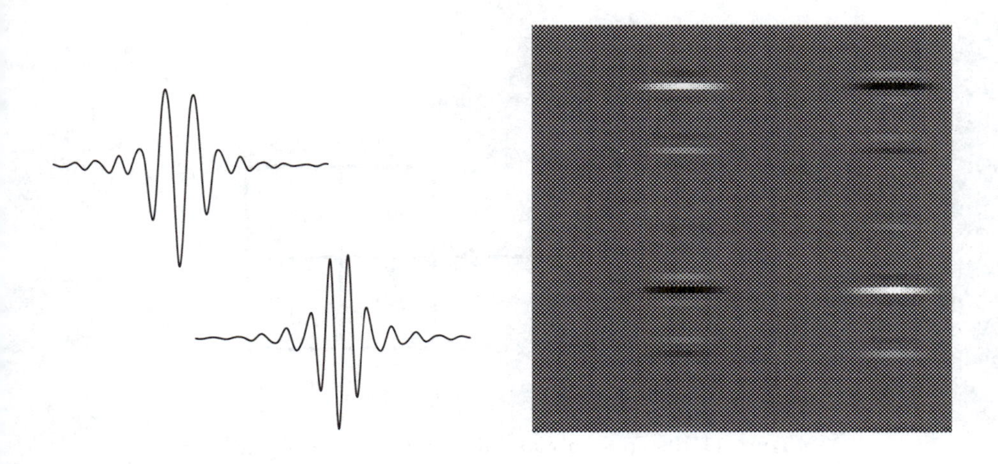

Figure 11.8: Wavelet packets in one and two dimensions.

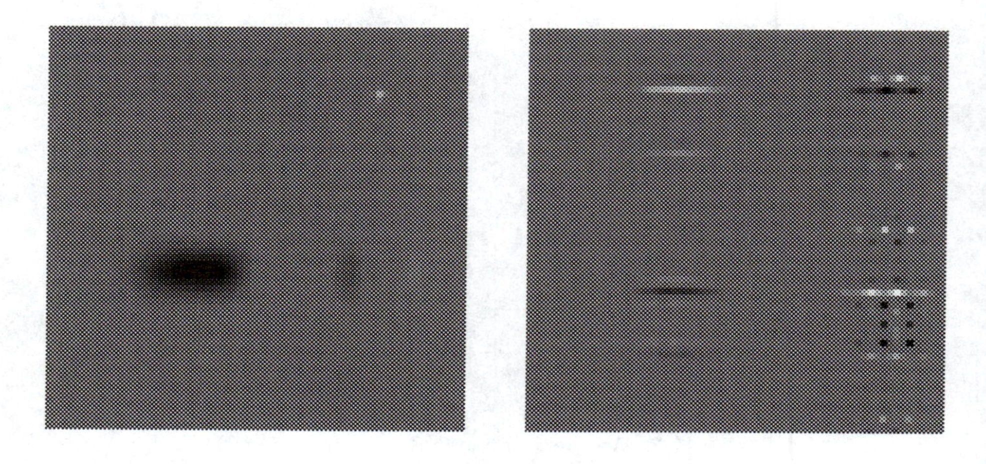

Figure 11.9: Superpositions: three wavelets, three wavelet packets.

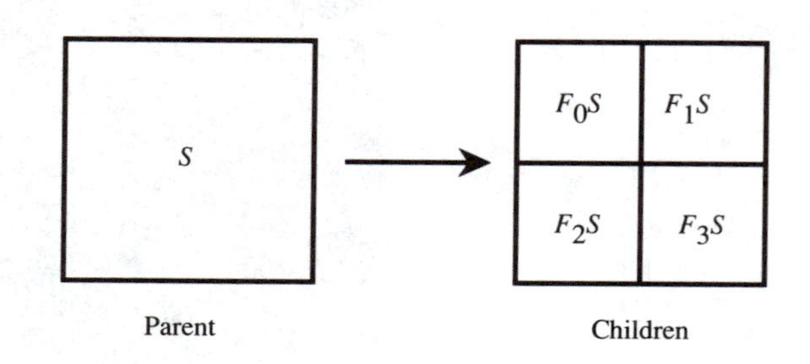

Figure 11.10: Four subband descendents of a picture.

All subbands together form a *quadtree*, in which each subspace forms a node and directed edges go from precursors to descendents. The orthogonality relation among the subspaces implies that every connected subtree which contains the root picture corresponds to an orthonormal subband decomposition of the original picture determined by the leaves of the subtree. Individual subbands are in one-to-one correspondence with rectangular regions in wavenumber space, and the quadtree stacks these regions one on top of the other. We can idealize various orthogonal subband bases as disjoint covers of wavenumber space by squares: they preserve their aspect ratio, since both axes are halved when we descend one level. A few such bases are schematically represented in Figure 9.14 in Chapter 9.

Best basis of wavelet packets. Note that the wavelet basis, the FBI WSQ basis, and the traditional bases of subband image coding are all custom wavelet packet bases giving custom subband coding schemes. It is possible to design a custom scheme for any class of pictures by using the two-dimensional best basis of wavelet packets for each element and then finding a joint best basis. In the case of fingerprints, the FBI expanded many sample pictures into individual best-bases before selecting the fixed WSQ basis, thus bolstering confidence in the effectiveness of the fixed choice. The low complexity and perceived freedom from patent protection of the resulting fixed-basis transform justified the risk of poor compression for exceptional images. But it is possible to use a wavelet packet best basis for each individual image, as discussed in Chapters 8 and 9. We will then be forced to include basis description data with the coefficients; if we use the levels list and encounter order convention, the description overhead can be kept to a minimum.

Quantization

The second ("Quantize") block of the transform coding compression flow chart replaces the real number coordinates with lower precision approximations which can be coded in a (small) finite number of digits. The output of this block is a stream of small integers. If the transform step was effective, then the output integers are mostly very small (namely zero) and can be ignored, while only a few are large enough to survive. The two sides of Figure 11.11 illustrate how coefficients fall into *quantization bins* indexed by small integers, before and after an effective transform step. If our goal is to reduce the rank of the representation, we can now stop and take only the surviving large values, which correspond to nonnegligible amplitudes, and tag them with some identifiers.

Not all coefficients need to be quantized into the same number of bins. For example, if the images are intended for the human eye, we can combine a knowledge of how distortion is perceived with a knowledge of the properties of our synthesis

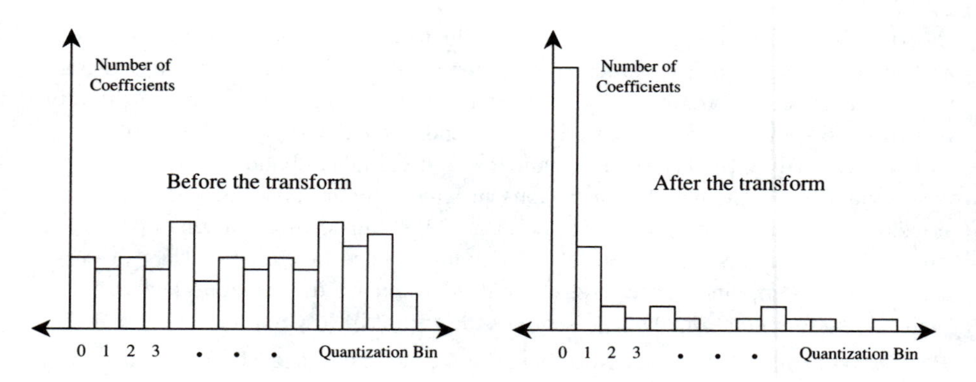

Figure 11.11: Number of coefficients per quantization bin, before and after an effective transform step.

functions in order to concentrate quantization error where it will be least visible. This is easy to do with a custom subband coding scheme, since synthesis functions within a subband are characterized by their spatial frequencies which in turn, to a large extent, determine their visibility. We can use fewer bins and coarser quantization in high-frequency subbands, where errors are less visible. But even for nonhuman audiences, such as automatic image detection, classification, and recognition machines, the quantization can be adjusted to preserve or even emphasize image features which are known to be important. In the FBI WSQ case, subbands which describe pore frequencies and ridge spacings are quantized more finely to assist subsequent automatic fingerprint identification systems, as well as human fingerprint examiners. For individual best-basis compression, it may be necessary to append a quantization table to the compressed image data. The JPEG standard [1] also allows individualizing the quantization table for each image, to improve its performance on exceptional images.

We may also vary the quantization by assigning different numbers of bins to different groups of coefficients, in proportion to how much variance those coefficients exhibit. We may compute the variance empirically for a group of coefficients from an individual image, such as all coefficients within one subband, or we may estimate the variance of each coefficient from a model of the class of images we plan to compress. In the first case, we must transmit a *bit allocation* table along with the compressed image, so we should take fairly large groups of coefficients to keep down the overhead. In the second case, the bit allocation scheme will be fixed and known to both transmitter and receiver, so there will be no overhead.

Redundancy removal

Our goal is to reduce the number of bits we must transmit or store, so after quantization we should code the sequence of bin indices for greatest efficiency. The third block ("Remove redundancy") replaces the stream of small integers with a more efficient alphabet of variable-length characters. In this alphabet the frequently occurring letters (like "0") are represented more compactly than rare letters.

The Huffman algorithm ([104], p.39ff) or one of its variants may be used for entropy coding. These treat the input as a sequence of independent Bernoulli trials with a known or empirically determined probability distribution function; they chose a new alphabet so as to minimize the expected length of the output. Our assumption is that the transform step has decorrelated the coefficients to the point that they are hard to distinguish from a sequence of independent random variables. This assumption is not valid if there are long runs of zeroes, but that can be fixed by introducing special code words for common run lengths. Such a technique is part of the FBI WSQ algorithm definition, since certain subbands are expected to be quantized to zero almost entirely.

11.2 Fast approximate factor analysis

There are many names for the algorithms described below: *factor analysis*, *principal component analysis*, *singular value decomposition*, and the *Karhunen–Loève transform* are some of the more common ones. It has both an algebraic interpretation, finding the eigenvalues of a matrix, and an analytic interpretation, finding the minimum of a cost function over the set of orthogonal matrices. The first interpretation is best when we need an exact solution, but it has high arithmetic complexity. The minimization point of view leads to a lower complexity approximate algorithm which gets close to the minimum but does not attain it.

Principal orthogonal decomposition can be used to solve two related problems: distinguishing elements from a collection by making d measurements, and inverting a complicated map from a p-parameter configuration space to a d-dimensional measurement space. In the case where d is more than 1000 or so, the classical $O(d^3)$ singular value decomposition algorithm becomes very costly, but it can be replaced with an approximate best basis method that has complexity $O(d^2 \log d)$. This can be used to compute an approximate Jacobian for a complicated map from $\mathbf{R}^p \rightarrow \mathbf{R}^d$ in the case $p \ll d$.

Consider the problem of most efficiently distinguishing elements from a collection by making d measurements. In general, we will need all d measured values to fully specify an element. However, it is possible to use less information if some of the measurements are correlated. For example, if the objects are parametrized by a small number $p \ll d$ of parameters, then the d measurements should separate them in a redundant fashion. That is, we can change basis locally in the d-dimensional measurement space to find just p combinations of measurements which change with the p parameters. This idea works even if there are many parameters but only p of them are relatively important.

Note the resemblance between the problem of distinguishing elements and the problem of inverting a complicated map from $\mathbf{R}^p$ to $\mathbf{R}^d$. In the first problem, we must find a discrete object given its description in $\mathbf{R}^d$. In the second, we must find the parameters in $\mathbf{R}^p$ from the description in $\mathbf{R}^d$. These problems are identical if the collection of objects is produced by evaluating the complicated map at discrete grid points in $\mathbf{R}^p$.

The combinations of measurements which root out the underlying parameters are called *principal (orthogonal) components* or *factors*; they have a precise meaning, and the well-known and well-behaved method of *singular value decomposition* or *SVD* produces them with arbitrary accuracy. However, SVD has a complexity that is asymptotically $O(d^3)$, making it impractical for problems larger than $d \approx 1000$. In this section, we will describe the classical principal factor algorithm, then give a lower complexity *approximate principal factor algorithm.* Finally, we will give example applications of the approximate algorithm to the two mentioned problems.

The Karhunen–Loève transform

The *principal orthogonal* or *Karhunen–Loève coordinates* for an ensemble $X = \{X_n \in \mathbf{R}^d : n = 1, \ldots, N\}$ correspond to the choice of axes in $\mathbf{R}^d$ which minimizes the volume of the variance ellipsoid defined by Figure 11.1. These axes should be the eigenvectors of the autocovariance matrix $A = E(\tilde{X} \otimes \tilde{X})$ of the ensemble, defined by Equation 11.8. Since this matrix is positive semidefinite, we can find an orthonormal basis for $\mathbf{R}^d$ consisting of eigenvectors, and these eigenvectors will have nonnegative real eigenvalues. Thus we are assured that the *Karhunen–Loève* basis exists for the ensemble X; this is the orthonormal basis of eigenvectors of A.

The Karhunen–Loève basis eigenvectors are also called *principal orthogonal components* or *principal factors*, and computing them for a given ensemble X is also called *factor analysis.* Since the autocovariance matrix for the Karhunen–Loève eigenvectors is diagonal, it follows that the Karhunen–Loève coordinates of the vectors in the sample space X are uncorrelated random variables. Let us denote these basis eigenvectors by $\{Y_n : n = 1, \ldots, N\}$, and let us denote by K the $d \times d$ matrix whose columns are the vectors $Y_1, \ldots, Y_N$. The adjoint of K, or K^*, is the matrix which changes from the standard coordinates into Karhunen–Loève coordinates;

this map is called the *Karhunen–Loève transform.*

Unfortunately, finding these eigenvectors requires diagonalizing a matrix of order d, which has complexity $O(d^3)$. In addition, even after already computing the Karhunen–Loève eigenvectors of an ensemble, updating the basis with some extra random vectors will cost an additional $O(d^3)$ operations since it requires another diagonalization.

Such a high order of complexity imposes a ceiling on the size of problem we can do by this method. In many cases of interest, d is very large and X spans $\mathbf{R}^d$, implying $N \geq d$. Thus even to find the coefficients of the autocovariance matrix requires $O(d^3)$ operations. At the present time, we are limited to $d \leq 10^3$ if we must use common desk top computing equipment, and $d \leq 10^4$ for the very most powerful computers.

So we shall take another perspective. We shall pose the problem of finding the Karhunen–Loève eigenvectors as an optimization over the set of orthogonal transformations of the original ensemble X. The quantity to be maximized will be a *transform coding gain*, or the amount of compression we achieve by using another basis to represent the ensemble. This gain is increased if we can decrease the volume of the variance ellipsoid for the signal ensemble, and is thus maximized by the Karhunen–Loève transform. We get an approximate Karhunen–Loève transform from any efficient transform which significantly increases the gain, even if does not attain the maximum.

Alternatively, we could introduce a distance function on the set of orthonormal bases for $\mathbf{R}^d$, treat the Karhunen–Loève basis as a distinguished point in this set, and then try to find an efficiently computable basis which is close to the Karhunen–Loève basis.

A metric on the orthogonal matrices

Consider Figure 11.12, which schematically depicts all the orthonormal bases of $\mathbf{R}^d$. These can be identified with certain transformations of $\mathbf{R}^d$, namely the orthogonal $d \times d$ matrices. The points marked by "x" represent bases in some library of fast transforms. The point marked "o" represents the optimal, or Karhunen–Loève basis, for a given ensemble of vectors. The point marked "xx" represents the fast transform closest to the Karhunen–Loève bases.

Let U be an orthogonal $d \times d$ matrix, and write $Y = UX$ to signify that $Y_n = UX_n$ for each $n = 1, 2, \ldots, N$. Since U is linear, $E(Y) = E(UX) = UE(X)$, which will be zero if we started with $E(X) = 0$. Since U is orthogonal, it preserves sums of squares, so $\mathrm{Var}(X) = \mathrm{Var}(Y)$. Using wavelet packets or adapted local trigonometric functions, it is possible to build a library of more than 2^d fast transforms U of $\mathbf{R}^d$

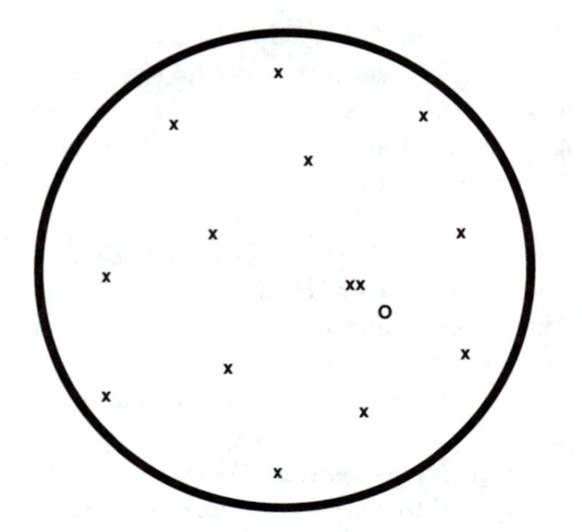

x = fast transform bases; o = Karhunen–Loève basis; xx = best fast basis.

Figure 11.12: Orthonormal bases for $\mathbf{R}^d$.

to use for the "x" points. We will illustrate the construction using wavelet packets. These transforms are arranged in a structure that permits us to search for the one closest to the "o" point in $O(d \log d)$ operations. We will use a notion of closeness that is derived from the function minimized by the Karhunen–Loève transform.

As in [61], define the *transform coding gain* for an orthogonal matrix by the formula

$$G_{TC}(U) = \text{Var}(UX)/\exp H(UX), \quad \text{where} \quad H(X) = \frac{1}{d}\sum_{i=1}^{d} \log \sigma(X)(i). \quad (11.11)$$

From this formula we see that $G_{TC}(UX)$ is maximized when $H(UX)$ is minimized. The quantity H has various interpretations. It is the entropy of the direct sum of d independent Gaussian random variables with variances $\sigma(X)(i)$, $i = 1, \ldots, d$. It is also equal to the logarithm of the volume of the variance ellipsoid (if we add $\log \omega_d$), so we see that minimizing $H(UX)$ or maximizing $G_{TC}(UX)$ is equivalent to minimizing the volume of the variance ellipsoid for the ensemble UX over all orthogonal matrices U.

Since the Karhunen–Loève transform is a global minimum for H, we will therefore say that the best approximation to the Karhunen–Loève transform from a library $\mathcal{U}$ of orthogonal matrices is the minimum of $H(UX)$ with the constraint $U \in \mathcal{U}$. We can define the *approximate factor analysis algorithm* to be the search through a library of orthonormal bases for the one whose H is closest to that of the Karhunen–Loève basis. If the library of bases is organized to facilitate a fast search, we will dare to call the result a *fast approximate Karhunen–Loève algorithm.*

The "closeness" of a basis U to the Karhunen–Loève basis K can be measured by computing the transform coding gain of U and subtracting that of K. This give us a *transform coding gain metric*:

$$dist_X(U,V) = |H(UX) - H(VX)|.$$

Notice that we get a different metric for each ensemble X. This is a degenerate metric on the orthogonal group, since it gives a distance of zero between bases which have the same transform coding gain for X. However, this technical point can be overcome by constructing a topological quotient space in which such bases are considered equivalent.

A minimum for $H(VX)$ is the Karhunen–Loève basis $V = K$, so minimizing $H(UX)$ over fast transforms U is equivalent to minimizing $dist_X(U,K)$: it finds the closest fast transform for this ensemble in the transform coding sense.

The entropy metric

If U is a $d \times d$ orthogonal matrix, then we can define its *entropy* to be

$$\mathcal{H}(U) = -\sum_{i=1}^{d}\sum_{j=1}^{d} |U(i,j)|^2 \log\left(|U(i,j)|^2\right). \tag{11.12}$$

We have the two inequalities

$$0 \leq \mathcal{H}(U) \leq d \log d, \tag{11.13}$$

which are elementary consequences of properties of the entropy of a probability distribution function and the fact that the squares of the elements of each of the d columns of U sum up to one. $\mathcal{H}$ is a functional on $\mathbf{O}(d)$, the compact Lie group of orthogonal linear transformations of $\mathbf{R}^d$. $\mathcal{H}$ can be used to define a distance function.

The *entropy metric* on the orthogonal group is the function

$$dist(U,V) = \mathcal{H}(U^*V)$$

defined for $U, V \in \mathbf{O}(d)$. This function adds up the entropy of the coordinates of the basis vectors for V when they are expanded in the basis U.

Note that $\mathcal{H}(I) = 0$ for the identity matrix I, and that $\mathcal{H}(U^*) = \mathcal{H}(U)$. These two properties of $\mathcal{H}$ imply that $dist(U,U) = 0$ and $dist(U,V) = dist(V,U)$. We also have the triangle inequality: $dist(U,V) \leq dist(U,W) + dist(W,V)$.

11.2.1 The approximate KL transform

We now turn to the large library of rapidly computable orthonormal wavelet packet bases, constructed to take advantage of the rapid growth of the number of subtrees of a binary tree. We will fix our attention on wavelet packets; it is relatively easy to generalize this example to the other adapted wavelet approximate Karhunen–Loève expansions.

Let H and G be conjugate orthogonal QFs. Starting with a signal $x = \{x(j) : j = 0, 1, \ldots, d-1\}$ of $d = M2^L$ samples, we recursively apply H and G a total of L times to get a complete wavelet packet analysis down to level L. We arrange the resulting sequences into a binary tree, which now contains very many basis subsets. Each graph, for example, gives a different orthonormal basis.

Suppose that the sequence of coefficients in block f of level s of the tree for the signal X_n is called $\{\lambda_{sf}^{(n)}(p)\}$. We then sum the coefficients of the N *signal* trees into two *accumulator* trees:

- The tree of *means*, which contains $\sum_{n=0}^{N-1} \lambda_{sf}^{(n)}(p)$ in location p of block f at level s, and so on;
- The tree of *squares*, which contains $\sum_{n=0}^{N-1} \left[\lambda_{sf}^{(n)}(p)\right]^2$ in location p of block f at level s, and so on.

Computing all the coefficients of all the blocks in an L-level tree starting from d samples takes $O(dL) = O(d \log d)$ operations per random vector, for a total of $O(Nd \log d)$ operations. After we do this for all the random vectors X_n in the ensemble X, we can produce the binary tree of *variances* by using Equation 11.2: at index p of block f at level s, for example, it contains

$$\sigma_{sf}^2(X)(p) \stackrel{\text{def}}{=} \frac{1}{N} \sum_{n=0}^{N-1} \left[\lambda_{sf}^{(n)}(p)\right]^2 - \left[\frac{1}{N} \sum_{n=0}^{N-1} \lambda_{sf}^{(n)}(p)\right]^2. \tag{11.14}$$

This is the variance of the wavelet packet coefficient $\lambda_{sf}(p)$ defined by the filters H, G. Forming this tree takes an additional $O(d \log d)$ operations.

The tree of variances may now be searched for the graph basis which maximizes transform coding gain. We use the $\log \ell^2$ information cost function:

$$\mathcal{H}(V_{jn}) \stackrel{\text{def}}{=} \sum_{k=0}^{d/2^j-1} \log \sigma_{jn}(X)(k). \tag{11.15}$$

Notice that each block is examined twice during the best basis search: once when it is a parent and once when it is a child. This means that the search requires as many comparison operations as there are blocks in the tree, which is $O(d)$. Computing $\mathcal{H}$ costs no more than a fixed number of arithmetic operations per coefficient in the tree, which is $O(d \log d)$. Thus the total cost of the search is $O(d \log d)$.

Let U be the best basis in the variance tree; call it the *joint best basis* for the ensemble X, in the wavelet packet library. We can also denote by U the $d \times d$ orthogonal matrix which corresponds to the orthonormal basis. Abusing notation, we write $\{U_i \in \mathbf{R}^d : i = 1, \ldots, d\}$ for the rows of U. We may suppose that these rows are numbered so that $\sigma(UX)$ is in decreasing order; this can be done by sorting all the d coefficients in all the blocks $V \in U$ into decreasing order, which can be done in $O(d \log d)$ operations.

If we fix $\epsilon > 0$ and let d' be the smallest integer such that

$$\sum_{n=1}^{d'} \sigma(UX)(n) \geq (1-\epsilon)\mathrm{Var}(X),$$

then the projection of X onto the row span of $U' = \{U_1 \ldots U_{d'}\}$ contains $1-\epsilon$ of the total variance of the ensemble X. Call this projected ensemble X'. The d' row vectors of U' are already a good basis for the ensemble X', but they may be further decorrelated by Karhunen–Loève factor analysis. The row vectors of U' are just $U'_i = U_i$ for $1 \leq i \leq d'$, and the autocovariance matrix for this new collection is given by

$$M'_{ij} = \frac{1}{N} \sum_{n=1}^{N} U'_i \tilde{X}'_n U'_j \tilde{X}'_n.$$

Here $\tilde{X}'_n = X'_n - E(X')$ is a vector in $\mathbf{R}^{d'}$, and $E(\tilde{X}') = 0$. Thus M' is a $d' \times d'$ matrix and can be diagonalized in $O(d'^3)$ operations. Let K' be the matrix of eigenvectors of M'. Then K'^* changes from the joint best basis coordinates (calculated from the standard coordinates by U') into coordinates with respect to these decorrelated eigenvectors. We may thus call the composition K'^*U' the *approximate Karhunen–Loève transform* with relative variance error ϵ.

Complexity

The approximate Karhunen–Loève or joint best basis algorithm is fast because we expect that even for small ϵ we will obtain $d' \ll d$. To count operations in the worst case, we make the following assumptions:

Assumptions for computing complexity

1. There are N random vectors;
2. They belong to a d-dimensional parameter space $\mathbf{R}^d$;
3. The autocovariance matrix has full rank, so $N \geq d$.

There are five parts to the algorithm to build the tree of variances and search it for the joint best basis, which *trains* the algorithm to choose a particular decomposition:

Finding the approximate Karhunen–Loève basis

- Expanding N vectors $\{X_n \in \mathbf{R}^d : n = 1, 2, \ldots, N\}$ into wavelet packet coefficients: $O(Nd \log d)$;
- Summing squares into the variance tree: $O(d \log d)$;
- Searching the variance tree for a best basis: $O(d + d \log d)$;
- Sorting the best basis vectors into decreasing order: $O(d \log d)$;
- Diagonalizing the autocovariance matrix of the top d' best basis vectors: $O(d'^3)$.

Adding these up, we see that the total complexity of constructing the approximate Karhunen–Loève transform K'^*U', is $O(Nd \log d + d'^3)$. This compares favorably with the complexity $O(Nd^2 + d^3)$ of the full Karhunen–Loève expansion, since we expect $d' \ll d$.

Depending on circumstances, the last step $U' \mapsto K'^*U'$ may not be necessary, since a large reduction in the number of parameters is already achieved by transforming into the orthonormal basis determined by U'. This reduces the complexity to $O(Nd \log d)$, with the penalty being less decorrelation of the factors.

After training to learn the joint best basis, the algorithm *classifies* new vectors by expanding them in the chosen basis and separating them by their principal components:

The approximate Karhunen–Loève transform of one vector

- Computing the wavelet packet coefficients of one vector: $O(d \log d)$.
- Applying the $d' \times d'$ matrix K'^*: $O(d'^2)$.

Since $d' \ll d$, this estimate compares favorably with the complexity of applying the full Karhunen–Loève transform to a vector, which is $O(d^2)$. Further savings are possible, notably because only a small fraction $d'' \ll d'$ of the Karhunen–Loève singular vectors are needed to capture almost all of the original ensemble variance. Hence we can take K'' to be just the first d'' of the columns of K', and then the total complexity of applying K''^*U' is bounded by $O(d \log d + d''d')$.

If we expect to update the Karhunen–Loève basis, then we might also expect to update the average vector and the average value of each coordinate in the library of bases, as well as the variance of the ensemble. But since we keep a sum-of-squares tree and a sum-of-coefficients or means tree rather than a variance tree, each additional random vector just contributes its wavelet packet coordinates into the means tree and the squares of its coordinates into the sum-of-squares tree. The variance tree is updated at the end using the correct new means. This results in the following update complexity:

Updating the approximate Karhunen–Loève basis

- Expanding one vector into wavelet packet coefficients: $O(d \log d)$.
- Adding the coefficients into the means tree: $O(d \log d)$.
- Adding the squared coefficients into the squares tree: $O(d \log d)$.
- Forming the variance tree and computing the new information costs: $O(d \log d)$.
- Searching the variance tree for the joint best basis: $O(d + d \log d)$.

So one new vector costs $O(d \log d)$, and updating the basis with $N > 1$ new vectors costs $O(Nd \log d)$.

11.2.2 Classification in large data sets

The Karhunen–Loève transform can be used to reparametrize a problem so as to extract prominent features with the fewest measurements. When the number of measurements is huge, the fast approximate algorithm must be used at least as a "front end" to reduce the complexity of the SVD portion of finding the Karhunen–Loève basis.

We list a few examples to give some indication of the size of a problem that can be treated by the approximate method on typical table top computing equipment.

The rogues' gallery problem

The "rogues' gallery" problem is to identify a face from among a collection of faces. This problem was first suggested to me by Lawrence Sirovich, who also provided the data used in this experiment. The random vectors were several thousand digitized 128×128 pixel, eight-bit gray-scale pictures of Brown University students, so $d =$

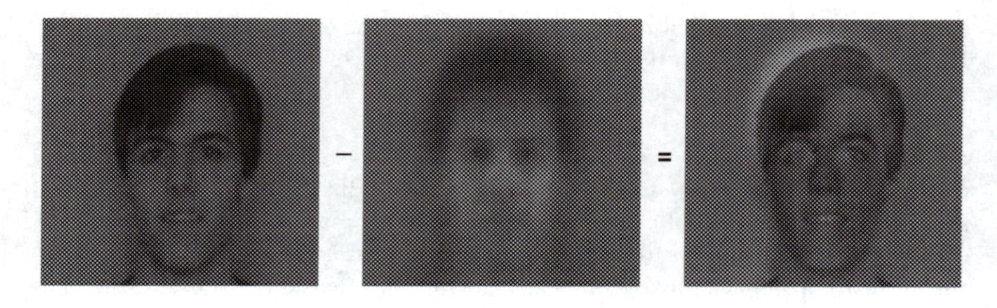

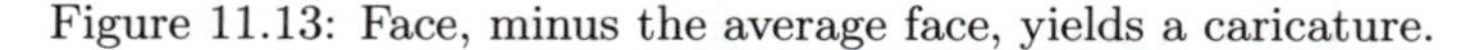

Figure 11.13: Face, minus the average face, yields a caricature.

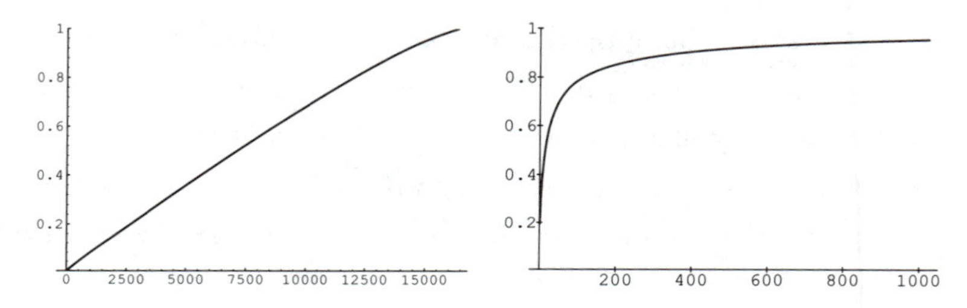

Figure 11.14: Accumulation of variance in the original basis and the joint best basis.

$128^2 = 16,384$. These were initially normalized with the pupils impaled on two fixed points near the center of the image. In [100, 62], a supercomputer was used to compute the Karhunen–Loève transform either of the complete set of pixels or else of an oval subset centered about the eyes. In the following, we will follow Sirovich's methodology and nomenclature, only we will replace the Karhunen–Loève transform with the lower complexity approximate algorithm.

For the experiment described below, we start with a more limited data set containing 143 pictures. Since the ensemble was fixed, we could subtract the average vector at the outset. Thus we transformed the data to floating point numbers, computed average values for the pixels, and then subtracted the average from each pixel to obtain "caricatures," or deviations from the average. Figure 11.13 is one of these caricatures.

The left graph in Figure 11.14 shows how the variance accumulates pixel by pixel, with the pixels sorted into decreasing order of variance.

Each caricature was treated as a picture and expanded into two-dimensional wavelet packets. The squares of the amplitudes were summed into a tree of variances, which was then searched for the joint best basis. In the joint best basis,

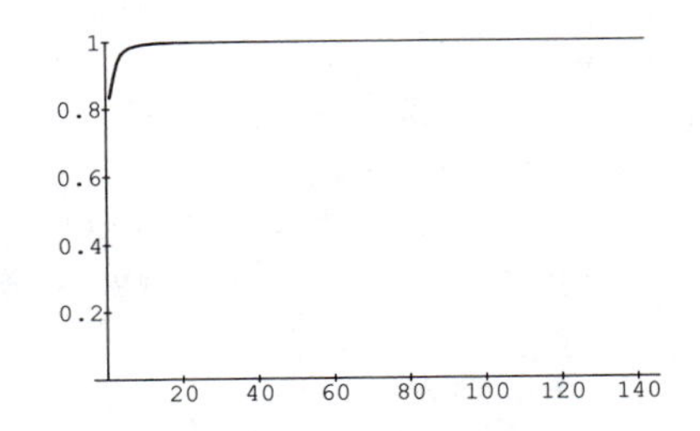

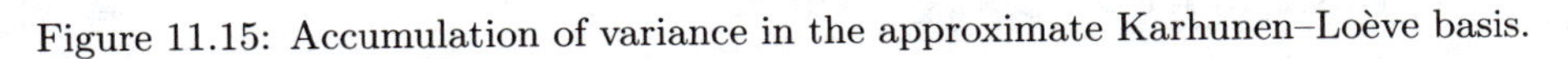
Figure 11.15: Accumulation of variance in the approximate Karhunen–Loève basis.

400 coordinates (of 16,384) contained more than 90 percent of the variance of the ensemble.

The right graph in Figure 11.14 shows the accumulation of total variance on the first d' coordinates in the joint best basis, sorted in decreasing order, as a fraction of the total variance of the ensemble, for $1 \leq d' \leq 1000$. Using 1000 parameters captures more than 95 percent of the ensemble variance, but requires somewhat more computer power than is readily available on a desk top. A 400 parameter system, on the other hand, can be analyzed on a typical workstation in minutes so we choose $d' = 400''$.

The top 400 coordinates were recomputed for each caricature and their auto-covariance matrix over the ensemble was diagonalized using the LINPACK [42] singular value decomposition routine.

Figure 11.15 shows the accumulation of total variance on the first d'' coordinates in the approximate Karhunen–Loève basis, sorted in decreasing order, as a fraction of the total variance of the 400 joint best basis coefficients, for $1 \leq d'' \leq 143$. The Karhunen–Loève post-processing for this small ensemble concentrates 98 percent of the retained variance from the top 400 joint best basis parameters into ten coefficients.

The fingerprint classification problem

Virtually the same method as the one applied to faces can be applied to fingerprints for identification purposes. The United States FBI uses eight bits per pixel to define the shade of gray and stores 500 pixels per inch, which works out to about 700,000 pixels and 0.7 megabytes per finger to store fingerprints in electronic form. This means that $d \approx 10^6$, so we are forced to use the fast approximate algorithm if we wish to compute the Karhunen–Loève expansion of a fingerprint.

There is no apparent relation between the parameters chosen by the Karhunen–Loève transform and the traditional parameters (location of "minutiæ" points in a fingerprint) which are used by police inspectors to describe fingerprints. Thus the additional classifying values would have to be stored alongside the more traditional values. But since the Karhunen–Loève parameters occupy only a few hundred bytes, which is a negligible amount of space compared to the million bytes of raw data, this subsidiary classification can be added to the fingerprint data base at a negligible cost.

Rank reduction for complex classifiers

The principle behind the fast approximate Karhunen–Loève transform is to employ a relatively low complexity $O(d^2 \log d)$ "front end" to reduce the rank of the subsequent high complexity $O(d^3)$ algorithm from d to $d' \ll d$.

If some measurements on an ensemble of random vectors are to be processed further by some other complex algorithm, we may similarly gain a significant speed advantage by preprocessing the data to reduce the number of parameters. A typical example would be processing for statistical classification from a large set of measurements. Classes, or regions in the measurement space $\mathbf{R}^d$, may have complicated boundaries which must be approximated by high-order polynomial hypersurfaces. Deciding at high orders whether a point lies in a particular region becomes very expensive when the dimension d grows large, so a reduction in the number of parameters will gain speed even if does not by itself simplify the geometry of the regions.

High complexity classifiers are used in speech recognition and machine vision systems. Adding a front end to reduce their workload is like adding a hearing aid or glasses so they can better focus on the most varying features. In some cases, the speed is desirable because we wish to perform the classification in "real time" or least fast enough to keep up with the inflowing data. Some examples are the following:

Ranks of various feature detection problems

- Mechanical failure detection from strain gauge data: $d \approx 10^2$;
- Target recognition from high-resolution radar range profiles: $d \approx 10^2$;
- Detection of irregular heartbeats from acoustic samples: $d \approx 10^2$;
- Phoneme detection: $d \approx 10^3$;
- Optical character recognition: $d \approx 10^3$;
- Detection of machine tool breakage from acoustic samples: $d \approx 10^3$.

11.2.3 Jacobians of complicated maps

Suppose that $T : \mathbf{R}^p \to \mathbf{R}^d$ is some smooth vector field with $p \ll d$. We may think of making plenty (d) of measurements of Tx for a variable x with only a few (p) degrees of freedom. This situation models one course of action to determine what a complicated map T is doing.

Approximating the tangent space in principal components

Recall that the *Jacobian* of T at a point $x \in \mathbf{R}^p$ is the $d \times p$ matrix $J = J_T[x]$ which gives the best linear approximation to T in a neighborhood of x. The coefficients of J are the various partial derivatives of T.

$$J_T[x](i,j) \stackrel{\text{def}}{=} \lim_{r\to 0} \left\langle e_i, \; \frac{T(x+re_j)-T(x)}{r} \right\rangle. \tag{11.16}$$

Here $1 \le i \le d$, $1 \le j \le p$, and e_i is the ith standard basis vector. However, the numerical computation of this Jacobian poses some difficulties because the difference quotient is ill-conditioned. Furthermore, the Jacobian might itself be an ill-conditioned matrix, but this difference quotient procedure offers no way of estimating the condition number of J. We will address these difficulties by replacing the difference quotient formula with an approximation based on the Karhunen–Loève transform for the positive matrix JJ^*. The error will lie solely in the approximation, since the Karhunen–Loève transform is orthogonal and thus perfectly conditioned. We will estimate the *condition number* of J from the singular value decomposition of J^*J. Then

$$cond(J) = \sqrt{cond(J^*J)} \approx \sqrt{\mu_1/\mu_p}, \tag{11.17}$$

where μ_1 and μ_p are respectively the first and n^{th} singular values of our estimate for J^*J.

Suppose first that T is a linear map, so that $T = J$ is its own Jacobian. Fix $x \in \mathbf{R}^p$, and consider the ball $B_r = B_r(x) \stackrel{\text{def}}{=} \{y \in \mathbf{R}^p : \|y - x\| \le r\}$ of radius $r > 0$, centered at x. We can consider the image $JB_r = \{Jy : y \in B_r(x)\} \subset \mathbf{R}^d$ of this ball under the transformation J to be an ensemble of random vectors. This will have expectation $E(JB_r) = JE(B_r) = Jx$, and we can compute the autocovariance matrix of the zero-mean ensemble $\widetilde{JB_r} \stackrel{\text{def}}{=} JB_r - Jx$ as follows:

$$\begin{aligned} E(\widetilde{JB_r} \otimes \widetilde{JB_r}) &= E_{y\in B_r(x)}(J\tilde{y}\,[J\tilde{y}]^*) \\ &= JE_{y\in B_r(x)}(\tilde{y}\tilde{y}^*)J^* = r^2 JJ^*. \end{aligned}$$

Here $\tilde{y} = y - x$ and the last equality holds since $E_{y\in B_r(x)}(\tilde{y}\tilde{y}^*) = r^2 I_d$ is just a constant times the $d \times d$ identity matrix and thus commutes out from between J and J^*. Thus $r^{-2}E(\widetilde{JB_r} \otimes \widetilde{JB_r}) = JJ^*$.

Proposition 11.3 *For every matrix J,*

$$\mathit{Rank}\ JJ^* = \mathit{Rank}\ J = \mathit{Rank}\ J^* = \mathit{Rank}\ J^*J.$$

Proof: To prove the first equality, notice that Range $JJ^* \subset$ Range J implies that Rank $J \geq$ Rank JJ^*. Now suppose that Range $JJ^* \neq$ Range J. There is some $y \neq 0$ with $y \in$ Range J and $\langle y, JJ^*z\rangle = \langle J^*y, J^*z\rangle = 0$ for all z. Putting $z = y$ we see that $J^*y = 0$. But by its definition, $y = Jx$ for some x, so we have $\|y\|^2 = \langle y, Jx\rangle = \langle J^*y, x\rangle = 0$, a contradiction. The third equality follows from the same argument if we substitute J^* for J. For the middle equality, note that Rank $AB \leq \min\{\text{Rank } A, \text{Rank } B\}$, so that the first equality gives Rank $J \leq$ Rank J^* while the third gives Rank $J \geq$ Rank J^*. □

Suppose J is an $d \times p$ matrix with $d \geq p$. If J has maximal rank p, then J^*J also has rank p. Now, the condition number of J is

$$\sup\{\frac{\|Jx\|}{\|x\|} : x \neq 0\} \Big/ \inf\{\frac{\|Jy\|}{\|y\|} : y \neq 0\}. \tag{11.18}$$

If J^*J has n nonzero singular values $\mu_1 \geq \cdots \geq \mu_p > 0$, counting multiplicities, then the supremum is $\sqrt{\mu_1}$ and the infimum is $\sqrt{\mu_p}$. To see this, let $z_1, \ldots, z_p$ be the orthonormal basis of $\mathbf{R}^p$ consisting of unit singular vectors for J^*J, which is guaranteed to exist because J^*J is a Hermitean matrix. Writing $x = \sum_i a_i z_i$ we have

$$\frac{\|Jx\|^2}{\|x\|^2} = \frac{\langle Jx, Jx\rangle}{\|x\|^2} = \frac{\langle J^*Jx, x\rangle}{\|x\|^2} = \frac{\sum_i a_i^2 \mu_i}{\sum_i a_i^2}. \tag{11.19}$$

This average is maximized when just a_1 is nonzero and minimized when just a_p is nonzero; it then equals μ_1 and μ_p, respectively. Thus we can compute the condition number of J using the formula in Equation 11.17.

Now suppose that T is any smooth vector field from $\mathbf{R}^p$ to $\mathbf{R}^d$, and x is some point in $\mathbf{R}^p$. We compute $z_r = E(TB_r)$, where $TB_r = TB_r(x) = \{Ty : \|y - x\| \leq r\}$; this is the expected value of Ty for y in the ball $B_r(x)$ of radius r centered at x. This average gives a second-order approximation to Tx:

Proposition 11.4 $\|z_r - Tx\| = O(r^2)$ *as* $r \to 0$.

Proof: We can use Taylor's theorem to write $T(x + y) = Tx + Jy + O(\|y\|^2)$ for $\|y\| \leq r$. But $E(Jy) = JE(y) = 0$ because we evaluate the expectation over $y \in B_r(0)$. Thus $E(TB_r) = Tx + O(r^2)$. □

We now define a positive matrix

$$A = A_r = E([TB_r - z_r] \otimes [TB_r - z_r]), \tag{11.20}$$

where the expectation is taken over the ball of radius r. Our main theorem is the following:

Theorem 11.5 $\displaystyle\lim_{r\to 0} \frac{1}{r^2} A_r = JJ^*.$

Proof: Using Proposition 11.4 we write $z_r = Tx + O(r^2)$. We then use Taylor's theorem to get the following estimate:

$$\begin{aligned}[T(x+y) - z] \otimes [T(x+y) - z] &= \left[Jy+O(\|y\|^2)\right] \otimes \left[Jy+O(\|y\|^2)\right] \\ &= (Jy)(Jy)^* + O(\|y\|^3).\end{aligned}$$

Taking the expectation of both sides over $y \in B_r(0)$ gives $A_r = r^2 JJ^* + O(r^3)$, yielding the desired result. The error estimate is $\|JJ^* - \frac{1}{r^2}A_r\| = O(r)$. □

Now suppose that $x \in \mathbf{R}^p$ is a point for which the Jacobian $J_T[x]$ has full rank p. Full rank is an open condition, *i.e.*, is it possessed by all matrices sufficiently close to J as well, so we have the following reassuring fact:

Corollary 11.6 *For all sufficiently small $r > 0$, Rank $A_r \geq$ Rank $J = p$.* □

Then we can approximate the map T in a neighborhood of Tx using the singular vectors for A_r:

Corollary 11.7 *If $\{z_1, \ldots, z_p\} \subset \mathbf{R}^d$ is a set of unit orthogonal singular vectors for A_r, then there are p linear functions $c_1, \ldots, c_p$ on $\mathbf{R}^p$ such that $T(x+y) = z + \sum_{i=1}^{p} c_i(y) z_i + O(r\|y\|^2)$.* □

We are not really concerned with the rank of A_r being too small, since A_r is a $d \times d$ matrix and $d \gg p$ is the interesting situation. Rather, we worry that choosing a too large value for r will result in Rank A_r being too large, so that we will not be able to identify the top few singular vectors which approximately span Range J. The problem is that if T has nonvanishing higher order derivatives, then the range of A_r will be higher-dimensional than the tangent space to T at Tx, which is the range of J. Schematically, this is depicted in Figure 11.16. The range of A_r is drawn as the two unit singular vectors z_1, z_2 multiplied by their respective singular values μ_1, μ_2. Notice that $\mu_2 \ll \mu_1$, illustrating what happens with smooth T: the variation μ_2 of TB_r in the nontangential direction z_2 is much smaller than the variation μ_1 in the tangent or z_1 direction. In practice, we will always have Rank $A_r = d$ because

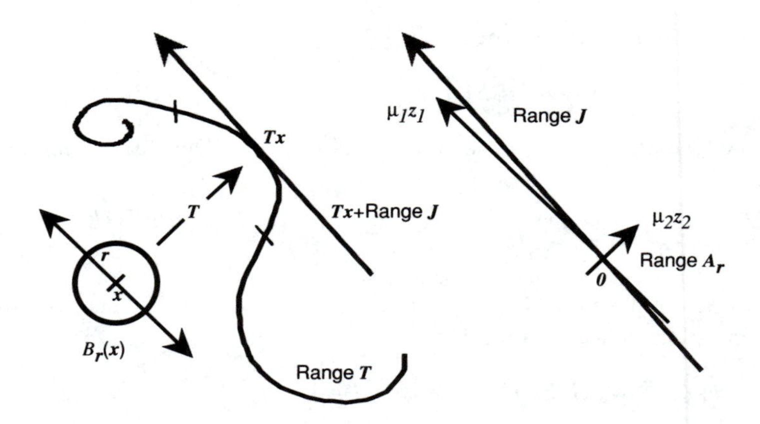

Figure 11.16: Tangent space (Range J) of T and its approximation (Range A_r).

to a finite precision machine every approximate matrix looks like it has full rank. However, if we arrange the singular values of A_r (with multiplicity) in decreasing order $\mu_1 \geq \cdots \mu_p \geq \cdots \geq 0$, then for small enough r we expect a steep drop between μ_p and μ_{p+1}. This then provides a method of choosing the largest r for which the singular vectors of A_r provide an accurate parametrization of T near x. Namely, we let r increase until $\sqrt{\mu_{p+1}/\mu_p}$ reaches some preset threshold of precision ϵ_μ. Then the nontangential components will contribute an error which is $1/\epsilon_\mu$ times smaller than the tangential components.

The functions $c_1, \ldots, c_p$ in Corollary 11.7 correspond to partial derivatives, but they are computed by orthogonal projection. We define matrix coefficients c_{ij} using the elementary basis vectors e_j of $\mathbf{R}^p$ as follows:

$$C_{ij} \stackrel{\text{def}}{=} \frac{1}{r} \langle z_i, T(x + re_j) - z \rangle . \tag{11.21}$$

Then we extend this to the superposition $y = \sum_{j=1}^{p} a_j e_j$ by taking linear combinations as follows:

$$c_i(y) \stackrel{\text{def}}{=} \sum_{j=1}^{p} C_{ij} a_j. \tag{11.22}$$

Comparing Equation 11.21 with Equation 11.16, we see that the $d \times p$ matrix J has been replaced with the $p \times p$ matrix C, the limit has been reduced to a single evaluation using the largest acceptable r, the initial point Tx is now an average z, and the standard basis $\{e_j : j = 1, \ldots, d\}$ in the range space has been replaced with a new orthonormal basis which is locally adapted to T.

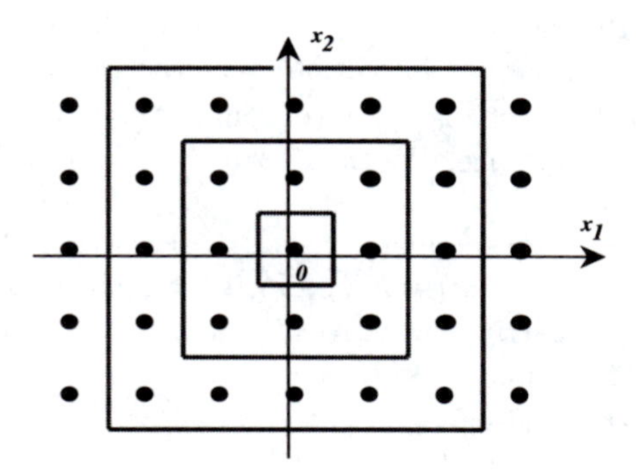

Figure 11.17: Cubes of various radii centered at zero.

Notice that the columns of the $p \times p$ matrix $C = (C_{ij})$ are given by the top p coordinates of the Karhunen–Loève transform of the secant vectors

$$\frac{1}{r}\left[T(x + re_1) - z\right], \frac{1}{r}\left[T(x + re_2) - z\right], \ldots, \frac{1}{r}\left[T(x + re_p) - z\right],$$

since the unit singular vectors $z_1, \ldots, z_p$ of A_r are the Karhunen–Loève eigenvectors for the ensemble TB_r. These secant vectors are approximations to the directional derivatives of T in the directions $e_1, e_2, \ldots, e_p$, and the Karhunen–Loève transform projects them onto the principal orthogonal components along Range T.

Fast approximate Jacobians

We can use the fast approximate Karhunen–Loève algorithm to compute the approximate Jacobian. Suppose that the domain of T includes a cube centered at the origin, namely $B_r = B_r(0) \stackrel{\text{def}}{=} \{x \in \mathbf{R}^p : |x_1| \leq r, \ldots, |x_p| \leq r\}$. Suppose we lay down a uniform grid of points of the form $x_i = k$ where $k = 0, \pm 1, \pm 2, \ldots$ and $i = 1, \ldots, p$. The intersection of the cube with this grid, which we will also call B_r, contains $(2r+1)^p$ points in all. We now compute Tx at all points $x \in B_r$, and call the resulting set TB_r. This will be our ensemble of "random" vectors. Each $x \in B_r$ produces a vector $Tx \in \mathbf{R}^d$ which requires d numbers to store, so TB_r will contain $|B_r|d = (2r+1)^p \times d$ floating-point numbers.

The approximation to T at zero using B_r will be computed by the wavelet packet best basis algorithm above. The mean vector $z = E(TB_r)$ is computed in all wavelet packet bases at once in the means tree. We form the variance tree from the squares tree and the means tree, and search it for the joint best basis of TB_r. We may assume that this basis is sorted into decreasing order of variance.

Now we take d' of the most varying terms out of the d in the joint best basis to retain $1-\epsilon$ of the total variance. We then form the $d' \times d'$ autocovariance matrix for these d' joint best basis vectors and diagonalize it, finding the singular vectors $z_1, \ldots, z_{d'}$ and corresponding singular values $\mu_1 \geq \ldots \geq \mu_{d'}$. We put the singular vectors into the columns of a matrix K', and get the *approximate Karhunen–Loève transform* K'^*, a $d' \times d'$ matrix which gives the approximate principal components when applied to the top d' joint best basis coordinates.

We now test whether the cube B_r is too large for the singular vectors to approximate the tangent space well. The rank of the Jacobian is at most p, so we must have $\epsilon_\mu(r) = \sqrt{\mu_{p+1}/\mu_p} \ll 1$. This gives the first parameter of the algorithm: we know that $\epsilon_\mu(r) \to 0$ as $r \to 0$, so if it exceeds a preset threshold we need to reduce r. However, if $\epsilon_\mu(r)$ meets our requirements, then we can discard all but the first p columns of K' to get the $d' \times p$ matrix K''. The range of K'' serves as the approximate tangent space at $T0$, and K''^* computes coordinates in this space from the top d' joint best basis coordinates.

Finally, we form the approximate Jacobian into this approximate tangent space by using Equation 11.21 with an approximate principal factor for z_i. One by one, we take the secant vectors $\frac{1}{r}\left[T(x+re_j)-z\right]$ for $j=1,2,\ldots,p$, which live in $\mathbf{R}^d$, and we find their joint best basis expansions. We then extract from each of those expansions the previously chosen d' coordinates which have most of the variance over the ensemble and apply K''^* to the vectors of these coordinates. That gives a list of p vectors in $\mathbf{R}^p$ which approximate the coefficients of the partial derivatives $\partial_1 T, \ldots, \partial_p T$ in the approximate tangent space basis.

The approximate Jacobian data consists of the following:

Data needed to approximate the Jacobian

- The joint best basis description down to d' coordinates (d' numbers);
- The p vectors of the approximate tangent space expressed as combinations of the first d' joint best basis vectors (pd' numbers);
- The $p \times p$ matrix C of partial derivatives expressed as combinations of the approximate tangent vectors.

Computing these quantities will cost us $O(|B_r| \times [d'^3 + d^2 \log d])$ arithmetic operations, where we expect $d' \ll d$.

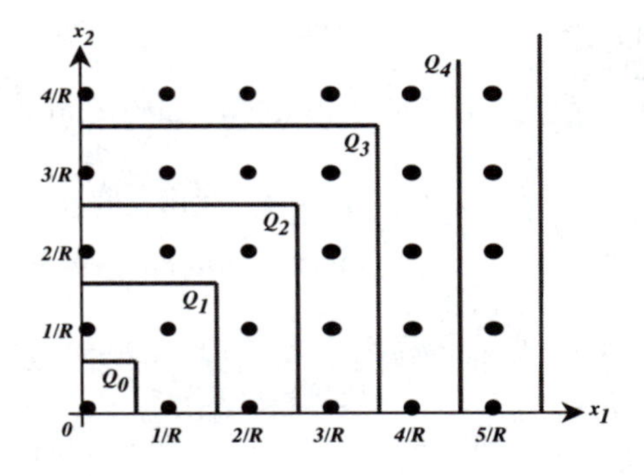

Figure 11.18: Patches of gridpoints at various distances from zero.

Efficient storage of complicated maps

Suppose for this application that the domain of T is the unit cube in $Q \subset \mathbf{R}^p$ defined by $Q = \{x \in \mathbf{R}^p : 0 \leq x_1 \leq 1, \ldots, 0 \leq x_p \leq 1\}$, and suppose we lay down a uniform grid of points of the form $x_i = r/R$ where $r = 0, 1, \ldots, R$ and $i = 1, \ldots, p$. We will call this grid G; it has mesh $1/R$ and contains $|G| = (R+1)^p$ points in all. We now compute Tx at all points $x \in G$, and call the resulting set TG. This is an enormous data set: each $x \in G$ produces a vector $Tx \in \mathbf{R}^d$ which requires d numbers to store, so TG will contain $|G|d = (R+1)^p \times d$ floating-point numbers.

We now use approximate Jacobians to reduce the size of the data set. We will do this by building up patches in the domain where T is well-approximated by its approximate Jacobian. With the fast update algorithm, we can segment the domain of T into patches on which we are sure that the approximation remains within a preset error. Suppose we start at $0 \in G$, at one corner of the cube Q. For each $r = 0, 1, 2, \ldots$ we define the set $Q_r = Q_r(0) \stackrel{\text{def}}{=} \{x \in G : 0 \leq x_i \leq r/R \text{ for } i = 1, \ldots, p\}$. We also define the set $P_r = P_r(0) \stackrel{\text{def}}{=} Q_r(0) \setminus Q_{r-1}(0)$, the partial cubical shell at radius r/R. Then $Q_{r+1} = Q_r \cup P_{r+1}$. This arrangement is depicted for the two-dimensional case $p = 2$ by the points enclosed in lightly colored boxes in Figure 11.18. Note that Q_r contains $|Q_r| = (r+1)^p$ points, while P_r contains $|P_r| = |Q_r| - |Q_{r-1}| = (r+1)^p - r^p \approx pr^{p-1}$ points.

The segmentation algorithm works by first initializing $r = 1$, and then iterating through an algorithm that enlarges a patch on which T is linearly approximated by its approximate Jacobian. We stop enlarging the patch Q_r when the variance of TQ_r along the approximate tangent vectors to T stops being much larger than the variance along approximate normal vectors. In following, we assume that the approximate Karhunen–Loève basis for TQ_{r-1} has already been determined, using the algorithm in Section 11.2.1 above. The update algorithm is also defined in that section.

Segmentation into regions of good linear approximation

- Compute the wavelet packet means and sums-of-squares trees for the extra vectors TP_r;
- Update the joint best basis for TQ_{r-1} by adding in the data from TP_r to get the joint best basis for TQ_r;
- Compute and store the approximate Karhunen–Loève basis K'_r for TQ_r and the singular values $\mu_1 \geq \ldots \geq \mu_{d'} \geq 0$;
- If $\epsilon_\mu(r) = \sqrt{\mu_{p+1}/\mu_p}$ is too large, then:
  - Compute and store $z = ETQ_{r-1}$ for the center;
  - Compute and store the approximate tangent vectors K''_{r-1} for the patch TQ_{r-1};
  - Compute and store the approximate Jacobian using K''_{r-1} and Equation 11.21;
  - Reset $k = 1$;
  - Move to the next free point in G;
- Else if $\epsilon_\mu(r)$ is still small enough, then increment r by one;
- Repeat.

This algorithm will eat away at the domain G, producing a covering of patches Q of various sizes, each with its center point $z_Q = ETQ$, its local approximate tangent space K''_Q, and its approximate Jacobian C_Q. These quantities will require d', pd', and p^2 real numbers to store, respectively. If there are a total of N patches, then the total amount of data to store is $N(d'+pd'+p^2) = O(Npd')$ numbers since $p \leq d'$. If p is small, $N \ll |G|$ and $d' \ll d$, then this compares favorably with the storage requirements for TG, namely $O(d|G|)$.

The complexity of computing these quantities on all of the patches can be estimated from the complexity of finding the approximate Jacobian for the single patch containing all of G, since this is the worst case. But from the previous section, we see that this is $O(|G| \times [d'^3 + d^2 \log d])$ arithmetic operations.

Applying this approximation of T to a vector $x \in \mathbf{R}^p$ involves first finding the patch Q with $x \in Q$. Suppose that x_Q is the center grid point of Q. Then in the first d' joint best basis coordinates,

$$\widetilde{Tx} = z_Q + K''_Q C_Q (x - x_Q). \tag{11.23}$$

We finally superpose the d' joint best basis vectors to get the coordinates of the point $Tx \in \mathbf{R}^d$ from $\widetilde{Tx}$. The complexity of computing Tx this way is $O(p + p^2 + pd' + d' + d \log d)$ which under our assumptions is bounded by $O(d \log d)$.

Precomputation for inverse problems

The final application is to use the local approximate Jacobians to invert the map $x \mapsto Tx$, which we suppose has already been computed at all points x on a finite grid G.

One classical way is to use linear interpolation: given $y \in \mathbf{R}^d$, we find the nearest points $Tx_k \in \mathbf{R}^d$ computed from grid points $x_k \in G$ and write $y = \sum_k a_k T x_k$. Then the linear approximation to $T^{-1}y$ is just $\sum_k a_k x_k$. This is exactly correct for linear maps T and has at least $O(h)$ accuracy for a differentiable map T on a grid with mesh h. However, it requires that we store the precomputed values $\{Tx : x \in G\}$ and that we search the whole list for the points close to y. The last step, in particular, requires computing $|G|$ distances for a grid G.

If we have invested the effort to compute the approximate Jacobian representation of T, then the inverse can be approximated from that data instead. Let N be the number of patches in the cover of G, and suppose that $N \ll |G|$. We also suppose for the sake of simplicity that we keep the same number d' of joint best basis components in each patch, although of course they may be different components in different patches. Finally, we suppose that we have already computed the inverses C_Q^{-1} of the approximate Jacobians on each patch Q, which requires a one-time investment of $O(Np^3)$. Then the necessary computations and their complexities for computing $T^{-1}y$ at a single $y \in \mathbf{R}^d$ are the following:

Approximate inverse via approximate Jacobian

- Find the complete wavelet packet expansion of y, which simultaneously computes all joint best basis expansions $\tilde{y}$: $O(d \log d)$;
- Compute the distances from $\tilde{y}$ to the means z_Q for each patch Q, and let Q henceforth be the patch with nearest mean: $O(Nd')$;
- Compute the approximate inverse,

$$T^{-1}y \approx x_Q + C_Q^{-1} {K''}^*_Q (\tilde{y} - z_Q),$$

for the nearest patch Q: $O(d' + pd' + p^2 + p) = O(pd')$.

11.3 Nonstandard matrix multiplication

By decomposing a matrix into its two-dimensional best basis, we reduce the number of nonnegligible coefficients and thus reduce the computational complexity of matrix application.

11.3.1 Two-dimensional best basis sparsification

Write $\mathcal{W}(\mathbf{R})$ for the collection of one-dimensional wavelet packets. Let ψ_{sfp} be a representative wavelet packet of frequency f at scale s and position p. Conjugate quadrature filters may be chosen so that $\mathcal{W}(\mathbf{R})$ is dense in many common function spaces. With minimal hypotheses, $\mathcal{W}(\mathbf{R})$ will be dense in $L^2(\mathbf{R})$. Using the Haar filters $\{1/\sqrt{2}, 1/\sqrt{2}\}$ and $\{1/\sqrt{2}, -1/\sqrt{2}\}$, for example, produces $\mathcal{W}(\mathbf{R})$ which is dense in $L^p(\mathbf{R})$ for $1 < p < \infty$. Longer filters can generate smoother wave packets, so we can also produce dense subsets of Sobolev spaces, etc.

Ordering wave packets

Wavelet packets ψ_{sfp} can be totally ordered. We say that $\psi < \psi'$ if $(s, f, p) < (s', f', p')$. The triplets are compared lexicographically, counting the scale parameters s, s' as most significant.

Write $\psi_X = \psi_{s_X f_X p_X}$, etc., and observe that tensor products of wavelet packets inherit this total order. We can say that $\psi_X \otimes \psi_Y < \psi'_X \otimes \psi'_Y$ if $\psi_X < \psi'_X$ or else if $\psi_X = \psi'_X$ but $\psi_Y < \psi'_Y$. This is just lexicographical comparison of the lists $(s_X, f_X, p_X, s_Y, f_Y, p_Y)$ and $(s'_X, f'_X, p'_X, s'_Y, f'_Y, p'_Y)$ from left to right.

The *adjoint order* $<^*$ just exchanges X and Y indices: $\psi_X \otimes \psi_Y <^* \psi'_X \otimes \psi'_Y$ if and only if $\psi_Y \otimes \psi_X <^* \psi'_Y \otimes \psi'_X$. It is also a total order.

Projections

Let $\mathcal{W}^1$ denote the space of bounded sequences indexed by the three wave packet indices s, f, p. With the ordering above, we obtain a natural isomorphism between ℓ^∞ and $\mathcal{W}^1$. There is also a natural injection $J^1 : L^2(\mathbf{R}) \hookrightarrow \mathcal{W}^1$ given by $J^1 x = \{\lambda_{sf}(p)\}$ for $x \in L^2(\mathbf{R})$, with $\lambda_{sf} = \langle x, \psi^<_{sfp} \rangle$ being the sequence of backwards inner products with functions in $\mathcal{W}(\mathbf{R})$. If B is a basis subset, then the composition J^1_B of J^1 with projection onto the subsequences indexed by B is also injective. J^1_B is an isomorphism of $L^2(\mathbf{R})$ onto $l^2(B)$, which is defined to be the square-summable sequences of $\mathcal{W}^1$ whose indices belong to B.

The inverse is a map $R^1 : \mathcal{W}^1 \to L^2(\mathbf{R})$ defined by

$$R^1\lambda(t) = \sum_{(s,f,p)\in\mathbf{Z}^3} \bar{\lambda}_{sf}(p)\psi'^{<}_{sfp}(t). \tag{11.24}$$

This map is defined and bounded on the closed subspace of $\mathcal{W}^1$ isomorphic to l^2 under the natural isomorphism mentioned above. In particular, R^1 is defined and bounded on the range of J^1_B for every basis subset B. The related restriction $R^1_B : \mathcal{W}^1 \to L^2(\mathbf{R})$ defined by $R^1_B\lambda(t) = \sum_{(s,f,p)\in B} \bar{\lambda}_{sfp}\psi'^{<}_{sfp}(t)$ is a left inverse for J^1 and J^1_B. In addition, $J^1R^1_B$ is a projection of $\mathcal{W}^1$. Likewise, if $\sum_i \alpha_i = 1$ and $R^1_{B_i}$ is one of the above maps for each i, then $J^1\sum_i \alpha_i R^1_{B_i}$ is also a projection of $\mathcal{W}^1$. It is an orthogonal projection on any finite subset of $\mathcal{W}^1$.

Similarly, writing $\mathcal{W}^2$ for $\mathcal{W}^1 \times \mathcal{W}^1$, the ordering of tensor products gives a natural isomorphism between ℓ^∞ and $\mathcal{W}^2$. Objects in the space $L^2(\mathbf{R}^2)$, *i.e.*, the Hilbert-Schmidt operators, inject into this sequence space $\mathcal{W}^2$ in the obvious way, namely $M \mapsto \langle M, \psi^{<}_{s_X f_X p_X} \otimes \psi^{<}_{s_Y f_Y p_Y}\rangle$. Call this injection J^2. If B is a basis subset of $\mathcal{W}^2$, then the composition J^2_B of J^2 with projection onto subsequences indexed by B is also injective. J^2_B is an isomorphism of $L^2(\mathbf{R}^2)$ onto $\ell^2(B)$, the square summable sequences of $\mathcal{W}^2$ whose indices belong to B.

The map $R^2 : \mathcal{W}^2 \to L^2(\mathbf{R}^2)$ given by $R^2c(x,y) = \sum \bar{c}_{XY}\psi'^{<}_X(x)\psi'^{<}_Y(y)$, is bounded on that subset of $\mathcal{W}^2$ naturally isomorphic to ℓ^2. In particular, it is bounded on the range of J^2_B for every basis subset B.

We may also define the restrictions R^2_B of R^2 to subsequences indexed by B, defined by $R^2_Bc(x,y) = \sum_{(\psi_X,\psi_Y)\in B} \bar{c}_{XY}\psi'^{<}_X(x)\psi'^{<}_Y(y)$. There is one for each basis subset B of $\mathcal{W}^2$. Then R^2_B is a left inverse of J^2 and J^2_B, and $J^2R^2_B$ is a projection of $\mathcal{W}^2$. As before, if $\sum_i \alpha_i = 1$ and B_i is a basis subset for each i, then $J^2\sum_i R^2_{B_i}$ is also a projection of $\mathcal{W}^2$. It is an orthogonal projection on any finite subset of $\mathcal{W}^2$.

11.3.2 Applying operators to vectors

For definiteness, let X and Y be two named copies of $\mathbf{R}$. Let $x \in L^2(X)$ be a vector, whose coordinates with respect to wave packets form the sequence $J^1x = \{\langle x, \psi^{<}_X\rangle : \psi_X \in \mathcal{W}(X)\}$.

Let $M : L^2(X) \to L^2(Y)$ be a Hilbert-Schmidt operator. Its matrix coefficients with respect to the complete set of tensor products of wave packets form the sequence $J^2M = \{\langle M, \psi^{<}_X \otimes \psi^{<}_Y\rangle : \psi_X \in \mathcal{W}(X), \psi_Y \in \mathcal{W}(Y)\}$. We obtain the identity

$$\langle Mv, \psi^{<}_Y\rangle = \sum_{\psi_X\in\mathcal{W}(X)} \langle M, \psi^{<}_X \otimes \psi^{<}_Y\rangle \,\langle x, \psi^{<}_X\rangle. \tag{11.25}$$

This identity generalizes to a linear action of $\mathcal{W}^2$ on $\mathcal{W}^1$ defined by

$$c(x)_{sfp} = \sum_{(s'f'p')} \lambda_{sfps'f'p'} v_{s'f'p'}. \tag{11.26}$$

Now, images of operators form a proper submanifold of $\mathcal{W}^2$. Likewise, images of vectors form a submanifold $\mathcal{W}^1$. We can lift the action of M on x to these larger spaces via the commutative diagram

$$\begin{array}{ccc} \mathcal{W}^1 & \stackrel{J_B^2 M}{\longrightarrow} & \mathcal{W}^1 \\ J^1 \uparrow & \bigcirc & \downarrow R^1 \\ L^2(\mathbf{R}) & \stackrel{M}{\longrightarrow} & L^2(\mathbf{R}). \end{array} \tag{11.27}$$

The significance of this lift is that by a suitable choice of B we can reduce the complexity of the map $J_B^2 M$, and therefore the complexity of the operator application.

Example

We illustrate the above ideas by considering nonstandard multiplication by square matrices of order 16. Using wavelet packets of isotropic dilations, we obtain the *best isotropic basis* action depicted in the Figure 11.19. The domain vector is first injected into a $16 \log 16$-dimensional space by expansion into the complete domain wavelet packet analysis tree at bottom. It then has its components multiplied by the best basis coefficients in the matrix square at center, as indicated by the arrows. The products are summed into a range wavelet packet synthesis tree at right. The range tree is then projected onto the 16-dimensional range space by the adjoint of the wavelet packet expansion. Notice that input blocks have the same size as the output blocks, so that this method reveals only how energy moves within scales. However, the best isotropic basis expansion of the matrix makes explicit how frequencies and positions are mixed. Of course it is possible to move energy between scales, but the action is not immediately readable from the nonstandard matrix coefficients.

Figure 11.20 depicts the best basis expansion for the matrix $m_{ij} = \sin \frac{\pi i j}{1024}$, with $0 \leq i, j < 1024$. All coefficients smaller in magnitude than one percent of the largest have been set to zero; the survivors are depicted as black dots in the location they would occupy within the tree. No information about the chosen basis is available in that picture; we have suppressed drawing the outlines of the subspace boxes in order to avoid clutter.

If we allow all tensor products of wavelet packets in the decomposition of the matrix, we get the *best tensor basis* nonstandard representation. Finding the best

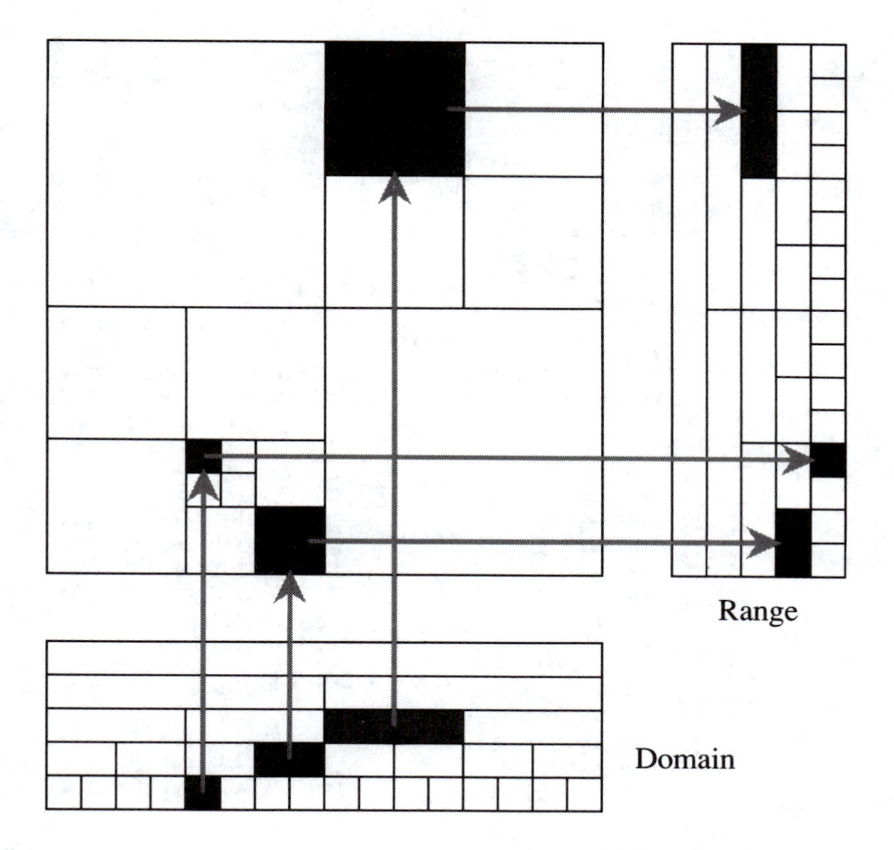

Figure 11.19: Nonstandard 16×16 matrix multiplication in isotropic best basis.

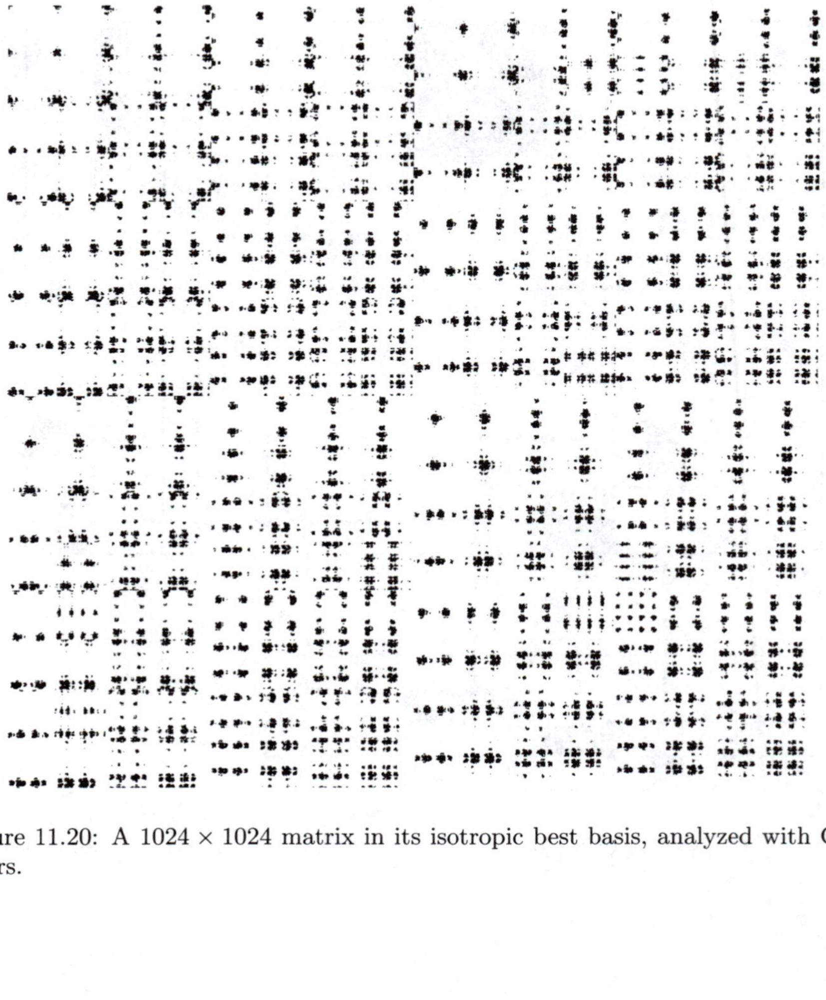

Figure 11.20: A 1024×1024 matrix in its isotropic best basis, analyzed with C30 filters.

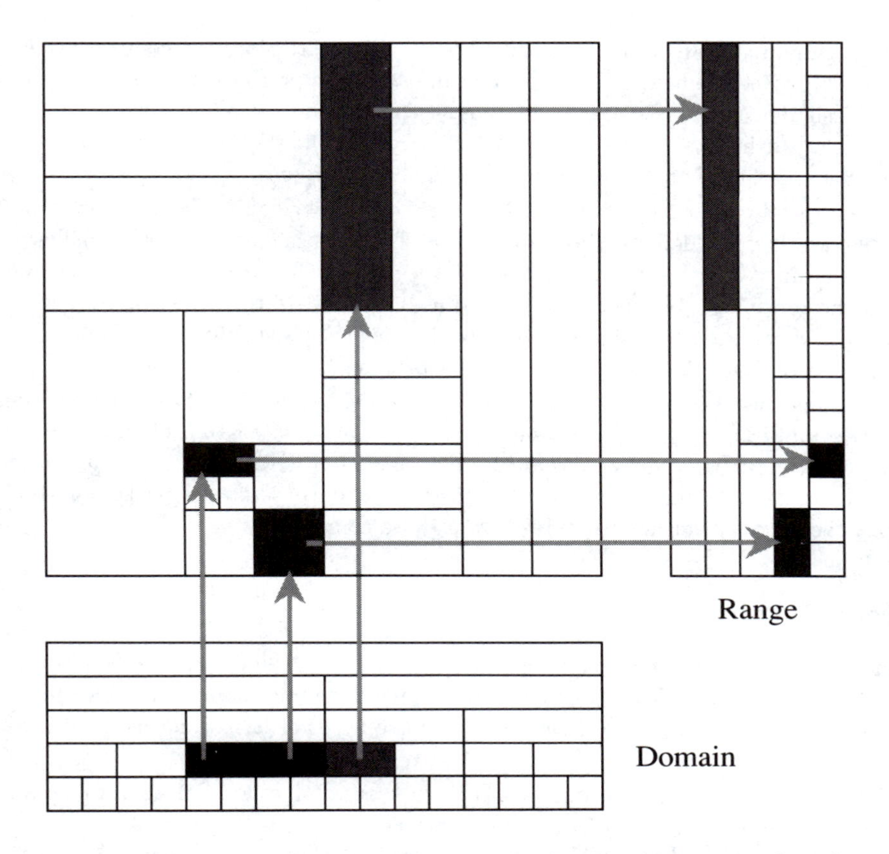

Figure 11.21: Nonstandard 16×16 matrix multiplication: anisotropic best basis.

basis for the matrix becomes harder, but the multiplication algorithm is practically the same. Again we must develop the complete wavelet packet analysis of the input to get the domain tree. Blocks in the nonstandard matrix send blocks in the domain tree to possibly larger or smaller blocks in the range tree, where they are synthesized into the output. This is depicted in Figure 11.21, for a 16×16 matrix. Such an expansion explicitly shows how energy moves from one scale to another, as well as how position and frequency are mixed.

Rather than choose a basis set among the two-dimensional wavelet packets, we may also consider matrix multiplication to be a list of inner products. The complexity of the operation may be reduced by expressing each row of the matrix in its individual best basis, expanding the domain vector into the complete tree of wavelet packet coefficients, then evaluating the inner products row by row, up to any desired accuracy. Schematically, this accelerated inner product may be depicted as in the left-hand side of Figure 11.22. Notice how the wavelet components of the input are gathered and summed to yield the output values.

A typical candidate for such a *best row basis* expansion is the 1024-point sine transform matrix, depicted in Figure 11.23. In this schematic, enough coefficients were retained to preserve 99 percent of the Hilbert–Schmidt norm of each row, and the surviving coefficients are plotted as black dots in the locations they would occupy. Approximately seven percent of the coefficients survived. No information is provided about the chosen bases in that picture.

We may also expand the columns in their own individual best-bases and consider matrix multiplication to be the superposition of a weighted sum of these wavelet packet expansions. This *best column basis* algorithm is depicted schematically in Figure 11.24. The input value is sprayed into the output wavelet packet synthesis tree, then the output is reassembled from those components.

Operation count

Suppose that M is a nonsparse operator of rank r. Ordinary multiplication of a vector by M takes at least $O(r^2)$ operations, with the minimum achievable only by representing M as a matrix with respect to the bases of its r-dimensional domain and range.

On the other hand, the injection J^2 will require $O(r^2[\log r]^2)$ operations, and each of J^1 and R^1 require $O(r \log r)$ operations. For a fixed basis subset B of $\mathcal{W}^2$, the application of $J^2_B M$ to $J^1 v$ requires at most $\#|J^2_B M|$ operations, where $\#|U|$ denotes the number of nonzero coefficients in U. We may choose our wavelet packet library so that $\#|J^2_B M| = O(r^2)$. Thus the multiplication method described above costs an initial investment of $O(r^2[\log r]^2)$, plus at most an additional $O(r^2)$ per right-hand side. Thus the method has asymptotic complexity $O(r^2)$ per vector in its exact form, as expected for multiplication by a matrix of order r.

We can obtain lower complexity if we take into account the finite accuracy of our calculation. Given a fixed matrix of coefficients C, write C_δ for the same matrix with all coefficients set to zero whose absolute values are less than δ. By the continuity of the Hilbert-Schmidt norm, for every $\epsilon > 0$ there is a $\delta > 0$ such that $\|C - C_\delta\|_{HS} < \epsilon$. Given M and ϵ as well as a library of wavelet packets, we can choose a basis subset $B \subset \mathcal{W}^2$ so as to minimize $\#|(J^2_B M)_\delta|$. The choice algorithm has complexity $O(r^2[\log r]^2)$, as shown above. For a certain class of operators, there

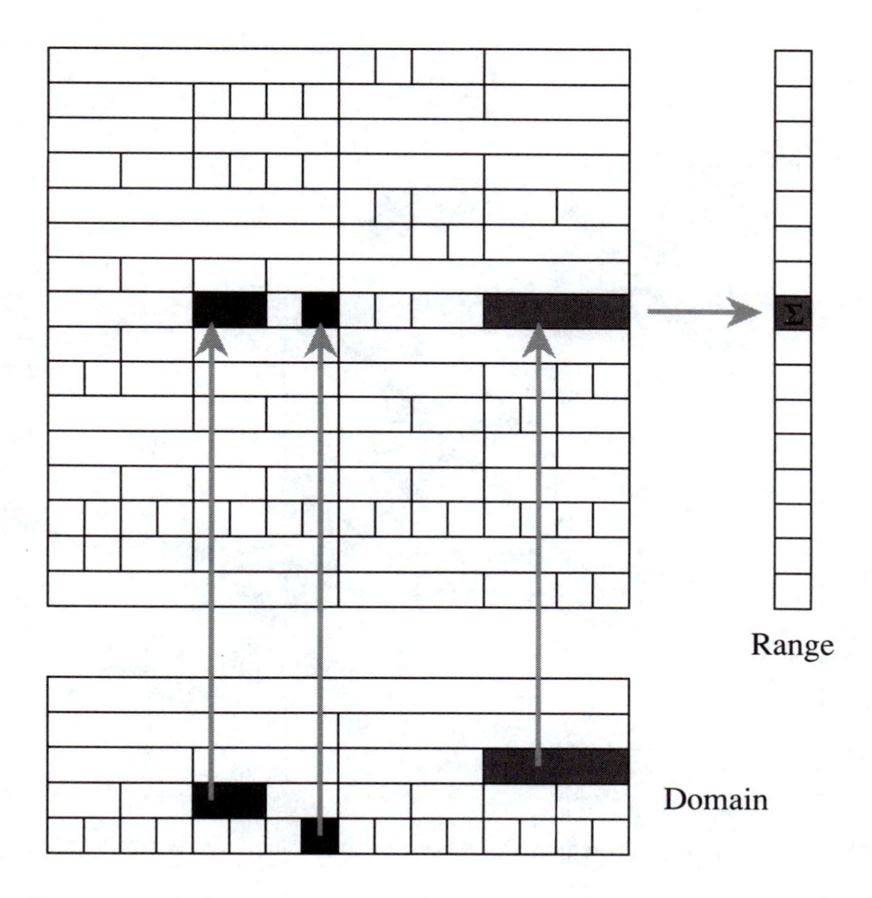

Figure 11.22: Nonstandard 16×16 matrix multiplication in best row basis.

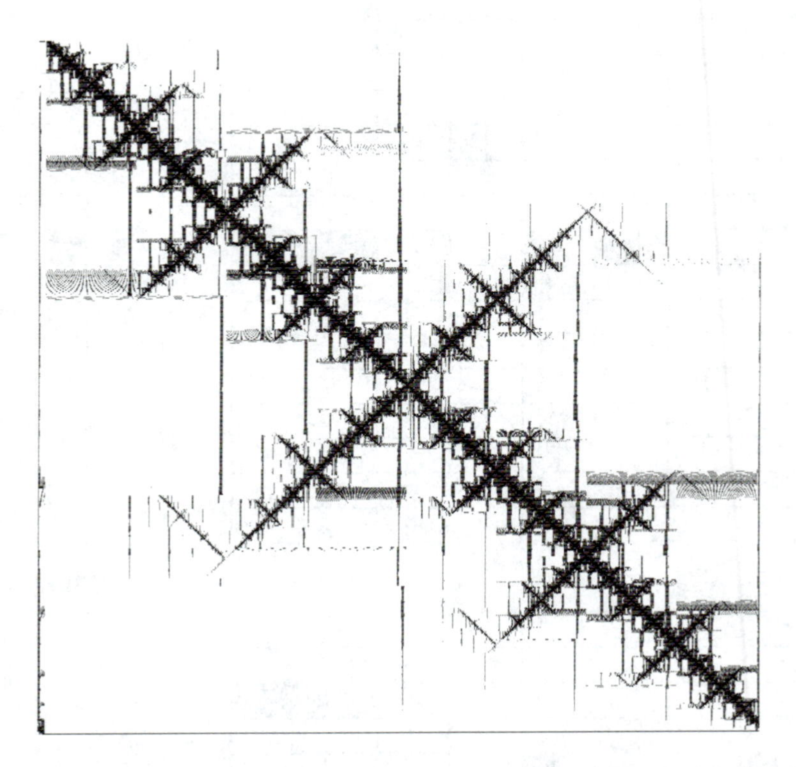

Figure 11.23: A 1024×1024 matrix in its best row basis, analyzed with C30 filters.

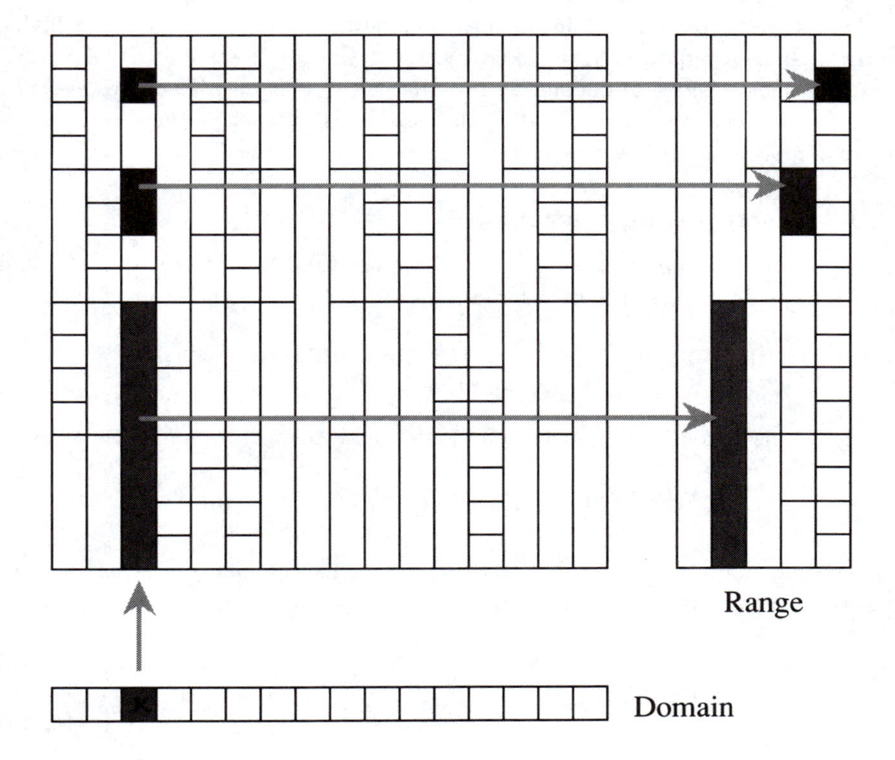

Figure 11.24: Nonstandard 16×16 matrix multiplication: best column basis.

is a library of wavelet packets such that for every fixed $\delta > 0$ we have

$$\#|(J_B^2 M)_\delta| = O(r \log r), \tag{11.28}$$

with the constant depending, of course, on δ. Call this class with property 11.28 the *sparsifiable* Hilbert-Schmidt operators $\mathcal{S}$. By the estimate above, finite-precision multiplication by sparsifiable rank-r operators has asymptotic complexity $O(r \log r)$.

The best row basis algorithm is also asymptotically $O(r \log r)$ for sparsifiable matrices, since it requires $O(r \log r)$ operations to find the complete tree of wavelet packet coefficients for the domain vector, then $O(\log r)$ multiply-adds to evaluate each of the r inner products. Finding the best basis for each of the r rows of the matrix requires an initial investment of $O(r^2 \log r)$ operations.

11.3.3 Composing operators

Let X, Y, Z be three named copies of $\mathbf{R}$. Suppose that $M : L^2(X) \to L^2(Y)$ and $N : L^2(Y) \to L^2(Z)$ are Hilbert-Schmidt operators. We have the identity

$$\langle NM, \psi_X^{\lessgtr} \otimes \psi_Z^{\lessgtr} \rangle = \sum_{\psi_Y^{\lessgtr} \in \mathcal{W}(Y)} \langle N, \psi_Y^{\lessgtr} \otimes \psi_Z^{\lessgtr} \rangle \, \langle M, \psi_X^{\lessgtr} \otimes \psi_Y^{\lessgtr} \rangle .$$

This generalizes to an action of $\mathcal{W}^2$ on $\mathcal{W}^2$, which is defined by the formula

$$\lambda(d)_{sfps'f'p'} = \sum_{s''f''p''} \mu_{sfps''f''p''} \, \lambda_{s''f''p''s'f'p'},$$

where λ and μ are sequences in $\mathcal{W}^2$. Using J^2, we can lift multiplication by N to an action on these larger spaces via the commutative diagram

$$\begin{array}{ccc} \mathcal{W}^2 & \xrightarrow{J_B^2 N} & \mathcal{W}^2 \\ J^2 \uparrow & \bigcirc & \downarrow R^2 \\ L^2(\mathbf{R}^2) & \xrightarrow{N} & L^2(\mathbf{R}^2). \end{array} \tag{11.29}$$

Again, by a suitable choice of B the complexity of the operation may be reduced to below that of ordinary operator composition.

Operation count

Suppose that M and N are rank-r operators. Standard multiplication of N and M has complexity $O(r^3)$. The complexity of injecting N and M into $\mathcal{W}^2$ is

$O(r^2[\log r]^2)$. The action of $J_B^2 N$ on $J^2 M$ has complexity $O(\sum_{sfp} \#|J_B^2 N_{YZ} : (s_Y, f_Y, p_Y) = (s, f, p)|\#|J^2 M_{XY} : (s_Y, f_Y, p_Y) = (s, f, p)|)$. The second factor is a constant $r \log r$, while the first when summed over all sfp is exactly $\#|J_B^2 N|$. Thus the complexity of the nonstandard multiplication algorithm, including the conjugation into the basis set B, is $O(\#|J_B^2 N|\, r \log r)$. Since the first factor is r^2 in general, the complexity of the exact algorithm is $O(r^3 \log r)$ for generic matrices, reflecting the extra cost of conjugating into the basis set B.

For the approximate algorithm, the complexity is $O(\#|(J_B^2 N)_\delta|\, r \log r)$. For the sparsifiable matrices, this can be reduced by a suitable choice of B to a complexity of $O(r^2[\log r]^2)$ for the complete algorithm. Since choosing B and evaluating J_B^2 each have this complexity, it is not possible to do any better by this method.

11.4 Speech signal segmentation

Here we apply the adapted local trigonometric transform to decompose digitized speech signals into orthogonal elementary waveforms in such a way that the choice of widths for the waveforms yields a segmentation of the signal. This algorithm leads to a local time-frequency representation which can also be used for compression and recognition. We present some experimental results of signal compression and automatic *voiced-unvoiced segmentation*, prepared by Eva Wesfreid [110]. The compressed signal has been simplified but still appears to be useful for detecting fundamental frequencies and characterizing *formants*.

We begin with a clean, digitized speech signal. The signal is decomposed by a complete adapted local trigonometric analysis, into cosines or sines multiplied by smooth cutoff functions. It is possible to compute several local cosine transforms all at once, recursively subdividing segments of the signal into halves. The basis functions on each subinterval are the orthogonal direct sum of the basis functions on its left and right halves, and this orthogonality propagates up through the multiple levels of the binary "family tree" in Figure 4.17. We then apply the best basis method obtained by entropy minimization [30]. This algorithm produces an adapted orthogonal elementary waveform decomposition, which is a local spectral representation for the speech signal. Roughly speaking, we get a windowed cosine transform of the signal, with the window size well-adapted to the spectrum it contains. A superposition of these functions may be depicted by a sequence of adjacent envelopes or windows, with vertical lines drawn between the nominal window boundaries. This is done in Figure 4.15.

The associated time partition, or choice of windows, appears to be useful for segmentation into voiced and unvoiced portions, which can be recognized by the number of peaks of the local spectrum. This number of peaks is related to the theoretical dimension of the decomposition. The time partition provides short segments where there is fast frequency variation and long segments where there is slow frequency variation. The spectral representation is invertible and allows both perfect reconstruction (analysis-synthesis) and lossy approximation (compression).

11.4.1 Adapted local spectral analysis

We introduce a *formant representation* as follows. We examine the spectrum in each segment and locate the centers-of-mass for the top few peaks. Keeping just the top few peaks, or just a few of the most energetic waveform components, is a kind of compression, and computing the centers of mass of the peaks is a drastic reduction of the amount of data to be used for subsequent recognition. The formant representation is the resulting set of locally constant spectral lines, or step function approximations to the time-frequency function.

For comparison and to suggest variations of this algorithm, we point out that other elementary waveform representations can be used in similar adapted decompositions and segmentations. Some of these are described in references [35], [66], and [95].

The transform can be computed using a standard fast discrete cosine transform, after preliminary "folding" step described in Chapter 4. This "folding" splits $S(t)$ into a set of local finite energy signals $S_j(t) \in L^2(I_j)$, $j \in Z$, such that applying a standard discrete cosine transform to the coefficients in $S_j(t)$ is equivalent to computing all inner products with the functions Ψ_k^j. In other words,

$$S_j(t) = \sum_{k \in Z} c_k^j \; \phi_k^j(t), \qquad \text{with} \quad c_k^j = \langle S_j(t), \phi_k^j(t) \rangle, \tag{11.30}$$

where

$$\phi_k^j(t) = \frac{\sqrt{2}}{\sqrt{|I_j|}} \cos \frac{\pi}{|I_j|}(k + \frac{1}{2})(t - a_j)\chi_{I_j}(t). \tag{11.31}$$

Discrete sampled cosines at half-integer frequencies are the basis functions of the *DCT-IV* transform. The characteristic function $\mathbf{1}_{I_j}(t)$ is equal to one if $t \in I_j$ and zero otherwise. If $S(t)$ is a sampled signal with $t \in \{0, 1, 2, \ldots, 2^N - 1\}$, then we can fold first at the boundaries which gives $S^0(t)$ and next recursively at the middle points in a few levels. Therefore, this folding splits each function $S_j^\ell(t) \in L^2(I_j^\ell)$ into $S_{2j}^{\ell+1}(t) \in L^2(I_{2j}^{\ell+1})$ and $S_{2j+1}^{\ell+1}(t) \in L^2(I_{2j+1}^{\ell+1})$. We then calculate the standard DCT-IV transform for each S_j^ℓ which gives the spectral tree. This computation is done by the functions `lcadf()` and `lcadf()` in Chapter 4.

The orthogonality of the lapped orthogonal transform implies the following energy conservation identities:

$$\|S\|^2 = \|S^0\|^2 = \|S^j\|^2 \stackrel{\text{def}}{=} \sum_k \|S_k^j\|^2; \qquad \|S_k^j\|^2 = \|d_k^j\|^2.$$

If $\{x_k\}$ belongs to l^2 and $l^2 \log l^2$ then we can define the *spectral entropy* of $\{x_k\}$ to be

$$H(x) = -\sum_k \frac{|x_k|^2}{\|x\|^2} \log \frac{|x_k|^2}{\|x\|^2} = \frac{\lambda(x)}{\|x\|^2} + \log \|x\|^2, \tag{11.32}$$

with $\lambda(x) = -\sum |x_k|^2 \log |x_k|^2$. Then $\exp\bigl(H(x)\bigr) = \|x\|^2 \exp(\frac{\lambda(x)}{\|x\|^2})$ may be called the *theoretical dimension* of the sequence $\{x\}$.

The *adapted local spectrum* a_j^ℓ over the time interval I_j^ℓ is just the best basis of local cosines defined by this information cost function. See Chapter 8 for a detailed description of this algorithm. It divides the signal into segments I_j^ℓ so as to minimize the total spectral entropy. We will call the chosen division into intervals the *adapted (entropy-minimizing) time partition* for the given signal.

11.4.2 Voiced-unvoiced segmentation

In this section we suppose that our signal $S = S(t)$ is speech sampled over a total time interval $[0, T]$, and that we have calculated its adapted time-partition:

$$[0, T] = \bigcup_{0 \le j < N} I_j.$$

The folding of S at the boundaries between subintervals can be viewed as a segmentation of the function:

$$S \longrightarrow \bigl(S_0, S_1, \ldots, S_{N-1}\bigr).$$

Using Equations 11.30 and 11.31, each S_j can be decomposed into a set of orthogonal elementary waveforms:

$$S_j(t) = \sum_{0 \le k < n_j} c_k^j \, \phi_k^j(t).$$

Here n_j is the number of samples in I_j, with c_k^j is a spectral coefficient computed via the standard DCT-IV transform [92]. The analysis-synthesis may be represented by the following scheme:

$$S \xrightarrow{\text{fold}} \begin{matrix} S_0 & \xrightarrow{\text{DCT-IV}} & \{c_k^0\} & \xrightarrow{\text{DCT-IV}} & S_0 \\ S_1 & \xrightarrow{\text{DCT-IV}} & \{c_k^1\} & \xrightarrow{\text{DCT-IV}} & S_1 \\ \vdots & & \vdots & & \vdots \\ S_{N-1} & \xrightarrow{\text{DCT-IV}} & \{c_k^{N-1}\} & \xrightarrow{\text{DCT-IV}} & S_{N-1} \end{matrix} \xrightarrow{\text{unfold}} S.$$

Each discrete local cosine coefficient c_k^j gives the amplitude of an elementary orthogonal waveform component. This elementary waveform's period is $T_k = \frac{2\pi}{\omega_k}$ with $\omega_k = \frac{\pi}{|I_j|}(k + \frac{1}{2})$, therefore its frequency is

$$F_k = \frac{\omega_k}{2\pi} = \frac{k + \frac{1}{2}}{2|I_j|}. \tag{11.33}$$

The maximum frequency that we can distinguish is about half the sampling rate. This sampling rate is $n_j/|I_j|$, and thus

$$0 \leq F_k < \frac{n_j + \frac{1}{2}}{2|I_j|}.$$

If we start with a signal that is sampled uniformly over the entire interval $[0, T]$, then n_j will be proportional to $|I_j|$ and each interval will have approximately the same top frequency. However, since the shorter intervals have fewer coefficients, their frequency resolution will be lower. In our experiments, the signal was sampled uniformly at 8 kHz so the maximum detectable frequency was about 4 kHz. The time subintervals ranged in length down to 32 samples, or 4 ms.

A voiced speech signal is produced by regular glottal excitation. Its fundamental frequency is typically in the range 140 to 250 Hz for a female speaker and 100 to 150 Hz for a male speaker [110]. The frequency $F_k = 250$Hz corresponds to about $k = n_j/16$. Using this estimate, we shall introduce a criterion to distinguish voiced from unvoiced speech segments. Define first the frequency index k_0 of the strongest spectral component:

$$|c_{k_0}| = \max_{0 \leq k < n} |c_k|.$$

We will call F_{k_0} the *first fundamental frequency* of the segment. We will say that the signal segment S_j over I_j is *voiced* if $k_0 < n/16$; otherwise it is *unvoiced.*

For more sophisticated recognition problems, we may wish to use more than one frequency to describe the spectrum in a segment. Having found a first fundamental frequency, we suppress all the coefficients at its nearest neighbor frequencies and then look for the strongest survivor. Consider for example a signal S_j over I_j with two fundamental frequencies F_{k_0} and F_{k_1}. To compute F_{k_0} and F_{k_1} we first calculate k_0 from the last equality, set c_k equal to zero if $|k - k_0| < T$ for a preset threshold T (*i.e.*, in a neighborhood of k_0), and then deduce k_1 from

$$|c_{k_1}| = \max_{0 \le k < n} |c_k|.$$

Finally the two frequencies are obtained via Equation 11.33. This procedure may be iterated as long as there are nonzero coefficients c_k. We can force the procedure to terminate earlier by using only the top a percent of the spectrum, so that relatively small peaks will not show up as fundamental frequencies in any segment. Then T and a are parameters of the algorithm, which can be set empirically, but which should depend upon on the signal-to-noise ratio of the signal. Our experiments typically used $a = 5$ and $T = 5$ or $a = 10$ and $T = 10$.

The coefficients c_k with $|k - k_0| < T$ are associated to the frequency F_{k_0} and contain some extra information. We can average this information into the parameter we extract. Consider the following notion of *center-of-frequency* associated to each fundamental frequency F_{k_i}:

$$\mu_i = \frac{M_i}{E_i}, \qquad \text{where} \quad E_i = \sum_{k=k_i-T}^{k=k_i+T} c_k^2 \quad \text{and} \quad M_i = \sum_{k=k_i-T}^{k=k_i+T} k c_k^2 \quad \text{for } i = 0, 1, 2, \ldots.$$

In this way, for each fundamental frequency F_{k_i}, we can describe its *(locally constant) formants* with the pair $\{\mu_i, E_i\}$. The *formant representation* for a speech signal consists of a list of intervals together with the top few most energetic locally constant formants for each interval. This data can be used for recognition.

11.4.3 Experimental results

We consider the signal corresponding to the first second of the French sentence "*Des gens se sont levés dans les tribunes*," uttered by a female speaker. Figure 11.25 shows the *original signal* on the top part, the *local spectrum* which minimizes entropy in the center and the *reconstructed signal* at the bottom. The reconstructed signal is obtained from the top five percent of the spectrum inside each subinterval. The *adapted time-partition* associated to the local spectrum is drawn with vertical lines.

After finding the adapted time-partition and testing whether the first fundamental frequency in each window is below the cutoff $n_j/16$. We then *merge* the adjacent voiced segments and the adjacent unvoiced segments to leave only the

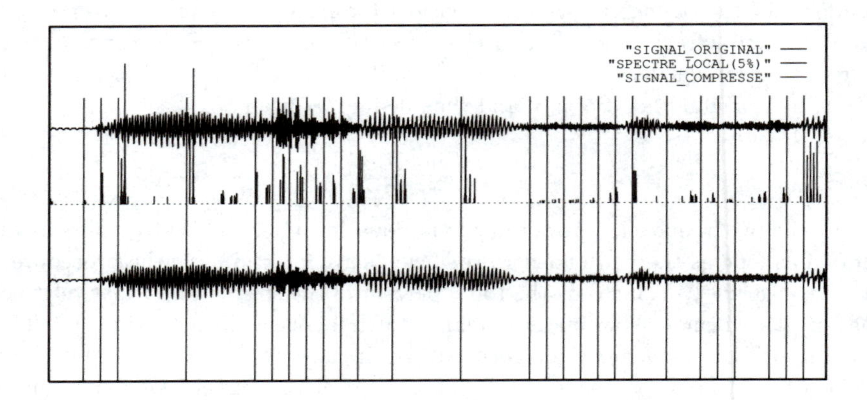

Figure 11.25: Entropy-minimizing local spectral decomposition.

windows of Figure 11.26. This highlights just the transitions between voiced and unvoiced segments.

We note that transitions are found where we expect. We extracted the time-partitions and listened to the sounds they contain. This gave us the labels in Figure 11.26, namely /d/ – /e/ – /g/ – /en/ – /s/ – /e/ – /s/. These are alternating voiced and unvoiced segments.

A more sophisticated criterion may be applied to merge adjacent voiced segments which have sufficiently similar formants. This can be done with by thresholding with a low-dimensional metric, since the formant representation of each segment has only four or five parameters in practice. Such a *phoneme recognition* algorithm needs to be well-engineered and would probably require some understanding of the speech content to resolve ambiguities, so it is beyond the scope of our discussion. The present algorithm may be considered as a "front end" to a speech recognition device, intended to simplify the representation of speech down to a few most relevant parameters.

Remark. On the horizontal axis of Figure 11.25, we must simultaneously display the sample number, the time in milliseconds, the locations of the window boundaries, and the local frequencies within each window. We solve this problem by not using any labels at all. On the topmost and bottommost traces, the horizontal axis represents time. All three traces are intersected by vertical lines indicating the window boundaries, or the endpoints of the intervals I_j in the adapted time-

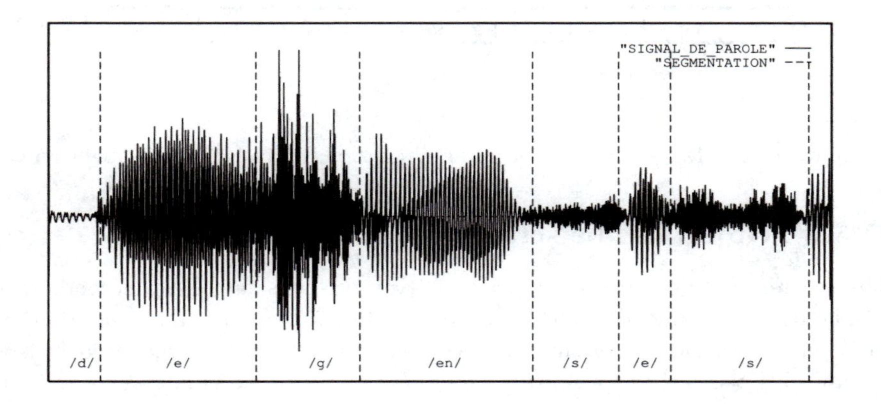

Figure 11.26: Adapted time-partition after voiced-unvoiced recognition and merging of adjacent similar segments.

partition. Within each window I_j, position along the horizontal axis of the middle trace of Figure 11.25 gives the frequency number n_j, which must be scaled by I_j to give the actual frequency. Thus the middle trace is mostly useful for counting the number of formants and gauging their relative strengths and frequencies. This style of presentation is due to Xiang Fang; similar graphs may be seen in [30].

11.5 Speech scrambling

A sequence $\{C_i : i = 1, 2, \ldots\}$ of TFA1s represents an encoded speech signal. The sequence has an associated distribution of values. One may transform the sequence with a permutation $\pi : \{C_i\} \to \{C'_i\}$ so as to preserve this distribution. Signals reconstructed from C' rather than from C will be garbled, but in a way that makes the garbled signal hard to distinguish automatically from ordinary speech. In effect, the scrambled speech will seem like some other speaker using a different language.

The map $\pi : \{C_i\} \to \{C'_i\}$ may be taken to be invertible (or even a permutation of the indices s, f, p). The map can be determined by a public key encryption method. Two individuals may each have a key, so that they may share a conversation in confidence, with eavesdroppers facing a difficult problem unscrambling either half, or even detecting that scrambling has occurred.

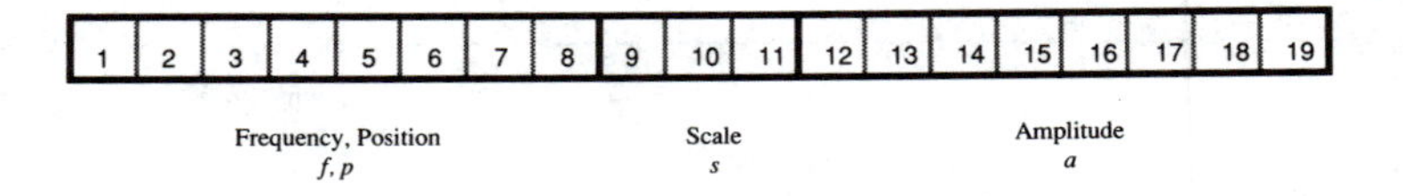

Figure 11.27: Tagged format for a one-dimensional wavelet packet coefficient.

11.5.1 Objects to be scrambled

Suppose we digitally sample speech at 8012 eight-bit samples per second, as in telephony. We divide this into 256-sample (31 ms) windows and find the best basis representation, quantizing and retaining the 13 most energetic coefficients (around five percent) per window. These may be packaged as follows: The result is $13 \times 19 = 247$ bits per window, or a rate of just under eight kbps without any entropy coding or removal of silences. This kind of packaging is needed to keep within the bandwidth limitations of telephone channels. Although the transmitted coefficients will in general be further encoded, we will assume that this is done in a lossless manner so that the parameters a, s, f, p may be extracted.

11.5.2 Feature-preserving permutations

Let C_i, $i = 1, 2, \ldots$ be TFA1s representing a speech signal in a time-frequency analysis down to level L. To the index space $\mathcal{C} = \{(s, f, p) : 0 \leq s \leq L, 0 \leq f < 2^s, 0 \leq p < 2^{L-s}\}$ of these atoms, we associate a probability measure $P : \mathcal{C} \rightarrow [0, 1]$. The probability of an index is the average energy per unit time found in coefficients with that index. This probability may be determined empirically from a sufficiently long sample of speech. Then the probability density may be depicted as a graph over the tree of wavelet packet coefficients.

Figure 11.28 is a freehand schematic of the situation, and is much less complicated than what we expect to find in practice. To create such a picture with actual data, we will accumulate $|a|^2$ in the coefficient position (s, f, p) for a reasonably long segment of real speech. Figure 11.29 depicts P for the author's recitation of part of *The Walrus and the Carpenter* by Lewis Carroll, compressed to eight kbps by discarding small coefficients of the best basis on windows of 512 samples in 62 ms.

We quantize the probability density into a small number of level regions with approximately equal numbers of coefficients in each region. Suppose that we choose five regions, as shown in the example. Encryption will consist of applying five permutations on subsets of indices, each fixing one of the five index subregions.

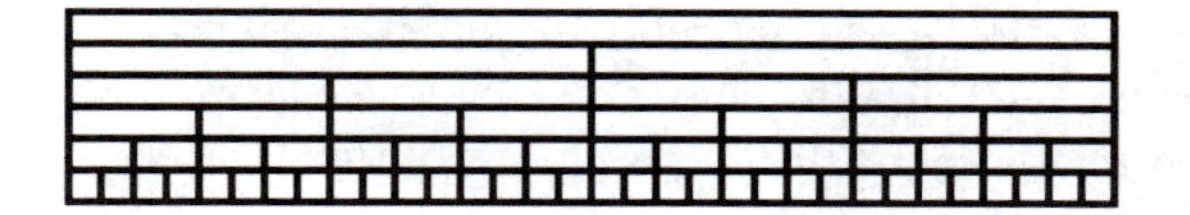

Wavelet Packet Coefficients

Example Coefficient Probability Density

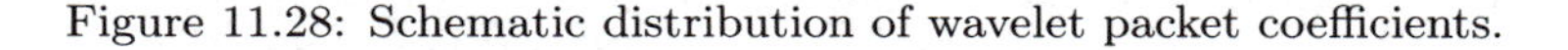

Figure 11.28: Schematic distribution of wavelet packet coefficients.

Figure 11.29: Actual distribution of wavelet packet coefficients.

The deviation between the statistical distribution of the original and encrypted coefficients may be made smaller by increasing the number of level regions. Of course, this number must remain small compared to the number of wavelet packet coefficients, otherwise there will be only trivial permutations within the regions.

11.6 Adapted waveform de-noising

We have considered expansions of a signal in several libraries of waveforms with the idea of choosing the library which best represents it. If no library does particularly well, we can peel off layers of a signal by taking one or a few waveforms out at a time, then re-analyzing the remainders. This is an example of a *meta-algorithm*, which can be used at a high level to choose an appropriate analysis for the given signal.

11.6.1 Coherency and noise

As an application combining these ideas we now focus on an algorithm for *de-noising* or, more precisely, *coherent structure extraction.* This is a difficult and ill-defined problem, not least because what is "noise" is often ill-defined. We chose instead to view an N-sample signal as being noisy or incoherent relative to a basis of waveforms if it does not correlate well with the waveforms of the basis, *i.e.*, if its entropy is of the same order of magnitude as

$$\log(N) - \epsilon \tag{11.34}$$

with small ϵ. From this notion, we are led to the following iterative algorithm based on the several previously defined libraries of orthonormal bases. We start with a signal f of length N, find the best basis in each library and select from among them the "best" best basis, the one minimizing the cost of representing f. We put the coefficients of f with respect to this basis into decreasing order of amplitude.

The rate at which the coefficients decrease controls the *theoretical dimension* N_0, which is a number between one and N describing how many of the coefficients are significant. We can define N_0 in several ways; the simplest is to count the coefficients with amplitudes above some threshold. Another is to exponentiate the entropy of the coefficient sequence, which matches the criterion in Equation 11.34.

Theoretical dimension is a kind of information cost. We will say that the signal is *incoherent* if its theoretical dimension is greater than a preset "bankruptcy" threshold $\beta > 0$. The threshold β is chosen to determine if unacceptably bad compression was obtained even with the best choice of waveforms. This condition terminates the iteration when further decompositions gain nothing.

If the signal is not incoherent, *i.e.*, if it contains components that we can recognize as signals, then we can pick a fixed fraction $0 < \delta \leq 1$ and decompose f into $c_1 + r_1$, where c_1 is reconstructed from the δN_0 big coefficients, while r_1 is the remainder reconstructed from the small ones. We proceed by using $r_1, r_2, \ldots$ as the signal and iterating the decomposition. The procedure is depicted in Figure 11.30 below. We can stop after a fixed number of decompositions, or else we can iterate until we are left with a remainder whose theoretical dimension exceeds β. We then superpose the coherent parts to get the coherent part of the signal. What remains qualifies as *noise* to us, because it cannot be well-represented by any sequence of our adapted waveforms. Thus the adapted waveform de-noising algorithm peels a particular signal into layers; we take as many of the top layers as we want, assured that the bottom layers are not cost-effective to represent.

The two parameters, β and δ, can be adjusted to match an *a priori* estimate of the signal-to-noise ratio, or can be adjusted by feedback to get the cleanest looking signal if no noise model is known.

Adapted waveform de-noising is a fast approximate version of the *matching pursuit* procedure described by Mallat [72]. There the waveforms are Gaussians, and just one component is extracted at each iteration. That procedure always produces the best decomposition, at the cost of many more iterations plus more work for each iteration. Mallat's stopping criterion is to test the amplitude ratio of successive extracted amplitudes; this is a method of recognizing remainders which have the statistics of random noise.

11.6.2 Experimental results

As an example, we start with a mechanical rumble masked by the noise of aquatic life, recorded through an underwater microphone. The calculations were performed by the program "denoise" [69, 22], using $\delta = 0.5$ and manually limiting the number of iterations to four. This application is also described in [31]; the data was provided through Ronald R. Coifman. Figure 11.31 shows the original signal paired with its de-noised version. Note that very little smoothing of the signal has taken place. Figures 11.32, 11.33, 11.34, and 11.35 respectively show the coherent parts and the remainders of the first four iterations. Notice how the total energy in each successive coherent part decreases, while the remainders continue to have roughly the same energy as the original.

Figures 11.36, 11.37, 11.38, and 11.39 respectively show the successive reconstructions from the coherent parts paired with a plot of the best basis coefficient amplitudes of the remainders, rearranged into decreasing order. A visual estimate of the theoretical dimension from these plots gives evidence after the fact that little is gained after four iterations.

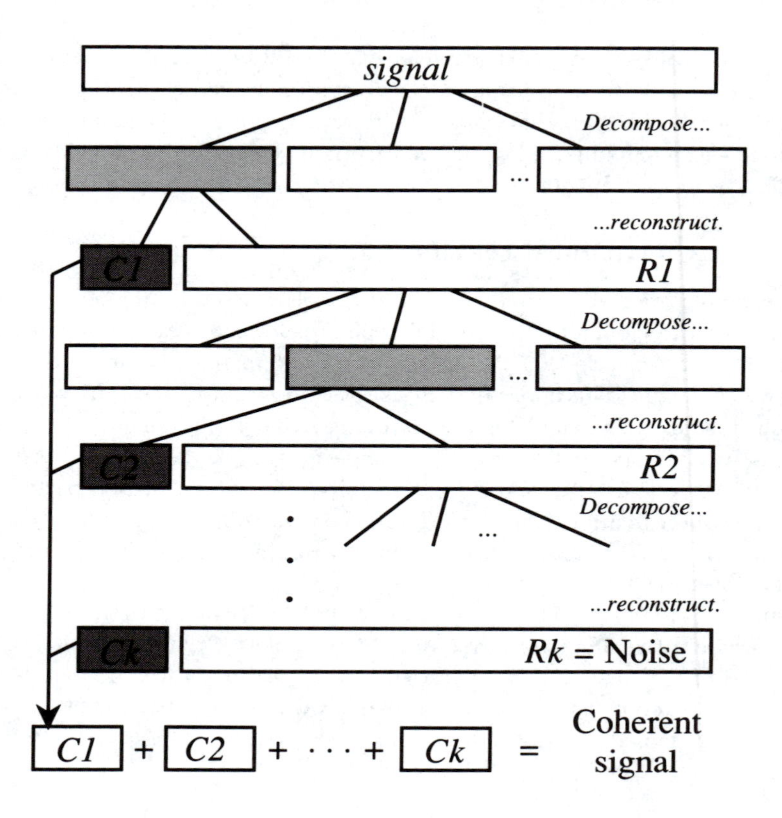

Figure 11.30: Schematic of adapted waveform de-noising.

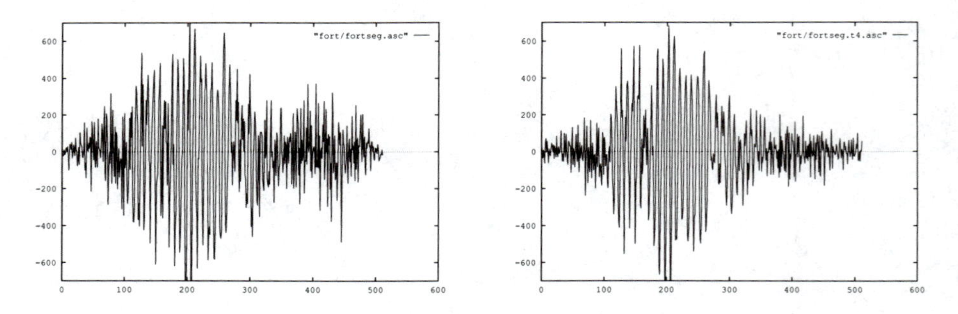

Figure 11.31: Original and de-noised signals.

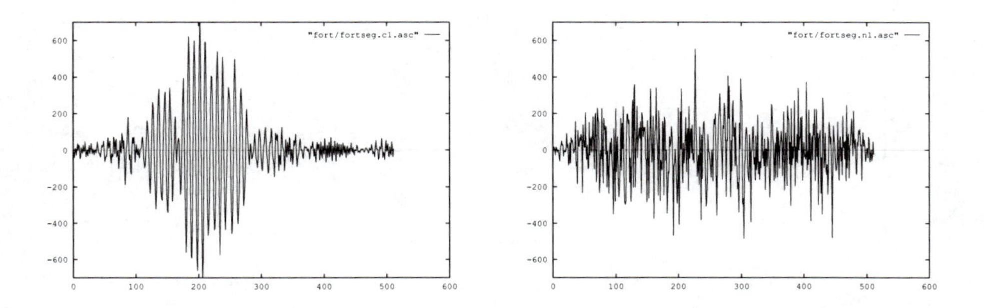

Figure 11.32: First coherent part and first remainder part.

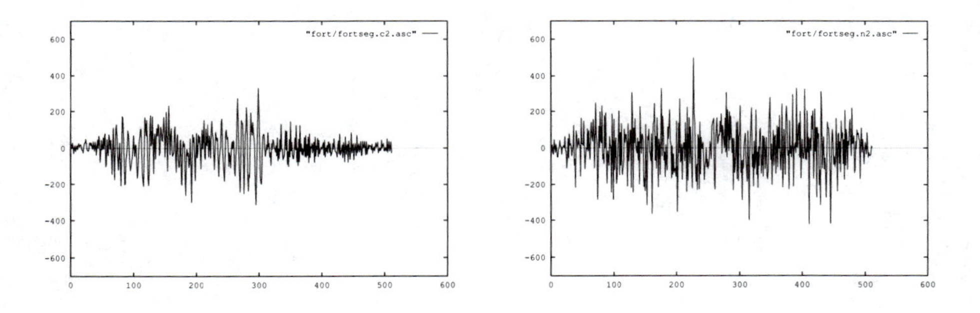

Figure 11.33: Second coherent part and second remainder part.

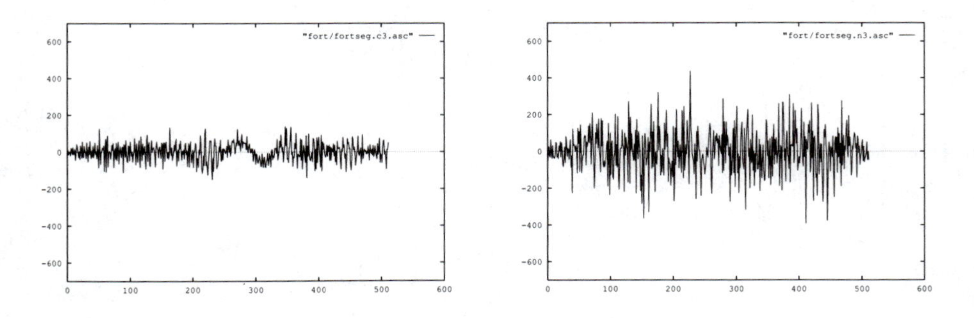

Figure 11.34: Third coherent part and third remainder part.

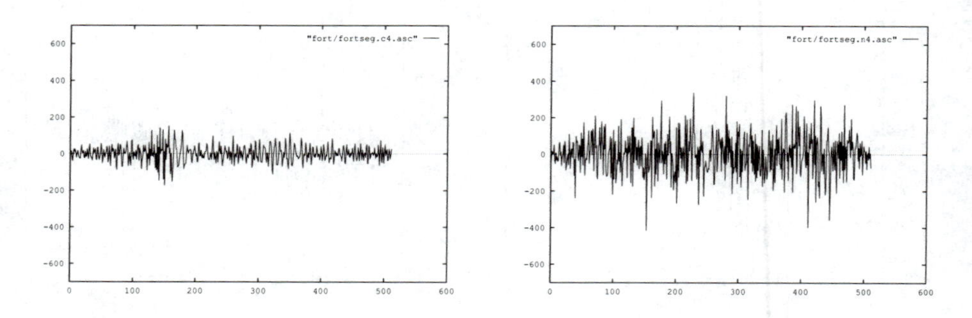

Figure 11.35: Fourth coherent part and fourth noise part.

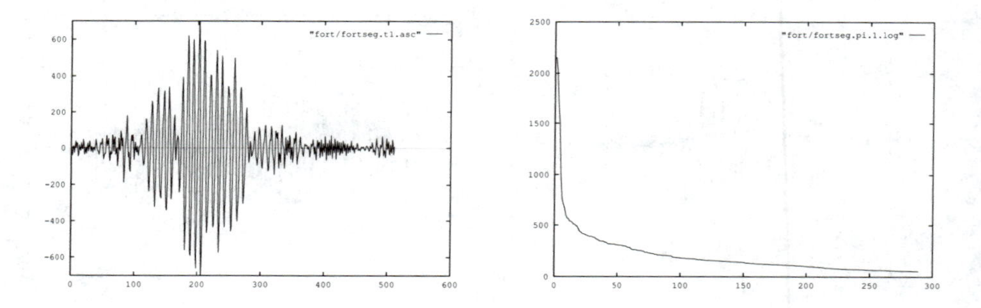

Figure 11.36: First reconstruction and its sorted remainder coefficients.

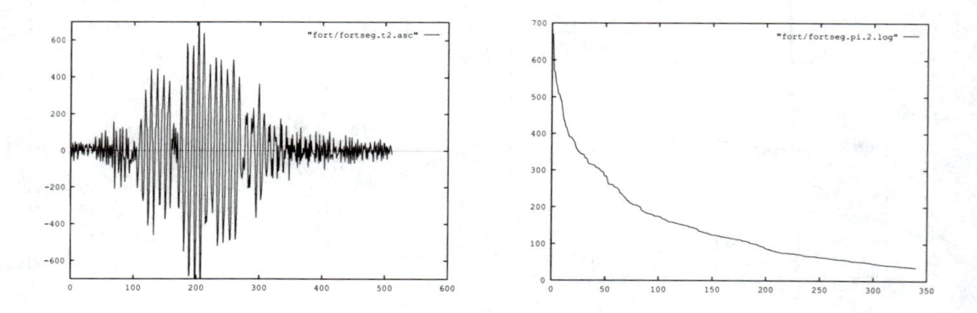

Figure 11.37: Second reconstruction and its sorted remainder coefficients.

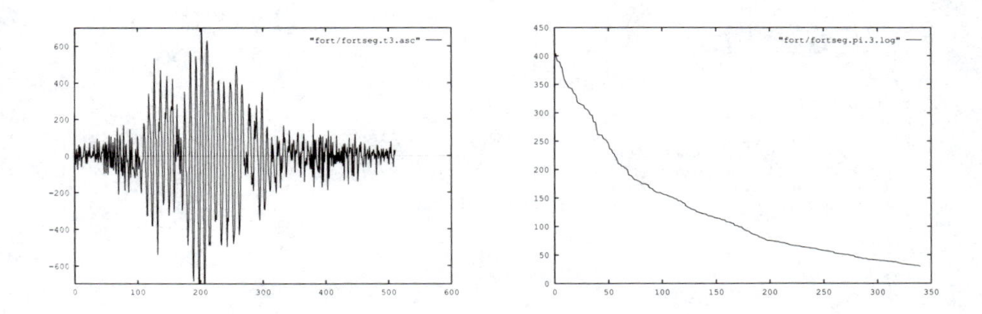

Figure 11.38: Third reconstruction and its sorted remainder coefficients.

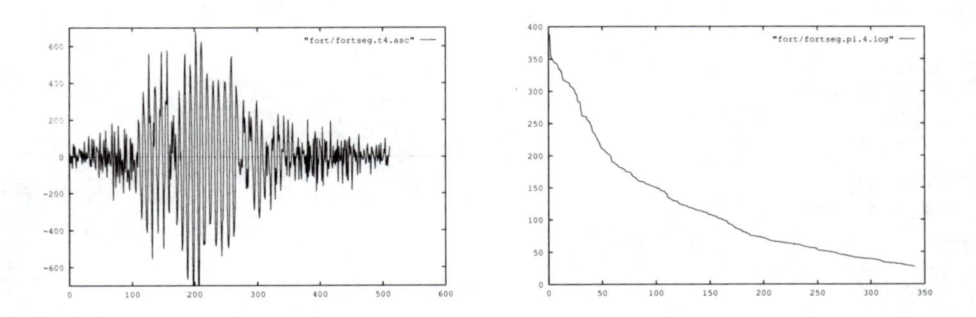

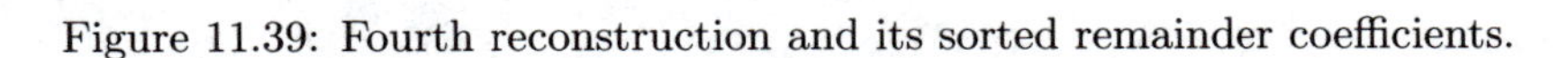

Figure 11.39: Fourth reconstruction and its sorted remainder coefficients.

Appendix A

Solutions to Some of the Exercises

Chapter 2: Programming Techniques

Chapter 2, Exercise 3. A BTN tree is more general than an array binary tree:

Convert an array binary tree into an BTN tree

```
abt2btnt( DATA, LENGTH, MAXLEVEL, LEVEL ):
  Let ROOT = makebtn( DATA, NULL, NULL, NULL )
  If LEVEL<MAXLEVEL then
    Let CHILD = DATA+LENGTH
    Let ROOT.LEFT = abt2btnt(CHILD, LENGTH, MAXLEVEL, LEVEL+1)
    Let CHILD = DATA+LENGTH+(LENGTH>>LEVEL)/2
    Let ROOT.RIGHT = abt2btnt(CHILD, LENGTH, MAXLEVEL, LEVEL+1)
  Return ROOT
```

Chapter 2, Exercise 4. Avoiding recursion, we get:

```
bisectn( ARRAY, N, U ):
   While N > 0
      N /= 2
      If U is odd then ARRAY += N
      U /= 2
   Return ARRAY
```

Chapter 4: Local Trigonometric Transforms

Chapter 4, Exercise 9. Notice that all we have to do is replace "c" with "s" in `lpic()` and `ilpic()`:

In place local periodization to N points, sine polarity

```
lpis( SIG, N, RISE ):
   fips( SIG, SIG, RISE )
   fips( SIG+N, SIG+N, RISE )
   uips( SIG+N, SIG, RISE )
```

Inverse local periodization from N points, sine polarity

```
ilpis( SIG, N, RISE ):
   fips( SIG+N, SIG, RISE )
   uips( SIG, SIG, RISE )
   uips( SIG+N, SIG+N, RISE )
```

Chapter 4, Exercise 10. We must invert the order of operations used by `lpica()`:

Inverse local periodizations of adjacent intervals, cosine polarity

```
ilpica( SIG, LENGTHS, NUM, RISE ):
   For I = 0 to NUM-1
      fipc( SIG+LENGTHS[I], SIG, RISE )
      uipc( SIG, SIG, RISE )
      SIG += LENGTHS[I]
   uipc( SIG, SIG, RISE )
```

To get the sine polarity version `ilpisa()`, we simply replace each `fipc()` with an `fips()` and each `uipc()` with a `uips()`.

Chapter 5: Quadrature Filters

Chapter 5, Exercise 1. Let e and f be two filter sequences from a quadruplet of biorthogonal quadrature filters. Then for $0 \le n < 2q$ we have

$$\begin{aligned}\sum_{k=0}^{2q-1} e'_{2q}(k)\bar{f}_{2q}(k+2n) &= \sum_{k=0}^{2q-1}\left(\sum_{x=-\infty}^{\infty} e'(k+2qx)\right)\left(\sum_{y=-\infty}^{\infty} \bar{f}(k+2n+2qy)\right) \\ &= \sum_{x=-\infty}^{\infty}\sum_{y=-\infty}^{\infty}\sum_{z=2qx}^{2q(x+1)-1} e'(z)\bar{f}(z+2[n-q(x-y)]) \\ &= \sum_{w=-\infty}^{\infty}\sum_{z=-\infty}^{\infty} e'(z)\bar{f}(z+2[n+qw]).\end{aligned}$$

Here we have substituted $k \leftarrow z - 2qx \Rightarrow 2qx \le z < 2q(x+1)$ and $y \leftarrow w + x \Rightarrow -\infty < w < \infty$. Then the sum over z and x can be combined into a sum over z with $-\infty < z < \infty$.

We now use Equations 5.6. If $e = g$ and $f' = h'$ or $e = h$ and $f' = g'$, then the sum in z vanishes for all n, w. If $e = h$ and $f' = h'$ or $e = g$ and $f' = g'$ are dual filters, then the sum in z is nonzero only if $n = qw$, which is possible only if $n = 0$ and $w = 0$. The sum in z yields 1, while the sum in w collapses to a single term, completeing the proof. □

Chapter 5, Exercise 4. The action of filters on functions yields combinations of samples at intermediate points:

$$\begin{aligned}H^*Hu(t) &= \frac{1}{2}u(t-1) + u(t) + \frac{1}{2}u(t+1), \\ G^*Gu(t) &= -\frac{1}{2}u(t-1) + u(t) - \frac{1}{2}u(t+1), \\ H^*Gu(t) &= \frac{1}{2}u(t-1) - \frac{1}{2}u(t+1), \\ G^*Hu(t) &= -\frac{1}{2}u(t-1) + \frac{1}{2}u(t+1).\end{aligned}$$

Chapter 6: The Discrete Wavelet Transform

Chapter 6, Exercise 3. We use a loop instead of a recursive function call to keep track of the decomposition level:

Nonrecursive disjoint L-level periodic DWT on $N = M * 2^L$ points

```
dwtpdn( DIFS, SUMS, IN, N, L, H, G ):
   While L > 0
      cdpi( DIFS+N/2, 1, IN, N, G )
      cdpi( SUMS+N/2, 1, IN, N, H )
      L -= 1
      N /= 2
      Let IN = SUMS + N
   For K = 0 to N-1
      DIFS[K] += IN[K]
```

We make the same disjointness and initialization assumptions about the three arrays `IN[]`, `SUMS[]`, and `DIFS[]` in this case as for the recursive implementation.

Chapter 6, Exercise 4. The main difference between this and the disjoint nonrecursive version is that we use `cdpe()` rather than `cdpo()`. After computing the sums in the scratch array, we can overwrite the combined input/output array with wavelet coefficients:

Nonrecursive in place L-level periodic DWT on $N = M * 2^L$ points

```
dwtpin( DATA, WORK, N, L, H, G ):
   While L > 0
      cdpe( WORK+N/2, 1, DATA, N, G )
      cdpe( WORK, 1, DATA, N, H )
      For I = 0 to N-1
         Let DATA[I] = WORK[I]
      N /= 2
      L -= 1
```

Chapter 6, Exercise 5. We use a recursive function call and retain a local variable in each instance of the function to keep track of the level of decomposition:

Recursive disjoint L-level periodic iDWT on $N = 2^L$ points

```
idwtpd0( OUT, SUMS, IN, N, H, G ):
   If N > 1 then
      N /=2
      idwtpd0( SUMS+N, SUMS, IN, N, H, G)
      acdpi( OUT, 1, SUMS+N, N, H )
      acdpi( OUT, 1,  IN +N, N, G )
   Else
      OUT[0] += IN[0]
   Return
```

Chapter 6, Exercise 6. We retain a local variable in each instance of the recursive function to keep track of the level of decomposition. Using `acdpe()` overwrites the combined input/output array with the reconstruction from the low-pass (H) child; then the reconstructed high-pass (G) child is added in with `acdpo()`:

Recursive in place L-level periodic iDWT on $N = 2^L$ points

```
idwtpi0( DATA, WORK, N, L, H, G ):
   If N > 1 then
      idwtpi0( DATA, WORK, N/2, H, G )
      acdpe( WORK, 1,     DATA,     N/2, H )
      acdpo( WORK, 1, DATA+(N/2), N/2, G )
      For I = 0 to N-1
         Let DATA[I] = WORK[I]
   Return
```

Chapter 6, Exercise 7. The following function generalizes `idwtpd0n()`. It keeps track of the decomposition level with a loop down to the desired depth.

Nonrecursive disjoint L-level periodic iDWT on $N = M * 2^L$ points

```
idwtpdn( OUT, SUMS, IN, N, L, H, G ):
   If L > 0 then
      M = N>>L
      SUMS += M
      For K=0 to M-1
         Let SUMS[K] = IN[K]
      While M < N/2
         acdpi( SUMS+M, 1, SUMS, M, H )
         acdpi( SUMS+M, 1,  IN,  M, G )
         SUMS += M
         IN   += M
         M  *= 2
      acdpi( OUT, 1, SUMS, N/2, H )
      acdpi( OUT, 1, IN,   N/2, G )
   Else
      For K=0 to N-1
         Let OUT[K] += IN[K]
```

Chapter 6, Exercise 8. The following function generalizes `idwtpi0n()` and uses a loop to keep track of the decomposition level.

Nonrecursive in place L-level periodic iDWT on $N = M * 2^L$ points

```
idwtpin( DATA, WORK, N, L, H, G ):
   Let M = N>>L
   While M < N
      acdpe( WORK, 1,  DATA,  M, H )
      acdpo( WORK, 1, DATA+M, M, G )
      For I = 0 to 2*M-1
         Let DATA[I] = WORK[I]
      M *= 2
```

Notice that we first use `acdpe()` to write the H reconstruction, then superpose the G reconstruction with `acdpo()`.

Chapter 6, Exercise 9. We use a loop to keep track of the depth of decomposition:

L-level nonrecursive aperiodic DWT from an INTERVAL

```
dwtan( V, W, IN, L, H, G ):
   For K = 0 to L-1
      cdao( V[K].ORIGIN, 1, IN.ORIGIN, IN.LEAST, IN.FINAL, H )
      cdao( W[K].ORIGIN, 1, IN.ORIGIN, IN.LEAST, IN.FINAL, G )
      Let IN = V[K]
   For N = IN.LEAST to IN.FINAL
      Let W[L].ORIGIN[N] = IN.ORIGIN[N]
```

The arrays `V[]` and `W[]` of INTERVALs hold the scaling and wavelet decomposition subspaces, respectively. They must be preallocated with $L+1$ INTERVALs in `W[]` and L INTERVALs in `V[]`.

Output values are actually written to two preallocated arrays `SUMS[]` and `DIFS[]`, which must be long enough to hold all the wavelet and scaling amplitudes produced by the aperiodic DWT. Their respective lengths may be computed by calling `intervalstotal(V,L)` and `intervalstotal(W,L+1)`, as defined in Chapter 2, after temporarily filling the endpoint members of the INTERVAL elements of `V[]` and `W[]` by calling `dwtaintervals()`. Similarly, the origin members of `V[]` and `W[]` are set by calling `dwtaorigins()`.

Chapter 6, Exercise 10. Let `shifttoorigin()` and `shifttonextinterval()` be defined so as to concatenate from the end. We pass the ends of the coefficient arrays `SUMS[]` and `DIFS[]` to the function `dwtaorigins()`.

L-level DWTA from an array

```
dwtacomplete( DATA, LENGTH, L, H, G ):
   Allocate L empty INTERVALs at V
   Allocate L+1 empty INTERVALs at W
   Let IN = makeinterval( DATA, 0, LENGTH-1 )
   dwtaintervals( V, W, IN, L, H, G )
   Let NS = intervalstotal( V, L )
   Allocate an array of NS zeroes at SUMS
   Let ND = intervalstotal( W, L+1 )
   Allocate an array of NS zeroes at DIFS
   dwtaorigins( V, W, SUMS+NS, DIFS+ND, L )
   dwta( V, W, IN, L, H, G )
   Deallocate SUMS[]
   Deallocate IN, but leave DATA[] alone
   Deallocate V[]
   Return W
```

Chapter 6, Exercise 11. Let `shifttoorigin()` and `shifttonextinterval()` concatenate from the end. We pass the end of the array `SUMS[]` to `idwtaorigins()`:

L-level iDWTA from an array of INTERVALs

```
idwtacomplete( W, L, H, G ):
   Allocate L empty INTERVALs at V
   Allocate 1 empty INTERVAL at OUT
   idwtaintervals( OUT, V, W, L, H, G )
   Let OUT = enlargeinterval( OUT, OUT.LEAST, OUT.FINAL )
   Let NS = intervalstotal( V, L )
   Allocate an array of NS zeroes at SUMS
   idwtaorigins( V, SUMS+NS, L )
   idwta( OUT, V, W, L, H, G )
   Deallocate SUMS[]
   Deallocate V[]
   Return OUT
```

Chapter 6, Exercise 12. For input, we need to specify the following:

- `V`, an array of L empty INTERVAL pointers for the scaling subspaces;
- `W`, an array of $L+1$ empty INTERVAL pointers for the wavelet subspaces and V_L;
- `IN[]`, the input INTERVAL; levels;
- L, the number of levels of decomposition;
- `H,G`, the low-pass and high-pass QFs;

The output is written to the progressively allocated V and W arrays by side-effect:

L-level DWTA from an INTERVAL, local allocation

```
dwtalocal( V, W, IN, L, H, G ):
   If L > 0
      Let LEAST = cdaleast( IN, H )
      Let FINAL = cdafinal( IN, H )
      Let V = makeinterval( NULL, LEAST, FINAL )
      cdai( V.ORIGIN, 1, IN.ORIGIN, IN.LEAST, IN.FINAL, H )
      Let LEAST = cdaleast( IN, G )
      Let FINAL = cdafinal( IN, G )
      Let W = makeinterval( NULL, LEAST, FINAL )
      cdai( W.ORIGIN, 1, IN.ORIGIN, IN.LEAST, IN.FINAL, G )
      dwtalocal( V+1, W+1, V, L-1, H, G )
   Else
      Let W = makeinterval( NULL, IN.LEAST, IN.FINAL )
      For N = IN.LEAST to IN.FINAL
         Let W.ORIGIN[N] = IN.ORIGIN[N]
   Return
```

Chapter 7: Wavelet Packets

Chapter 7, Exercise 3. We combine sprinkling amplitudes from a hedge into an array binary tree with a previously written periodic wavelet packet synthesis function:

Periodic DWPS from a hedge to an array binary tree

```
hedge2dwpspabt( GRAPH, N, HQF, GQF ):
   Let MAXLEVEL = GRAPH.LEVELS[0]
   For I = 1 to GRAPH.BLOCKS-1
      Let MAXLEVEL = max( MAXLEVEL, GRAPH.LEVELS[I] )
   Allocate an array of N*(MAXLEVEL+1) 0s at DATA
   hedge2abt( DATA, GRAPH, N )
   abt2dwpsp( DATA, N, MAXLEVEL, HQF, GQF )
   Return DATA
```

Chapter 7, Exercise 4. The following modification to `btnt2dwpsa()` deallocates each BTN as soon as its parent is reconstructed:

Aperiodic DWPS from a tree, immediate deallocation

```
btnt2dwpsa0( ROOT, HQF, GQF ):
  If ROOT != NULL then
    btnt2dwpsa0( ROOT.LEFT, HQF, GQF )
    If ROOT.LEFT != NULL then
      Let ROOT.CONTENT = acdaparent( ROOT.CONTENT,
                                     ROOT.LEFT.CONTENT, HQF )
      Let ROOT.LEFT = freebtn( ROOT.LEFT, freeinterval, free )
    btnt2dwpsa0( ROOT.RIGHT, HQF, GQF )
    If ROOT.RIGHT != NULL then
      Let ROOT.CONTENT = acdaparent( ROOT.CONTENT,
                                   ROOT.RIGHT.CONTENT, GQF )
      Let ROOT.RIGHT = freebtn( ROOT.RIGHT, freeinterval, free )
  Return
```

Chapter 8: The Best Basis Algorithm

Chapter 8, Exercise 5. We assume that `GRAPH.BLOCKS` is zero at the outset:

Recursively find the best basis nodes in a complete BTN tree

```
btnt2bbhedge( GRAPH, ROOT, S, L ):
  Let MYCOST = infocost( ROOT.CONTENT.ORIGIN,
                    ROOT.CONTENT.LEAST, ROOT.CONTENT.FINAL )
  If S == L then
    Let GRAPH.CONTENT[GRAPH.BLOCKS] = ROOT.CONTENT
    Let GRAPH.LEVELS[GRAPH.BLOCKS] = L
    GRAPH.BLOCKS += 1
    Let BESTCOST = MYCOST
  Else
    Let BLOCKS = GRAPH.BLOCKS
    Let LCOST = btnt2bbhedge( GRAPH, ROOT.LEFT, S+1, L )
    Let RCOST = btnt2bbhedge( GRAPH, ROOT.RIGHT, S+1, L )
    If MYCOST > LCOST + RCOST then
       Let BESTCOST = LCOST + RCOST
    Else
       Let BESTCOST = MYCOST
       Let GRAPH.BLOCKS = BLOCKS
       Let GRAPH.CONTENTS[GRAPH.BLOCKS] = ROOT.CONTENT
       Let GRAPH.LEVELS[GRAPH.BLOCKS] = S
       GRAPH.BLOCKS += 1
  Return BESTCOST
```

Chapter 8, Exercise 6. We call `costs2blevel()` after `btnt2costs()`:

Return a best level hedge for a BTN tree of INTERVALs

```
btnt2blevel( ROOT, MINLEVEL, MAXLEVEL ):
  btnt2costs( ROOT )
  Let GRAPH = makehedge( 1<<MAXLEVEL, NULL, NULL, NULL )
  Let GRAPH.BLOCKS = 0
  costs2blevel( GRAPH, ROOT, MINLEVEL, MAXLEVEL )
  Return GRAPH
```

Chapter 8, Exercise 7. We first implement a function that computes and returns the information cost of a level in the original BTN tree of INTERVALs:

Information cost of one level in a BTN tree of INTERVALs

```
btntlevelcost( ROOT, LEVEL ):
  Let COST = 0
  For BLOCK = 0 to (1<<LEVEL)-1
    Let NODE = btnt2btn( ROOT, LEVEL, BLOCK )
    Let I = NODE.CONTENT
    COST += infocost( I.ORIGIN, I.LEAST, I.FINAL )
  Return COST
```

Then we use this function in an analog of `btnt2blevel()`:

Find the best level basis in a BTN tree of INTERVALs

```
btnt2blhedge( GRAPH, ROOT, MINLEVEL, MAXLEVEL ):
  Let BESTLEVEL = MINLEVEL
  Let BESTCOST = btntlevelcost( ROOT, MINLEVEL )
  For LEVEL = MINLEVEL+1 to MAXLEVEL
    Let COST = btntlevelcost( ROOT, LEVEL )
    If COST<BESTCOST then
      Let BESTCOST = COST
      Let BESTLEVEL = LEVEL
  Let GRAPH.BLOCKS = 1<<BESTLEVEL
  For BLOCK = 0 to GRAPH.BLOCKS-1
      Let NODE = btnt2btn( ROOT, LEVEL, BLOCK )
      Let GRAPH.CONTENTS[BLOCK] = NODE.CONTENT
      Let GRAPH.LEVELS[BLOCK] = BESTLEVEL
  Return BESTCOST
```

Chapter 8, Exercise 8. We call `costs2blevel()` after preparation with `abt2costs()`:

Return the best level hedge from an array binary tree

```
abt2blevel( DATA, LENGTH, MINLEVEL, MAXLEVEL ):
  Let COSTS = abt2costs( DATA, LENGTH, MAXLEVEL )
  Let GRAPH = makehedge( 1<<MAXLEVEL, NULL, NULL, NULL )
  Let GRAPH.BLOCKS = 0
  costs2blevel( GRAPH, COSTS, MINLEVEL, MAXLEVEL )
  freebtnt( COSTS, NULL, free )
  Return GRAPH
```

Chapter 8, Exercise 9. Since the amplitudes in an array binary tree are stored contiguously by level, we can search for the level with least information cost without introducing any additional data structures:

Find the best level basis in an array binary tree

```
abt2blhedge( GRAPH, DATA, LENGTH, MINLEVEL, MAXLEVEL ):
  Let BESTLEVEL = MINLEVEL
  Let BESTCOST = infocost(DATA+LENGTH*MINLEVEL, 0, LENGTH-1)
  For LEVEL = MINLEVEL+1 to MAXLEVEL
    Let COST =  infocost(DATA+LENGTH*LEVEL, 0, LENGTH-1)
    If COST<BESTCOST then
      Let BESTCOST = COST
      Let BESTLEVEL = LEVEL
  Let GRAPH.BLOCKS = 1<<BESTLEVEL
  For BLOCK = 0 to GRAPH.BLOCKS-1
      Let GRAPH.CONTENTS[BLOCK] =
             DATA + abtblock( LENGTH, BESTLEVEL, BLOCK )
      Let GRAPH.LEVELS[BLOCK] = BESTLEVEL
  Return BESTCOST
```

The outermost loop computes the cost of each level and bubbles up the minimum; it is then trivial to write the output hedge.

Chapter 9: Multidimensional Library Trees

Chapter 9, Exercise 1. We first need to perform the following initializations:

- Allocate an array of integer lengths `LEN[]` with d members and assign it the dimensions of the initial data array;
- Allocate an array of small integer depths `LVL[]` with d members, and choose a depth of decomposition for each dimension which is no more than the number of times that two divides that dimension;
- Allocate an array of $2d$ PQF data structures and fill it with pairs of conjugate QFs, one pair for each dimension.
- Allocate a scratch array at `WORK[]` and fill it with zeroes;
- Allocate an input/output array at `DATA[]`, and read the input into it.

The following pseudocode implements a d-dimensional tensor wavelet transform on data whose dimensions are `LEN[0]` $\times \cdots \times$ `LEN[D-1]`, using the filters `H=QFS[2*K]`, `G=QFS[2*K+1]`, down to level `LVL[K]` along the axis indexed by $\mathtt{K}^{\text{th}}$ variable:

In place d-dimensional periodic DWT with varying levels and filters in each dimension

```
dwtpvd( DATA, D, LEN, LVL, QFS, WORK ):
  Let VOLUME = LEN[0]
  For K = 1 to D-1
    VOLUME *= LEN[K]
  Let K = D-1
  While K >= 0
    Let I = 0
    While I < VOLUME
      dwtpi(DATA+I,WORK,LEN[D-1],LVL[K],QFS[2*K],QFS[2*K+1])
      I += LEN[D-1]
    xpid( DATA, LEN, D )
    K -= 1
```

Chapter 9, Exercise 2. The same ideas used in Problem 1 work to produce the isotropic two-dimensional wavelet transform. It is necessary to vary the level of decomposition for a row depending upon the row index, and likewise to adjust the level of decomposition for a column depending upon the column index. The resulting output does not follow the indexing convention of our quadtree two-dimensional wavelet packet bases, but for some applications it may be more efficient. The following implementation works for rectangular arrays and allows different decomposition depths in the two dimensions:

In place isotropic two-dimensional periodic wavelet basis

```
dwtpi2(DATA, IX,IY, LX,LY, HX,GX, HY,GY, WORK):
   Let M = IX>>LX
   Let I = 0
   While I < M*IY
      dwtpi( DATA, WORK, IY, LY, HY, GY )
      I += IY
   While M < IX
      While I < 2*M*IY
         dwtpi( DATA, WORK, IY, LY, HY, GY )
         I += IY
      M *= 2
   xpi2( DATA, IX, IY )
   Let M = IY>>LY
   Let I = 0
   While I < M*IX
      dwtpi( DATA, WORK, IX, LX, HX, GX )
      I += IX
   While M < IY
      While I < 2*M*IX
         dwtpi( DATA, WORK, IX, LX, HX, GX )
         I += IX
      M *= 2
   xpi2( DATA, IY, IX )
```

Chapter 9, Exercise 3. Inverses `idwtpt2()`, `idwtptd()`, `idwtpvd()`, and `idwtpi2()` are obtained by replacing each call to the function `dwtpi()` with a call to `idwtpi()`. Note that the factors in a tensor product of transforms commute.

Chapter 9, Exercise 4. With disjoint transposition, it is not necessary to use a temporary variable, nor is it necessary to check whether the (i, j) coefficient has a lower offset than the (j, i) coefficient.

Two-dimensional transposition and copying between disjoint arrays

```
xpd2(OUT, IN, X, Y):
   For I = 0 to X-1
      For J = 0 to Y-1
         Let OUT[J*X + I] = IN[I*Y + J]
```

Chapter 9, Exercise 7. The following implementation can be made more efficient with a few extra variables:

Disjoint separable two-dimensional cosine unfolding

```
udc2( OUT, IN0, IN1, IN2, IN3, X0, X1, Y0, Y1, WORK, RX, RY ):
   Let WPTR = WORK + Y0*(X0+X1)
   For I = 0 to X0-1
     udcn( WPTR+I, X0+X1, IN0+Y0+I*Y0, IN1+I*Y1, Y0, RY )
   Let WPTR = WORK + Y0*(X0+X1) + X0
   For I = 0 to X0-1
     udcn( WPTR+I, X0+X1, IN2+Y0+I*Y0, IN3+I*Y1, Y0, RY )
   Let WPTR = WORK + X0
   OUT += (Y0+Y1)*X0
   For I = 0 to Y0-1
     udcn( OUT+I, Y0+Y1, WPTR, WPTR, X0, RX )
     udcp( OUT+I, Y0+Y1, WPTR, WPTR, X1, RX )
     WPTR += X0+X1
   For I = 0 to X0-1
     udcp( WORK+I, X0+X1, IN0+Y0+I*Y0, IN1+I*Y1, Y1, RY )
   Let WPTR = WORK + X0
   For I = 0 to X1-1
     udcp( WPTR+I, X0+X1, IN2+Y0+I*Y0, IN3+I*Y1, Y1, RY )
   Let WPTR = WORK + X0
   OUT += Y0
   For I = 0 to Y1-1
     udcn( OUT+I, Y0+Y1, WPTR, WPTR, X0, RX )
     udcp( OUT+I, Y0+Y1, WPTR, WPTR, X1, RX )
     WPTR += X0+X1
```

Appendix B

List of Symbols

Symbol	Definition
$\Rightarrow$	Implies.
$\mathbf{Z}$	Integers $\{\ldots,-2,-1,0,1,2,3,\ldots\}$.
$\mathbf{N}$	Natural numbers $\{0,1,2,\ldots\}$.
$\#S$	The number of elements in the set S.
$\gcd(p,q)$	Greatest common divisor of p and q.
$\lceil x \rceil$	Least integer greater than or equal to x.
$\lfloor x \rfloor$	Greatest integer less than or equal to x.
$\mathbf{R}$	Real numbers.
$\mathbf{R}^+$	Nonnegative real numbers; $[0,\infty[$.
$\mathbf{T}$	The circle of unit circumference; also, the interval $[0,1]$.
$\mathbf{C}$	Complex numbers; the complex plane.
$]a,b[$	The open interval $\{x : a < x < b\}$.
$[a,b]$	The closed interval $\{x : a \leq x \leq b\}$.
$]a,b]$	The half-open interval $\{x : a < x \leq b\}$.
$[a,b[$	The half-closed interval $\{x : a \leq x < b\}$.
$B_\epsilon(\alpha)$	The ball of radius ϵ centered at α.
$\lvert E\rvert$	Lebesgue measure of a set E, $E \subset \mathbf{R}^d$.
$\operatorname{diam} E$	Diameter of a set E; $\inf\{R : s,t \in E \Rightarrow \lvert s-t\rvert \leq R\}$.
$\sup X$	Supremum of a set X; the least $L \in \mathbf{R}$ with $x \leq L$ for all $x \in X$.
$\inf X$	Infimum of a set X; the greatest $L \in \mathbf{R}$ with $x \leq L$ for all $x \in X$.
$\bar{z}$	Complex conjugate; if $z = a + ib$ for a,b real, then $\bar{z} = a - ib$.
$\Re z$	Real part of z; $\Re z = \frac{1}{2}(z+\bar{z})$.
$\Im z$	Imaginary part of z; $\Im z = \frac{1}{2i}(z-\bar{z})$.

Symbol	**Definition**
A^T	Transpose of the matrix A; $A^T(n,m) = A(m,n)$.
A^*	Adjoint of the matrix A; $A^*(n,m) = \overline{A(m,n)}$.
$\oplus$	Direct sum; for matrices A and B, $A \oplus B = \begin{pmatrix} A & 0 \\ 0 & B \end{pmatrix}$
$cond(A)$	Condition number of a matrix; $\|A\|\,\|A^{-1}\|$.
$\langle u, v\rangle$	Hermitean inner product; $u, v \in \ell^2 \Rightarrow \langle u, v\rangle = \sum_k \bar{u}(k)v(k)$.
$\|u\|$	Root-mean-square norm of u; $\sqrt{\langle u, u\rangle}$.
$\operatorname{supp} f$	Support of f; smallest closed set S with $t \notin S \Rightarrow f(t) = 0$.
$dist_X(u, v)$	Distance between u and v in the metric space X.
$C(\mathbf{T})$	Continuous one-periodic functions.
L^1	Lebesgue integrable functions.
L^2	Square-integrable functions.
L^p	Functions f such that $\lvert f\rvert^p$ is Lebesgue integrable.
ℓ^1	Absolutely summable sequences.
ℓ^2	Square-summable sequences.
ℓ^p	Sequences u such that $\{\lvert u(k)\rvert^p\}$ is summable.
$\ll$	Much less than.
`==`	Logical equality (EQUALS).
$\mathtt{\vert\vert}$	Logical disjunction (OR).
`&&`	Logical conjunction (AND).
`^`	Bitwise exclusive-disjunction (XOR).
A `^=` B	Replace A with its bitwise exclusive-disjunction with B.
A `<<` B	Bit-shift A toward the left by B bits; $2^B A$.
A `>>` B	Bit-shift A to the right by B bits; $\lfloor A/2^B \rfloor$.
A `+=` B	Increment A by B and store the result in A.
A `-=` B	Decrement A by B and store the result in A.
A `*=` B	Multiply A by B and store the result in A.
A `/=` B	Divide A by B and store the result in A.
$\mathbf{1}$	The characteristic function of the interval $[0, 1]$.
$\mathbf{1}_I$	The characteristic function of some interval I.
f_e	Even terms of the sequence f; $f_e(n) = f(2n)$.
f_o	Odd terms of the sequence f; $f_o(n) = f(2n + 1)$.
$\psi^<$	Backwards reflection of ψ; $\psi^<(t) \stackrel{\text{def}}{=} \psi(-t)$.
$r_{[n]}$	Iteration n for a recursively-defined function r.
f_q	q-periodization of the function f; $f_q(t) = \sum_{k\in\mathbf{Z}} f(t + kq)$.
$c[f]$	Center of energy of the sequence f.
$d[f]$	Deviation from linear phase of the filter sequence f.

Appendix C

Quadrature Filter Coefficients

Here we give the coefficients of some of the quadrature filters mentioned in the text. These are also listed on the diskette, in a form suitable for inclusion in Standard C programs.

C.1 Orthogonal quadrature filters

We first define a few quantities useful in getting full-precision values on arbitrary machines. They are, respectively, $\sqrt{2}$, $\sqrt{3}$, $\sqrt{10}$ and $\sqrt{15}$.

```
#define SR2  (1.4142135623730950488)
#define SR3  (1.7320508075688772935)
#define SR10 (3.1622776601683793320)
#define SR15 (3.8729833462074168852)
```

In addition, we precompute the quantities $A = \frac{1}{4}\sqrt{2}\left[1 + \sqrt{10} + \sqrt{5 + 2\sqrt{10}}\right]$ and $B = \dfrac{3}{4A}$:

```
#define A  (2.6613644236006609279)
#define B  (0.2818103350856762928)
```

C.1.1 Beylkin filters

The "Beylkin 18" filter was designed by placing roots for the frequency response polynomial close to the Nyquist frequency on the real axis, thus concentrating power spectrum energy in the desired band.

Beylkin 18

Low-pass	High-pass
9.93057653743539270 E-02	6.40485328521245350 E-04
4.24215360812961410 E-01	2.73603162625860610 E-03
6.99825214056600590 E-01	1.48423478247234610 E-03
4.49718251149468670 E-01	-1.00404118446319900 E-02
-1.10927598348234300 E-01	-1.43658079688526110 E-02
-2.64497231446384820 E-01	1.74604086960288290 E-02
2.69003088036903200 E-02	4.29163872741922730 E-02
1.55538731877093800 E-01	-1.96798660443221200 E-02
-1.75207462665296490 E-02	-8.85436306229248350 E-02
-8.85436306229248350 E-02	1.75207462665296490 E-02
1.96798660443221200 E-02	1.55538731877093800 E-01
4.29163872741922730 E-02	-2.69003088036903200 E-02
-1.74604086960288290 E-02	-2.64497231446384820 E-01
-1.43658079688526110 E-02	1.10927598348234300 E-01
1.00404118446319900 E-02	4.49718251149468670 E-01
1.48423478247234610 E-03	-6.99825214056600590 E-01
-2.73603162625860610 E-03	4.24215360812961410 E-01
6.40485328521245350 E-04	-9.93057653743539270 E-02

C.1.2 Coifman or "Coiflet" filters

The "Coifman" filters are designed so that both the scaling function and the mother wavelet will have vanishing moments.

Coifman 6

```
   Low-pass
   --------
((SR15-3.0)/32.0)*SR2      =    3.8580777747886749O E-02
((1.0-SR15)/32.0)*SR2      =   -1.26969125396205200 E-01
((3.0-SR15)/16.0)*SR2      =   -7.71615554957734980 E-02
((SR15+3.0)/16.0)*SR2      =    6.07491641385684120 E-01
((SR15+13.0)/32.0)*SR2     =    7.45687558934434280 E-01
((9.0-SR15)/32.0)*SR2      =    2.26584265197068560 E-01

   High-pass
   ---------
((9.0-SR15)/32.0)*SR2       =   2.26584265197068560 E-01
(-(SR15+13.0)/32.0)*SR2     =  -7.45687558934434280 E-01
((SR15+3.0)/16.0)*SR2       =   6.07491641385684120 E-01
((SR15-3.0)/16.0)*SR2       =   7.71615554957734980 E-02
((1.0-SR15)/32.0)*SR2       =  -1.26969125396205200 E-01
((3.0-SR15)/32.0)*SR2       =   3.85807777478867490 E-02
```

Coifman 12

```
          Low-pass                       High-pass
          --------                       ---------
 1.63873364631797850 E-02      -7.20549445368115120 E-04
-4.14649367819664850 E-02       1.82320887091009920 E-03
-6.73725547222998740 E-02       5.61143481936598850 E-03
 3.86110066823092900 E-01      -2.36801719468767500 E-02
 8.12723635449606130 E-01      -5.94344186464712400 E-02
 4.17005184423777600 E-01       7.64885990782645940 E-02
-7.64885990782645940 E-02       4.17005184423777600 E-01
-5.94344186464712400 E-02      -8.12723635449606130 E-01
 2.36801719468767500 E-02       3.86110066823092900 E-01
 5.61143481936598850 E-03       6.73725547222998740 E-02
-1.82320887091009920 E-03      -4.14649367819664850 E-02
-7.20549445368115120 E-04      -1.63873364631797850 E-02
```

Coifman 18

Low-pass	High-pass
-3.79351286437787590 E-03	-3.45997731974026950 E-05
7.78259642567078690 E-03	7.09833025057049280 E-05
2.34526961421191030 E-02	4.66216959820144030 E-04
-6.57719112814312280 E-02	-1.11751877082696180 E-03
-6.11233900029556980 E-02	-2.57451768812796920 E-03
4.05176902409616790 E-01	9.00797613673228960 E-03
7.93777222625620340 E-01	1.58805448636159010 E-02
4.28483476377618690 E-01	-3.45550275733444640 E-02
-7.17998216191705900 E-02	-8.23019271063202830 E-02
-8.23019271063202830 E-02	7.17998216191705900 E-02
3.45550275733444640 E-02	4.28483476377618690 E-01
1.58805448636159010 E-02	-7.93777222625620340 E-01
-9.00797613673228960 E-03	4.05176902409616790 E-01
-2.57451768812796920 E-03	6.11233900029556980 E-02
1.11751877082696180 E-03	-6.57719112814312280 E-02
4.66216959820144030 E-04	-2.34526961421191030 E-02
-7.09833025057049280 E-05	7.78259642567078690 E-03
-3.45997731974026950 E-05	3.79351286437787590 E-03

Coifman 24

Low-pass	High-pass
8.92313668220275710 E-04	-1.78498455869993380 E-06
-1.62949201311084900 E-03	3.25968044485761290 E-06
-7.34616632765623490 E-03	3.12298760780433580 E-05
1.60689439640692360 E-02	-6.23390338657646180 E-05
2.66823001556288040 E-02	-2.59974552319421750 E-04
-8.12666996803130540 E-02	5.89020756811437840 E-04
-5.60773133164719500 E-02	1.26656192867951870 E-03
4.15308407030430150 E-01	-3.75143615692490270 E-03
7.82238930920498790 E-01	-5.65828668594603800 E-03
4.34386056491468390 E-01	1.52117315272391490 E-02
-6.66274742630007520 E-02	2.50822618451469330 E-02
-9.62204420335636970 E-02	-3.93344271229132190 E-02
3.93344271229132190 E-02	-9.62204420335636970 E-02
2.50822618451469330 E-02	6.66274742630007520 E-02
-1.52117315272391490 E-02	4.34386056491468390 E-01
-5.65828668594603800 E-03	-7.82238930920498790 E-01
3.75143615692490270 E-03	4.15308407030430150 E-01
1.26656192867951870 E-03	5.60773133164719500 E-02
-5.89020756811437840 E-04	-8.12666996803130540 E-02
-2.59974552319421750 E-04	-2.66823001556288040 E-02
6.23390338657646180 E-05	1.60689439640692360 E-02
3.12298760780433580 E-05	7.34616632765623490 E-03
-3.25968044485761290 E-06	-1.62949201311084900 E-03
-1.78498455869993380 E-06	-8.92313668220275710 E-04

Coifman 30

Low-pass	High-pass
-2.12080863336306810 E-04	-9.51579170468293560 E-08
3.58589677255698600 E-04	1.67408293749300630 E-07
2.17823630484128470 E-03	2.06380639023316330 E-06
-4.15935878160399350 E-03	-3.73459674967156050 E-06
-1.01311175380455940 E-02	-2.13150140622449170 E-05
2.34081567615927950 E-02	4.13404844919568560 E-05
2.81680290621414970 E-02	1.40541148901077230 E-04
-9.19200105488064130 E-02	-3.02259519791840680 E-04
-5.20431632162377390 E-02	-6.38131296151377520 E-04
4.21566206728765440 E-01	1.66286376908581340 E-03
7.74289603740284550 E-01	2.43337320922405380 E-03
4.37991626228364130 E-01	-6.76418541866332000 E-03
-6.20359639056089690 E-02	-9.16423115304622680 E-03
-1.05574208705835340 E-01	1.97617790117239590 E-02
4.12892087407341690 E-02	3.26835742832495350 E-02
3.26835742832495350 E-02	-4.12892087407341690 E-02
-1.97617790117239590 E-02	-1.05574208705835340 E-01
-9.16423115304622680 E-03	6.20359639056089690 E-02
6.76418541866332000 E-03	4.37991626228364130 E-01
2.43337320922405380 E-03	-7.74289603740284550 E-01
-1.66286376908581340 E-03	4.21566206728765440 E-01
-6.38131296151377520 E-04	5.20431632162377390 E-02
3.02259519791840680 E-04	-9.19200105488064130 E-02
1.40541148901077230 E-04	-2.81680290621414970 E-02
-4.13404844919568560 E-05	2.34081567615927950 E-02
-2.13150140622449170 E-05	1.01311175380455940 E-02
3.73459674967156050 E-06	-4.15935878160399350 E-03
2.06380639023316330 E-06	-2.17823630484128470 E-03
-1.67408293749300630 E-07	3.58589677255698600 E-04
-9.51579170468293560 E-08	2.12080863336306810 E-04

C.1.3 Standard Daubechies filters

For a given length, the "Daubechies" filters [37] maximize the smoothness of the associated scaling function by maximizing the rate of decay of its Fourier transform.

Daubechies 2, or Haar–Walsh

```
           Low-pass                          High-pass
           --------                          ---------
SR2 = 7.07106781186547 E-01      SR2 =  7.07106781186547 E-01
SR2 = 7.07106781186547 E-01     -SR2 = -7.07106781186547 E-01
```

Daubechies 4

```
          Low-pass
          --------
 (1.0+SR3)/(4.0*SR2)     =     4.82962913144534160 E-01
 (3.0+SR3)/(4.0*SR2)     =     8.36516303737807940 E-01
 (3.0-SR3)/(4.0*SR2)     =     2.24143868042013390 E-01
 (1.0-SR3)/(4.0*SR2)     =    -1.29409522551260370 E-01

          High-pass
          ---------
 (1.0-SR3)/(4.0*SR2)     =    -1.29409522551260370 E-01
 (SR3-3.0)/(4.0*SR2)     =    -2.24143868042013390 E-01
 (3.0+SR3)/(4.0*SR2)     =     8.36516303737807940 E-01
-(1.0+SR3)/(4.0*SR2)     =    -4.82962913144534160 E-01
```

Daubechies 6

```
         Low-pass
         --------
 0.125*A                =     3.32670552950082630 E-01
 0.125*(SR2+2.0*A-B)    =     8.06891509311092550 E-01
 0.125*(3.0*SR2-2.0*B)  =     4.59877502118491540 E-01
 0.125*(3.0*SR2-2.0*A)  =    -1.35011020010254580 E-01
 0.125*(SR2+2.0*B-A)    =    -8.54412738820266580 E-02
 0.125*B                =     3.52262918857095330 E-02

         High-pass
         ---------
 0.125*B                =    3.52262918857095330 E-02
 0.125*(A-SR2-2.0*B)    =    8.54412738820266580 E-02
 0.125*(3.0*SR2-2.0*A)  =   -1.35011020010254580 E-01
 0.125*(2.0*B-3.0*SR2)  =   -4.59877502118491540 E-01
 0.125*(SR2+2.0*A-B)    =    8.06891509311092550 E-01
-0.125*A                =   -3.32670552950082630 E-01
```

Daubechies 8

```
       Low-pass                  High-pass
       --------                  ---------
  2.303778133090 E-01      -1.059740178500 E-02
  7.148465705530 E-01      -3.288301166700 E-02
  6.308807679300 E-01       3.084138183700 E-02
 -2.798376941700 E-02       1.870348117190 E-01
 -1.870348117190 E-01      -2.798376941700 E-02
  3.084138183600 E-02      -6.308807679300 E-01
  3.288301166700 E-02       7.148465705530 E-01
 -1.059740178500 E-02      -2.303778133090 E-01
```

Daubechies 10

```
     Low-pass                High-pass
     --------                ---------
 1.601023979740 E-01     3.335725285000 E-03
 6.038292697970 E-01     1.258075199900 E-02
 7.243085284380 E-01    -6.241490213000 E-03
 1.384281459010 E-01    -7.757149384000 E-02
-2.422948870660 E-01    -3.224486958500 E-02
-3.224486958500 E-02     2.422948870660 E-01
 7.757149384000 E-02     1.384281459010 E-01
-6.241490213000 E-03    -7.243085284380 E-01
-1.258075199900 E-02     6.038292697970 E-01
 3.335725285000 E-03    -1.601023979740 E-01
```

Daubechies 12

```
     Low-pass                High-pass
     --------                ---------
 1.115407433500 E-01    -1.077301085000 E-03
 4.946238903980 E-01    -4.777257511000 E-03
 7.511339080210 E-01     5.538422010000 E-04
 3.152503517090 E-01     3.158203931800 E-02
-2.262646939650 E-01     2.752286553000 E-02
-1.297668675670 E-01    -9.750160558700 E-02
 9.750160558700 E-02    -1.297668675670 E-01
 2.752286553000 E-02     2.262646939650 E-01
-3.158203931800 E-02     3.152503517090 E-01
 5.538422010000 E-04    -7.511339080210 E-01
 4.777257511000 E-03     4.946238903980 E-01
-1.077301085000 E-03    -1.115407433500 E-01
```

Daubechies 14

Low-pass	High-pass
7.785205408500 E-02	3.537138000000 E-04
3.965393194820 E-01	1.801640704000 E-03
7.291320908460 E-01	4.295779730000 E-04
4.697822874050 E-01	-1.255099855600 E-02
-1.439060039290 E-01	-1.657454163100 E-02
-2.240361849940 E-01	3.802993693500 E-02
7.130921926700 E-02	8.061260915100 E-02
8.061260915100 E-02	-7.130921926700 E-02
-3.802993693500 E-02	-2.240361849940 E-01
-1.657454163100 E-02	1.439060039290 E-01
1.255099855600 E-02	4.697822874050 E-01
4.295779730000 E-04	-7.291320908460 E-01
-1.801640704000 E-03	3.965393194820 E-01
3.537138000000 E-04	-7.785205408500 E-02

Daubechies 16

Low-pass	High-pass
5.441584224300 E-02	-1.174767840000 E-04
3.128715909140 E-01	-6.754494060000 E-04
6.756307362970 E-01	-3.917403730000 E-04
5.853546836540 E-01	4.870352993000 E-03
-1.582910525600 E-02	8.746094047000 E-03
-2.840155429620 E-01	-1.398102791700 E-02
4.724845740000 E-04	-4.408825393100 E-02
1.287474266200 E-01	1.736930100200 E-02
-1.736930100200 E-02	1.287474266200 E-01
-4.408825393100 E-02	-4.724845740000 E-04
1.398102791700 E-02	-2.840155429620 E-01
8.746094047000 E-03	1.582910525600 E-02
-4.870352993000 E-03	5.853546836540 E-01
-3.917403730000 E-04	-6.756307362970 E-01
6.754494060000 E-04	3.128715909140 E-01
-1.174767840000 E-04	-5.441584224300 E-02

Daubechies 18

Low-pass	High-pass
3.807794736400 E-02	3.934732000000 E-05
2.438346746130 E-01	2.519631890000 E-04
6.048231236900 E-01	2.303857640000 E-04
6.572880780510 E-01	-1.847646883000 E-03
1.331973858250 E-01	-4.281503682000 E-03
-2.932737832790 E-01	4.723204758000 E-03
-9.684078322300 E-02	2.236166212400 E-02
1.485407493380 E-01	-2.509471150000 E-04
3.072568147900 E-02	-6.763282906100 E-02
-6.763282906100 E-02	-3.072568147900 E-02
2.509471150000 E-04	1.485407493380 E-01
2.236166212400 E-02	9.684078322300 E-02
-4.723204758000 E-03	-2.932737832790 E-01
-4.281503682000 E-03	-1.331973858250 E-01
1.847646883000 E-03	6.572880780510 E-01
2.303857640000 E-04	-6.048231236900 E-01
-2.519631890000 E-04	2.438346746130 E-01
3.934732000000 E-05	-3.807794736400 E-02

Daubechies 20

Low-pass	High-pass
2.667005790100 E-02	-1.326420300000 E-05
1.881768000780 E-01	-9.358867000000 E-05
5.272011889320 E-01	-1.164668550000 E-04
6.884590394540 E-01	6.858566950000 E-04
2.811723436610 E-01	1.992405295000 E-03
-2.498464243270 E-01	-1.395351747000 E-03
-1.959462743770 E-01	-1.073317548300 E-02
1.273693403360 E-01	-3.606553567000 E-03
9.305736460400 E-02	3.321267405900 E-02
-7.139414716600 E-02	2.945753682200 E-02
-2.945753682200 E-02	-7.139414716600 E-02
3.321267405900 E-02	-9.305736460400 E-02
3.606553567000 E-03	1.273693403360 E-01
-1.073317548300 E-02	1.959462743770 E-01
1.395351747000 E-03	-2.498464243270 E-01
1.992405295000 E-03	-2.811723436610 E-01
-6.858566950000 E-04	6.884590394540 E-01
-1.164668550000 E-04	-5.272011889320 E-01
9.358867000000 E-05	1.881768000780 E-01
-1.326420300000 E-05	-2.667005790100 E-02

C.1.4 Vaidyanathan filters

The "Vaidyanathan" coefficients correspond to the filter sequence #24B constructed by Vaidyanathan and Huong in [106]. These coefficients give an exact reconstruction scheme, but they do not satisfy any moment condition, including the normalization: the sum of the low-pass filter coefficients is close but not equal to $\sqrt{2}$. The function one obtains after iterating the reconstruction procedure nine times looks continuous, but is of course not differentiable. This filter has been optimized, for its length, to satisfy the standard requirements for effective speech coding.

Vaidyanathan 24

Low-pass	High-pass
-6.29061181907475230 E-05	4.57993341109767180 E-02
3.43631904821029190 E-04	-2.50184129504662180 E-01
-4.53956619637219290 E-04	5.72797793210734320 E-01
-9.44897136321949270 E-04	-6.35601059872214940 E-01
2.84383454683556460 E-03	2.01612161775308660 E-01
7.08137504052444710 E-04	2.63494802488459910 E-01
-8.83910340861387800 E-03	-1.94450471766478170 E-01
3.15384705589700400 E-03	-1.35084227129481260 E-01
1.96872150100727140 E-02	1.31971661416977720 E-01
-1.48534480052300990 E-02	8.39288843661128300 E-02
-3.54703986072834530 E-02	-7.77097509019694100 E-02
3.87426192934114400 E-02	-5.58925236913735480 E-02
5.58925236913735480 E-02	3.87426192934114400 E-02
-7.77097509019694100 E-02	3.54703986072834530 E-02
-8.39288843661128300 E-02	-1.48534480052300990 E-02
1.31971661416977720 E-01	-1.96872150100727140 E-02
1.35084227129481260 E-01	3.15384705589700400 E-03
-1.94450471766478170 E-01	8.83910340861387800 E-03
-2.63494802488459910 E-01	7.08137504052444710 E-04
2.01612161775308660 E-01	-2.84383454683556460 E-03
6.35601059872214940 E-01	-9.44897136321949270 E-04
5.72797793210734320 E-01	4.53956619637219290 E-04
2.50184129504662180 E-01	3.43631904821029190 E-04
4.57993341109767180 E-02	6.29061181907475230 E-05

C.2 Biorthogonal quadrature filters

All of the filters below are derived in [37]. They would have rational coefficient sequences if we normalized them to have sum two rather than sum $\sqrt{2}$. For convenience, we introduce the constants $\sqrt{2}/2^{17}$, $\sqrt{2}/2^{15}$, and $\sqrt{2}/2^{14}$:

```
#define SR2OVER2E17 (SR2/131072.0)  /* sqrt(2.0)/(1<<17) */
#define SR2OVER2E15 (SR2/32768.0)   /* sqrt(2.0)/(1<<15) */
#define SR2OVER2E14 (SR2/16384.0)   /* sqrt(2.0)/(1<<14) */
```

The low-pass and high-pass filters indexed by a single number (like 2) play the role of H and G', and are dual to any of the filters indexed by a corresponding pair (like 2,4), which play the role of H' and G.

C.2.1 Symmetric/antisymmetric, one moment

1 symmetric/antisymmetric, or Haar–Walsh

```
      Low-pass                         High-pass
      --------                         ---------
SR2 = 7.07106781186547 E-01     SR2 =  7.07106781186547 E-01
SR2 = 7.07106781186547 E-01    -SR2 = -7.07106781186547 E-01
```

1,3 symmetric/antisymmetric

```
      Low-pass                         High-pass
      --------                         ---------
  (-1.0/16.0)*SR2                  (-1.0/16.0)*SR2
  ( 1.0/16.0)*SR2                  (-1.0/16.0)*SR2
  ( 1.0/2.0)*SR2                   ( 1.0/2.0)*SR2
  ( 1.0/2.0)*SR2                   (-1.0/2.0)*SR2
  ( 1.0/16.0)*SR2                  ( 1.0/16.0)*SR2
  (-1.0/16.0)*SR2                  ( 1.0/16.0)*SR2
```

1,5 symmetric/antisymmetric

```
     Low-pass                     High-pass
     --------                     ---------
(  3.0/256.0)*SR2            (  3.0/256.0)*SR2
( -3.0/256.0)*SR2            (  3.0/256.0)*SR2
(-11.0/128.0)*SR2            (-11.0/128.0)*SR2
( 11.0/128.0)*SR2            (-11.0/128.0)*SR2
(  1.0/  2.0)*SR2            (  1.0/  2.0)*SR2
(  1.0/  2.0)*SR2            ( -1.0/  2.0)*SR2
( 11.0/128.0)*SR2            ( 11.0/128.0)*SR2
(-11.0/128.0)*SR2            ( 11.0/128.0)*SR2
( -3.0/256.0)*SR2            ( -3.0/256.0)*SR2
(  3.0/256.0)*SR2            ( -3.0/256.0)*SR2
```

C.2.2 Symmetric/symmetric, two moments

2 symmetric/symmetric

```
     Low-pass                     High-pass
     --------                     ---------
  (1.0/4.0)*SR2                (-1.0/4.0)*SR2
  (1.0/2.0)*SR2                ( 1.0/2.0)*SR2
  (1.0/4.0)*SR2                (-1.0/4.0)*SR2
```

2,2 symmetric/symmetric

```
     Low-pass                     High-pass
     --------                     ---------
  (-1.0/8.0)*SR2               (-1.0/8.0)*SR2
  ( 1.0/4.0)*SR2               (-1.0/4.0)*SR2
  ( 3.0/4.0)*SR2               ( 3.0/4.0)*SR2
  ( 1.0/4.0)*SR2               (-1.0/4.0)*SR2
  (-1.0/8.0)*SR2               (-1.0/8.0)*SR2
```

2,4 symmetric/symmetric

```
     Low-pass                       High-pass
     --------                       ---------
( 3.0/128.0)*SR2              (  3.0/128.0)*SR2
(-3.0/ 64.0)*SR2              (  3.0/ 64.0)*SR2
(-1.0/  8.0)*SR2              ( -1.0/  8.0)*SR2
(19.0/ 64.0)*SR2              (-19.0/ 64.0)*SR2
(45.0/ 64.0)*SR2              ( 45.0/ 64.0)*SR2
(19.0/ 64.0)*SR2              (-19.0/ 64.0)*SR2
(-1.0/  8.0)*SR2              ( -1.0/  8.0)*SR2
(-3.0/ 64.0)*SR2              (  3.0/ 64.0)*SR2
( 3.0/128.0)*SR2              (  3.0/128.0)*SR2
```

2,6 symmetric/symmetric

```
       Low-pass                        High-pass
       --------                        ---------
(   -5.0/1024.0)*SR2            (   -5.0/1024.0)*SR2
(    5.0/ 512.0)*SR2            (   -5.0/ 512.0)*SR2
(   17.0/ 512.0)*SR2            (   17.0/ 512.0)*SR2
(  -39.0/ 512.0)*SR2            (   39.0/ 512.0)*SR2
(-123.0/1024.0)*SR2             (-123.0/1024.0)*SR2
(   81.0/ 256.0)*SR2            (  -81.0/ 256.0)*SR2
(  175.0/ 256.0)*SR2            (  175.0/ 256.0)*SR2
(   81.0/ 256.0)*SR2            (  -81.0/ 256.0)*SR2
(-123.0/1024.0)*SR2             (-123.0/1024.0)*SR2
(  -39.0/ 512.0)*SR2            (   39.0/ 512.0)*SR2
(   17.0/ 512.0)*SR2            (   17.0/ 512.0)*SR2
(    5.0/ 512.0)*SR2            (   -5.0/ 512.0)*SR2
(   -5.0/1024.0)*SR2            (   -5.0/1024.0)*SR2
```

2,8 symmetric/symmetric

```
     Low-pass                         High-pass
     --------                         ---------
  35.0*SR2OVER2E15               35.0*SR2OVER2E15
 -70.0*SR2OVER2E15               70.0*SR2OVER2E15
-300.0*SR2OVER2E15             -300.0*SR2OVER2E15
 670.0*SR2OVER2E15             -670.0*SR2OVER2E15
1228.0*SR2OVER2E15             1228.0*SR2OVER2E15
-3126.0*SR2OVER2E15            3126.0*SR2OVER2E15
-3796.0*SR2OVER2E15           -3796.0*SR2OVER2E15
10718.0*SR2OVER2E15          -10718.0*SR2OVER2E15
22050.0*SR2OVER2E15           22050.0*SR2OVER2E15
10718.0*SR2OVER2E15          -10718.0*SR2OVER2E15
-3796.0*SR2OVER2E15           -3796.0*SR2OVER2E15
-3126.0*SR2OVER2E15            3126.0*SR2OVER2E15
1228.0*SR2OVER2E15             1228.0*SR2OVER2E15
 670.0*SR2OVER2E15             -670.0*SR2OVER2E15
-300.0*SR2OVER2E15             -300.0*SR2OVER2E15
 -70.0*SR2OVER2E15               70.0*SR2OVER2E15
  35.0*SR2OVER2E15               35.0*SR2OVER2E15
```

C.2.3 Symmetric/antisymmetric, three moments

3 symmetric/antisymmetric

```
   Low-pass                 High-pass
   --------                 ---------
(1.0/8.0)*SR2           (-1.0/8.0)*SR2
(3.0/8.0)*SR2           ( 3.0/8.0)*SR2
(3.0/8.0)*SR2           (-3.0/8.0)*SR2
(1.0/8.0)*SR2           ( 1.0/8.0)*SR2
```

3,1 symmetric/antisymmetric

```
    Low-pass               High-pass
    --------               ---------
(-1.0/4.0)*SR2         ( 1.0/4.0)*SR2
( 3.0/4.0)*SR2         ( 3.0/4.0)*SR2
( 3.0/4.0)*SR2         (-3.0/4.0)*SR2
(-1.0/4.0)*SR2         (-1.0/4.0)*SR2
```

3,3 symmetric/antisymmetric

```
    Low-pass               High-pass
    --------               ---------
( 3.0/64.0)*SR2        ( -3.0/64.0)*SR2
(-9.0/64.0)*SR2        ( -9.0/64.0)*SR2
(-7.0/64.0)*SR2        (  7.0/64.0)*SR2
(45.0/64.0)*SR2        ( 45.0/64.0)*SR2
(45.0/64.0)*SR2        (-45.0/64.0)*SR2
(-7.0/64.0)*SR2        ( -7.0/64.0)*SR2
(-9.0/64.0)*SR2        (  9.0/64.0)*SR2
( 3.0/64.0)*SR2        (  3.0/64.0)*SR2
```

3,5 symmetric/antisymmetric

```
    Low-pass               High-pass
    --------               ---------
( -5.0/512.0)*SR2      (   5.0/512.0)*SR2
( 15.0/512.0)*SR2      (  15.0/512.0)*SR2
( 19.0/512.0)*SR2      ( -19.0/512.0)*SR2
(-97.0/512.0)*SR2      ( -97.0/512.0)*SR2
(-13.0/256.0)*SR2      (  13.0/256.0)*SR2
(175.0/256.0)*SR2      ( 175.0/256.0)*SR2
(175.0/256.0)*SR2      (-175.0/256.0)*SR2
(-13.0/256.0)*SR2      ( -13.0/256.0)*SR2
(-97.0/512.0)*SR2      (  97.0/512.0)*SR2
( 19.0/512.0)*SR2      (  19.0/512.0)*SR2
( 15.0/512.0)*SR2      ( -15.0/512.0)*SR2
( -5.0/512.0)*SR2      (  -5.0/512.0)*SR2
```

3,7 symmetric/antisymmetric

```
          Low-pass                      High-pass
          --------                      ---------
   35.0*SR2OVER2E14            -35.0*SR2OVER2E14
 -105.0*SR2OVER2E14           -105.0*SR2OVER2E14
 -195.0*SR2OVER2E14            195.0*SR2OVER2E14
  865.0*SR2OVER2E14            865.0*SR2OVER2E14
  363.0*SR2OVER2E14           -363.0*SR2OVER2E14
-3489.0*SR2OVER2E14          -3489.0*SR2OVER2E14
 -307.0*SR2OVER2E14            307.0*SR2OVER2E14
11025.0*SR2OVER2E14          11025.0*SR2OVER2E14
11025.0*SR2OVER2E14         -11025.0*SR2OVER2E14
 -307.0*SR2OVER2E14           -307.0*SR2OVER2E14
-3489.0*SR2OVER2E14           3489.0*SR2OVER2E14
  363.0*SR2OVER2E14            363.0*SR2OVER2E14
  865.0*SR2OVER2E14           -865.0*SR2OVER2E14
 -195.0*SR2OVER2E14           -195.0*SR2OVER2E14
 -105.0*SR2OVER2E14            105.0*SR2OVER2E14
   35.0*SR2OVER2E14             35.0*SR2OVER2E14
```

3,9 symmetric/antisymmetric

```
     Low-pass                         High-pass
     --------                         ---------
  -63.0*SR2OVER2E17               63.0*SR2OVER2E17
  189.0*SR2OVER2E17              189.0*SR2OVER2E17
  469.0*SR2OVER2E17             -469.0*SR2OVER2E17
-1911.0*SR2OVER2E17            -1911.0*SR2OVER2E17
-1308.0*SR2OVER2E17             1308.0*SR2OVER2E17
 9188.0*SR2OVER2E17             9188.0*SR2OVER2E17
 1140.0*SR2OVER2E17            -1140.0*SR2OVER2E17
-29676.0*SR2OVER2E17          -29676.0*SR2OVER2E17
  190.0*SR2OVER2E17             -190.0*SR2OVER2E17
87318.0*SR2OVER2E17            87318.0*SR2OVER2E17
87318.0*SR2OVER2E17           -87318.0*SR2OVER2E17
  190.0*SR2OVER2E17              190.0*SR2OVER2E17
-29676.0*SR2OVER2E17           29676.0*SR2OVER2E17
 1140.0*SR2OVER2E17             1140.0*SR2OVER2E17
 9188.0*SR2OVER2E17            -9188.0*SR2OVER2E17
-1308.0*SR2OVER2E17            -1308.0*SR2OVER2E17
-1911.0*SR2OVER2E17             1911.0*SR2OVER2E17
  469.0*SR2OVER2E17              469.0*SR2OVER2E17
  189.0*SR2OVER2E17             -189.0*SR2OVER2E17
  -63.0*SR2OVER2E17              -63.0*SR2OVER2E17
```

Bibliography

[1] ISO/IEC JTC1 Draft International Standard 10918-1. Digital compression and coding of continuous-tone still images, part 1: Requirements and guidelines. Available from ANSI Sales, (212)642-4900, November 1991. ISO/IEC CD 10918-1 (alternate number SC2 N2215).

[2] Tom M. Apostol. *Calculus*, volume II. John Wiley & Sons, New York, second edition, 1969.

[3] Tom M. Apostol. *Introduction to Mathematical Analysis.* Addison–Wesley, Reading, Massachusetts, second edition, January 1975.

[4] Pascal Auscher. Remarks on the local Fourier basis. In Benedetto and Frazier [7], pages 203–218.

[5] Pascal Auscher, Guido Weiss, and Mladen Victor Wickerhauser. Local sine and cosine bases of Coifman and Meyer and the construction of smooth wavelets. In Chui [15], pages 237–256.

[6] Michael F. Barnsley and Lyman P. Hurd. *Fractal Image Compression.* AK Peters, Ltd., Wellesley, Mass., 1993.

[7] John J. Benedetto and Michael Frazier, editors. *Wavelets: Mathematics and Applications.* Studies in Advanced Mathematics. CRC Press, Boca Raton, Florida, 1992.

[8] Gregory Beylkin. On the representation of operators in bases of compactly supported wavelets. *SIAM Journal of Numerical Analysis*, 6-6:1716–1740, 1992.

[9] Gregory Beylkin, Ronald R. Coifman, and Vladimir Rokhlin. Fast wavelet transforms and numerical algorithms I. *Communications on Pure and Applied Mathematics*, XLIV:141–183, 1991.

[10] Richard E. Blahut. *Fast Algorithms for Digital Signal Processing.* Addison-Wesley, Reading, Massachusetts, 1985.

[11] Brian Bradie, Ronald R. Coifman, and Alexander Grossmann. Fast numerical computations of oscillatory integrals related to acoustic scattering, I. *Applied and Computational Harmonic Analysis*, 1(1):94–99, December 1993.

[12] P. J. Burt and E. H. Adelson. The Laplacian pyramid as a compact image code. *IEEE Transactions on Communication*, 31(4):532–540, 1983.

[13] Lennart Carleson. On convergence and growth of partial sums of Fourier series. *Acta Mathematica*, 116:135–157, 1966.

[14] Charles K. Chui. *An Introduction to Wavelets.* Academic Press, Boston, 1992.

[15] Charles K. Chui, editor. *Wavelets–A Tutorial in Theory and Applications.* Academic Press, Boston, 1992.

[16] Mac A. Cody. The fast wavelet transform. *Dr. Dobb's Journal*, 17(4):44, April 1992.

[17] Mac A. Cody. A wavelet analyzer. *Dr. Dobb's Journal*, 18(4):44, April 1993.

[18] Mac A. Cody. The wavelet packet transform. *Dr. Dobb's Journal*, 19(4):44, April 1994.

[19] Albert Cohen, Ingrid Daubechies, and Jean-Christophe Feauveau. Biorthogonal bases of compactly supported wavelets. *Communications on Pure and Applied Mathematics*, 45:485–500, 1992.

[20] Albert Cohen, Ingrid Daubechies, and Pierre Vial. Wavelets on the interval and fast wavelet transforms. *Applied and Computational Harmonic Analysis*, 1:54–81, December 1993.

[21] Donald L. Cohn. *Measure Theory.* Birkhäuser, Boston, 1980.

[22] Ronald R. Coifman, Fazal Majid, and Mladen Victor Wickerhauser. Denoise. Available from Fast Mathematical Algorithms and Hardware Corporation, 1020 Sherman Ave., Hamden, CT 06514 USA, 1992.

[23] Ronald R. Coifman and Yves Meyer. Nouvelles bases othonormées de $L^2(\mathbf{R})$ ayant la structure du système de Walsh. Preprint, Department of Mathematics, Yale University, New Haven, 1989.

[24] Ronald R. Coifman and Yves Meyer. Remarques sur l'analyse de Fourier à fenêtre. *Comptes Rendus de l'Académie des Sciences de Paris*, 312:259–261, 1991.

[25] Ronald R. Coifman, Yves Meyer, Stephen R. Quake, and Mladen Victor Wickerhauser. Signal processing and compression with wavelet packets. In Meyer and Roques [84], pages 77–93.

[26] Ronald R. Coifman, Yves Meyer, and Mladen Victor Wickerhauser. Numerical adapted waveform analysis and harmonic analysis. In *Proceedings of the Symposium in Honor of Elias Stein*, Princeton, New Jersey, July 1991. Princeton University, Princeton University Press.

[27] Ronald R. Coifman, Yves Meyer, and Mladen Victor Wickerhauser. Adapted waveform analysis, wavelet-packets and applications. In *Proceedings of ICIAM '91*, pages 41–50, Washington, DC, 1991, 1992. SIAM, SIAM Press.

[28] Ronald R. Coifman, Yves Meyer, and Mladen Victor Wickerhauser. Size properties of wavelet packets. In Ruskai et al. [97], pages 453–470.

[29] Ronald R. Coifman, Yves Meyer, and Mladen Victor Wickerhauser. Wavelet analysis and signal processing. In Ruskai et al. [97], pages 153–178.

[30] Ronald R. Coifman and Mladen Victor Wickerhauser. Entropy based algorithms for best basis selection. *IEEE Transactions on Information Theory*, 32:712–718, March 1992.

[31] Ronald R. Coifman and Mladen Victor Wickerhauser. Wavelets and adapted waveform analysis. In Benedetto and Frazier [7], pages 399–423.

[32] Ronald R. Coifman and Mladen Victor Wickerhauser. Wavelets and adapted waveform analysis: A toolkit for signal processing and numerical analysis. In Daubechies [38], pages 119–153. Minicourse lecture notes.

[33] James W. Cooley and John W. Tukey. An algorithm for the machine calculation of complex Fourier series. *Mathematics of Computation*, 19:297–301, 1965.

[34] Christophe D'Alessandro, Xiang Fang, Eva Wesfreid, and Mladen Victor Wickerhauser. Speech signal segmentation via Malvar wavelets. In Meyer and Roques [84], pages 305–308.

[35] Christophe D'Alessandro and Xavier Rodet. Synthèse et analyse–synthèse par fonction d'ondes formantiques. *Journal d'Acoustique*, 2:163–169, 1989.

[36] Ingrid Daubechies. Orthonormal bases of compactly supported wavelets. *Communications on Pure and Applied Mathematics*, XLI:909–996, 1988.

[37] Ingrid Daubechies. *Ten Lectures on Wavelets*, volume 61 of *CBMS-NSF Regional Conference Series in Applied Mathematics.* SIAM Press, Philadelphia, Pennsylvania, 1992.

[38] Ingrid Daubechies, editor. *Different Perspectives on Wavelets*, number 47 in Proceedings of Symposia in Applied Mathematics, San Antonio, Texas, 11-12 January 1993. American Mathematical Society.

[39] Ingrid Daubechies, Alexander Grossmann, and Yves Meyer. Painless nonorthogonal expansions. *Journal of Mathematical Physics*, 27:1271–1283, 1986.

[40] Ingrid Daubechies, Stéphane Jaffard, and Jean-Lin Journé. A simple Wilson orthonormal basis with exponential decay. *SIAM Journal of Mathematical Analysis*, 22(2):554–573, 1991.

[41] Ron Devore, Bjørn Jawerth, and Bradley J. Lucier. Image compression through wavelet transform coding. *IEEE Transactions on Information Theory*, 38:719–746, March 1992.

[42] J. J. Dongarra, J. R. Bunch, C. B. Moler, and G. W. Stewart. *LINPACK User's Guide.* SIAM Press, Philadelphia, 1979.

[43] D. Esteban and C. Galand. Application of quadrature mirror filters to split band voice coding systems. In *Proceedings of IEEE ICASSP-77*, pages 191–195, Washington, D.C., May 1977.

[44] Lawrence C. Evans and Ronald F. Gariepy. *Measure Theory and Fine Properties of Functions.* Studies in Advanced Mathematics. CRC Press, Boca Raton, Florida, 1992.

[45] Xiang Fang and Eric Seré. Multiple folding local sine transform. *Applied and Computational Harmonic Analysis*, 1994. To appear.

[46] Marie Farge, Eric Goirand, Yves Meyer, Frédéric Pascal, and Mladen Victor Wickerhauser. Improved predictability of two-dimensional turbulent flows using wavelet packet compression. *Fluid Dynamics Research*, 10:229–250, 1992.

[47] Jean-Christophe Feauveau. Filtres miroirs conjugés: un théorie pour les filtres miroirs en quadrature et l'analyse multiresolution par ondelettes. *Comptes Rendus de l'Académie des Sciences de Paris*, 309:853–856, 1989.

[48] Gerald B. Folland. *Harmonic Analysis in Phase Space*. Number 122 in Annals of Mathematics Studies. Princeton University Press, Princeton, New Jersey, 1989.

[49] P. Franklin. A set of continuous orthogonal functions. *Mathematische Annalen*, 100:522–529, 1928.

[50] Michael Frazier, Bjørn Jawerth, and Guido Weiss. *Littlewood–Paley Theory and the Study of Function Spaces*. Number 79 in CBMS Regional Conference Lecture Notes. American Mathematical Society, Providence, Rhode Island, 1990.

[51] Eric Goirand, Mladen Victor Wickerhauser, and Marie Farge. A parallel two dimensional wavelet packet transform and its application to matrix-vector multiplication. In Babu Joseph and Rudolph Motard, editors, *Applications of Wavelets to Chemical Engineering*. Kluwer Academic Publishers, Norwell, Massachusetts, 1994. To appear.

[52] Ramesh A. Gopinath and C. Sidney Burrus. Wavelet transforms and filter banks. In Chui [15], pages 603–654.

[53] Philippe Guillemain, Richard Kronland-Martinet, and B. Martens. Estimation of spectral lines with the help of the wavelet transform—applications in NMR spectroscopy. In Meyer [82], pages 38–60.

[54] A. Haar. Zur theorie der orthogonalen funktionensysteme. *Mathematische Annalen*, 69:331–371, 1910.

[55] Nikolaj Hess-Nielsen. *Frequency Localization of Wavelet Packets*. PhD thesis, University of Aalborg, 1992.

[56] Nikolaj Hess-Nielsen. A comment on the frequency localization of wavelet packets. *Applied and Computational Harmonic Analysis*, 1994. To appear.

[57] Frédéric Heurtaux, Fabrice Planchon, and Mladen Victor Wickerhauser. Scale decomposition in Burgers' equation. In Benedetto and Frazier [7], pages 505–523.

[58] P. G. Hjorth, Lars F. Villemoes, J. Teuber, and R. Florentin-Nielsen. Wavelet analysis of 'double quasar' flux data. *Astronomy and Astrophysics*, 255:20–23, 1992.

[59] David Holzgang. *Understanding PostScript Programming.* Sybex, San Francisco, 1987.

[60] IAFIS-IC-0110v2. WSQ gray-scale fingerprint image compression specification. Version 2, US Department of Justice, Federal Bureau of Investigation, 16 February 1993.

[61] Nuggehally S. Jayant and Peter Noll. *Digital Coding of Waveforms: Principles and Applications to Speech and Video.* Prentice-Hall, Englewood Cliffs, New Jersey, 1984.

[62] Michael Kirby and Lawrence Sirovich. Application of the Karhunen–Loève procedure for the characterization of human faces. *IEEE Transactions on Pattern Analysis and Machine Intelligence*, 12:103–108, January 1990.

[63] Richard Kronland-Martinet, Jean Morlet, and Alexander Grossmann. Analysis of sound patterns through wavelet transforms. *International Journal of Pattern Recognition and Arificial Intelligence*, 1(2):273–302, 1987.

[64] Enrico Laeng. Une base orthonormale de $L^2(\mathbf{R})$, dont les éléments sont bien localisés dans l'espace de phase et leurs supports adaptés à toute partition symétrique de l'espace des fréquences. *Comptes Rendus de l'Académie des Sciences de Paris*, 311:677–680, 1990.

[65] Wayne M. Lawton. Necessary and sufficient conditions for existence of orthonormal wavelet bases. *Journal of Mathematical Physics*, 32:57–61, 1991.

[66] J. S. Liénard. Speech analysis and reconstruction using short time, elementary waveforms. In *Proceedings of IEEE ICASSP-87*, pages 948–951, Dallas, Texas, 1987.

[67] Pierre-Louis Lyons, editor. *Problémes Non-Linéaires Appliqués, Ondelettes et Paquets D'Ondes*, Roquencourt, France, 17–21June 1991. INRIA. Minicourse lecture notes.

[68] Wolodymyr R. Madych. Some elementary properties of multiresolution analyses of $L^2(\mathbf{R}^n)$. In Chui [15], pages 259–294.

[69] Fazal Majid. Applications des paquets d'ondelettes au débruitage du signal. Preprint, Department of Mathematics, Yale University, 28 July 1992. Rapport d'Option, Ecole Polytechnique.

[70] Stéphane G. Mallat. Multiresolution approximation and wavelet orthonormal bases of $L^2(\mathbf{R})$. *Transactions of the AMS*, 315:69–87, 1989.

[71] Stéphane G. Mallat. A theory for multiresolution signal decomposition: The wavelet decomposition. *IEEE Transactions on Pattern Analysis and Machine Intelligence*, 11:674–693, 1989.

[72] Stéphane G. Mallat and Zhifeng Zhang. Matching pursuits with time-frequency dictionaries. *IEEE Transactions on Signal Processing*, 41(12):3397–3415, December 1993.

[73] Stéphane G. Mallat and Sifen Zhong. Wavelet transform maxima and multi-scale edges. In Lyons [67], pages 141–177. Minicourse lecture notes.

[74] Henrique Malvar. Lapped transforms for efficient transform/subband coding. *IEEE Transactions on Acoustics, Speech, and Signal Processing*, 38:969–978, 1990.

[75] Henrique Malvar. *Signal Processing with Lapped Transforms.* Artech House, Norwood, Massachusetts, 1992.

[76] David Marr. *Vision: A Computational Investigation Into the Human Representation and Processing of Visual Information.* W. H. Freeman, San Francisco, 1982.

[77] P. Mathieu, M. Barlaud, and M. Antonini. Compression d'images par transformée en ondelette et quantification vectorielle. *Traitment du Signal*, 7(2):101–115, 1990.

[78] Yves Meyer. De la recherche pétrolière à la géometrie des espaces de Banach en passant par les paraproduits. Technical report, École Polytechnique, Palaiseau, 1985–1986.

[79] Yves Meyer. *Ondelettes et Opérateurs*, volume I: Ondelettes. Hermann, Paris, 1990.

[80] Yves Meyer. *Ondelettes et Opérateurs*, volume II: Opérateurs. Hermann, Paris, 1990.

[81] Yves Meyer. Méthodes temps-fréquences et méthodes temps-échelle en traitement du signal et de l'image. In Lyons [67], pages 1–29. Minicourse lecture notes.

[82] Yves Meyer, editor. *Wavelets and Applications*, Procedings of the International Conference "Wavelets and Applications" Marseille, May 1989, Paris, 1992. LMA/CNRS, Masson.

[83] Yves Meyer. *Wavelets: Algorithms and Applications.* SIAM Press, Philadelphia, 1993.

[84] Yves Meyer and Sylvie Roques, editors. *Progress in Wavelet Analysis and Applications*, Procedings of the International Conference "Wavelets and Applications", Toulouse, France, 8–13 June 1992. Observatoire Midi-Pyrénées de l'Université Paul Sabatier, Editions Frontieres.

[85] Charles A. Micchelli. Using the refinement equation for the construction of pre-wavelets IV: Cube spline and elliptic splines united. IBM research report number 76222, Thomas J. Watson Research Center, New York, 1991.

[86] F. Mintzer. Filters for distortion-free two-band multirate filter banks. *IEEE Transactions on Acoustics, Speech and Signal Processing*, 33:626–630, 1985.

[87] J. M. Nicolas, J. C. Delvigne, and A. Lemer. Automatic identification of transient biological noises in underwater acoustics using arborescent wavelets and neural network. In Meyer [82].

[88] R. E. Paley. A remarkable series of orthogonal functions I. *Proceeedings of the London Mathematical Society*, 34:241–279, 1932.

[89] Stefan Pittner, Josef Schneid, and Christoph W. Ueberhuber. Wavelet literature survey. Technical report, Institute for Applied and Numerical Mathematics, Technical University in Vienna, Austria, 1993.

[90] William H. Press, Saul A. Teukolsky, William T. Vetterling, and Brian P. Flannery. *Numerical Recipes in C: The Art of Scientific Computing.* Cambridge University Press, New York, second edition, 1992.

[91] John P. Princen and Alan Bernard Bradley. Analysis/synthesis filter bank design based on time domain aliasing cancellation. *IEEE Transactions on Acoustics, Speech and Signal Processing*, 34(5):1153–1161, October 1986.

[92] K. R. Rao and P. Yip. *Discrete Cosine Transform.* Academic Press, New York, 1990.

[93] Robert D. Richtmyer. *Principles of Advanced Mathematical Physics*, volume I of *Texts and Monographs in Physics.* Springer-Verlag, New York, 1978.

[94] B. Riemann. Über die darstellbarkeit einer funktion durch eine trigonometrische reihe. In *Gessamelte Werke*, pages 227–271. Reprinted by Dover Books, 1953, First published in Leipzig, 1892, 1854.

[95] Xavier Rodet. Time domain formant-wave-function synthesis. In J. C. Simon, editor, *Spoken Language Generation and Understanding.* D. Reidel Publishing Company, Dordrecht, Holland, 1980.

[96] Halsey L. Royden. *Real Analysis.* Macmillan Publishing Company, 866 Third Avenue, New York, New York 10022, third edition, 1988.

[97] Mary Beth Ruskai et al., editors. *Wavelets and Their Applications.* Jones and Bartlett, Boston, 1992.

[98] Herbert Schildt. *The Annotated ANSI C Standard: ANSI/ISO 9899-1990.* Osborne Mcgraw Hill, Berkeley, California, 1993.

[99] Claude E. Shannon and Warren Weaver. *The Mathematical Theory of Communication.* The University of Illinois Press, Urbana, 1964.

[100] Lawrence Sirovich and Carole H. Sirovich. Low dimensional description of complicated phenomena. *Contemporary Mathematics*, 99:277–305, 1989.

[101] M. J. T. Smith and T. P. Barnwell III. A procedure for designing exact reconstruction filter banks for tree-structured sub-band coders. In *Proceedings of IEEE ICASSP-84*, pages 27.1.1–27.1.4, San Diego, CA, March 1984.

[102] Henrik V. Sorensen, Douglas L. Jones, C. Sidney Burrus, and Michael T. Heideman. On computing the discrete Hartley transform. *IEEE Transactions on Acoustics, Speech, and Signal Processing*, ASSP-33(4):1231–1238, October 1985.

[103] Elias M. Stein and Guido Weiss. *Introduction to Fourier Analysis on Euclidean Spaces.* Number 32 in Princeton Mathematical Series. Princeton University Press, Princeton, New Jersey, 1971.

[104] James Andrew Storer. *Data Compression: Methods and Theory.* Computer Science Press, 1803 Research Boulevard, Rockville, Maryland 20850, 1988.

[105] J. Strömberg. A modified Haar system and higher order spline systems on $\mathbf{R}^n$ as unconditional bases for Hardy spaces. In William Beckner and others., editors, *Conference in Harmonic Analysis in Honor of Antoni Zygmund II*, pages 475–493, Belmont, California, 1981. Wadsworth.

[106] P. P. Vaidyanathan and Phuong-Quan Huong. Lattice structures for optimal design and robust implementation of two-channel perfect-reconstruction QMF banks. *IEEE Transactions on Acoustics, Speech, and Signal Processing*, 36(1):81–94, January 1988.

[107] Martin Vetterli. Filter banks allowing perfect reconstruction. *Signal Processing*, 10(3):219–244, 1986.

[108] James S. Walker. *Fast Fourier Transforms.* CRC Press, Boca Raton, Florida, 1992.

[109] Gregory K. Wallace. The JPEG still picture compression standard. *Communications of the ACM*, 34:30–44, April 1991.

[110] Eva Wesfreid and Mladen Victor Wickerhauser. Adapted local trigonometric transform and speech processing. *IEEE Transactions on Signal Processing*, 41(12):3596–3600, December 1993.

[111] Mladen Victor Wickerhauser. Fast approximate factor analysis. In *Curves and Surfaces in Computer Vision and Graphics II*, volume 1610, pages 23–32, Boston, October 1991. SPIE.

[112] Mladen Victor Wickerhauser. INRIA lectures on wavelet packet algorithms. In Lyons [67], pages 31–99. Minicourse lecture notes.

[113] Mladen Victor Wickerhauser. Acoustic signal compression with wavelet packets. In Chui [15], pages 679–700.

[114] Mladen Victor Wickerhauser. Computation with adapted time-frequency atoms. In Meyer and Roques [84], pages 175–184.

[115] Mladen Victor Wickerhauser. High-resolution still picture compression. *Digital Signal Processing: a Review Journal*, 2(4):204–226, October 1992.

[116] Mladen Victor Wickerhauser. Best-adapted wavelet packet bases. In Daubechies [38], pages 155–171. Minicourse lecture notes.

[117] Mladen Victor Wickerhauser. Smooth localized orthonormal bases. *Comptes Rendus de l'Académie des Sciences de Paris*, 316:423–427, 1993.

[118] Mladen Victor Wickerhauser. Large-rank approximate principal component analysis with wavelets for signal feature discrimination and the inversion of complicated maps. *Journal of Chemical Information and Computer Science*, 1994. Proceedings of Math-Chem-Comp 1993, Rovinj, Croatia. To appear.

Index

a.e.x, 4
absolutely integrable, 6
absolutely summable, 2
absval(), 46
abt2bbasis(), 293
abt2blevel(), 437
abt2blhedge(), 437
abt2btnt(), 425
abt2costs(), 292
abt2dwpsp(), 263
abt2hedge(), 58
abt2tfa1(), 61
abt2tfa1s(), 62
abtblength(), 51, 260
abtblock(), 51, 260
abthedge2tfa1s(), 64
accumulator trees, 382
acdae(), 198
acdafinal(), 269
acdai(), 198
acdaleast(), 269
acdao(), 197
acdaparent(), 269
acdpe(), 208
acdpi(), 208
acdpo(), 207
action region, 107
adapted local cosines, 131
    analysis, 146
    synthesis, 151
adapted waveform analysis, 286, 331
adjacent compatible intervals, 113
adjoint, 15, 153–154, 160
aliasing, 23
almost everywhere, 4
analysis, 1
    functions, 33
    objective, 286
analytic signal, 333
angular Fourier transform, 336
antisymmetric about $-\frac{1}{2}$, 192
aperiodic, 191, 254
    DWPA, 252
    DWT, 215
    wavelet analysis and synthesis, 231
arborescent wavelets, 237
array2tfa1s(), 63
assert(), 43
assignments, 45
autocovariance matrix, 368

B-splines, 35
band-limited, 35
basis, 13
    best, 275
        column, 404
        isotropic, 400
        LCT, 371
        level, 293
        row, 404
        tensor, 400

choice overhead, 237
compactly supported, 133
description, 289
Dirac, 336
fixed-level, 345
graph, 254
Hilbert, 13
Karhunen–Loève, 378
of adapted waveforms, 273
orthonormal, 13
Riesz, 14
splines, 35
standard, 15, 336
subband, 244
subset, 299
wavelet, 217, 244
wavelet packet, 244
bbwp2(), 323
bell, 109
Bessel's inequality, 33
best basis, 275, 371, 400
joint, 383
best branch, 294
binary tree node, 51, 259
biorthogonal, 154, 254
DWPA, 254
DWT, 215
quadrature filters, 156
bit allocation, 376
bit-reversal, 72
bitrevd(), 74
bitrevi(), 74
block cosine, 109, 331
block sines, 331
BQFs, 156
br(), 74
bra and ket notation, 165
branch, 54
BTN (data structure), 51
BTN.CONTENT, 51
BTN.LEFT, 52
BTN.RIGHT, 52
BTN.TAG, 52
btn2branch(), 54
btnt2bbasis(), 292
btnt2bbhedge(), 435
btnt2blevel(), 435
btnt2blhedge(), 436
btnt2btn(), 53
btnt2costs(), 292
btnt2dwpsa(), 270
btnt2dwpsa0(), 434
btnt2hedge(), 59
btnt2tfa1(), 63
btnt2tfa1s(), 63
btntlevelcost(), 436

cas(), 78, 80
Cauchy, 1
sequence, 7
criterion, 1
Cauchy–Schwarz inequality, 8
cbwp2(), 324
CCMULIM(), 48
CCMULRE(), 48
cdachild(), 265
cdae(), 196
cdafinal(), 227, 231, 265
cdai(), 196
cdaleast(), 227, 231, 265
cdao(), 196
cdmi(), 202
cdmo(), 200
cdpe(), 206
cdpi(), 203
cdpo(), 206
cdpo1(), 204

cdpo2(), 206
center of energy, 164
center-of-frequency, 413
centered frequency , 260
child, 147, 265
children, 246
chirp, 349
classifies, 384
coe(), 194
coefficients, 55, 289
coherent structure extraction, 417
compactly supported, 10
  basis, 133
  distribution, 13
  function, 10
  wavelets, 214
complementary subspaces, 217
complete metric space, 7
completeness, 13
COMPLEX (data structure), 47
  COMPLEX.IM, 47
  COMPLEX.RE, 47
complicated function, 286
compression, 339
concentration in ℓ^p, 276
condition number, 42, 389
conditional, 45
  expression, 46
conjugate, 160
  quadrature filter, 154
    sequences, 157
containment, 216
conventional indexing, 192
conventional normalization, 158, 160
convergence, 2
  pointwise, 2
  uniform, 2
convex, 280
convolution, 25
  and decimation, 154, 160
  formulas, 246
  of distributions, 13
  with long signals, 29
  with short signals, 29
correlation coefficient, 364
cos(), 47
cosine
  polarity, 143
  transform, 23
costs2bbasis(), 291
costs2blevel(), 293
countable subadditivity, 4
countably additive, 4
CQFs, 154
crooks, 375
CRRMULIM(), 48
CRRMULRE(), 48
cubic splines, 33
cyclic transposition, 315

DCT(), 96
DCT-IV, 410
de-noising, 417
decimation, 31
decrease, 216
decreasing rearrangement, 280
decrements, 45
def, 354
degree, 37
degree d decay, 10
deperiodize, 174
depth of decomposition, 244
description sequence, 286
DFT, 23
dft(), 77
dht(), 83
dhtcossin(), 81

dhtnormal(), 81
dhtproduct(), 82
diameter, 10
dilation, 31, 216, 331
Dirac
  basis, 336
  delta function, 12
  mass, 12
discrete
  Fourier transform, 23, 68
  wavelet packets, 237
    analysis (DWPA), 237
    synthesis (DWPS), 237
    transform (DWPT), 237
disjoint, 107
  dyadic cover, 243
distribution function, 5
distributions, 12
double precision, 47
doubly infinite sequences, 2
doubly stochastic, 279
DST(), 96
dual
  index, 11
  set, 33
  space, 11
  wavelet packets, 238
duality, 156
DWPA, 237
  on an interval, 254
dwpa0(), 258
dwpaa2btnt(), 266
dwpaa2btntr(), 266
dwpaa2hedge(), 267
dwpaa2hedger(), 267
dwpap2abt(), 261
dwpap2abt0(), 261
dwpap2hedge(), 262
dwpap2hedger(), 262
DWPS, 237
dwps0(), 259
DWPT, 237
DWT, 214
  on the interval, 215
dwta(), 230
dwtacomplete(), 432
dwtaintervals(), 228, 431
dwtalocal(), 433
dwtan(), 431
dwtaorigins(), 229
dwtpd(), 223
dwtpd0(), 221
dwtpd0n(), 222
dwtpdn(), 428
dwtpi(), 224
dwtpi2(), 327, 439
dwtpi0(), 222
dwtpi0n(), 222
dwtpin(), 428
dwtpt2(), 326
dwtptd(), 326
dwtpvd(), 327, 438
dyadic
  interval, 243
  dyadic cover, 243
    disjoint, 107

efficient representatives, 286
eigenfunctions, 17
eigenvalue, 16
eigenvector, 16
elementary sequences, 15
encounter order, 310
enlargeinterval(), 50, 269
entropy, 274, 276, 277, 381
  coding, 366
  metric, 381

epsepilogue(), 355
epsfrect(), 357
epslineto(), 355
epsmoveto(), 355
epsprologue(), 354
epsstroke(), 355
epstranslate(), 355
exp(), 47
expectation, 363
exponential decay, 10

factor analysis, 377–378
    algorithm, 380
factors, principal, 378
    approximate, 378
fdc2(), 327
fdcn(), 141
fdcp(), 141
fds2(), 328
fdsn(), 141
fdsp(), 142
FFT, 69
fftnormal(), 76
fftomega(), 75
fftproduct(), 76
fill, 354
filter, 153
    finite impulse response, 153
    high-pass, 157
    infinite impulse response, 153
    low-pass, 157
    multiplier, 176
    periodized, 153
    variable length, 191
final, 48
finitely supported, 10
fipc(), 143
fips(), 143
FIR, 153
first fundamental frequency, 412
fixed folding, 129, 147
fixed-level basis, 345
float, 47
folding, 106
for almost every $x \in \mathbf{R}$, 4
For loops, 45
formants, 409
representation, 410, 413
Fourier coefficients, 18, 20
Fourier integral, 19
Fourier series, 20
Fourier transform, 18
    angular, 336
    inverse, 20
    of a sequence, 21
    of a tempered distribution, 13
fractal image synthesis, 363
frame, 14
    bounds, 14
    tight, 14
freebtn(), 53
freebtnt(), 54
freehedge(), 56
freeinterval(), 49
frequency, 125, 330
    index, 242
    response, 176

Galerkin's method, 33
Gaussian function, 25, 335
Gaussian random variables, 380
generator, 216
Gram–Schmidt, 37
graph bases, 244
graph theorem, 244
Gray code, 186
gray-scale, 362
graycode(), 187

Hölder condition, 176
Haar scaling function, 184
Haar–Walsh functions, 331
  wavelet packets, 242
Hartley transform, 23
hat functions, 33
Heaviside function, 13
hedge, 55
HEDGE (data structure), 55
  HEDGE.BLOCKS, 55
  HEDGE.CONTENTS, 55
  HEDGE.LEVELS, 55
  HEDGE.TAG, 55
hedge2abt(), 59
hedge2btnt(), 60
hedge2dwpsa(), 271
hedge2dwpsp(), 264
hedge2dwpspabt, 434
hedge2dwpspr(), 264
hedgeabt2tfa1s(), 64
Heisenberg product, 330
Heisenberg's inequality, 24
Hermite group, 336
Hermitean, 15
  inner product, 7
  symmetry, 7
Hilbert space, 8

ICH(), 195
idwta(), 234
idwtacomplete(), 432
idwtaintervals(), 232
idwtaorigins(), 232
idwtpd(), 225
idwtpd0(), 429
idwtpd0n(), 224
idwtpdn(), 430
idwtpi(), 226
idwtpi0(), 429
idwtpi0n(), 225
idwtpi2(), 328, 440
idwtpin(), 430
idwtpt2(), 328, 440
idwtptd(), 328, 440
idwtpvd(), 328, 440
IFH(), 195
igraycode(), 187
IIR, 153
ilct(), 144
ilpic(), 145
ilpica(), 146, 152, 426
ilpis(), 145, 152, 426
ilpisa(), 146, 152, 426
ilst(), 144
incoherent, 418
increase, 127, 216
increments, 45
independence, 156, 158
index intervals, 243
indexing, conventional, 192
infocost()
  $\ell^{1/4}$, 290
  entropy, 290
information cell, 333
information cost, 274, 275
  functions, 274
  metrics, 275
ininterval(), 49
initlcabtn(), 148
initrcf(), 148
initrcfs(), 148
inner product space, 8
instrument response, 218
interpolation, quadrature filter, 182
INTERVAL (data structure), 48
  INTERVAL.FINAL, 48
  INTERVAL.LEAST, 48

INTERVAL.ORIGIN, 48
interval2tfa1s(), 65
intervalhedge2tfa1s(), 65
intervalstotal(), 50, 431
invariant under translation, 7

Jacobian, 389
joint best basis, 383

Karhunen–Loève, 378
basis, 276
transform, 369, 377, 379
approximate, 383, 394
fast approximate, 381
kernel, 301

l1norm(), 288
Laplacian pyramid scheme, 214
lcadf(), 147
lcadf2hedge(), 150
lcadm(), 149
lcadm2hedge(), 150
lcsdf(), 151
lct(), 144
leaf, 214
least, 48
leaves, 291
Lebesgue, 3
L^p-norms, 9
integrable, 6
integral, 6
measurable, 4
function, 4
set, 4
measure, 3
outer measure, 4
space, 9
Legendre polynomials, 37
level, 215
levelcost(), 293
levels, 55
levels list, 312
library, 282
limit, 7
linear phase response, 166
linear space, 6
linearity, 7
lineto, 354
LLXS, 356
LLYS, 356
localized exponential functions, 103
localized, 164
in time and frequency, 330
trigonometric functions, 103
cosines, 110
sines, 110
well, 330
locally constant formants, 413
locally integrable, 6
log(), 47
logarithm of energy, 276
logl2(), 289
long signal convolution, 29
lossy, 367
lossless, 367
lpdc(), 359
lpds(), 359
lphdev(), 194
lpic(), 145
lpica(), 146
lpis(), 145, 152, 426
lpisa(), 146
lpnormp(), 288
lst(), 144

makebtn(), 52
makebtnt(), 53
makehedge(), 56

makeinterval(), 49
matching pursuit, 418
max(), 46
measure, 4
  of an interval, 4
  positive, 4
merge, 413
mesh, 5
meta-algorithm, 286, 417
metric, 7
  space, 7
min(), 46
MINGRAY, 357
ml2logl2(), 288
modulation, 331
moment, 37
  vanishing, 37
    k^{th}, 37
    zero, 37
monotonicity, 4
more concentrated pdf, 280
mother wavelet, 213, 217, 238
moveto, 354
mul, 354
multiple folding, 130
multipliers, 45
multiresolution analysis, 216

natural order, 250
noise, 418
nominal frequency, 242
  intervals, 343
nominal position, 242
nondegeneracy, 7
norm, 7
normalization, 13, 156, 158
normalizers, 45
numerically smooth, 365
Nyquist, 336
  frequency, 336
  theorem, 336

octave subband decomposition, 214
operator norm, 11
OQF, 159
origin, 48
orthogonal, 214, 254
  components, principal, 378
  decomposition, principal, 369
  DWPA, 254
  DWT, 215
  factors, principal, 378
  filters, 154
  MRA, 216
  quadrature filter, 158
orthogonality, 13
orthonormal, 14
  basis, 13
  graph bases, 254
  wavelet packets, 242

Paley order, 250
parent, 147
Parseval's formula, 14
partial sum sequence, 280
pdf, 164, 276
periodic, 191, 254
  adjoint, 155
  convolution-decimation, 155
  DWPA, 253
  DWT, 215
periodization, 30, 118
periodize(), 199
periodized filters, 153
phase response, 164, 165
phoneme recognition, 414
PI (π), 45, 75
plotsig(), 356

pointwise convergence, 2
polarization, 8
position, 125, 215, 330
  index, 242
  nominal, 242
positive measure, 4
positivity, 7
PostScript, 354
PQF (data structure), 193
  PQF.ALPHA, 193
  PQF.CENTER, 193
  PQF.DEVIATION, 193
  PQF.F, 193
  PQF.FP, 193
  PQF.OMEGA, 193
PQFL(), 202
PQFO(), 202
preperiodize, 30
principal components, 378
  analysis, 377
  orthogonal, 378
product, infinite, 3

QF, 154
qfcirc(), 203
QMF, 154
quadrature, 301
quadrature filter, 153
  conjugate, 154
  interpolation, 182
  mirror, 154
  sequences, 157
  conventional indexing, 192
  conventional normalization, 158, 160
quadtree, 299, 375
quantization, 366
  bins, 375

rapid decrease, 10
rcf(), 139
rcfgrid(), 139
rcfis(), 138
rcfmidp(), 140
rcfth(), 138
REAL, 47
reconstruction
  error of, 367
  exact, 154, 156, 158
redundancy removal, 366
registration, 296
point, 295
registrationpoint(), 296
regular sampling, 35
regularity of wavelets, 214
remainder, 37
Return, 46
Riemann integrable, 5
Riemann integral, 5
  improper, 5
Riemann sums, 5
Riesz basis, 14
rising cutoff function, 104
root, 214, 283, 291
row-by-row scan, 310

sacdpd2(), 320
sacdpe2(), 321
sacdpi2(), 321
sample point, 33
sampling, 33
  function, 218
  interval, 32
    scale, 244
scale, 215, 330
  index, 242
scaling, 7
  function, 182

scan order, 310
  row-by-row, 310
  zig-zag scan, 310
scdpd2(), 318
scdpe2(), 319
scdpi2(), 319
Schwartz class, 10
self-duality, 158
self-similarity, 213
selfadjoint, 15
separable, 299
sequency order, 250
sequential input, 196, 198, 200, 202, 203, 207, 208
sequential output, 196, 197, 200, 204, 206, 207
sesquilinearity, 7
setgray, 354
Shannon, 35
  scaling function, 183
  synthesis function, 35
  wavelet, 211
shift, 31
  cost function, 297
shiftscosts(), 295
shifttonext(), 228
  from beginning, 229
  from end, 228
shifttoorigin(), 228
  from beginning, 229
  from end, 228
short signal convolution, 29
showpage, 354
signal trees, 382
Simpson's rule, 67
sin(), 47
sinc function, 183
sine polarity, 143
sine transform, 23
single precision, 47
singular value decomposition, 377, 378
smooth, 10, 214
  local periodization, 103
  periodic interval restriction, 122
  projections, 112
  to degree d, 10
space, 6
  dual, 11
  Hilbert, 8
    separable, 299
  inner product, 8
  Lebesgue, 8
  linear, 6
  metric, 7
    complete, 7
  vector, 6
sparsifiable, 408
spectral entropy, 411
spectral information, 18
SQ2 ($\sqrt{2}$), 75
SQH ($\sqrt{\frac{1}{2}}$), 75
sqrt(), 47
SQUARE(), 45
square-summable, 2
standard deviation, 363
stationary, 368
stochastic matrix, 279
stroke, 354
Sturm–Liouville, 16
  boundary conditions, 17
  differential operator, 16
  eigenvalue problem, 17
  regular operator, 16
subbands, 302, 372
  basis, 244
  coding, 214, 237, 303

superalgebraic decay, 10
support, 10
  intervals, 29
  widths, 29
SVD, 378
symmetric, 192
  about $-\frac{1}{2}$, 192
  about 0, 192
symmetry, 7
synthesis functions, 32
synthesize, 247

tdim(), 289
tempered distributions, 12
tensor products, 299
  of QFs, 301
test functions, 11
TFA1 (data structure), 57
  TFA1.AMPLITUDE, 57
  TFA1.BLOCK, 57
  TFA1.LEVEL, 57
  TFA1.OFFSET, 57
tfa12btnt(), 62
tfa1inabt(), 61
tfa1s2abt(), 61
tfa1s2btnt(), 62
tfa1s2dwpsa(), 271
tfa1s2dwpsp(), 263
tfa1s2ps(), 358
tfa1sinabt(), 61, 260
TFA2 (data structure), 57
  TFA2.AMPLITUDE, 57
  TFA2.XBLOCK, 57
  TFA2.XLEVEL, 57
  TFA2.XOFFSET, 57
  TFA2.YBLOCK, 57
  TFA2.YLEVEL, 57
  TFA2.YOFFSET, 57
TFAD (data structure), 57, 58
  TFAD.AMPLITUDE, 58
  TFAD.BLOCKS, 58
  TFAD.DIMENSION, 58
  TFAD.LEVELS, 58
  TFAD.OFFSETS, 58
theoretical dimension, 410–411, 418
theta1(), 138
theta3(), 138
thresh(), 287
threshold, 275
tight frame, 14
time-frequency
  atom, 57, 329–330
  molecule, 330
  plane, 329
training, 384
transform coding, 366
  gain, 379, 380
    metric, 381
translation, 31, 331, 354
  invariance of DWTP, 294
tree, 283
  accumulator, 382
  of means, 382
  quad, 299, 375
  signal, 382
  of squares, 382
  of variances, 382
triangle inequality, 7
trigonometric
  functions, localized, 103
  polynomial, 37
  series, 21
  waveforms, windowed, 331
truncated infinite algorithm, 1
two-scale equation, 216

udc2(), 328, 440
udcn(), 142

udcp(), 142
uds2(), 328
udsn(), 142
udsp(), 142
uipc(), 143
uips(), 143
unboundedness, 127
uncertainty, 25
    Heisenberg principle, 25
unfolding, 106
uniform, 2
    convergence, 2
    norm, 8
unit mass, 36
unstable, 1
unvoiced, 412
URXS, 356
URYS, 356

Vandermonde matrix, 68
variance ellipsoid, 363
vector space, 6
voiced, 412
voiced-unvoiced segmentation, 409

Walrus and the Carpenter, The, 416
Walsh functions, 189
Walsh-type, 244
warp, 115
wavelet, 213, 331
    arborescent, 237
    basis, 217, 244
    compactly supported, 214
    decomposition, 217
    equation, 217
    mother, 213, 217, 238
    on an interval, 254
    registration, 294
    regular, 214
    Shannon, 211
    subspaces, 217
    with vanishing moments, 214
wavelet packet, 237–238, 242, 331
    analysis, 247
    basis, 244
    coefficients, 259
    Haar–Walsh, 242
    orthonormal, 242
    synthesis, 248
waveletregistration(), 296
well-localized, 164
While loops, 46
width, 10
window, 109
windowed trigonometric waveforms, 331

xpi2(), 315
xpd2(), 328, 440
xpid(), 317

Z?X:Y, 46
zero measure, 4
zig-zag scan, 310